**Recent Developments
in Switching Theory**

ELECTRICAL SCIENCE
A Series of Monographs and Texts

Editors: Henry G. Booker and Nicholas DeClaris
A complete list of titles in this series appears at the end of this volume

Recent Developments in Switching Theory

Edited by
Amar Mukhopadhyay
Department of Computer Science
University of Iowa
Iowa City, Iowa

1971

ACADEMIC PRESS   New York and London

ACADEMIC PRESS, INC.
111 Fifth Avenue, New York, New York 10003

United Kingdom Edition published by
ACADEMIC PRESS, INC. (LONDON) LTD.
Berkeley Square House, London W1X 6BA

LIBRARY OF CONGRESS CATALOG CARD NUMBER: 70-137618

PRINTED IN THE UNITED STATES OF AMERICA

CONTENTS

List of Contributors		ix
Preface		xi
Acknowledgments		xiii

I. COMPLETE SETS OF LOGIC PRIMITIVES
Amar Mukhopadhyay

I.	Introduction	1
II.	Iteratively Closed System of Functions	3
III.	Characterization of Weak Complete Set of Logic Primitives	5
IV.	Reduction Theorems	7
V.	Theorem of Post	13
VI.	Bases and Simple Bases	15
VII.	Almost Complete Sets of Logic Primitives	16
	Appendix. Proof of Theorem 7.1	21
	References	26

II. COMBINATIONAL CIRCUITS WITH FEEDBACK
David A. Huffman

I.	Introduction	28
II.	Circuit Visualization of Markov's Result	31
III.	A Circuit with a Single Not-Element Which Inverts Two Variables	39
IV.	The Design of "Multi-Inversion" Circuits Which Use Only One Inverter	43
V.	Proof of the Necessity of Unstable Circuit Equilibria	48
VI.	A "Multi-Inversion" Circuit Which Is Stable	51
VII.	Summary and Conclusions	54
	References	55

III. LUPANOV DECODING NETWORKS
Amar Mukhopadhyay

I.	Introduction	57
II.	Disjunctive and Nondisjunctive Complete Decoding Networks	58

III. The Case When $r \neq 2^k$ 64
IV. The Optional Terms   68
V. Toward a General Theory   72
VI. Conclusions   82
    References   83

IV. COUNTING THEOREMS AND THEIR APPLICATIONS TO CLASSIFICATION OF SWITCHING FUNCTIONS

    Michael A. Harrison

I. Introduction to Boolean Functions and Classification Problems   86
II. Group Theory and Pólya's Theorem   87
III. Some Applications of Pólya's Theorem to Switching Functions   98
IV. Structure Theorems for Permutation Groups and the Determination of Cycle Indices   101
V. Operations on the Range, Genera, and a Lower Bound   113
    Appendix 1. Cycle Index Polynomials for S_n 116
    Appendix 2. Cycle Index Polynomials for G_n 117
    Appendix 3. Cycle Index Polynomials for $GL_n(W_2)$ 118
    Appendix 4. Cycle Index Polynomials for $A_n(Z_2)$ 119
    References   120

V. HARMONIC ANALYSIS OF SWITCHING FUNCTIONS

    Robert J. Lechner

I. Summary   122
II. Survey of Abstract Harmonic Analysis   125
III. Combinatorial Applications   156
IV. Analysis of the Prototype Equivalence Relation   173
V. Synthesis of Encoded Input Logic   196
    References   225

VI. UNIVERSAL LOGIC MODULES

    Harold S. Stone

I. Statement of the Problem   230
II. Bounds for $M(n)$ 231
III. The Construction of ULM'S for Small n 241
IV. Other Approaches to the Universal Module Problem   249
V. Historical References   252
    References   252

VII. CELLULAR LOGIC

    Amar Mukhopadhyay and Harold S. Stone

I. Introduction   256
II. Single-Rail Cascades   258

III. Two-Rail Cascades 281
IV. Two-Dimensional Arrays 285
V. Minimization of Cellular Arrays 300
VI. Review of Other Works in Cellular Area 307
References 311

VIII. THE THEORY OF MULTIRAIL CASCADES
Bernard Elspas

I. Introduction 316
II. Decomposition Theory of Group Functions 320
III. Synthesis of Multirail Cascades 336
References 367

IX. PROGRAMMABLE CELLULAR LOGIC
William H. Kautz

I. Introduction 369
II. Programmable Cellular Arrays 373
III. Arrays for Arbitrary Logic 374
IV. Special-Purpose Arrays 394
V. Conclusion 420
References 421

Author Index 423
Subject Index 427

List of Contributors

Numbers in parentheses indicate the pages on which the authors' contributions begin.

BERNARD ELSPAS, Information Science Laboratory, Stanford Research Institute, Menlo Park, California (315)

MICHAEL A. HARRISON, Department of Computer Science, University of California, Berkeley, California (85)

DAVID A. HUFFMAN, Board of Studies in Information and Computer Science, University of California, Santa Cruz, California (27)

WILLIAM H. KAUTZ, Information Science Laboratory, Stanford Research Institute, Menlo Park, California (369)

ROBERT J. LECHNER, Sylvania Electronic Systems, Inc., Needham Heights, Massachusetts; present address Applied Research Division, Honeywell EDP, Waltham, Massachusetts (121)

AMAR MUKHOPADHYAY, Department of Computer Science, University of Iowa, Iowa City, Iowa (1, 57, 255)

HAROLD S. STONE, Departments of Electrical Engineering and Computer Science, Stanford University, Stanford, California (229, 255)

ix

PREFACE

During the last decade a number of basic and significant advances have been made in switching theory. Some of these have their roots in classical switching theory while others are of recent origin, primarily from the impact of large-scale-integrated semiconductor technology in digital circuit design. To keep abreast of the recent developments in switching theory and to help maintain a sense of unity in this area, there is a need for presenting these results in a comprehensive book.

A summary of topics included in this book is as follows: Chapters I–IV deal with classical switching theory problems. Chapter I develops a new and simplified proof of Post's theorem on completeness of logic primitives and presents some recent results on simple bases and almost complete sets of logic primitives. Chapter II provides a deep insight into the role of feedback in combinational switching circuits and proves the "minimal-NOT" result for circuits with feedback. Chapter III presents a systematic procedure for the design of Lupanov decoding networks and gives the best results known so far on this subject. Chapter IV gives a review of classical results on counting theorems and their application to the classification of switching functions under different notions of equivalence, including linear and affine equivalences.

Chapter V is a self-contained development of abstract harmonic analysis of combinational switching functions. This chapter adds new conceptual insights and unifying principles to the traditional approaches to switching theory and also provides synthesis algorithms suitable for the new technology.

Chapters VI–IX deal with problems whose motivations primarily come from large-scale-integration technology. Chapter VI presents the theory of universal logic modules, methods of their construction, and upper bounds on the input terminals. Chapter VII is concerned with cellular logic: theory of single-rail cascades, elementary theory of two-rail cascades, and logical design techniques for combinational functions using two-dimensional cellular arrays. A discussion of minimization problems in cellular arrays and a review of research in the macrocellular area are also included. Chapter VIII is concerned with systematic techniques for the realization of multi-output logic functions by means of multirail cellular cascades. A mathematical theory of multirail cascades—the decomposition theory of group functions—is developed which permits systematic techniques of great generality to be applied to the logic synthesis problem. Chapter IX presents programmable cellular logic, a form of versatile cellular logic whose behavior can be electronically programmed. The design and operation of several types of programmable cellular arrays for the realization of arbitrary logical behavior are described. Four special-purpose cellular arrays are offered to illustrate the theoretical techniques employed in the logical design of programmable arrays.

This book will be of general interest to all English-reading research workers in the major fields of computer science and discrete information processing systems. The book can also be used as a textbook for a graduate course on switching theory in computer science and electrical engineering as well as a supplementary reference book to already-established courses in switching theory in the universities. As far as possible, each chapter has been made self-contained, but a little background in modern algebra and classical switching theory will be helpful.

We believe that this book will create new interest in switching theory research and will provide stimulation for organizing new switching theory courses in universities.

ACKNOWLEDGMENTS

I would like to express my appreciation to all the contributors for their participation and prompt cooperation in the project. The book includes materials derived from a large number of published and unpublished sources which have been acknowledged in the text. I would like to thank the Computer Science faculty of the University of Iowa, especially Professor Gerard P. Weeg, for moral support; Professor Sudhakar M. Reddy of the Electrical Engineering Department for his careful reading of Chapter I; and many of my students for their help in different stages of preparation of the book. I would also like to acknowledge the help from the offices of the Mathematical Sciences Division of the College of Liberal Arts and the University Computer Center, particularly the help of Ada Burns, Lois Friday, and Marsha Paxton in typing portions of the manuscript.

I would like to dedicate my contributions in this book to the memory of my late father, Surendra Nath Mukhopadhyay, and to my affectionate mother. I thank my wife, Pampa, for her understanding, and my daughters, Mita and Paula, for their love.

Recent Developments
in Switching Theory

Chapter I

COMPLETE SETS OF LOGIC PRIMITIVES

AMAR MUKHOPADHYAY

I. INTRODUCTION   1

II. ITERATIVELY CLOSED SYSTEM OF FUNCTIONS   3

III. CHARACTERIZATION OF A WEAK COMPLETE SET OF LOGIC PRIMITIVES   5

IV. REDUCTION THEOREMS   7

V. THEOREM OF POST   13

VI. BASES AND SIMPLE BASES   15

VII. ALMOST COMPLETE SETS OF LOGIC PRIMITIVES   16

APPENDIX. PROOF OF THEOREM 7.1   21

REFERENCES   26

ABSTRACT. This chapter is concerned with the problems of building up arbitrarily complex combinational switching circuits by using an interconnection of a set of simpler combinational circuits called primitives. A new and simplified proof of Post's theorem on completeness is presented. Some of the recent results on simple bases by Shestopal and on almost complete sets of logic primitives by Kobayashi are also given.

I. INTRODUCTION

This chapter is concerned with the problem of building up arbitrarily complex combinational switching circuits by using an interconnection of a set of simpler combinational circuits called *primitives*. We assume that the circuit is to handle binary inputs $x_1, x_2, \ldots, x_n$ each of which can have values 1 or 0, and the output of the circuit is a logic function (switching function, Boolean function) $f(x_1, \ldots, x_n)$ of the inputs. To avoid logical inconsistencies, the primitives are assumed to be interconnected to form a well-formed combinational circuit (Burks and Wright, 1953) which is defined recursively as: (i) each

1

primitive is a logic circuit; (ii) if N_1 and N_2 are two distinct logic circuits, then identification of some of the inputs of N_2 with an output of N_1 is a logic circuit; (iii) identification of any number of inputs of a logic circuit which are not outputs results in a logic circuit; (iv) each input of the primitive is connected to some input x_i ($1 \leq i \leq n$) of the circuit or to constant signals 1 or 0 or to the output of another primitive; the circuit has only one output which is not connected to any primitive input.

One should note that the interconnection rules do not allow closed feedback loops to exist within the circuit. Incorporation of closed feedback loops typically produces a sequential rather than a combinational circuit, but there are circuits which are combinational even with closed feedback loops. This problem will be treated in Chapter II.

A set of primitives is said to be *strong complete* if any arbitrary logic function $f(x_1, \ldots, x_n)$ can be realized as the output of a logic circuit which contains a finite number of primitives from the given set and whose inputs are identified with the set of variables $x_1, \ldots, x_n$. The definition implies that, in particular, the constant functions $f(x_1, \ldots, x_n) = 1$ and $f(x_1, \ldots, x_n) = 0$ must also be producible.

If the requirement for the realizability of constant functions is removed and if it can be assumed that the constants 1 and 0 can be applied to the inputs whenever necessary, the set of primitives is said to be *weak complete*. Note that a strong complete set of primitives is weak complete but the converse is not necessarily true.

An example of a strong complete set of primitives is provided by the well-known canonic expansion of a logic function (Post, 1921; Shannon, 1949).

$$f(x_1, \ldots, x_n) = \sum \dot{x}_1 \dot{x}_2 \cdots \dot{x}_n f(\varphi_1, \varphi_2, \ldots, \varphi_n) \qquad (1)$$

where $\dot{x}_i = x_i$ or $\bar{x}_i$ (the negation or complement of x_i), $\varphi_i = 1$ if $\dot{x}_i = x_i$, $\varphi_i = 0$ if $\dot{x}_i = \bar{x}_i$, $1 \leq i \leq n$, and $\sum$ denotes the extended logical sum operation. The canonic expansion says that the set of primitives consisting of the 2-input logical AND operation, the 2-input logical OR operation[1], and the single input negation or NOT operation can be used to synthesize any arbitrary logic function including the constant functions 0 and 1 (since $x_i + \bar{x}_i = 1$ and $x_i \bar{x}_i = 0$) and, therefore, form a strong complete set of primitives. In fact, (AND, NOT) or (OR, NOT) form a strong complete set since OR can be synthesized using AND and NOT, and AND can be synthesized using

---

[1] Since AND, OR, and EXCLUSIVE OR operations are associative, k-input AND operation, k-input OR operation, and k-input EXCLUSIVE OR operation for $k > 2$ can be realized using only 2-input AND, 2-input OR, and 2-input EXCLUSIVE OR operations, respectively,

OR and NOT as

$$x_1 x_2 = (\overline{\overline{x}_1 + \overline{x}_2}) \tag{2}$$

$$x_1 + x_2 = \overline{\overline{x}_1 \overline{x}_2} \tag{3}$$

An example of a weak complete set of logic primitives is provided by the well-known complement-free ring sum canonic expansion (Zhegalkin, 1927; Muller, 1954; Reed, 1954)

$$f(x_1, \ldots, x_n) = \sum a_i x_{j_1} x_{j_2} \cdots x_{j_k} \tag{4}$$

where a_i is 0 or 1, $0 \leq i \leq 2^n - 1$, $0 \leq k \leq n$, $j_i < j_{i+1}$ and the set of variables $(x_{j_1}, x_{j_2}, \ldots, x_{j_k})$ denotes a subset of k variables (this subset is the empty set when $k = 0$) out of the input variables $(x_1, \ldots, x_n)$ and $\sum$ denotes the extended logical EXCLUSIVE OR operation. This expansion says the set of primitives consisting of the AND operation and the 2-input EXCLUSIVE OR operation form a weak complete set but not a strong complete set since the constant function $a_0 = 1$ cannot be realized using these operations.

Each primitive in the logic circuit may be looked upon as performing a mathematical operation on its inputs. Thus, the function of the entire circuit might be interpreted as performing a mathematical *composition* with operations corresponding to the primitives. Thus, completeness of logical primitives is mathematically equivalent to logical universality of the set of operations by which it is possible to represent any arbitrary two-valued function of a finite number of two-valued variables as a composition or well-formed formula involving the basic operations and the logic arguments $x_1, \ldots, x_n$ and possibly the constant arguments 1 and 0. A complete solution to this problem was presented by Post (1941).

Although Post's theorem has been known to the switching theorists by word of mouth, its proof never appeared in any switching theory textbooks. One of the reasons for this may be that Post's original proof involves the development of a full hierarchy of an "iteratively closed system of two-valued mathematical logic" using a language and terminology which is not commonly used by switching theorists. The proof of Post's theorem by Yablonskii (1958) is not readily available to the English-reading worker. In this chapter we propose to give a simplified proof of Post's theorem on completeness. We shall then give some of the recent results on simple bases by Shestopal (1961) and on almost complete sets of logic primitives by Kobayashi (1967). The reference by Ibuki, Naemura, and Nozaki (1963) came to the author's notice after this chapter was written.

II. ITERATIVELY CLOSED SYSTEM OF FUNCTIONS

Basic to the understanding of Post's theorem is the concept of an "iteratively closed system of functions." Following Post, we shall consider only a *contracted* system of functions in which all n-variable logic functions are assumed to have arguments $x_1, \ldots, x_n$. Given some property P, a finite set of functions $F = \{f(x_1, \ldots, x_n)\}$ is said to be iteratively closed with respect to the property P if any function obtained as a composition of functions in F belongs in F and hence preserves the property P. Some of the relevant properties of logic functions which we shall use quite often in later developments are:

PROPERTY 1 (P1) (Monotonicity). Let $A = (a_1, a_2, \ldots, a_n)$ and $B = (b_1, b_2, \ldots, b_n)$ be two Boolean n-tuples where a_i and b_i $(1 \le i \le n)$ are binary constants 0 or 1. We write $A \le B$ if for all i, $a_i \le b_i$, where $0 \le 0, 0 \le 1$, and $1 \le 1$. Let $f(A)$ denote the value of the function $f(x_1, \ldots, x_n)$ when $x_i = a_i$. Similarly, we define $f(B)$. The function $f(x_1, \ldots, x_n)$ is said to be monotonic if for all A such that $f(A) = 1$, it is true that $f(B) = 1$ whenever $A \le B$. Monotonic functions are more commonly known in switching theory literature as *positive unate functions* (see Chapter II).

PROPERTY 2 (P2) (Linearity). A function $f(x_1, \ldots, x_n)$ is said to be linear if its canonic expansion given by Eq. (4) has the form

$$f(x_1, \ldots, x_n) = a_0 \oplus a_1 x_1 \oplus a_2 x_2 \oplus \cdots \oplus a_n x_n \tag{5}$$

where a_i $(0 \le i \le n)$ is 0 or 1 and $\oplus$ denotes the 2-input EXCLUSIVE OR operation.

PROPERTY 3 (P3) (Self-Duality). A function $f(x_1, \ldots, x_n)$ is said to be self-dual if complementing its inputs $x_1, \ldots, x_n$ results in the complementary function, i.e.,

$$f(x_1, \ldots, x_n) = \bar{f}(\bar{x}_1, \ldots, \bar{x}_n) \tag{6}$$

PROPERTY 4 (P4) (Zero Preservation). A function $f(x_1, \ldots, x_n)$ is said to be a function preserving zero if

$$f(0, 0, \ldots, 0) = 0 \tag{7}$$

PROPERTY 5 (P5) (One Preservation). A function $f(x_1, \ldots, x_n)$ is said to be a function preserving one if

$$f(1, 1, \ldots, 1) = 1 \tag{8}$$

The proof of the following theorem will be left as an exercise.

THEOREM 2.1.   The set of *n*-variable logic functions having the single property P*I* ($1 \leq I \leq 5$) forms an iteratively closed system of functions with respect to P*I*.

Stated in other words, the above theorem says that a monotonic function of monotonic functions is a monotonic function, a linear function of linear functions is a linear function, etc. The classification of all 1-variable and 2-variable functions according to the Properties P1 through P5 is given in Table I. In this table each function occupies a row and if it does *not* possess

TABLE 1

Classification of 1- and 2-Variable Functions

	FUNCTION	P1	P2	P3	P4	P5
	0	0	0	1	0	1
	1	0	0	1	1	0
	x_1	0	0	0	0	0
NOT	$(\bar{x}_1)$	1	0	0	1	1
AND	$(x_1 x_2)$	0	1	1	0	0
OR	$(x_1 + x_2)$	0	1	1	0	0
NAND	$(\bar{x}_1 + \bar{x}_2)$	1	1	1	1	1
NOR	$(\bar{x}_1 \bar{x}_2)$	1	1	1	1	1
IMP	$(x_1 + \bar{x}_2,\ \bar{x}_1 + x_2)$	1	1	1	1	0
NIMP	$(x_1 \bar{x}_2,\ \bar{x}_1 x_2)$	1	1	1	0	1
EXOR	$(x_1 \oplus x_2)$	1	0	1	0	1
EQUIV	$(x_1 \oplus \bar{x}_2 = x_1 \odot x_2)$	1	0	1	1	0

the property P*I*, 1 is entered in the P*I* column position for the row; otherwise the entry is 0.

III. CHARACTERIZATION OF A WEAK COMPLETE SET OF LOGIC PRIMITIVES

In this section, the necessary and sufficient conditions for a set of logic primitives to be weak complete will be given. The following two theorems have been adopted from Glushkov (1963).

THEOREM 3.1. If the constant functions 0 and 1 are available, the operation of negation can be synthesized by means of any nonmonotonic logic function.

Proof: Let us define a relation $A \leftarrow B$ between any two Boolean n-tuples $A = (a_1, \ldots, a_n)$ and $B = (b_1, \ldots, b_n)$ if there exists an index $j (1 \leq j \leq n)$ such that $a_j = 0$ and $b_j = 1$ and $a_i = b_i$ for all $i \neq j$, $1 \leq i \leq n$. Let $f(x_1, \ldots, x_n)$ be nonmonotonic. This means that there exist Boolean n-tuples A and B where $A \leq B$ such that $f(A) = 1$ and $f(B) = 0$. Thus, there must exist n-tuples $A_0, A_1, A_2, \ldots, A_k$ such that $A = A_0 \leftarrow A_1 \leftarrow A_2 \leftarrow \cdots \leftarrow A_k = B$ for some k, $1 \leq k \leq n$. Therefore for some t, $0 \leq t < k$, there must exist A_t and A_{t+1} such that $f(A_t) = 1$ and $f(A_{t+1}) = 0$. Since $A_t \leftarrow A_{t+1}$, let $A_t = (a_1{}^t, a_2{}^t, \ldots, a_{s-1}^t, 0, a_{s+1}^t, \ldots, a_n{}^t)$ and $A_{t+1} = (a_1{}^t, a_2{}^t, \ldots, a_{s-1}^t, 1, a_{s+1}, \ldots, a_n{}^t)$ for some s, $1 \leq s \leq n$. This means that $f(a_1{}^t, a_2{}^t, \ldots, a_{s-1}^t, x_s, a_{s+1}^t, \ldots, a_n{}^t) = \bar{x}_s$, which shows that negation can be synthesized.

THEOREM 3.2. If the constant functions 1 and 0 are available, the AND and OR operations can be synthesized by means of a nonlinear logic function.

Proof: Let $f(x_1, \ldots, x_n)$ be a nonlinear function. Then the canonic expansion of f given by Eq. (4) must contain at least one term of the form $x_i x_j p$, where $1 \leq i, j \leq n$, $i \neq j$ and p is a logical product of some of the variables from $(x_1, \ldots, x_{i-1}, x_{i+1}, \ldots, x_{j-1}, x_{j+1}, \ldots, x_n)$ or $p = 1$. Let us pick up one of such terms containing the fewest number of variables in p. Assign value 1 to all the variables in p so that $p = 1$ and assign value 0 to all of the remaining variables but x_i and x_j. Then f is reduced to a function $h(x_i, x_j)$ given by

$$h(x_i, x_j) = a_0 \oplus a_i x_i \oplus a_j x_j \oplus x_i x_j \qquad (9)$$

The expressions for h for eight different assignments of values to a_0, a_i, and a_j are shown in Table II. If $(a_0, a_i, a_j) = (0, 0, 0)$, we get an AND operation $x_i x_j$. If $(a_0, a_i, a_j) = (0, 1, 1)$, we get an OR operation $x_i + x_j$. In all other cases, in view of the classification of Table I, h is nonmonotonic, so that negation can be synthesized by means of h; thus it is possible to synthesize either the AND or the OR operation. Corollary 1 follows from the above proof.

COROLLARY 3.1. If the operation of negation is available, the operation of AND or OR can be synthesized by means of a 2-variable nonlinear function.

TABLE II

$h(x_i, x_j)$ for Different Values of (a_0, a_i, a_j)

a_0	a_1	a_j	$h(x_i, x_j)$
0	0	0	$x_i x_j$
0	0	1	$x_j \oplus x_i x_j = \bar{x}_i x_j$
0	1	0	$x_i \oplus x_i x_j = x_i \bar{x}_j$
0	1	1	$x_i \oplus x_j \oplus x_i x_j = x_i + x_j$
1	0	0	$1 \oplus x_i x_j = \bar{x}_i + \bar{x}_j$
1	0	1	$1 \oplus x_j \oplus x_i x_j = x_i + \bar{x}_j$
1	1	0	$1 \oplus x_i \oplus x_i x_j = \bar{x}_i + x_j$
1	1	1	$1 \oplus x_i \oplus x_j \oplus x_i x_j = \bar{x}_i \bar{x}_j$

THEOREM 3.3 (Weak Completeness Theorem). A set of functions is weak complete if and only if the set contains at least one nonmonotonic function and at least one nonlinear function.

Proof: The necessity of the conditions follows from Theorem 2.1. Assuming the availability of constants 1 and 0, we can synthesize the negation operation by using the nonmonotonic operation (by Theorem 3.1) and either the AND or the OR operation by using the nonlinear operation (by Theorem 3.2) and hence from Eqs. (1)–(3), it follows that the set is weak complete.

IV. REDUCTION THEOREMS

In this section we shall derive a set of theorems which will give the effect of identifying (i.e., connecting together) some of the inputs of a primitive so as to reduce it to a primitive of a smaller number of variables with the preservation of certain properties. Theorem 4.2 is new and all other theorems have been obtained by Shestopal (1961), wherein proofs are either not given or given in the form of hints or sketches. We shall give complete proofs. These theorems will be used in Section V to prove the theorem of Post.

THEOREM 4.1. If $f(x_1, \ldots, x_n)$ is a nonmonotonic function of $n > 3$ variables, then it can be reduced to a nonmonotonic function of not more than three variables by identifying some of the inputs.

Proof: Since $f(x_1, \ldots, x_n)$ is nonmonotonic, we can assume, without loss of generality, that there exist product terms $p_1 = x_1 x_2 \cdots x_k \bar{x}_{k+1} \bar{x}_{k+2} \cdots \bar{x}_{n-1} x_n$ and $p_2 = x_1 x_2 \cdots x_k \bar{x}_{k+1} \bar{x}_{k+2} \cdots \bar{x}_{n-1} \bar{x}_n$, $0 \le k \le n-1$, such that $f = 0$ when $p_1 = 1$ and $f = 1$ when $p_2 = 1$. If $k > 0$, identify $x_1, \ldots, x_{k-1}$ and x_k with the variable u and $x_{k+1}, \ldots, x_{n-2}$ and x_{n-1} with the variable v. Then $f(x_1, \ldots, x_n)$ is reduced to a 3-variable function $F(u, v, x_n)$ which is nonmonotonic. If $k = 0$, identify $x_1, \ldots, x_{n-2}$ and x_{n-1} with the variable u. Then $f(x_1, \ldots, x_n)$ is reduced to a 2-variable function $G(u, x_n)$ which is also obviously nonmonotonic.

THEOREM 4.2. All 3-variable non-self-dual nonlinear functions can be reduced by identification of variables to either 2-variable nonlinear functions or nontrivial 2-variable linear functions.

Proof: The theorem is proved by exhaustion. Table III gives a list of all

TABLE III

Table of 3-Variable Functions

(d_1, d_2, d_3)	NUMBER OF FUNCTIONS	REPRESENTATIVE FUNCTION	IDENTIFICATION OR COMMENT	THE REDUCED FUNCTION
$(3,3,1)$	2	$x_1 \oplus x_2 \oplus x_3 \oplus x_1 x_2 \oplus x_1 x_3 \oplus x_2 x_3 \oplus x_1 x_2 x_3$	$x_2 = x_3$	$x_1 \oplus x_2 \oplus x_1 x_2$
$(3,3,0)$	2	$x_1 \oplus x_2 \oplus x_3 \oplus x_1 x_2 \oplus x_1 x_3 \oplus x_2 x_3$	$x_2 = x_3$	$x_1 \oplus x_2$
$(3,2,1)$	6	$x_1 \oplus x_2 \oplus x_3 \oplus x_1 x_2 \oplus x_1 x_3 \oplus x_1 x_2 x_3$	$x_2 = x_3$	$x_1 \oplus x_1 x_2$
$(3,2,0)$	6	$x_1 \oplus x_2 \oplus x_3 \oplus x_1 x_2 \oplus x_1 x_3$	$x_1 = x_3$	$x_1 \oplus x_2 \oplus x_1 x_2$
$(3,1,1)$	6	$x_1 \oplus x_2 \oplus x_3 \oplus x_1 x_3 \oplus x_1 x_2 x_3$	$x_1 = x_3$	$x_1 \oplus x_2 \oplus x_1 x_2$
$(3,1,0)$	6	$x_1 \oplus x_2 \oplus x_3 \oplus x_1 x_2$	$x_2 = x_3$	$x_1 \oplus x_1 x_2$
$(3,0,1)$	2	$x_1 \oplus x_2 \oplus x_3 \oplus x_1 x_2 x_3$	$x_2 = x_3$	$x_1 \oplus x_1 x_2$
$(3,0,0)$	2	$x_1 \oplus x_2 \oplus x_3$	Linear	
$(2,3,1)$	6	$x_1 \oplus x_2 \oplus x_1 x_2 \oplus x_1 x_3 \oplus x_2 x_3 \oplus x_1 x_2 x_3$	$x_2 = x_3$	$x_1 \oplus x_1 x_2$
$(2,3,0)$	6	$x_1 \oplus x_2 \oplus x_1 x_2 \oplus x_1 x_3 \oplus x_2 x_3$	Selfdual[a]	
$(2,2,1)_1$	12	$x_1 \oplus x_2 \oplus x_1 x_2 \oplus x_1 x_3 \oplus x_1 x_2 x_3$	$x_2 = x_3$	$x_1 \oplus x_2 \oplus x_1 x_2$
$(2,2,1)_2$	6	$x_1 \oplus x_3 \oplus x_1 x_2 \oplus x_2 x_3 \oplus x_1 x_2 x_3$	$x_1 = x_3$	$x_1 x_2$
$(2,2,0)_1$	12	$x_1 \oplus x_3 \oplus x_1 x_2 \oplus x_1 x_3$	$x_1 = x_3$	$x_1 \oplus x_1 x_2$
$(2,2,0)_2$	6	$x_1 \oplus x_2 \oplus x_1 x_3 \oplus x_2 x_3$	$x_2 = x_3$	$x_1 \oplus x_1 x_2$
$(2,1,1)_1$	12	$x_1 \oplus x_2 \oplus x_1 x_3 \oplus x_1 x_2 x_3$	$x_1 = x_3$	$x_2 \oplus x_1 x_2$
$(2,1,1)_2$	6	$x_1 \oplus x_3 \oplus x_1 x_3 \oplus x_1 x_2 x_3$	$x_2 = x_3$	$x_1 \oplus x_1 x_2$
$(2,1,0)_1$	12	$x_1 \oplus x_2 \oplus x_1 x_3$	$x_2 = x_3$	$x_1 \oplus x_2 \oplus x_1 x_2$
$(2,1,0)_2$	6	$x_1 \oplus x_2 \oplus x_1 x_2$	None	$x_1 \oplus x_2 \oplus x_1 x_2$
$(2,0,1)$	6	$x_1 \oplus x_2 \oplus x_1 x_2 x_3$	$x_1 = x_3$	$x_1 \oplus x_2 \oplus x_1 x_2$

TABLE III (continued)

$(2,0,0)$	6	$x_1 \oplus x_2$	Linear	
$(1,3,1)$	6	$x_1 \oplus x_1 x_2 \oplus x_1 x_3 \oplus x_2 x_3 \oplus x_1 x_2 x_3$	$x_1 = x_3$	$x_1 x_2$
$(1,3,0)$	6	$x_1 \oplus x_1 x_2 \oplus x_1 x_3 \oplus x_2 x_3$	$x_2 = x_3$	$x_1 \oplus x_2$
$(1,2,1)_1$	12	$x_1 \oplus x_1 x_2 \oplus x_2 x_3 \oplus x_1 x_2 x_3$	$x_1 = x_3$	$x_1 \oplus x_1 x_2$
$(1,2,1)_2$	6	$x_1 \oplus x_1 x_2 \oplus x_1 x_3 \oplus x_1 x_2 x_3$	$x_2 = x_3$	$x_1 \oplus x_1 x_2$
$(1,2,0)_1$	12	$x_1 \oplus x_1 x_2 \oplus x_1 x_3$	$x_2 = x_3$	$x_1 \oplus x_2 \oplus x_1 x_2$
$(1,2,0)_2$	6	$x_1 \oplus x_1 x_2 \oplus x_2 x_3$	$x_2 = x_3$	$x_1 \oplus x_2 \oplus x_1 x_2$
$(1,1,1)_1$	12	$x_3 \oplus x_1 x_3 \oplus x_1 x_2 x_3$	$x_1 = x_3$	$x_1 x_2$
$(1,1,1)_2$	6	$x_1 \oplus x_2 x_3 \oplus x_1 x_2 x_3$	$x_2 = x_3$	$x_1 \oplus x_2 \oplus x_1 x_2$
$(1,1,0)_1$	12	$x_1 \oplus x_1 x_2$	None	$x_1 \oplus x_1 x_2$
$(1,1,0)_2$	6	$x_1 \oplus x_2 x_3$	$x_1 = x_3$	$x_1 \oplus x_1 x_2$
$(1,0,1)$	6	$x_1 \oplus x_1 x_2 x_3$	$x_1 = x_3$	$x_1 \oplus x_1 x_2$
$(1,0,0)$	6	x_1	Linear	
$(0,3,1)$	2	$x_1 x_2 \oplus x_1 x_3 \oplus x_2 x_3 \oplus x_1 x_2 x_3$	$x_2 = x_3$	$x_2 \oplus x_1 x_2$
$(0,3,0)$	2	$x_1 x_2\ \ x_1 x_3\ \ x_2 x_3$	Selfdual[a]	
$(0,2,1)$	6	$x_1 x_2 \oplus x_1 x_3 \oplus x_1 x_2 x_3$	$x_2 = x_3$	$x_1 x_2$
$(0,2,0)$	6	$x_1 x_2 \oplus x_1 x_3$	$x_1 = x_3$	$x_1 \oplus x_1 x_2$
$(0,1,1)$	6	$x_1 x_3 \oplus x_1 x_2 x_3$	$x_1 = x_3$	$x_1 \oplus x_1 x_2$
$(0,1,0)$	6	$x_1 x_2$	None	$x_1 x_2$
$(0,0,1)$	2	$x_1 x_2 x_3$	$x_2 = x_3$	$x_1 x_2$
$(0,0,0)$	2	0	Linear	

[a] When the function is self-dual, all identifications lead to the single-variable function x_1 or $\bar{x}_1$.

three variable functions along with self-dual and linear functions. A class of functions in this table is designated by a triplet (d_1, d_2, d_3) in the first column of the table where d_1, d_2, and d_3 denote that any function in this class will have d_1 terms which are single variable $(0 \leq d_1 \leq 3)$, d_2 terms which are products of two variables $(0 \leq d_2 \leq 3)$, and d_3 terms which are products of three variables $(0 \leq d_3 \leq 1)$ in its canonical expansion of Eq. (4). Column three of this table gives a representative function in this class and any function which can be obtained from the representative function by permutation of variables or by function complementation will have the same triplet (d_1, d_2, d_3). Note that the triplet (d_1, d_2, d_3) is not a unique set of invariants for the class.[2] In some cases, two classes have the same triplet and they are designated as $(d_1, d_2, d_3)_1$ and $(d_1, d_2, d_3)_2$. The second column gives the

[2] The problem of classification of switching functions will be treated in further depth in Chapters IV and V.

number of functions in the class. The fourth column gives one of the possible identifications of the variables which reduce the function according to the theorem or contains a remark. The resulting reduced functions are shown in the fifth column. Note that the nonlinear function whose class designations are $(2, 3, 0)$ and $(0, 3, 0)$ are not reducible according to the theorem and the functions in these classes are self-dual. All other nonlinear 3-variable functions are non-self-dual. Thus, the theorem is proved.

THEOREM 4.3. All 4-variable nonlinear functions can be reduced to nonlinear functions of three variables by identification of inputs.

Proof: Consider a 4-variable nonlinear function $f(x_1, x_2, x_3, x_4)$. Let us express f as

$$f(x_1, x_2, x_3, x_4) = x_4 \, g(x_1, x_2, x_3) \oplus h(x_1, x_2, x_3) \tag{10}$$

where

$$g(x_1, x_2, x_3) = f(x_1, x_2, x_3, 1) \oplus f(x_1, x_2, x_3, 0) \tag{11}$$

$$h(x_1, x_2, x_3) = f(x_1, x_2, x_3, 0) \tag{12}$$

First, assume $g = 0$. Then h must itself be nonlinear and f is trivially a 3-variable nonlinear function. Then, assume $g = 1$. Since f is nonlinear, h must be nonlinear and identification of x_4 with any one of the variables x_1, x_2, or x_3 keeps the resulting 3-variable function nonlinear. Now, assume g is not 0 or 1. Let us identify some of the variables in (x_1, x_2, x_3) such that g is reduced to either a nonlinear function of two variables or a nontrivial linear function of two variables or a single variable or complement of a single variable. If g is nonlinear, such a reduction is always possible as is shown in Table III. Also, if g is nonconstant linear, such a reduction is obviously possible. Since h does not involve the variable x_4, such an identification would reduce f to a 3-variable nonlinear function because of the presence of the term $x_4 \, g$ independent of the effect of this identification on the function h.

THEOREM 4.4. Nonlinear functions of $n > 3$ variables can be reduced to nonlinear functions of not more than three variables by identification of inputs.

Proof: Theorem 4.3 is a special case of this theorem for $n = 4$. The theorem will now be proved by induction. Assume that all nonlinear functions of $k > 3$ variables can be reduced to nonlinear functions of not more than $(k - 1)$

variables by identifying inputs. We then prove that all nonlinear functions of $(k + 1)$ variables can be reduced to nonlinear functions of not more than k variables by identifying inputs. We have

$$f(x_1, \ldots, x_{k+1}) = x_{k+1} g(x_1, \ldots, x_k) \oplus h(x_1, \ldots, x_k) \qquad (13)$$

where

$$g = f(x_1, \ldots, x_k, 1) \oplus f(x_1, \ldots, x_k, 0) \qquad (14)$$

$$h = f(x_1, \ldots, x_k, 0) \qquad (15)$$

First assume g is linear. If $g = 0$, h must be a nonlinear function of k variables since f is nonlinear. If $g = 1$, identification of x_{k+1} with any one of the variables $(x_1, \ldots, x_k)$ will reduce f to a nonlinear function of k variables. If g is a nonconstant linear function, we can always identify some variables in $(x_1, \ldots, x_k)$ such that g becomes a linear function of less than k variables. Thus $x_{k+1} g$ becomes a nonlinear function of less than $(k + 1)$ variables. Since x_{k+1} does not occur in h, f becomes a nonlinear function of less than $(k + 1)$ variables independent of the effect of the identification on h. Now, suppose g is nonlinear. By induction hypothesis, g can be reduced to a nonlinear function of not more than $(k - 1)$ variables. Again, $x_{k+1} g$ becomes a nonlinear function of at most k variables and since x_{k+1} does not occur in h, f becomes a nonlinear function of not more than k variables.

THEOREM 4.5. If $f(x_1, \ldots, x_2)$ is a non-self-dual function of $n > 2$ variables, then it can be reduced to a non-self-dual function of two variables by identifying its inputs.

Proof: Since f is non-self-dual, there exist two fundamental products $p_1 = \dot{x}_1 \dot{x}_2 \cdots \dot{x}_n$ and $p_2 = \bar{\dot{x}}_1 \bar{\dot{x}}_2 \cdots \bar{\dot{x}}_n$ where $\dot{x}_i$ is x_i or $\bar{x}_i$ such that if $f = 1$ when $p_1 = 1$, f is also 1 when $p_2 = 1$ or if $f = 0$ when $p_1 = 1$, f is also 0 when $p_2 = 1$. Assume first that $f = 1$. Without loss of generality, let $p_1 = x_1 x_2 \cdots x_k$ $\bar{x}_{k+1} \bar{x}_{k+2} \cdots \bar{x}_n$ where $1 \leq k \leq n$. Identify $x_1, x_2, \ldots, x_k$ to X and $x_{k+1}, \ldots, x_n$ to Y. Then $f(x_1, \ldots, x_n)$ is transformed to a function $F(X, Y)$ whose truth table looks like

X	Y	F
0	0	φ_1
0	1	1
1	0	1
1	1	φ_2

where φ_1 and φ_2 are each 0 or 1. When $\varphi_1 = \varphi_2 = 1$, $F(X, Y) = 1$, which is a non-self-dual function. For all other values of φ_1 and φ_2, $F(X, Y)$ is reduced to a nontrivial 2-variable function and hence a non-self-dual function (see Table I). For the special case when $p_1 = x_1 x_2 \cdots x_n$, identify the first k variables $x_1, x_2, \ldots, x_k$ ($1 \le k < n$) with X and $x_{k+1}, x_{k+2}, \ldots, x_n$ with Y. Then $f(x_1, \ldots, x_n)$ is transformed to a function $G(X, Y)$ whose truth table is

X	Y	G
0	0	1
0	1	φ_1
1	0	φ_2
1	1	1

Again, if both φ_1 and φ_2 are 1, $G = 1$. For all other values of φ_1 and φ_2, G is a nontrivial 2-variable function and hence a non-self-dual function. In a similar way, it can be proved that f can be reduced to a 2-variable non-self-dual function, assuming $f = 0$ for both p_1 and p_2.

THEOREM 4.6. A nonzero (one) preserving function $f(x_1, \ldots, x_n)$ of $n > 1$ variables remains a nonzero (one) preserving function under any identification of inputs.

The proof of this theorem is simple and is left as an exercise.

Before concluding this section, we shall prove another interesting theorem attributed to Yablonskii (1958). This theorem and its corollary will also be used to prove the theorem of Post in the next section.

THEOREM 4.7. Every operation that does not preserve zero also either does not preserve one or is non-self-dual.

Proof: Suppose $f(x_1, \ldots, x_n)$ does not preserve zero and does preserve one, i.e., $f(0, 0, \ldots, 0) = 1$ and $f(1, 1, \ldots, 1) = 1$, which means that f is non-self-dual since complementing the inputs does not complement the function value. Next, suppose that $f(x_1, \ldots, x_n)$ does not preserve zero and is self-dual. Then, $f(0, 0, \ldots, 0) = 1$ and $f(1, 1, \ldots, 1) = 0$, which implies that f does not preserve one.

COROLLARY 4.1. The operation of negation or the constant function 1 can be synthesized from a nonzero preserving operation.

Proof: Let $f(x_1, \ldots, x_n)$ be the nonzero preserving operation. Identify all its inputs to the single input x_1. If f is nonzero preserving and also non-one preserving, then the output function after identification is obviously $\bar{x}_1$. If f is nonzero preserving and non-self-dual, it means that $f(0, 0, \ldots, 0) = 1$ and $f(1, 1, \ldots, 1) = 1$, so that the output is a constant function 1 after identification.

V. THEOREM OF POST

THEOREM 5.1. A set of functions is strong complete if and only if it contains (1) at least one function not zero preserving, (2) at least one function not one preserving, (3) at least one non-self-dual function, (4) at least one nonlinear function, (5) at least one nonmonotonic function.

Proof: The necessity of these conditions follows from Theorem 2.1. To prove sufficiency, we shall show constructively that it is possible to synthesize the negation operation and either the AND or OR operation if the conditions of the theorem are satisfied.

Let us consider the primitive which is nonzero preserving. We have to consider two cases according to Theorem 4.7.

CASE A: The nonzero preserving primitive is also non-self-dual. By Corollary 4.1, we can synthesize the constant function 1 out of this primitive by identifying all its inputs. This constant 1 function is then connected to all the inputs of the nonone preserving primitive of the set yielding a constant 0 output. Thus, the constant functions 1 and 0 have been produced. Now, we can use the nonmonotonic primitive to synthesize the negation operation (Theorem 3.1) and the nonlinear primitive to synthesize either the AND or the OR operation (Theorem 3.2). Hence the given set of primitives is strong complete.

CASE B: The nonzero preserving primitive is also nonpreserving one. The proof in this case should proceed in the following steps.

Step 1. By identifying all the inputs of the primitive, the negation operation can be synthesized (Corollary 4.1).

Step 2. Consider now the nonlinear primitive. This primitive must produce a function of more than one variable since all single variable functions are linear. If this is a 2-variable function, we can synthesize either the AND or the OR operation since the negation operation has already been synthesized in Step 1 (Corollary 3.1). The constant functions 1 and 0 can be synthesized by using the relations $1 = x_1 + \bar{x}_1$ and $0 = x_1\bar{x}_1$, respectively.

Step 3. Now, suppose the nonlinear primitive produces a 3-variable function. We have to consider three subcases according to Theorem 4.2.

Subcase 1: The 3-variable nonlinear function can be reduced to a 2-variable nonlinear function by identification of appropriate inputs. We can then synthesize either AND or OR as in Step 2.

Subcase 2: The 3-variable nonlinear function can be reduced to a nontrivial 2-variable linear function (see the classes $(3, 3, 0)$ and $(1, 3, 0)$ in Table III). With the use of the negation operation synthesized in Step 1, we can now obtain either 1 or 0 functions since $1 = x_1 \oplus \bar{x}_1$ and $0 = x_1 \odot \bar{x}_1$. Applying the negation operation again, we can obtain 0(1) from 1(0), and hence both the constant functions have been obtained. We can now use another copy of the original nonlinear primitive and synthesize the AND or OR operations (Theorem 3.2).

Subcase 3: The 3-variable nonlinear function is self-dual and is not reducible as in Subcases 1 and 2 (see classes $(2, 3, 0)$ and $(0, 3, 0)$ in Table III). We need help from the non-self-dual primitive in the set which we have not used so far. If this non-self-dual primitive produces a function of $n > 2$ variables, we reduce it such that the function is a 2-variable non-self-dual function by identifying inputs (Theorem 4.5). We have to consider three sub-subcases: (a) The reduced non-self-dual function is a constant function 1(0). We use the negation synthesized in Step 1 and then use another copy of the original nonlinear function in the set to synthesize either the AND or OR operation. (b) The reduced non-self-dual function is a nontrivial 2-variable linear function. We proceed as in Subcase 2 in Step 3 to synthesize 1 and 0 and either AND or OR. (c) The reduced non-self-dual function is a 2-variable nonlinear function. We proceed as in Step 2 to synthesize 1 and 0 and either the AND or OR operation.

Step 4. Now, suppose the nonlinear primitive in the set produces a nonlinear function of more than three variables. Then we reduce it to a nonlinear function of not more than three variables (Theorem 4.4). If the reduced function is a 2-variable nonlinear function, then we proceed as in Step 2. If the reduced function is a 3-variable nonlinear function, we reduce it further by identifying inputs such that the resulting function is: (i) A nonlinear 2-variable function. Then we proceed as in Step 2. (ii) A nontrivial linear function of 2 variables. Then we proceed as in Subcase 2 in Step 3. (iii) A self-dual function. If the original nonlinear function from which this self-dual function has been derived is also self-dual then we pick up the non-self-dual primitive in the set and proceed as in Subcase 3 in Step 3. But if the original nonlinear function is also non-self-dual, or if there is any other non-self-dual primitive in the set, we can work with either one of them as in Subcase 3 in Step 3 and synthesize constant functions 1 and 0 and either the AND or OR operation.

This exhausts all possible cases and the theorem is proved.

VI. BASES AND SIMPLE BASES

A strong (weak) complete set of logic primitives is called a strong (weak) *basis* if no proper subset of it forms a strong (weak) complete set. While it can be easily seen that the total number of bases of any kind is infinite, the maximum number of primitives in a basis is bounded. According to Theorems 4.7 and 5.1, a strong basis can contain no more than four primitives; also, since there are nonlinear monotonic functions the maximum number of primitives in a weak basis is two.

A basis is called a *simple basis* if it is impossible to replace some of the primitives in the set by functions obtained by identifying some of the inputs of one or more primitives of the basis without destroying the completeness property.

Given a property P, a function is said to be *simple* (Shestopal, 1961; Kautz, 1966) with respect to P if any identification of its inputs destroys P for the function. Functions simple with repsect to at least one of the properties mentioned in the theorem of Post (Theorem 5.1) will be called *simple functions* in this section. It follows from the definition that a simple basis can contain only simple functions.

It is interesting that the number of simple functions is finite, as a result of which it follows that the number of simple bases is also finite. To see this, we note the following:

(1) From Theorem 4.1, functions simple with respect to nonmonotonicity must be nonmonotonic functions of at most three variables. By exhaustion, it can be shown that the simple nonmonotonic functions are: NOT, IMP, NIMP, EXOR, EQUIV, the 3-variable representative functions in Table III whose class designations are $(3, 2, 0)$, $(3, 1, 1)$, $(3, 0, 0)$, $(2, 3, 0)$, $(2, 2, 1)_1$, $(2, 2, 1)_2$, $(2, 1, 0)_1$, $(2, 0, 1)$, $(1, 3, 1)$, $(1, 2, 0)_1$, $(1, 2, 0)_2$, $(1, 1, 1)$, and the functions which can be obtained from these by permutation of inputs only.

(2) From Theorem 4.4, functions simple with respect to nonlinearity are nonlinear functions of at most three variables. These are: AND, OR, IMP, NIMP, NOR, NAND, and all the 3-variable functions in Table III whose class designations are $(3, 3, 0)$, $(2, 3, 0)$, $(1, 3, 0)$, and $(0, 3, 0)$.

(3) From Theorem 4.5, functions simple with respect to non-self-duality are 0, 1, AND, OR, NAND, and NOR.

(4) From Theorem 4.6, functions simple with respect to nonzero (one) preservation are 1 and $\bar{x}$ (0 and $\bar{x}$).

To obtain all possible simple and weak bases it is only necessary to take all possible irredundant combinations of simple nonmonotonic and simple nonlinear functions. Similarly, to obtain all possible simple and strong bases it is only necessary to take all possible irredundant combinations of simple functions—preserving the five properties of Theorem 5.1. Simple bases having single functions are NOR and NAND; one having a maximum of four primitives is $\{0, 1, x_1 x_2, x_1 \oplus x_2 \oplus x_3\}$. This basis was also mentioned by Loomis and Wyman (1965) as an example of a complete set.

EXERCISES

1. Determine the strong and weak basis consisting of logic primitives of at most two variables.
2. (Shestopal) Show that there are 48 simple bases; two bases consisting of one function, 22 bases consisting of two functions, 21 bases consisting of three functions, three consisting of four functions.

VII. ALMOST COMPLETE SETS OF LOGIC PRIMITIVES

Given a complete set, an arbitrarily complex combinational function may need potentially infinite number of copies of the primitives for its realization. Kobayashi (1967) posed the following interesting question: Given a set of primitives $F = \{f_1, \ldots, f_p\}$ which is not complete, if a set of primitives $G = \{g_1, g_2, \ldots, g_q\}$ can be added to F, is it possible to realize any arbitrary function by using each of g_i a bounded number, say k_i times, and using each of f_j a potentially infinite number of times, if necessary? If this is possible, F will be called an *almost complete set of logic primitives*.

The problem of almost complete set of logic primitives seems to have originated from a more intriguing problem in switching theory, called the "minimal-NOT" problem. This problem will be discussed in great detail in Chapter II, but we will briefly mention the following: in Chapter II, Fig. 3 gives a circuit for complementing three inputs by using only two NOT operations. Note that the T-elements in this circuit can be replaced by circuits consisting of only AND and OR primitives. In an arbitrary combinational circuit using AND, OR, and NOT primitives, by applying this "three-NOT-to-two-NOT" transformation repeatedly, it should be possible to get an equivalent circuit consisting of a large number of AND and OR primitives and at most two NOT primitives. One is therefore tempted to conclude that (AND, OR) is an almost complete set of primitives. The fallacy in the above argument is that the "three-NOT-to-two-NOT" transformations might create closed feedback loops in the circuit and the behavior of the circuit may correspond to a sequential circuit rather than a combinational circuit. Markov (1958) proved that m NOT primitives are sufficient for generating the complements of $2^m - 1$ input variables in a circuit without any closed feedback loops. Hence, any function of $n = 2^m - 1$ variables may be realized with m NOT's. But, here the number of NOT primitives grows with an increase in the number of input variables and hence is not bounded. Huffman proves in Chapter II that a single NOT primitive is sufficient to synthesize any arbitrary combinational function if and only if closed feedback loops are allowed in the network. Thus, {AND, OR} is almost an almost complete set! It seems that Kobayashi's general formulation of the problem has been motivated from the above observation.

A complete characterization of an almost complete set of logic primitives is not known, but Kobayashi (1967) gives the following necessary condition, whose proof is given in the Appendix.

THEOREM 7.1. Let $C(n)$ denote the number of n-variable functions that can be realized in an n-input network using the primitives $F = \{f_1, \ldots, f_p\}$, each of which can be used a potentially infinite number of times, if necessary. If for any nonnegative integers a and b

$$\lim_{n \to \infty} \frac{[C(n + a)n]^{n^b}}{2^{2^n}} = 0$$

then the set $F = \{f_1, \ldots, f_p\}$ is not almost complete.

Based on this theorem, it is shown that the characterization problem of an almost complete set is equivalent to deciding whether the set (AND, OR) is almost complete or not. We need to prove a number of theorems for this purpose.

THEOREM 7.2. Let H be the set of all the reduced functions of not more than two variables that are obtained from $f(x_1, \ldots, x_n)$ by identification of some inputs $x_1, \ldots, x_n$ to constants 0 or 1. Then $f(x_1, \ldots, x_n)$ is nonlinear monotonic if and only if $n \geq 2$, each of the reduced functions in H is monotonic, and at least one reduced function in H is the AND or OR function.

Proof: Suppose f is nonlinear and monotonic. Since f is nonlinear, $n \geq 2$. Also, since 0 and 1 functions are monotonic, all the reduced functions must also be monotonic (by Theorem 2.1). The theorem is obviously true for $n = 2$, since $x_1 + x_2$ and $x_1 x_2$ are the only nonlinear monotonic functions. To prove the theorem for $n \geq 2$, we shall utilize a unique normal form representation for a monotonic or positive unate function (Semon *et al.*, 1955)

$$f = \sum_{i=1}^{k} P_i \qquad (16)$$

where P_i is an essential prime implicant[3] (Quine, 1952) of f expressed as a product of uncomplemented variables. The prime implicants of f also possess the following property: for any arbitrary pair of prime implicants P_i and P_j, $i \neq j$, there must be at least one variable which does not occur in both P_i and P_j. We have to consider two cases: (i) all the prime implicants of f are single variable, that is, f has the form $x_1 + x_2 + \cdots + x_n$. In this case we set $x_3 = x_4 = \cdots = x_n = 0$ and we get an OR function $x_1 + x_2$; (ii) at least one of the prime implicants of f is a product of two or more variables. Let this prime implicant be, without loss of generality, $x_1 x_2 \cdots x_k$ ($2 \leq k \leq n$). Now, set $x_{k+1} = x_{k+2} = \cdots = x_n = 0$; this will reduce the function to a k-input AND function $x_1 x_2 \cdots x_k$, since at least one of the variables from

$$(x_{k+1}, x_{k+2}, \ldots, x_n)$$

must occur in all other prime implicants. Then we set $x_3 = x_4 = \cdots = x_k = 1$ and the function is reduced to AND function $x_1 x_2$. Thus the reduced set of functions includes at least one of the AND or OR functions. In most of the cases, both AND and OR functions can be realized as reduced functions (see Theorem 7.3).

To prove the converse, we note that since one of the reduced functions is nonlinear and a nonlinear function cannot be derived from a linear function by assigning constants 0 or 1 to some of the inputs, f must be nonlinear (Theorem 2.1). The function f must also be monotonic, otherwise we could have derived the negation function $\bar{x}_1$ from f (by Theorem 3.1) which contradicts the assumption that all the reduced functions are monotonic.

A nonlinear monotonic function f will be called an AND-type *function*, an OR-*type function*, or an AND-OR-type *function* if the reduced functions contain the AND but not the OR operation, the OR but not the AND operation, both the AND and the OR operations, respectively. Theorem 7.2 classifies all nonlinear monotonic functions into three disjoint classes: the AND-type functions, the OR-type functions, and the AND-OR-type functions. Furthermore, it is true that:

---

[3] A *prime implicant* of a function f is a product term such that all the fundamental products contained in the term are true fundamental products of f and no variable can be deleted from the term without violating this condition. It is *essential* if it contains at least one fundamental product not contained in any other prime implicant. See Chapter V for further details.

THEOREM 7.3.   A function $f(x_1, \ldots, x_n)$ is an AND-type (OR-type) function if and only if f is an m-input AND(OR) function for $2 \leq m \leq n$.

The proof of this theorem will be left as an exercise.

THEOREM 7.4.   An almost complete set of functions $F = \{f_1, \ldots, f_p\}$ includes a nonlinear function. If all nonlinear functions in F are AND-type functions or OR-type functions, then F must contain a linear function other than 0, 1, or x_i $(1 \leq i \leq n)$.

Proof: Suppose all the functions in F are linear. Then all n-input functions realized from $f_1, \ldots, f_p$ are linear. The number of all n-variable linear functions is $C_1(n) = 2^{n+1}$.

$$\lim_{n \to \infty} \frac{(C_1(n + a)n)^{n^b}}{2^{2^n}} \leq \lim_{n \to \infty} \frac{(2^{n+a+1}n)^{n^b}}{2^{2^n}} = 0$$

holds for any nonnegative integers a and b. Thus, F is not almost complete—a contradiction. Hence, F must include a nonlinear function. Then suppose all nonlinear functions in F are AND-type (OR-type) functions and all linear functions (if any) in F are 0, 1, or x_i. Therefore, the functions that can be realized from F are the 0, 1, x_i, and the m-input AND functions (OR functions), $2 \leq m \leq n$. The total number of such functions is obviously $C_2(n) = 2^n + 1$. Hence,

$$\lim_{n \to \infty} \frac{(C_2(n + a)n)^{n^b}}{2^{2^n}} \leq \lim_{n \to \infty} \frac{(2^{n+a} + 1)^{n^b}}{2^{2^n}} = 0$$

holds for any nonnegative integer a and b. Thus, F is not almost complete which is a contradiction. This proves the theorem.

An almost complete set F is called trivially almost complete if

$$\{f_1, \ldots, f_p, 0, 1\}$$

is complete; otherwise it is *nontrivially almost complete*. A *minimal nontrivially almost complete* set is a nontrivially almost complete set such that none of its proper subsets is a nontrivially almost complete set.

THEOREM 7.5.   Each function in a nontrivially almost complete set is one of 0, 1, x_i, or a nonlinear monotonic function. The set must include either an AND-OR-type function or both an AND-type and an OR-type function.

Proof: Suppose $F = \{f_1, \ldots, f_p\}$ is nontrivially almost complete. By Theorem 7.4, it contains a nonlinear function. All functions in the set must be monotonic since if one of the functions is nonmonotonic by using constants and the nonlinear function, F becomes complete (Theorem 3.3); but this contradicts the assumption that F is nontrivially almost complete. Hence, all nonlinear functions in the set are monotonic. Now, suppose that all nonlinear functions in the set are either AND-type or OR-type. Then, by Theorem 7.4, the set must contain a linear function other than 0, 1, x_i which must necessarily be a nonmonotonic function. Applying Theorem 3.3, we again come to a contradiction. Hence the theorem is proved.

Theorem 7.6. A minimal nontrivially almost complete set of functions consists of either an AND-OR-type function or both AND- and OR-type functions.

Proof: By Theorem 7.5, the minimal nontrivially almost complete set is the union of four disjoints sets: (i) a set of AND-OR-type functions $(g_1, g_2, \ldots, g_a)$, (ii) a set of AND-type functions $(h_1, h_2, \ldots, h_b)$; (iii) a set of OR-type functions $(u_1, u_2, \ldots, u_c)$, and (iv) a set of functions $(v_1, v_2, \ldots, v_d)$ of the form 0, 1, x_i. By Theorem 7.5, either $a \geq 1$ or $b \geq 1$ and $c \geq 1$.

Suppose $a \geq 1$. With a single element, say g, from set (i) and 0 and 1, all functions in (i), (ii), and (iii), (iv) can be synthesized. Hence g_1 itself forms a nontrivially almost complete set and hence a minimal set. Suppose that $b \geq 1$ and $c \geq 1$. With two elements h_1 and u_1, say, and constant functions 0 and 1, we can synthesize $x_1 x_2$ and $x_1 + x_2$, and hence all functions in (i), (ii), (iii), and (iv). Hence, (h_1, u_1) form a nontrivially almost complete set and hence a minimal set.

It follows from Theorem 7.6 that if (AND, OR) form an almost complete set then there are two types of minimal nontrivial almost complete set: a set consisting of an AND-OR-type function or a set consisting of an AND-type function and an OR-type function. If (AND, OR) is not almost complete, then there is no nontrivially almost complete set.

There are no 0-, 1-, 2-variable AND-OR-type functions. There are three essentially different 3-variable AND-OR-type functions: $x_1(x_2 + x_3)$, $x_1 + x_2 x_3$ and $x_1 x_2 + x_1 x_3 + x_2 x_3$. The only AND- and OR-type set (AND, OR) is mentioned in Muller (1954) and Markov (1958).

APPENDIX. PROOF OF THEOREM 7.1

Let S denote an arbitrary logic circuit with inputs $x_1, \ldots, x_n$ and consisting of the primitives from $F = \{f_1, \ldots, f_p\}$, called f-type primitives, each occurring unbounded number of times if necessary, and primitives from $G = \{g_1, \ldots, g_q\}$, called g-type primitives, occurring only bounded number of times such that each element g_t $(1 \leq t \leq q)$ is used r_t times where for some positive integer k_t, $r_t \leq k_t$ and $\sum r_t \leq \sum k_t = K$. We attach a label $\langle t, u \rangle$ with the uth element in these r_t elements g_t.

According to our interconnection rules S cannot have a closed feedback loop. But S might have a closed path which is nondirected, which occurs when the output of a certain primitive P is connected with inputs $I_1, \ldots, I_s$ of other primitives. We wish to transform S into an equivalent circuit free from such closed paths, thus essentially obtaining a treelike circuit equivalent to S. We start with one of the primitives P at the lowest level of the circuit. (If all the inputs to P are from the primary inputs $x_1, \ldots, x_n$, then its level is 1. An element belongs to the $g + 1$ level if there is one input to it from another primitive whose level number is g.) We construct s copies of the subcircuit whose output is P and connect the output of the jth subcircuit with the input I_j $(1 \leq j \leq s)$. This process is repeated in the resulting circuit iteratively until the output of each primitive in all the levels is connected with only one input. The circuit S has thus been transformed to a treelike circuit producing the same output function.

To be able to calculate the total number of functions realizable by such a circuit, it is convenient to group some of the f-type primitives into a single primitive $f_{m,j}$ that is defined later. Qualitatively, the grouping consists of replacing the maximal subcircuit containing only f-type primitives by a single primitive $f_{m,j}$. More precisely, in the set of all f-type primitives an equivalence relation of connectedness is defined. Two f-type primitives P and P' are connected if either (i) P and P' are the same element, or (ii) there exists a sequence of f-type primitives $P = P_1, P_2, \ldots, P_v = P'$ such that for each i, $1 \leq i \leq v$, the output of one of P_i, P_{i+1} is connected with an input of the other. The primitives of one equivalence class form a maximal subcircuit S_m and each f-type primitive falls in exactly one subcircuit S_m. Each subcircuit has only one output which is the output of the circuit or is connected to a g-type primitive and the inputs of the subcircuit are either $x_1, \ldots, x_n$ or the outputs of g-type primitives. Consider now each S_m to be a basic block and let I_1 and I_2 be two inputs to some S_m, which are connected to two g-type elements having the same label $\langle t, u \rangle$. The circuit can be further reduced by identifying I_1 and I_2 within the subcircuit S_m, deleting I_2 and also deleting the part of the circuit which produces I_2. This is repeated until no two inputs of an S_m are connected to outputs of two g-type elements having the same label number $\langle t, u \rangle$. Further, if there are two inputs I_1 and I_2 of an S_m connected to the same primary input x_i ($1 \le i \le n$), then these two inputs are identified inside the subcircuit and the input I_2 is deleted. Each subcircuit is now an m-input ($0 \le m \le n + K$) circuit consisting of f-type elements which produces at the output a function $f_{m,j}$ where $f_{m,j}$ is the jth function in the list of all $C(m)$ functions that can be produced by an m-input circuit using primitive $F = \{f_1, \dots, f_p\}$ unbounded number of times. Replace the subcircuit S_m by a single primitive $f_{m,j}$.

Thus the original circuit S has been transformed to an equivalent circuit having the following properties:

(i) The circuit consists of g-type primitives (each primitive g_t may have m_t number of inputs $0 \le m_t$) and $f_{m,j}$-type primitives ($1 \le j \le C(m)$, $0 \le m \le n + K$), and has a tree-like structure because no output of a primitive is connected to more than one input of other primitives.

(ii) The maximum level of the circuit is $2K + 1$. This is because input of one $f_{m,j}$-type primitive cannot be derived from the output of another $f_{m,j}$-type primitive; hence maximum level number occurs when $f_{m,j}$-type primitives and g-type primitives alternate from the output level to the first level, giving a total level number of $2K + 1$ since there are at most K distinct g-type primitives in the transformed circuit.

Let $N(n)$ denote the number of different circuits satisfying (i) and (ii). Then we will prove[4] that for sufficiently large n

$$N(n) \le (C(n + K)n)^{n^{4K+2}}$$

To prove this, let $B(n, \lambda)$ denote the set of different n-input circuits satisfying condition (i) and the condition that the level number of the circuit does not exceed λ. Let $N(n. \lambda)$ denote the number of circuits in $B(n, \lambda)$.

The output element of a circuit in $B(n, \lambda)$ could be either a g-type primitive or an $f_{m,j}$-type primitive. The input to this element could be either one of the n inputs x_i or the output of a circuit in $B(n, \lambda - 1)$. Hence for $\lambda \ge 2$ the following inequality holds:

$$N(n, \lambda) \le \sum_{t=1}^{q} (n + N(n, \lambda - 1))^{m_t} + \sum_{m=0}^{n+K} C(m)(n + N(n, \lambda - 1))^m \quad \text{(A.1)}$$

Also, since $N(n, 0) = 0$, we have

$$N(n, 1) \le \sum_{t=1}^{q} n^{m_t} + \sum_{m=0}^{n+K} C(m)n^m \quad \text{(A.2)}$$

---

[4] The detailed proof given here is from Kobayashi (1970). Reddy and Mukhopadhyay (1971) obtained a simplified proof of Kobayashi's theorem after this chapter was written.

Let M denote $\max\{m_1, \ldots, m_q\}$ and $M(n, \lambda)$ denote $n + N(n, \lambda)$. For $n \geq 0$ and $\lambda \geq 1$, we have

$$n \leq M(n, \lambda) \tag{A.3}$$

Also, from definition of M, for any t

$$m_t \leq M \tag{A.4}$$

Since any n-variable function may be regarded as an $(n + 1)$-variable function that does not depend on the $(n + 1)$th variable, we have

$$C(0) \leq C(1) \leq C(2) \leq \cdots \tag{A.5}$$

We assume $F = \{f_1, \ldots, f_p\}$ to be nonempty, otherwise the theorem is trivially true. Let f_1 be an n_1-variable function. Then the n_1-input circuit realizing f_1 is one of the members of $C(n_1)$. Hence $C(n_1) \geq 1$. Using (A.5) we have for any $n \geq n_1$

$$1 \leq C(n + K) \tag{A.6}$$

Let $n_2 = \max\{M - K, 1 + q, 3, K + 2, 2K + 1\}$. Then for any $n \geq n_2$ we have the following inequalities.

$$M \leq n + K \tag{A.7}$$

$$1 + q \leq n \tag{A.8}$$

$$3n \leq n^2 \tag{A.9}$$

$$2n \leq n^2 \tag{A.10}$$

$$1 \leq n \tag{A.11}$$

$$1 \leq 2n \tag{A.12}$$

$$1 \leq n + K \tag{A.13}$$

$$1 \leq M(n, \lambda) \quad \text{(by A.3 and A.11)} \tag{A.14}$$

$$n + K + 2 \leq 2n \tag{A.15}$$

$$n + K + 1 \leq 2n \tag{A.16}$$

$$2K + 1 \leq n \tag{A.17}$$

Let n_0 denote $\max\{n_1, n_2\}$. If $n \geq n_0$ and $\lambda \geq 2$ we have the following inequalities:

$$M(n, \lambda) = n + N(n, \lambda)$$

$$\leq n + \sum_{t=1}^{q} (n + N(n, \lambda - 1))^{m_t} + \sum_{m=0}^{n+K} C(m)(n + N(n, \lambda - 1))^m \quad \text{(by A.1)}$$

$$= n + \sum_{t=1}^{q} M(n, \lambda - 1)^{m_t} + \sum_{m=0}^{n+K} C(m)M(n, \lambda - 1)^m$$

$$\leq M(n, \lambda - 1) + qM(n, \lambda - 1)^M + (n + K + 1)C(n + K)$$
$$M(n, \lambda - 1)^{n+K}$$
$$\text{(by A.3, A.14, A.4, and A.5)}$$

$$\leq M(n, \lambda - 1)^{n+K} + qM(n, \lambda - 1)^{n+K} + (n + K + 1)C(n + K)$$
$$M(n, \lambda - 1)^{n+K}$$
$$\text{(by A.14, A.13, and A.7)}$$

$$= (1 + q + (n + K + 1)C(n + K))M(n, \lambda - 1)^{n+K}$$
$$\leq (n + 2nC(n + K))M(n, \lambda - 1)^{n+K} \quad \text{(by A.8 and A.16)}$$
$$\leq (nC(n + K) + 2nC(n + K))M(n, \lambda - 1)^{n+K} \quad \text{(by A.6)}$$
$$= 3nC(n + K)M(n, \lambda - 1)^{n+K}$$
$$\leq n^2 C(n + K)M(n, \lambda - 1)^{n+K} \quad \text{(by A.9)}$$
$$\leq M(n, \lambda - 1)^2 C(n + K)M(n, \lambda - 1)^{n+K} \quad \text{(by A.3)}$$
$$= C(n + K)M(n, \lambda - 1)^{n+K+2}$$
$$\leq C(n + K)M(n, \lambda - 1)^{2n} \quad \text{(by A.14 and A.15)} \quad \text{(A.18)}$$

If $n \geq n_0$ we have also the following inequalities:

$$M(n, 1) = n + N(n, 1)$$

$$\leq n + \sum_{t=1}^{q} n^{m_t} + \sum_{m=0}^{n+K} C(m)n^m \quad \text{(by A.2)}$$

$$\leq n + qn^M + (n + K + 1)C(n + K)n^{n+K} \quad \text{(by A.11, A.4, and A.5)}$$
$$\leq n^{n+K} + qn^{n+K} + (n + K + 1)C(n + K)n^{n+K}$$
$$\text{(by A.11, A.13, and A.7)}$$

$$= (1 + q + (n + K + 1)C(n + K))^{n+K}$$
$$\leq (n + 2nC(n + K))n^{n+K} \quad \text{(by A.8 and A.16)}$$
$$\leq (nC(n + K) + 2nC(n + K))n^{n+K} \quad \text{(by A.6)}$$
$$= 3nC(n + K)n^{n+K}$$
$$\leq n^2 C(n + K)n^{n+K} \quad \text{(by A.9)}$$
$$= C(n + K)n^{n+K+2}$$
$$\leq C(n + K)n^{2n} \quad \text{(by A.11 and A.15)} \quad \text{(A.19)}$$

If $n \geq n_0$ and $\lambda \leq 2K + 1$ we have the following inequalities:

$M(n, \lambda) \leq C(n + K)M(n, \lambda - 1)^{2n}$

$\qquad \leq C(n + K)^{1+2n}M(n, \lambda - 2)^{(2n)^2}$

$\qquad \leq C(n + K)^{1+2n+(2n)^2}M(n, \lambda - 3)^{(2n)^3}$

$\qquad \vdots$

$\qquad \leq C(n + K)^{1+2n+(2n)^3+\cdots+(2n)^{\lambda-1}}n^{(2n)^\lambda}$

$\qquad \leq C(n + K)^{\overbrace{(2n)^{\lambda-1}+(2n)^{\lambda-1}+\cdots+(2n)^{\lambda-1}}^{\lambda \text{ times}}}n^{(2n)^\lambda}$ $\qquad$ (by A.6 and A.12)

$\qquad = C(n + K)^{\lambda(2n)^{\lambda-1}}n^{(2n)^\lambda}$

$\qquad \leq C(n + K)^{n(2n)^{\lambda-1}}n^{(2n)^\lambda}$ $\qquad$ (by A.6 the condition $\lambda \leq 2K + 1$, and A.17)

$\qquad \leq C(n + K)^{n(n^2)^{\lambda-1}}n^{(n^2)^\lambda}$ $\qquad$ (by A.6, A.10, and A.11)

$\qquad = C(n + K)^{n^{2\lambda-1}}n^{n^{2\lambda}}$ $\qquad\qquad\qquad\qquad$ (A.20)

Obviously $N(n) = N(n, 2K + 1)$. Hence for any $n \geq n_0$

$\qquad N(n) = N(n, 2K + 1)$

$\qquad\qquad \leq n + N(n, 2K + 1)$

$\qquad\qquad = M(n, 2K + 1)$

$\qquad\qquad \leq C(n + K)^{n^{4K+1}}n^{n^{4K+2}}$ $\qquad$ (by A.20)

$\qquad\qquad \leq C(n + K)^{n^{4K+2}}n^{n^{4K+2}}$ $\qquad$ (by A.6 and A.11)

$\qquad\qquad = (C(n + K)n)^{n^{4K+2}}$ $\qquad\qquad\qquad$ (A.21)

which proves the required inequality.

Let $L(n)$ denote the number of n-variable functions realized by n-input circuits using elements of F unbounded number of times and the elements of G bounded number of times. Then for large n,

$$L(n) \leq N(n) \leq (C(n + K)n)^{n^{4K+2}} \qquad\qquad (A.22)$$

Since K does not grow with n, we have

$$\lim_{n \to \infty} \frac{L(n)}{2^{2^n}} \leq \lim_{n \to \infty} \frac{(C(n + K)n)^{n^{4K+2}}}{2^{2^n}} = 0 \qquad\qquad (A.23)$$

where 2^{2^n} is the total number of n-variable functions. Therefore, Eq. (A.23) shows that there exist n-variable functions that cannot be realized by using the elements of F unbounded number of times and the elements of G bounded number of times. Hence $F = \{f_1, \ldots, f_p\}$ is not almost complete.

REFERENCES

BURKS, A. W., and WRIGHT, J. B. (1953). *Proc. IRE* **41**, No. 10, 1357.

GLUSHKOV, V. M. (1963). "Introduction to Cybernetics." Academic Press, New York.

IBUKI, K., NAEMURA, K., and NOZAKI, A. (1963). *Electron Commun. Japan* **46**, No. 7, 55–75.

KAUTZ, W. H. (1966). *IEEE Trans. Electron. Comput.* **EC-15**, 164.

KOBAYASHI, K. (1967). *Electron. Commun. Japan* **50**, No. 12.

KOBAYASHI, K. (1970). Private Communication.

LOOMIS, H. H., JR., and WYMAN, R. H., JR. (1965). *IEEE Trans. Electron. Comput.* **EC-14**, 173.

MARKOV, A. A. (1958). *J. Ass. Comput. Mach.* **5**, No. 4, 331–334.

MULLER, D. E. (1954). *IRE Trans. Electron. Comput.* **EC-3**, 6–12.

POST, E. L. (1921). *Amer. J. Math.* **43**, 163–185.

POST, E. L. (1941). "Two-Valued Iterative Systems of Mathematical Logic." Princeton Univ. Press, Princeton, New Jersey.

QUINE, W. V. (1952). *Amer. Math. Mon.* **59**, No. 8, 521–531.

REDDY, S. M., and MUKHOPADHYAY, A. (1971). "Simple Proofs of Two Theorems on Complete Sets of Logic Primitives," unpublished manuscript.

REED, I. S. (1954). *IRE Trans. Inform. Theory*, **IT-4**, 38–49.

SEMON, W. L., and GOULD, R. (1955). "A Chart for Unate Functions." Progress Rept. BL-13, July 1–Nov. 1, Computation Lab., Harvard Univ., Cambridge, Massachusetts.

SHANNON, C. E. (1949). *Bell Syst. Tech. J.* **28**, 59.

SHESTOPAL, G. A. (1961). *Dokl. Akad. Nauk. SSSR* **140**, No. 2, 314.

YABLONSKII, S. V. (1958). *Trudy Mat. Inst. Steklov.* **51**, 5.

ZHEGALKIN, I. I. (1927). *Mater. Sb. Statni Vyzk. Ustav Mater. Technol.* **34**, 9–22.

Chapter II

COMBINATIONAL
CIRCUITS
WITH
FEEDBACK

DAVID A. HUFFMAN

I. INTRODUCTION	28
A. The Work of Kautz	28
B. The Work of Markov	29
C. The Work of Huffman (A Preview)	30
D. The Importance of Feedback Techniques to Logical Circuit Design	30
II. CIRCUIT VISUALIZATION OF MARKOV'S RESULT	31
A. Threshold Elements	31
B. The Equivalence of Inversion and Inhibition	33
C. A Circuit for the Inversion of Three Variables	33
D. Circuits for the Inversion of $2^k - 1$ variables	35
E. The Nesting of " Markov Circuits "	37
III. A CIRCUIT WITH A SINGLE NOT-ELEMENT WHICH INVERTS TWO VARIABLES	39
A. The First Version of the Circuit	39
B. An Improved Version of the Circuit	40
C. The Signal-Replication Trick	43
IV. THE DESIGN OF "MULTI-INVERSION" CIRCUITS WHICH USE ONLY ONE INVERTER	43
A. The Feedback Concept Applied to the Production of the Excitation E^*	43
B. The Design of the Output-Incrementing Circuit	45
C. Testing and Subsequent Modification of the Circuit	47
V. PROOF OF THE NECESSITY OF UNSTABLE CIRCUIT EQUILIBRIA	48
VI. A "MULTI-INVERSION" CIRCUIT WHICH IS STABLE	51
VII. SUMMARY AND CONCLUSIONS	54
REFERENCES	55

ABSTRACT. This chapter is concerned with the design of switching circuits which can invert arbitrarily large numbers of variables using the absolute minimum possible number of NOT-elements (one or two, depending on the stability requirement). The work of Markov proves that without using feedback techniques k NOT-elements can be used to invert as many as $2^k - 1$ variables and the inversion of 2^k variables requires $k + 1$ NOT-elements. However no constant upper bound could be placed on the number of NOT-elements required for the inversion of arbitrarily many variables if feedback was not used. The results of this chapter together with those of others such as Kautz increase our perspective on what might be possible in designing logical networks and emphasize the role of feedback in the design of combinational circuits.

I. INTRODUCTION

A. THE WORK OF KAUTZ

Feedback, for a switching theorist, is usually a concept firmly associated with sequential circuits, i.e., circuits having memory. The reasons for this are apparent. On the one hand, the logical operations AND, OR, and NOT correspond directly to easily realized gate elements. A Boolean-algebraic expression, as a statement of requirements for an arbitrary combinational circuit, is simply and directly associated with a corresponding logical network of these three types of gate elements using feedforward signal paths only. Thus, combinational circuits do not require feedback. To date, essentially all minimization problems for combinational circuits are or have been approached with an implicit or explicit assumption that only feedforward networks are to be considered. On the other hand, the requirements for a sequential circuit (a circuit having two or more circuit states corresponding to at least one state of the input variables) can be realized only by a logical network having feedback paths.

It is perhaps only natural, therefore, that the use of feedback is not usually considered in the design of combinational circuits. An example of an exception to that general rule is the recent work of Kautz (1970). In a short note he described networks of the type shown in Fig. 1. This network has three identical stages of two NOR-elements each. When the number of stages in such a circuit is odd and when all inputs (X_1, X_2, ...) are held fixed at the value 1, the circuit will oscillate. However, under these conditions all outputs ($Z_1, Z_2, ...$) will have the value 0, in spite of the oscillation. When the number of stages is even there will be no oscillation, but for one combination of values of the input variables the circuit may be in either one or the other of two stable states. For either of these two states the output values will be the same.

Each member of the class of circuits described by Kautz produces, therefore, a unique set of output values for each combination of input values

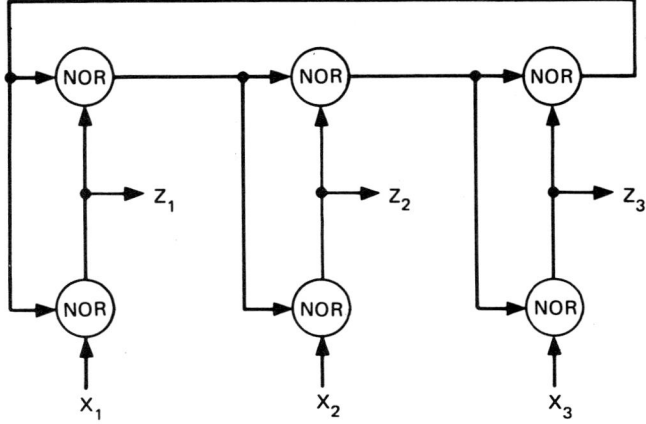

FIG. 1. A combinational circuit which makes use of feedback.

(in the circuit shown: $Z_1 = \bar{X}_1(X_2 + \bar{X}_3)$, $Z_2 = \bar{X}_2(X_3 + \bar{X}_1)$, and $Z_3 = \bar{X}_3(X_1 + \bar{X}_2)$) in spite of the fact that it has a feedback loop. Kautz proved that circuits which would produce the same functions without using feedback would require more NOR-elements than his circuit require. These circuits demonstrate that feedback can lead to more economical combinational circuits than would be the case if feedback were not permitted.

B. THE WORK OF MARKOV

When a circuit is composed entirely of one type of switching element (for example, entirely of 2-input NOR-elements) the meaning of "minimization" can be defined in an unambiguous way. It is then clear that a realization with a given number of elements is "better" in nearly any sense of the word than any realization requiring more elements. When several different types of elements are used, the situation is not so straightforward and the relative "costs" of the various types of elements must be considered.

An example of a problem in this area is that of minimizing the number of NOT elements in a circuit regardless of the numbers of AND- and OR-elements which may then be necessary. This problem is equivalent to that of minimizing the cost of networks under the assumption that NOT-elements are, by comparison, infinitely more costly than are the AND and OR elements. This problem is particularly important from a theoretical point of view, for if we knew how to generate economically the complements of a set of input variables, then the cost of an arbitrary set of functions on those input variables would need to be no more than that required to generate the complemented variables. This conclusion follows immediately from the fact that any function of n variables can be expressed as a set of input variables and their complements appropriately combined using the operations AND and OR.

The most significant contribution to date to the problem of minimizing NOT-elements was made by Markov (1957). He concluded that k NOT-elements (inversions) were sufficient for generating the complements of $2^k - 1$ Boolean variables and demonstrated, in a rather concise algebraic formulation, how this could be done. In an earlier paper Gilbert (1954) had considered the question of inversion complexity but offered no proof of minimality.

Shortly after Markov's paper was published Muller wrote an unpublished memorandum (Muller, 1958). Muller's paper included a schematic diagram for the case $k = 2$; that is, a diagram showing how to design a network containing only two NOT-elements which produced as its three outputs the complements of its three input variables.

C. THE WORK OF HUFFMAN (A PREVIEW)

The major result of this chapter was first published by this author in 1969. It depends heavily on the work of Markov mentioned above. However, instead of restricting myself to networks with feedforward signal paths only, as did Markov, I considered instead what might be accomplished if feedback paths were allowed. A constructive proof will follow which shows that *an arbitrarily large number of input variables may be complemented with a network containing only one NOT-element*. Since it is easy to demonstrate that the EXCLUSIVE OR function, for example, requires at least one NOT-element, the result described here will never be improved upon.

This chapter will show that any conceivable logical network whatsoever which produces from its two or more input variables the corresponding complemented variables and which contains only *one* NOT-element must have unstable equilibria (must oscillate). Another constructive proof will demonstrate that when *two* NOT-elements are used the oscillation can be avoided.

D. THE IMPORTANCE OF FEEDBACK TECHNIQUES TO LOGICAL CIRCUIT DESIGN

The circuits of both Huffman and Kautz constitute examples which demonstrate that feedback can allow combinational circuits to be designed which are more economical, in some sense of that word, than would be possible without feedback. At this time these examples are isolated ones. They do however provide tantalizing glimpses into an imaginable area of future research. As a result of that research the future logical circuit designer may begin his work with the assumption that feedback is likely to be necessary in achieving an economical circuit; cases in which feedback is not profitable may be the pathological ones. Will oscillation in these circuits prove to be the rule rather than the exception? Or will the perspective of time show us that the two examples above were atypical unstable artifacts in a generally stable population of circuits?

One more point before ending this introduction: if an arbitrary set of combinational circuits can be realized with one NOT-element, then an arbitrary sequential circuit can also be realized. This conclusion follows from the fact that any sequential circuit can be thought of as a set of combinational circuits with appropriate feedback. Thus, in theory at least, even the largest general purpose computer could be built with one NOT-element!

II. CIRCUIT VISUALIZATION OF MARKOV'S RESULT

A. THRESHOLD ELEMENTS

The essential nature of Markov's main result can probably best be appreciated if it is seen in the form of the diagram of a logical circuit. The diagrams shown here will contain threshold elements exclusively.

A commonly used definition of a threshold element (hereafter called a T-element) is the following one:

DEFINITION.   Associated with a T-element having q binary inputs X_1, X_2, ..., X_q are corresponding integers $W_1, W_2, \ldots, W_q$ called its "weights" and an integer T called its "threshold." For a given T-element each of these integers is fixed but each may be positive, negative, or zero. Each of the q inputs and the single output Z of the element shall be assumed here to have one of two complementary values, 0 or 1, at a given moment. The weighted sum $W_1 X_1 + W_2 X_2 + \cdots + W_q X_q$ will be called here the "excitation" E of the T-element. The output will be defined to be 1 if and only if the excitation of the element is equal to or exceeds the threshold value. That is,

$$\text{if} \quad E = W_1 X_1 + W_2 X_2 + \cdots + W_q X_q \geq T, \quad \text{then} \quad Z = 1$$

and

$$\text{if} \quad E = W_1 X_1 + W_2 X_2 + \cdots + W_q X_q < T, \quad \text{then} \quad Z = 0$$

Not every Boolean function can be realized by a single T-element. Those which can be will be called here "T-functions." An example of a T-function is

$Z = X_1 X_3 + X_3 \bar{X}_4 + X_1 \bar{X}_2 \bar{X}_4$. The schematic diagram of Fig. 2a shows a realization for this function by a T-element with a threshold $T = 2$ and with weights $W_a = 2$, $W_b = -1$, $W_c = 3$, and $W_d = -2$.

The input variables which are associated with negative weights may be said to have an inhibitory effect on the production of the output of a T-element. These variables are those which appear complemented in the Boolean algebraic expression. The positively weighted input variables have an excitatory effect on the element and they appear uncomplemented in the expression. It can easily be recognized that any T-function must be able to be expressed by some corresponding expression in which each variable appears either complemented or uncomplemented but not both, because no given input of a T-element can at the same time be both inhibitory and excitatory.

The converse of the above argument is not valid. Any function which can be written with each variable either complemented or uncomplemented but not requiring both is called an "unate" function. However, not all unate functions are T-functions. An illustrative example is $F = X_1 X_2 + X_3 X_4$. (That function is an example of a "positive unate" function.) If this were a T-function each weight would have to be positive. Because of the symmetry apparent in the expression each variable could have the same weight associated with it as do the other variables. But in that case the function would have the value 1 whenever *any* pair of the four variables had the value 1, not merely when one of the two pairs (X_1, X_2) or (X_3, X_4) had that value. Thus there is no method of assigning weights to a single T-element so that it will produce the function $F = X_1 X_2 + X_3 X_4$.

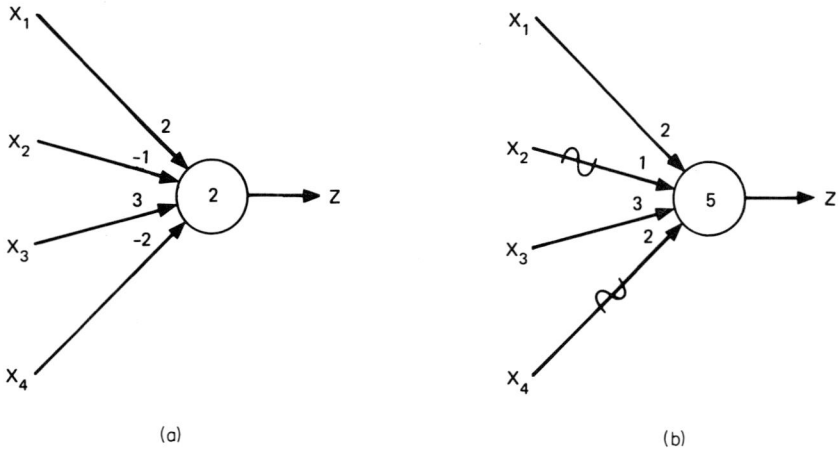

(a)   (b)

FIG. 2.   Equivalent realizations of $Z = X_1 X_3 + X_3 \bar{X}_4 + X_1 \bar{X}_2 \bar{X}_4$: (a) with negative weights, (b) with NOT-elements.

B. THE EQUIVALENCE OF INVERSION AND INHIBITION

It is always possible to replace an arbitrary T-element which has negative weights associated with some of its inputs by a T-element having only positive weights if those input variables are complemented before they are applied to the T-element and if the threshold value is increased accordingly. The example in Fig. 2a is derived from the equivalent T-element described in Fig. 2b. The symbol ⌣ is descriptive of the necessary complementation (or inversion) of the second and fourth binary variables. The required positive change in threshold value is the sum of the magnitudes of the weights assigned to those two variables.

The inversion (or complementation) of a binary signal in a network of logical elements corresponds to the NOT operator of logic. That inversion can itself be realized by a simple T-element having only one input and one output. The appropriate weight for the input is $W = -1$ and an appropriate threshold value is $T = 0$. Thus the transformation illustrated in Fig. 2 did not *eliminate* the negative weights. It merely removed them to locations away from the multi-input T-element. It is apparent that when we derive networks of T-elements for which we claim a minimum number of NOT-elements we must not allow other T-elements to have any negative weights for, as we have seen, these negative weights are equivalent to additional NOT-elements. Therefore, in the networks which follow, we shall allow only the required number of NOT-elements, and additional T-elements having only positive weights; we shall call these additional elements "positive" T-elements.

C. A CIRCUIT FOR THE INVERSION OF THREE VARIABLES

A network of positive T-elements and two NOT-elements which has as its outputs the complements of its three input variables is shown in Fig. 3. This circuit was inspired by and implicit in some of the equations found in the previously mentioned work of Markov (1957). The function g_2 (the reason for this subscript nomenclature will be apparent below) is a "majority" function of the input variables. One of the NOT-elements is used to invert this function (creating, thereby, the "minority" function). This new function h_2 is "high" (has the value 1) if and only if either none or one of the variables X_1, X_2, or X_3 is high.

Whenever none or one of the three X-values is high a contribution of 2 is made to the excitation of the T-element which produces the function g_1. Equivalently, for those cases, the effective value of the threshold of that element is reduced from 3 to 1 insofar as its three remaining inputs (X_1, X_2, and

X_3) are concerned. That is, the element producing g_1 acts as if it had only the three X-inputs, and the obvious threshold value 3, when two or three of these X-values are high; when none or only one of these values is high, the threshold value is effectively reduced to 1. The net effect of this is that the output g_1 will be high whenever either any one or all three of (i.e., whenever an odd number of) the X-inputs is high, and will be "low" (have the value 0) otherwise. The second NOT-element is used to invert this function. The resulting function $h_1 = \bar{g}_1$ is high if and only if none or two of (if an even number of) the three X-inputs is high.

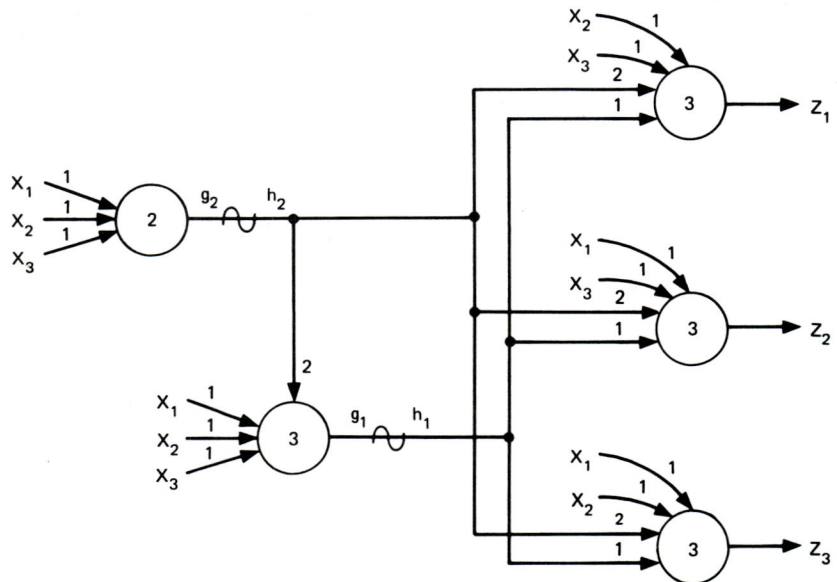

FIG. 3.   A circuit for the inversion of three variables.

A component of the excitation of each of the three remaining T-elements is common to each. This component is the binary function h_1 weighted by 1 plus the binary function h_2 weighted by 2. It can easily be deduced that this component is three when none of the X-inputs is high, is two when one of the X-inputs is high, is one when two of the X-inputs are high, and is zero when all three X-inputs are high. It is apparent that this component of excitation depends in a symmetric way upon the three X-inputs. Thus, it is convenient to let the symbol $|X|$ refer to the number of the X-inputs which have the value 1. That is, $|X|$ is equal to the ordinary sum $X_1 + X_2 + X_3$. With this nomenclature the common component of the excitation of the three rightmost T-elements may be expressed as $3 - |X|$.

The T-element which has as its output the function Z_1 is typical of the rightmost three elements. Its total excitation is $3 - |X| + X_2 + X_3 = 3 - |X| + |X| - X_1 = 3 - X_1$. This excitation therefore depends only upon the X_1-input. If X_1 is 0 the total excitation is 3; if X_1 is 1 that excitation is 2. Since the threshold value is 3, the output Z_1 of the element will be 1 when $X_1 = 0$ and will be 0 when $X_1 = 1$. In other words, $Z_1 = \overline{X}_1$, the complement of the X_1 input. It may be proved similarly that $Z_2 = \overline{X}_2$ and $Z_3 = \overline{X}_3$.

This remarkable and nonintuitive construction leads us to conclude that three independently-acting binary variables may be inverted in a logical network which contains only two NOT-elements and five additional T-elements which do not use inhibition. However interesting a result this is from a theoretical viewpoint, it might not seem to lead to any direct reduction in "cost" unless we can imagine a circuit technology in which the cost of five T-elements is less than the cost of a single inversion. This easy conclusion is not entirely correct unless we assume that the end goal of the design is to obtain the inverted variables for their own sake.

Recall that once a set of input variables and their complements are available, any function or set of functions whatsoever of these variables can be produced without further need either of NOT-elements or of elements with inhibitory inputs. Consequently, the incremental cost of saving one NOT-element in producing the complemented variables should properly be compared with the possible reduction in cost of not requiring further inhibitory inputs in *any* of the set of networks producing the desired set of functions.

D. CIRCUITS FOR THE INVERSION OF 2^k—1 VARIABLES

The circuit shown in Fig. 3 can be generalized to yield a class of circuits which use k inverters to invert as many as $2^k - 1$ variables. We shall illustrate with a circuit having $k = 4$ inverters which can invert $2^4 - 1 = 15$ variables. It will be described in two parts. The first part produces an excitation having the value $15 - |X|$. The second part consists of fifteen similarly acting T-elements which produce the inverted variables themselves.

In the diagram of Fig. 4a a shorthand notation has been used to describe the set of fifteen input variables. Each of these variables is given a weight of 1 at each of the four T-elements. Since each of the inputs contributes an equal excitation, it is again appropriate to use the symbol $|X|$ to represent the number of X-variables which have the value 1. Because there are fifteen X-variables, the possible value of $|X|$ ranges between 0 and 15, inclusive.

The table of Fig. 4b shows how the four g-functions and the four h-functions depend upon $|X|$. Immediately after each of the four g-functions is

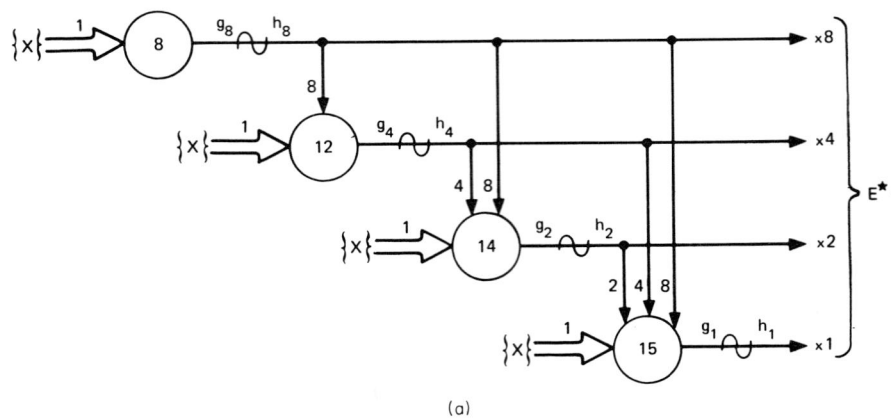

(a)

$\lvert X \rvert$	g_8	$8h_8$	g_4	$4h_4$	g_2	$2h_2$	g_1	$1 \cdot h_1$	$8h_8 + 4h_4 + 2h_2 + 1 \cdot h_1$
0	0	8	0	4	0	2	0	1	15
1	0	8	0	4	0	2	1	0	14
2	0	8	0	4	1	0	0	1	13
3	0	8	0	4	1	0	1	0	12
4	0	8	1	0	0	2	0	1	11
5	0	8	1	0	0	2	1	0	10
6	0	8	1	0	1	0	0	1	9
7	0	8	1	0	1	0	1	0	8
8	1	0	0	4	0	2	0	1	7
9	1	0	0	4	0	2	1	0	6
10	1	0	0	4	1	0	0	1	5
11	1	0	0	4	1	0	1	0	4
12	1	0	1	0	0	2	0	1	3
13	1	0	1	0	0	2	1	0	2
14	1	0	1	0	1	0	0	1	1
15	1	0	1	0	1	0	1	0	0

(b)

FIG. 4.   A circuit which produces $E^* = 15 - \lvert X \rvert$.

produced by a T-element, it is inverted by one of the four NOT-elements to generate the corresponding h-function. Each h-function is weighted by an integer value corresponding to the subscripted index of that function wherever else that h-function is used in the network.

The first (leftmost) T-element of the circuit has an excitation $\lvert X \rvert$ which ranges between 0 and 15 and it has a threshold value of 8. The majority function it generates is called g_8. The second T-element has an excitation $\lvert X \rvert + 8h_8$ which ranges between 8 and 15, and it has a threshold value of 12. The third T-element has an excitation $\lvert X \rvert + 8h_8 + 4h_4$ which ranges between

12 and 15, and it has a threshold value of 14. The final T-element has an excitation $|X| + 8h_8 + 4h_4 + 2h_2$ which can be either 14 or 15, and it has a threshold value of 15. The output of that final T-element is a function g_1 which is high whenever an odd number of the fifteen X-variables are high. The fourth, and final, NOT-element is used to invert this function.

The components of the excitation E^* are obtained by weighting the binary functions h_8, h_4, h_2, and h_1 by 8, 4, 2, and 1, respectively. This excitation has the value $15 - |X|$. Figure 5 shows how E^* is used as part of the

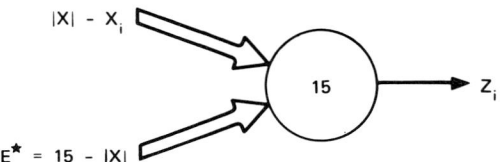

FIG. 5.   A typical one of the T-elements which generate $Z_i = \overline{X}_i$.

total excitation of a typical T-element which produces one of the circuit outputs Z_i. The remainder of the excitation is $|X|$ *minus* that of the appropriate input X_i. In circuit terms this could be done by applying as inputs to that T-element all 15 of the inputs *except* the one denoted by X_i. The *total* excitation to that element is therefore $15 - X_i$.

Whenever $X_i = 0$ the total excitation is 15 and, since the threshold is 15, the output $Z_i = 1$. Whenever $X_i = 1$ the total excitation is 14 and the resulting output is $Z_i = 0$. Consequently $Z_i = \overline{X}_i$, the value complementary to X_i.

It is apparent that the general scheme shown in the circuit of Figs. 4 and 5 can, in theory, be generalized to an arbitrary number $k \geq 2$ of NOT-elements. Such circuits would have, in addition to those k inverters, k T-elements for the formation of the excitation E^*, and $2^k - 1$ T-elements the outputs of which would be the complemented values of the $2^k - 1$ input variables.

E. THE NESTING OF "MARKOV CIRCUITS"

The net effect of the circuit formulation of Markov's results is that such a circuit containing only k inverters can be substituted for a set of $2^k - 1$ inverters. Thus, instead of using $127 = 2^7 - 1$ inverters, we can imagine using a circuit containing only seven inverters (and $134 = 7 + 127$ other T-elements). As another example, we can also think of replacing $7 = 2^3 - 1$ inverters by a circuit containing only three inverters (and $10 = 3 + 7$ other T-elements). Likewise, we can replace three inverters by the circuit of Fig. 3 which contains two inverters (and $5 = 2 + 3$ other T-elements). The idea of

" nesting" these circuits naturally arises and would lead immediately to the conclusion that 127 inverters (or even $2^{127} - 1$ inverters!) could ultimately be replaced by two inverters. This nesting idea is represented in Fig. 6. As attractive as this concept is, the actual analysis of such nested circuits is nearly beyond the realm of practicality.

What is *not* apparent from the diagram of Fig. 6 is that when the circuits are nested, feedback loops are formed. Therefore, the resulting circuit must be analyzed by sequential circuit techniques. Even that technique is not practical beyond the case of the two-layer circuit which results from the nesting of the "$3 \rightarrow 2$" circuit inside the "$7 \rightarrow 3$" circuit. The author has performed this analysis on that latter example but it is too complicated to give here, especially in view of the main result of this chapter which comes from an alternate approach which is simpler to describe.

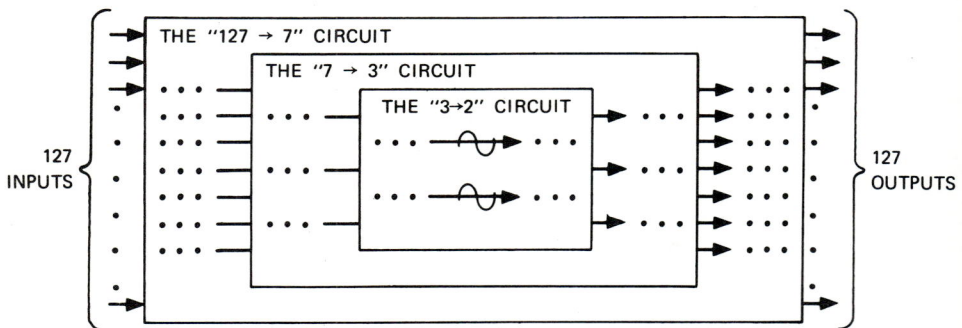

FIG. 6.   The basic nesting idea.

Before giving the results of that analysis, a few comments are in order. Imagine that several layers of these circuits are nested and that the inputs at each circuit layer are set to the values which would be appropriate for the inputs of the inverters which the circuit replaces. The outputs of each circuit would then be the same as the outputs of the inverters themselves. It is clear that however deeply the circuits are nested, there will exist, for each combination of values of the inputs to the overall circuit, such a circuit equilibrium and that it will be stable. Furthermore, at this equilibrium the circuit outputs will, in fact, be the complements of the corresponding input variables. The vital questions are whether or not there are *other* equilibria (either stable or unstable) as well and whether, when the circuit input values are changed, the circuit might enter one of these other equilibria.

These questions can be answered in the negative, at least for the two-layer circuit. That circuit contains only two feedback loops. The author has shown that for each of the 2^7 possible combinations of values of the seven input variables there is exactly one equilibrium which is stable and which therefore leads to the correctly complemented outputs.

The author has good reason to believe that the other more deeply nested circuits also have a unique equilibrium and thus that they also produce the hoped-for terminal action, particularly if the delays in the feedback loops in an "outer" layer of the next are significantly greater than delays in the loops within that layer.

III. A CIRCUIT WITH A SINGLE NOT-ELEMENT WHICH INVERTS TWO VARIABLES

A. THE FIRST VERSION OF THE CIRCUIT

In this section we shall demonstrate a simple sequential circuit which has only one NOT-element and which generates as its outputs the complements of its two input variables. Thus this circuit could be used to replace two NOT-elements. The circuit is shown in Fig. 7. It contains, in addition to the single NOT-element, three identical T-elements. Each of them has three equally weighted inputs and a threshold value of 2 and therefore each acts as a majority element. A signal-delay element is inserted into the single feedback loop as shown. We shall assume that the amount of this delay is much larger than any other delay to which signals may be subjected in other parts of the circuit.

Consider the action of the circuit when both X_1 and X_2 have the value 0. In that case the signal s is 0, independent of the value of q which is fed back to the first T-element. The inverted signal $\bar{s} = p$ is 1. After the circuit has reached

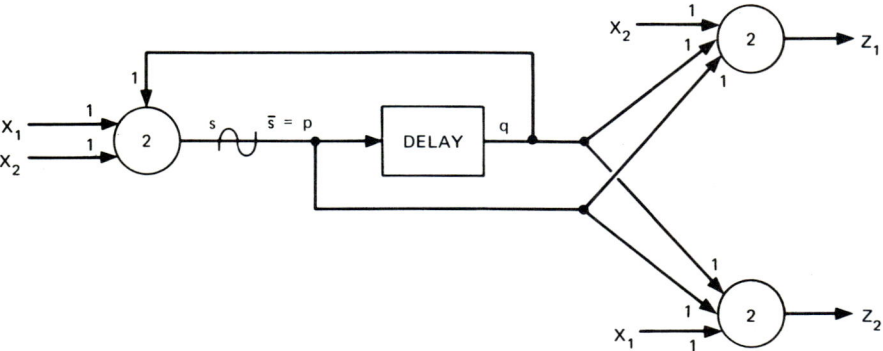

FIG. 7.   A circuit which inverts two variables.

its (stable) equilibrium the signal q is also 1. Since $p = q = 1$ the excitation of each of the other T-elements is 2, the same value as the thresholds of these elements. Consequently $Z_1 = Z_2 = 1$, the value which is complementary to the common value of X_1 and X_2.

Consider what happens if $X_1 = X_2 = 1$. Then we can deduce that $s = 1$, $\bar{s} = p = 0$, and, at equilibrium, $q = 0$. Since $p = q = 0$, the excitation of each of the two output T-elements is 1. Therefore $Z_1 = Z_2 = 0$, the value which is complementary to the common value of X_1 and X_2.

A more interesting situation arises if $X_1 = 0$ and $X_2 = 1$ or if $X_1 = 1$ and $X_2 = 0$. In those cases the excitation of the first T-element is $q + 1$. Since the threshold of that element is 2, its output depends directly upon the value of q. When $q = 1$ it follows that $s = 1$; when $q = 0$, $s = 0$. Thus $s = q$ and $\bar{s} = p = \bar{q}$.

Both p and q are inputs to the pair of T-elements on the right. Since p and q always have complementary values, they at all times make a joint contribution of exactly 1 to the excitation of those two elements. Because each of these two elements has a threshold value 2, their outputs will be high if and only if their X-input is high. That is $Z_1 = X_2$ and $Z_2 = X_1$. There are two cases to be considered. When $X_1 = 0$ and $X_2 = 1$, it follows that $Z_1 = X_2 = 1$ and $Z_2 = X_1 = 0$. When $X_1 = 1$ and $X_2 = 0$ it follows that $Z_1 = X_2 = 0$ and $Z_2 = X_1 = 1$. In either of these two events $Z_1 = \bar{X}_1$ and $Z_2 = \bar{X}_2$.

Our conclusion is that whichever of the four combinations of values of X_1 and X_2 may be present the circuit outputs Z_1 and Z_2 assume, respectively, the complements of these values. This was the desired result.

The situation in which p and q have complementary values (it occurs when X_1 and X_2 have complementary values) is worth investigating in more detail. If we assume that $q = 0$, the first T-element and the NOT-element produce the signal value $p = 1$. After a time corresponding to the delay of the delay element the value of q changes to 1. This value, in turn, causes p to change to 0. After another equal interval of time the value of q becomes 0 again and p then returns to 1. This cycle is repeated as long as the values of X_1 and X_2 are not changed. As we have already seen, this circuit oscillation does not affect the outputs. They remain constant at the desired values.

B. AN IMPROVED VERSION OF THE CIRCUIT

If the circuit of Fig. 7 were actually constructed we would undoubtedly notice variations from the ideal behavior described above. In particular, when the circuit was oscillating, the signals p and q which ideally have complementary values, might arrive at one or the other of the rightmost T-elements at slightly different times. This could occur, for instance, if the propagation times of p and q to the element which generates Z_1 were slightly different. In that event their joint contribution to the excitation of that element would not be absolutely constant at 1. Instead, the excitation might momentarily be either 0 or 2. The effect on Z_1 would be to make it change from its theoretical value of $\overline{X}_1$ for brief fractions of the period of oscillation. The larger the amount of the delay in the delay element the less that fraction of time would be.

If it became a matter of practical importance this nonideal circuit action could be corrected. The technique will be called the "signal-replication trick." It is the basis of the modified circuit shown in Fig. 8a. That circuit is the same as the previous one except that additional delay elements have been added. We assume that these three elements delay their signals by roughly equal amounts. Their exact values are not critical. We also assume that each of these delays is larger than any other signal delay in the circuit.

From the analysis of the earlier circuit it is easy to see that when X_1 and X_2 have complementary values, the circuit will oscillate at a frequency which depends on the magnitude of the delay between the signal a and the signal c. The signals b and d are delayed *replicas* of a and c, respectively. The behavior of these four signals as functions of time is shown in Fig. 8b.

In drawing this diagram it was assumed that all three delay elements were identical. In that idealized case exactly two of the signals $a, b, c,$ and d would be high (when the circuit was oscillating) and their contribution to the excitations of the rightmost T-elements would be constant at 2.

Departures from the ideal can occur. First, the three delays will not be exactly equal; second, the signal propagation times of these four signals to the T-elements will not be precisely the same. As long as the three inserted delays are nearly matched and each is significantly larger than the signal propagation times through the remainder of the network, we may be assured that the excitation these four signals provide will vary by no more than one unit from the nominal value of 2. That is, the excitation from the four signals will be in the range 1 to 3 when the circuit is oscillating.

When the circuit is *not* oscillating, the values of $a, b, c,$ and d will be the same; either all of them will be low or all of them will be high. That is, the excitation they yield in those cases will be either 0 or 4.

Now let us examine the action of the T-element from which Z_1 is generated. If $X_1 = X_2 = 0$, then $a = b = c = d = 1$ and $Z_1 = 1$. If $X_1 = X_2 = 1$, then $a = b = c = d = 0$ and $Z_1 = 0$. If X_1 and X_2 have complementary values, then $a + b + c + d$ will be in the range 1 to 3. If $X_2 = 0$ the total excitation will therefore be below the threshold value of 4, and Z_1 will be 0. If, however, $X_2 = 1$ the total excitation will be in the range 4 to 6, and Z_1 will be 1. In summary, Z_1 will always be complementary in value to X_1. A similar argument holds for Z_2 and X_2; the equilibrium value of Z_2 is $\overline{X}_2$.

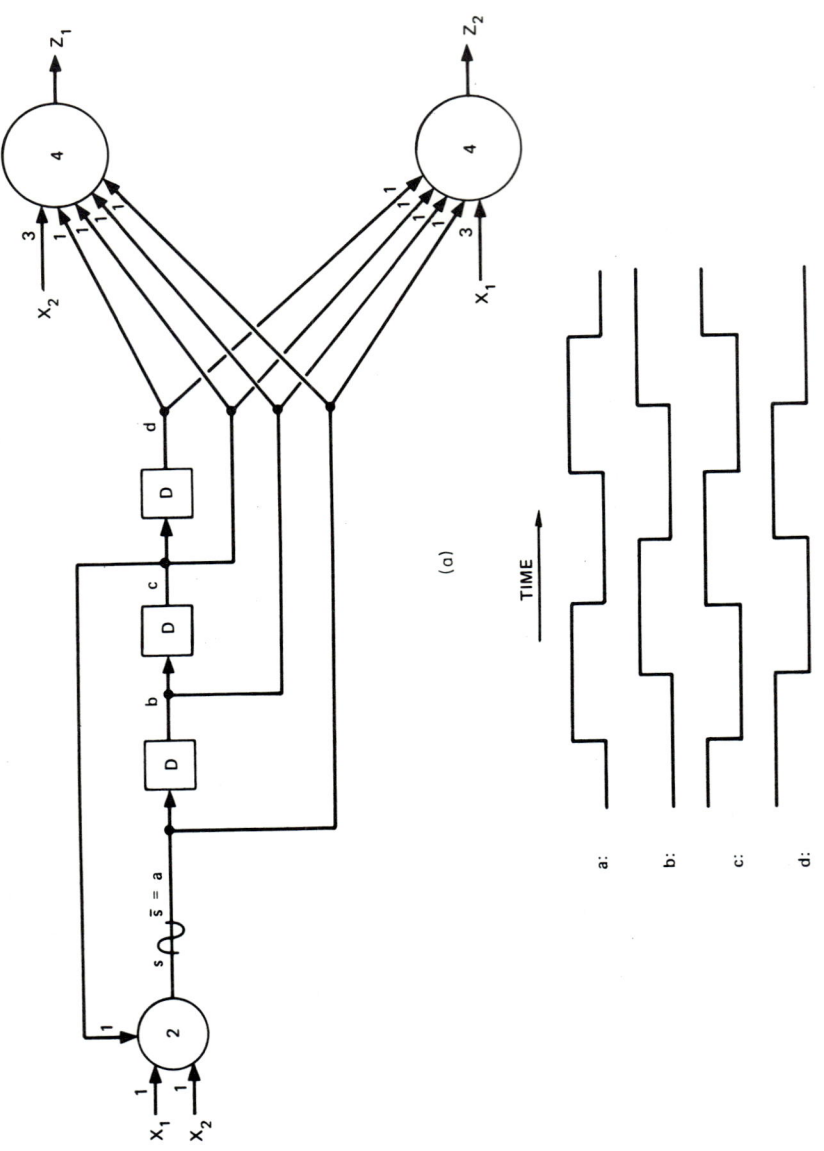

FIG. 8.   An improved version of the circuit of Fig. 7: (a) the circuit, (b) the behavior of the four signals a, b, c, and d when $X_1 = X_2$.

C. THE SIGNAL-REPLICATION TRICK

The "signal-replication trick" is one to which we shall refer later in this chapter. It was used here to compensate for the fact that p and q in the original circuit of Fig. 8, although ideally having complementary values when the circuit was oscillating, would in practice not actually provide an absolutely constant excitation component to the final T-elements. This ideal excitation component had the value 1. The transformation illustrated in Fig. 8 in effect provided that the a and c signals were each replicated (with a delay D) in the b and d signals. If we had been able to weight each of these four signals by the multiplicative factor $\frac{1}{2}$ then there would have been no need to change the original weight (1) associated with the X's and no need to change the threshold values (2). The component of excitation from the a, b, c, and d signals would have nominally been 2 but in the nonideal situation we wanted to be able to handle might have been $1\frac{1}{2}$, 2, or $2\frac{1}{2}$. Consequently, we can see that the changed values of the weights of the X-variables and of the thresholds were not fundamental. They were necessary only because, in the traditional mathematical model of a T-element, it was not possible to weight an input by a noninteger factor.

IV. THE DESIGN OF "MULTI-INVERSION" CIRCUITS WHICH USE ONLY ONE INVERTER

A. THE FEEDBACK CONCEPT APPLIED TO THE PRODUCTION OF THE EXCITATION E*

We saw in the last section that a sequential circuit allowed us to create two independent signal inversions with only one NOT-element. This is a result which it is not possible to achieve using the usual combinational circuit techniques. This fact will be proved in Section V. Indeed we shall prove there an even stronger result: a circuit which inverts two or more variables with only one inverter *must oscillate* (have an unstable equilibrium).

If we examine Fig. 4 from the viewpoint of someone familiar with the concept of feedback in a somewhat more conventional context, we might notice that the excitation E^* which the circuit produces plus the excitation associated with the input is constant. Any other circuit which would maintain the output at a value such that it plus the value of the input is constant at the desired value would serve the same purpose.

A schematic representation of the plan that a circuit designer might envision for the problem of inverting seven variables is shown in Fig. 9a. (The generalization to 15 or to any other number of variables will be obvious.) A T-element with a threshold of 7 is used for the required comparison. Whenever the sum of the excitations for the set of X-signals and the set of y-signals is less than the required value, the signal s will be equal to 0; whenever this sum is equal to or exceeds that threshold value, s will be equal to 1.

The value of the signal s is, therefore, an indicator of whether or not the output of the incrementing circuit is too low. The designer of more conventional feedback circuits might ordinarily expect that the "error" signal would tell him by exactly *how much* the output had deviated from the desired value. We shall see that the binary indication given by the signal s will suffice.

A single NOT-element will be used. Its purpose is to invert the signal s. The yet-to-be-designed part of the circuit in Fig. 9a must use only T-elements having positive weights if we are to fulfill the promise about constructing a circuit with only one inverter. (Recall that negative weights are equivalent to signal inversions.)

The terminal requirements of that part of the circuit are as follows:

(i)   When $\bar{s} = 1$ ($s = 0$), the value of $|y|$ is too low and it should be made to increase.

(ii)   When $\bar{s} = 0$ ($s = 1$), the value of $|y|$ *may* be too high and it should be made to decrease.

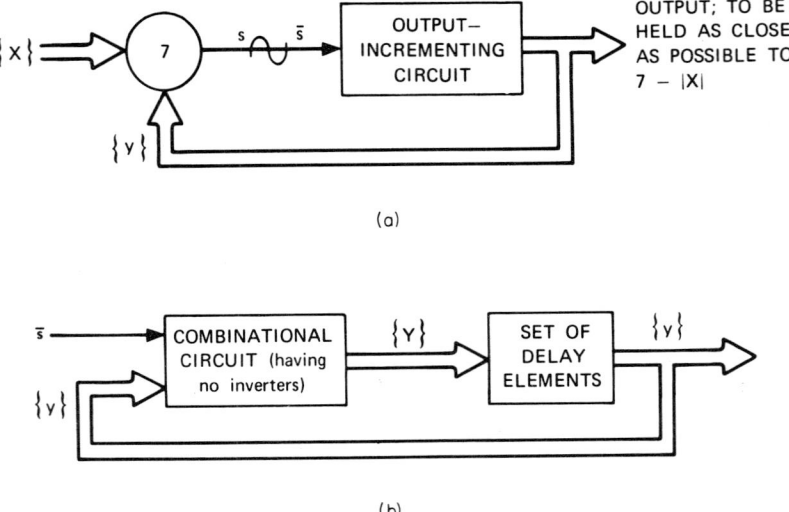

(a)

(b)

FIG. 9.   A design for the "multi-inversion" circuit: (a) a plan for circuits to replace circuits like that of Fig. 4, (b) a more detailed plan of the output-incrementing circuit.

The reader may have noticed the careful wording of (ii). The designer of logic circuits is constrained by the fact that the signals in his circuits are binary-valued. A single signal cannot be used to differentiate among the three obviously desirable types of error indication (too low; too high; exactly right). Only two kinds of indication are possible. We have chosen these to be "too low" and "too high *or* exactly right." The reader should have faith at this point that it *will* be possible to work around this only slightly troublesome limitation on what the signal $\bar{s}$ can, by its value, indicate.

B. THE DESIGN OF THE OUTPUT-INCREMENTING CIRCUIT

We call the circuit which has $\bar{s}$ as its input and the set of signals $\{y\}$ as its output the "output-incrementing" circuit. When $\bar{s} = 0$, we will decrease $|y|$ (the number of y-signals which are high), and when $\bar{s} = 1$, we will increase $|y|$. There is no *necessity* in deciding that the appropriate increment in $|y|$ is *one* unit, but that is, for convenience, the choice we shall make. (We could imagine, for instance, a more elaborate plan in which the size of the increment changed with time. Initially, it might be large, decreasing to one unit only after the signal s had changed value.)

The output-incrementing circuit will itself be a sequential circuit having one input $\bar{s}$ and a set of six feedback loops. The outputs of these loops are the signals in the set we have already called $\{y\}$ and the set of inputs to these loops will be called $\{Y\}$. The combinational component of the circuit will have feedforward paths only and must be designed so that the $\{Y\}$ signals have the proper dependence upon $\bar{s}$ and the $\{y\}$ signals. The table in Fig. 10a shows this dependence. There we have required that $|Y|$ (which is the "next" value of $|y|$) be less than $|y|$ when $\bar{s} = 0$ and greater than $|y|$ when $\bar{s} = 1$, whenever that is possible. Of course, the least possible value of $|Y|$ is 0 and its greatest possible value is 6.

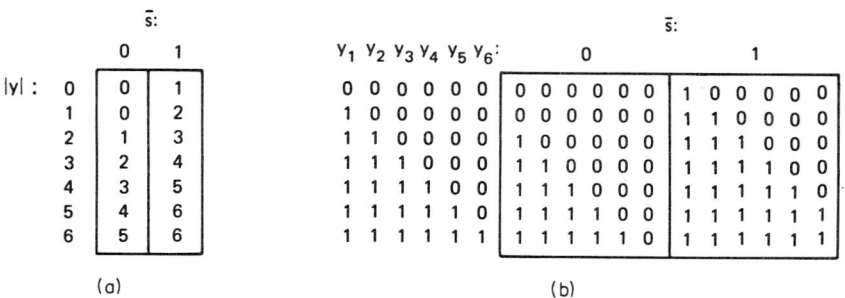

FIG. 10. Design requirements for an output-incrementing circuit: $|Y|$ as a function of $|y|$ and $\bar{s}$, (b) requirements for the Y-signals.

Another more detailed description of how the signals in the set $\{Y\}$ depend upon those in the set $\{y\}$ and upon $\bar{s}$ is given in Fig. 10b. A simple unary encoding of $|y|$ has been used. For instance, when $|y| = 4$, we shall represent that situation by letting $y_1 = y_2 = y_3 = y_4 = 1$ and $y_5 = y_6 = 0$. By observing the entries in the table we can determine how each of the six Y-functions depend upon the y-variables and $\bar{s}$. The functions Y_1 and Y_6 are both special cases. The other four functions illustrate the general rule for production of these Y-functions.

The leftmost components of the entries in the table tell us that Y_1 should be equal to 1 if and only if $\bar{s} = 1$ *or* $y_2 = 1$ (or both). The rightmost components of the table entries tell us that Y_6 should be equal to 1 if and only if $\bar{s} = 1$ *and* $y_5 = 1$. Each of these two functions requires a T-element with only two inputs, as is shown in Fig. 11.

As an example of the more general case we analyze the entries corresponding to Y_3. We see that under the appropriate conditions an increase (from 0 to 1) in the value of either $\bar{s}$ or y_2 or y_4 can cause an increase (from 0 to 1) in the value of Y_3. More explicitly, when none or one of the variables $\bar{s}$, y_2, and y_4 are equal to 1, the required value of Y_3 is 0. Whenever two or three of those variables are 1, the required value of Y_3 is 1. The function Y_3 can be produced by a T-element with these three variables as inputs, each weighted by 1. The threshold should have the value 2. This same type of dependence holds for the remaining Y-functions. Each function Y_i ($2 \leq i \leq 5$) is equal to 1 if and only if a *majority* of the variables $\bar{s}$, y_{i-1}, and y_{i+1} are equal to 1.

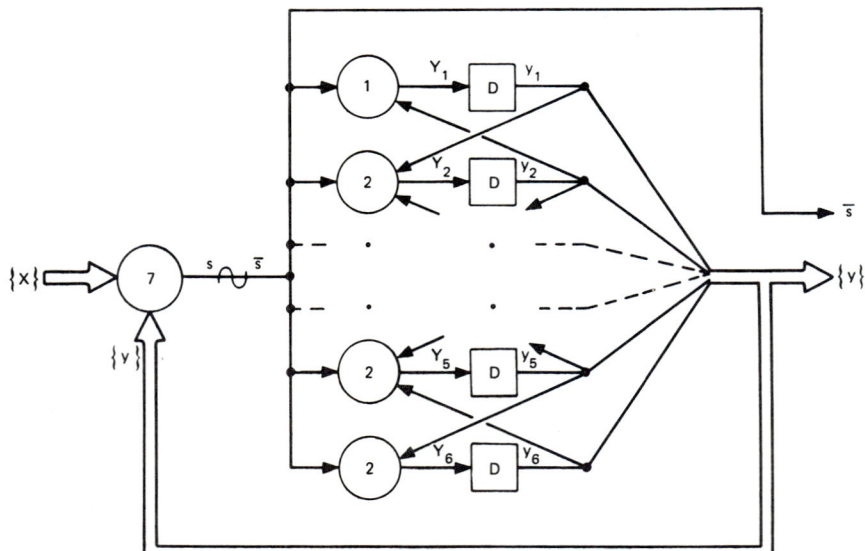

FIG. 11.   An unstable circuit which produces $E^* = \bar{s} + |y| = 7 - |X|$.

All of these requirements have been incorporated into the completed circuit diagram shown in Fig. 11. Since all inputs to each of the T-elements are weighted by 1, none of these unit weights have been shown. The reason for the adding of the $\bar{s}$ signal to the output will be given shortly.

C. TESTING AND SUBSEQUENT MODIFICATION OF THE CIRCUIT

It is now the appropriate time for us to judge how effectively the circuit we have designed would actually work. Rather than deal with the circuit itself we shall return to the original block diagram of Fig. 9a and the tabular description of the output-incrementing circuit given in Fig. 10a. As a first example for analysis we shall assume that $|X| = 2$ and that the value of $|y| = 3$. Since $|X| + |y| = 5$, it follows that $s = 0$ and $\bar{s} = 1$. From the table we see that when $\bar{s} = 1$ and $|y| = 3$, the produced value of $|Y|$ is 4. Therefore, after a delay corresponding to the signal propagation time around one of the feedback loops in the output-incrementing circuit, $|y|$ becomes 4. The excitation provided by $|X| + |y|$ still is less than the critical threshold value of 7, and consequently $s = 0$ and $\bar{s} = 1$ again. When $\bar{s} = 1$ and $|y| = 4$, it follows that $|Y| = 5$ and, after an interval of delay, $|y| = 5$.

With $|X| = 2$ and $|y| = 5$, the value of s changes to 1 and the value of $\bar{s}$ to 0. With $\bar{s} = 0$ and $|y| = 5$, the produced value of $|Y|$ is 4. After another delay $|y|$ becomes equal to 4. Almost coincident with that change, the value of s again becomes equal to 0 and that of $\bar{s}$ becomes equal to 1. Thus, when $|X| = 2$, the equilibrium reached is an unstable one. The circuit oscillates between two states: one in which $|y| = 5$ and $\bar{s} = 0$ and the other in which $|y| = 4$ and $\bar{s} = 1$. This action was not exactly what we had originally intended. However, even though $|y|$ has not been held constant at the value $7 - |X| = 5$, the *sum* of $|y|$ and $\bar{s}$ has been.

A more complete analysis would show that whenever $|X|$ has any value between 1 and 6, and whatever the initial value of $|y|$ may be, the circuit will ultimately reach an unstable equilibrium in which the sum of $|y|$ and $\bar{s}$ is constant at $7 - |X|$. The cases for which $|X| = 0$ or $|X| = 7$ are special ones. When $|X| = 0$, the final equilibrium is a *stable* one for which $|y| = 6$ and $\bar{s} = 1$. When $|X| = 7$, the equilibrium condition is also stable; at that equilibrium $|y| = 0$ and $\bar{s} = 0$. In each of these two special cases, however, the sum of $|y|$ and $\bar{s}$ has the desired value, $7 - |X|$, at equilibrium.

We conclude that whatever the value of $|X|$ may be, the equilibrium value of $|y| + \bar{s}$ is *constant* at $7 - |X|$. For this reason we have in Fig. 11 added the signal $\bar{s}$ to the output of that circuit. That circuit then has a terminal action which, after an equilibrium is reached, is exactly the same as that of a " Markov circuit" designed according to the general plan illustrated in Fig. 4a.

Consequently the output of our circuit could be used in combination with T-elements similar to the one shown in Fig. 5 to produce the complemented values of the seven X-variables.

At the unstable equilibria, there are nearly simultaneous changes of $\bar{s}$ and $|y|$ and their sum is constant. The situation is essentially that which we saw before in the circuit of Fig. 7. The signal-replication trick can be used here if we want to assure that the values of the inverted variables remain absolutely constant. What would be necessary would be the creation of a set of replicas of the $\bar{s}$ signal and of each of the y-signals, each to be delayed by an amount roughly equal to one half of the delays shown in Fig. 11. These signals, in addition to the originals from which they were produced, would then be available for use as inputs to the T-elements from which the complemented X-variables are produced.

We have, at last, produced a sequential circuit which has only one NOT-element and which at equilibrium will produce the complements of its seven input variables at its output terminals. We have called this circuit the "multi-inversion" circuit. It is obvious that multi-inversion circuits can be designed for an arbitrarily large number of input variables.

V. PROOF OF THE NECESSITY OF UNSTABLE CIRCUIT EQUILIBRIA

In the preceding section we developed the design for a "multi-inversion" circuit which used only one NOT-element and which nevertheless inverted an arbitrarily large number of X-variables. This particular circuit had a number of unstable equilibria. The resulting oscillations made no difference as far as the production of the complemented X-variables was concerned. However, a natural question to ask is whether or not the oscillation was necessary. That is, is it possible to design a circuit which has only one NOT-element and gives the terminal action we want but which has only stable equilibria? This section will show that that is not possible; the oscillation is necessary.

The method of proof is by contradiction. We shall assume that there is a multi-inversion circuit with one NOT-element which is stable at each of the set of equilibria which can be reached by changing the X-variables. Then we shall show that such a circuit could not produce the desired set of outputs.

The most general possible plan for a multi-inversion circuit having only one NOT-element is that pictured in Fig. 12. We already know that a sequential circuit is necessary in order to accomplish the desired result. We assume that all of the feedback loops have been shown in the diagram and that the network of logic elements contains no additional loops.

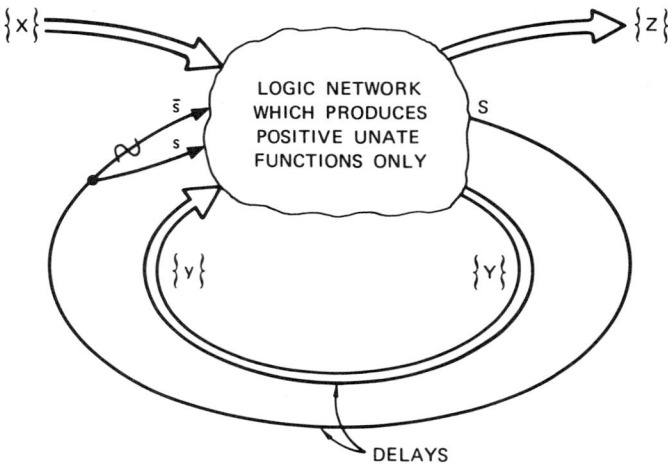

FIG. 12.   Plan for a hypothetical sequential circuit.

The inputs to the circuit are the set of X-variables which are to be inverted; we call this set $\{X\}$. The set $\{Z\}$ of outputs will be assumed, at the stable equilibria, to have constant values which are the complements of the corresponding input variables. The signal S which provides the input to the single NOT-element is also derived from the logic network as are the signals of the set $\{Y\}$ which are used to excite the feedback loops.

We allow the *possibility* that $\{Z\}$, $\{Y\}$, and S may all depend upon $\{X\}$, $\{y\}$, s, and $\bar{s}$ in some arbitrarily complicated way. One could imagine, for example, that several different (stable) equilibria were associated with some of the states of the input set $\{X\}$. We might, as another example, imagine that one could obtain some novel circuit action by careful control of the magnitudes of the signal delays in the circuit, perhaps even letting them change with time.

One restriction is, however, essential to our later arguments. That restriction is that all of the functions produced by the network of logic elements are *positive unate* functions. This is implicit in our earlier assumption that the only element with inhibition is the single NOT-element which is outside the network. An alternate way of stating the same thing is that no input to the network can by increasing its value (from 0 to 1) cause any function value to decrease, or by decreasing its value (from 1 to 0) cause any function value to increase.

If the number of inputs in the set $\{X\}$ is two or more (the only interesting case), it will be possible to imagine that the input variables are set so that some but not all of the X-variables are equal to 1, and that the circuit has come to rest at one of its assumed stable equilibria. At this equilibrium s and $\bar{s}$ have constant complementary values. With that as the assumed initial state of the circuit we shall conduct a pair of *gedanken* experiments In the first experiment we shall increase the value of one of the X-variables (from 0 to 1). In the second experiment we shall decrease the value of one of the X-variables (from 1 to 0).

At the initial equilibrium state either $s = 1$ and $\bar{s} = 0$ or $s = 0$ and $\bar{s} = 1$. Assume, first, that $s = 1$ and $\bar{s} = 0$. Let one of the X-variables *increase* in value from 0 to 1. If the circuit is to accomplish its assigned task, a corresponding Z-signal must (eventually) decrease from 1 to 0. This decrease cannot occur unless there is some way for either s or $\bar{s}$ or one of the signals in the set $\{y\}$ to decrease. Because s has been assumed to have a high value already it cannot increase further. Therefore, neither s nor $\bar{s}$ can change until S decreases. In addition no signal in the set $\{y\}$ can decrease until some corresponding signal in the set $\{Y\}$ decreases. Thus we have an impasse:

(i)   Neither s nor $\bar{s}$ or any signal in $\{y\}$ can decrease until either S or some signal in $\{Y\}$ decreases.

(ii) Neither S nor any signal in $\{Y\}$ or $\{Z\}$ can decrease until s, $\bar{s}$, or some signal in $\{y\}$ decreases.

If we assume that the initial circuit equilibrium is one in which $s = 1$ and $\bar{s} = 0$, there is no way in which the circuit can produce a decrease in one of the Z-signals when the corresponding X-signal is increased.

If we were to assume instead that the initial circuit equilibrium was one in which $s = 0$ and $\bar{s} = 1$, a similar impasse would result. For that case we would assume that one of the X-variables was decreased in value from 1 to 0. The impasse is:

(i)   Neither s nor $\bar{s}$ or any signal in $\{y\}$ can increase until either S or some signal in $\{Y\}$ increases.

(ii) Neither S nor any signal in $\{Y\}$ or $\{Z\}$ can increase until s, $\bar{s}$, or some signal in $\{y\}$ increases.

Our conclusion is that neither $s = 1$ (and $\bar{s} = 0$) nor $s = 0$ (and $\bar{s} = 1$) is appropriate for the assumed stable equilibrium. Therefore the assumed stable equilibrium *cannot* exist. *Any* circuit, however cleverly designed, which acts like a "multi-inversion" circuit and which has only one NOT-element must have unstable equilibria. We can understand now that the oscillations which occurred in the circuits of Fig. 7 and Fig. 11 were unavoidable. In the next section we shall prove that oscillations can be avoided *if* we allow ourselves to use *two* NOT-elements.

We also note that the proof given above does *not* eliminate the possibility that stable equilibria can exist when all of the X-variables are high or when all of them are low. When all of the inputs are high, it would be impossible to increase the value of one of the X-variables. Consequently the first *gedanken* experiment could be not performed. Thus, it might conceivably be possible for a stable equilibrium with $s = 1$ and $\bar{s} = 0$ to exist when all X-variables have the value 1.

Similarly, when all of the X-variables are low, it would be impossible to conduct the second *gedanken* experiment. It might, therefore, be possible for the circuit to have a stable equilibrium with $s = 0$ and $\bar{s} = 1$ when all X-variables have the value 0.

The circuits in Figs. 7 and 11 both illustrate that these two *stable* equilibria can exist.

VI. A "MULTI-INVERSION" CIRCUIT WHICH IS STABLE

The multi-inversion circuit which we designed earlier (see Figs. 9–11) used only one NOT-element. Within that circuit the signal s, since it was *binary*-valued, could indicate only whether the value of $|y|$ was too low *or not*. We hinted at that time that it would have been desirable to have a *ternary*-valued signal; that is, one which could differentiate among three possibilities: too low, too high, or exactly right.

The circuit described in this section resembles the earlier one but does have these three levels of indication of error. This effect is achieved by having *two* signals, s_1 and s_2. "Too low" will correspond to $s_1 = s_2 = 0$. "Too high" will correspond to $s_1 = s_2 = 1$. "Exactly right" will be indicated by $s_1 = 0, s_2 = 1$. *Each* of these two signals will be inverted by its own NOT-element. The use of these *two* NOT-elements will enable us to build a multi-inversion circuit which has only stable equilibria.

In Fig. 13a we show how the number of Y-functions which have the value

(a) $\bar{s}_1, \bar{s}_2$:

$|y|$:	00	10	11
0	0	0	1
1	0	1	2
2	1	2	3
3	2	3	4
4	3	4	5
5	4	5	5

(b) $\bar{s}_1, \bar{s}_2$:

Y_1 Y_2 Y_3 Y_4 Y_5:	00	10	11
0 0 0 0 0	0 0 0 0 0	0 0 0 0 0	1 0 0 0 0
1 0 0 0 0	0 0 0 0 0	1 0 0 0 0	1 1 0 0 0
1 1 0 0 0	1 0 0 0 0	1 1 0 0 0	1 1 1 0 0
1 1 1 0 0	1 1 0 0 0	1 1 1 0 0	1 1 1 1 0
1 1 1 1 0	1 1 1 0 0	1 1 1 1 0	1 1 1 1 1
1 1 1 1 1	1 1 1 1 0	1 1 1 1 1	1 1 1 1 1

FIG. 13. Design requirements for another output-incrementing circuit: (a) $|Y|$ as a function of $|y|$, $\bar{s}_1$, and $\bar{s}_2$, (b) requirements for the Y-signal.

1 are to depend upon the two complemented signals, $\bar{s}_1$ and $\bar{s}_2$. When $\bar{s}_1 = 0$ and $\bar{s}_2 = 1$ that is an indication that the value of $|y|$ is exactly right. For that situation we make $|Y| = |y|$. When the value of $|y|$ is too high (when $\bar{s}_1 = \bar{s}_2 = 0$) we make $|Y| = |y| - 1$, if that is possible. When the value of $|y|$ is too low (when $\bar{s}_1 = \bar{s}_2 = 1$) we make $|Y| = |y| + 1$, if that is possible. A circuit built to these specifications would change its value of $|y|$, one unit at time, until the circuit was at a stable equilibrium with $|Y| = |y|$.

The table of Fig. 13b shows in more detail how the five Y-functions can be generated; Y_3 is a typical function. It can, under appropriate circumstances, change its value from 0 to 1 when one of the variables $\bar{s}_1, \bar{s}_2, y_2, y_3$, or y_4 changes from 0 to 1. More explicitly, Y_3 is equal to 1 when a majority (three or more) of the five variables are equal to 1 and is 0 otherwise. Consequently, Y_3 can be generated by a single T-element which has a threshold of 3 and has those five inputs, each weighted by the factor 1. Y_2 and Y_4 are produced in a similar way.

The productions of the functions Y_1 and Y_5 are special cases of the general rule illustrated above. $Y_1 = 1$ if and only if at least two of the variables $\bar{s}_1, \bar{s}_2$, y_1, and y_2 are equal to 1. $Y_5 = 1$ if and only if at least three of the variables $\bar{s}_1, \bar{s}_2, y_4$, and y_5 are equal to 1.

Diagrammatic representations of how these five functions are produced are shown in Fig. 14. In that same circuit diagram we show also how the "error" signals s_1 and s_2 are generated. For $|X|$ in the range 1 to 6, inclusive the circuit will reach a stable equilibrium at which $|X| + |y| = 6$ and there-fore, at these equilibria, $\bar{s}_1 = 1$ and $\bar{s}_2 = 0$. Consequently the output of the circuit, $E^* = \bar{s}_1 + s_2 + |y|$, will have the value $7 - |X|$.

The remaining cases, $|X| = 0$ or $|X| = 7$, deserve special attention. When $|X| = 0$, there is no way for either s_1 or s_2 to have the value 1 even if $|y| = 5$. With $\bar{s}_1 = \bar{s}_2 = 1$ the output-incrementing portion of the circuit will increase $|y|$ step-by-step as far as is possible (to $|y| = 5$). Therefore, when $|X| = 0$, the the sum $\bar{s}_1 + \bar{s}_2 + |y| = 7$. This is the desired value.

Similarly, when $|X| = 7$, there is no way for either s_1 or s_2 to have the value 0, even if $|y| = 0$. With $\bar{s}_1 + \bar{s}_2 = 0$ the value of $|y|$ will be decreased, one unit at a time, until $|y| = 0$. Therefore, when $|X| = 7$, the sum $\bar{s}_1 + \bar{s}_2 + |y| = 0$. This, also, is the desired value.

We have shown that there is a method for building a multi-inversion circuit which has only stable equilibria and which uses only two NOT-elements. This is the minimum possible number since we have already proved that when only one such element is used some of the equilibria must be unstable.

The main result of this section might have been anticipated in view of the possibility of nesting several "Markov circuits." As it has been mentioned earlier, however, such nested circuits are at best difficult to analyze, and it

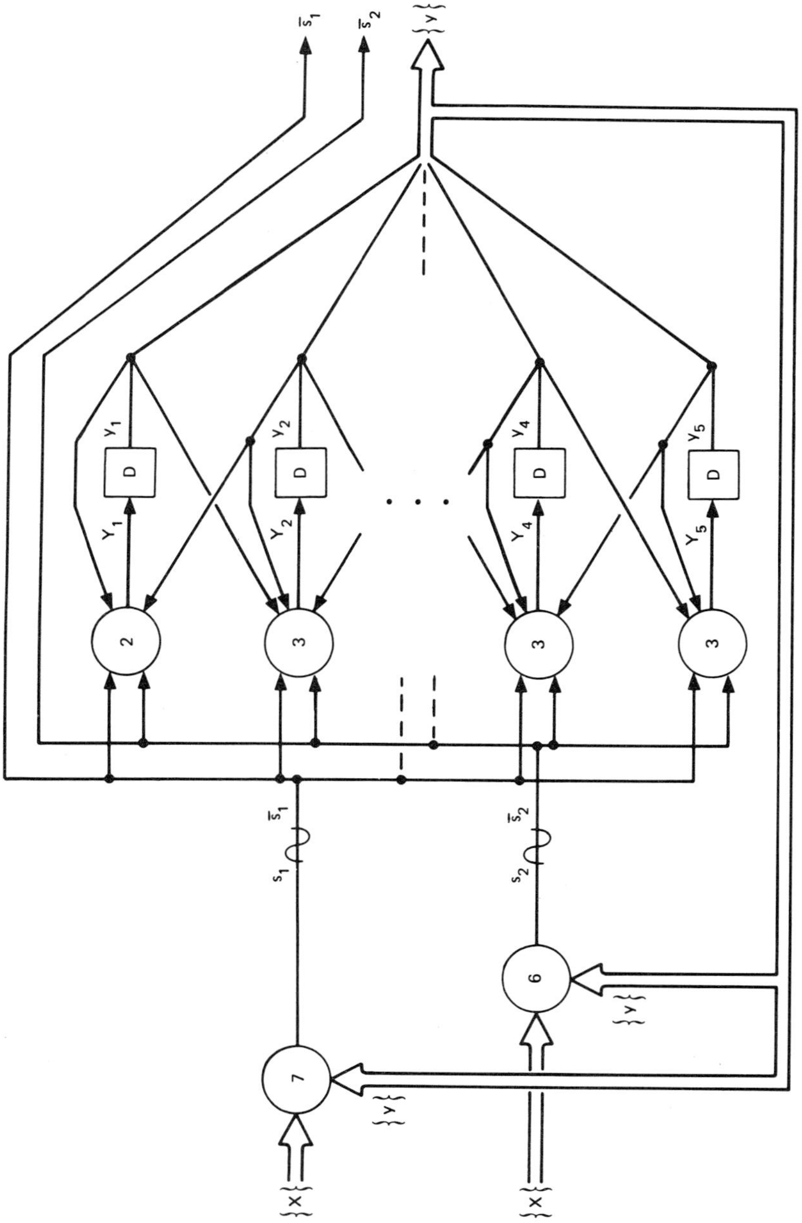

FIG. 14. A stable circuit which produces $E^* = s_1 + s_2 + |y| = 7 - |X|$.

probably would be required that the various layers of feedback loops have different orders of magnitude of signal delay. Circuits of the type illustrated in Fig. 14 are comparatively easy to analyze and their feedback delays can be approximately equal in magnitude even though this is not a necessary restriction.

VII. SUMMARY AND CONCLUSIONS

The major portion of this chapter has been dedicated to the design of circuits which can invert arbitrarily large numbers of variables using the absolute minimum possible number of NOT-elements (one or two, depending on the stability requirement). The work of Markov constituted proof that, without using feedback techniques, k NOT-elements could be used to invert as many as $2^k - 1$ variables. The inversion of 2^k variables required $k + 1$ NOT-elements. As remarkable as his result was, it still showed that no upper bound could be placed on the number of NOT-elements required for the inversion of arbitrarily many variables if feedback was not used. Thus the result of this chapter is proof that feedback may allow significant reduction in the "cost" of producing combinational functions.

The reader may argue that the "price" (large numbers of other elements) has, in this particular case, been enormous. Whether or not this is true depends upon the relative costs of NOT-elements and the other logical elements. But the real point is that the results of this chapter increase our perspective on what *might* be possible in designing logical networks. These results, together with those of others such as Kautz, dramatize that feedback should be considered whenever combinational circuits are to be designed. How this additional freedom can best be exploited cannot now be said. Undoubtedly there will in the future exist techniques which are more subtle than merely inverting all of the set of uncomplemented input variables and then realizing desired functions as positive unate functions of the full set of both complemented and uncomplemented variables.

Is feedback a technique which should be considered only for producing functions which are not themselves positive unate, or do there exist feedback networks having positive unate functions only which are minimal? Just how closely is the optimal use of feedback related to the "inversion complexity" of a function? All of these questions are tantalizing but, for the moment, we must wait for the answers until more research is done.

ACKNOWLEDGMENT

The research reported in this chapter was partially supported by the Stanford Research Institute.

REFERENCES

GILBERT, E. N. (1954). Lattice theoretic properties of frontal switching functions. *J. Math. Phys.* **33**, 57–67.

HUFFMAN, D. A. (1969). Logical design with one NOT-element. *In* " Proceedings Second Hawaii International Conference of Systems Science," pp. 735–738.

KAUTZ, W. H. (1970). The necessity of closed circuit loops in minimal combinational circuits. *IEEE Trans. Comput.* **C-19**, 162–166.

MARKOV, A. A. (1957). On the inversion complexity of a system of functions. *Dokl. Acad. Nauk SSSR* **116**, 917–919. (Also published in *J. Ass. Comput. Machinery* **5**, 331–334.)

MULLER, D. E. (1958). Minimizing the Number of NOT-Elements in Combinational Circuits. Unpublished memorandum.

Chapter III

LUPANOV DECODING NETWORKS

AMAR MUKHOPADHYAY

I. INTRODUCTION   57

II. DISJUNCTIVE AND NONDISJUNCTIVE COMPLETE DECODING
NETWORKS   58

III. THE CASE WHEN $r \neq 2^k$ 64

IV. THE OPTIONAL TERMS   68

V. TOWARD A GENERAL THEORY   72

VI. CONCLUSIONS   82

REFERENCES   83

ABSTRACT. Decoding networks find applications in computer memories for the selection of a particular item of data addressed by a binary code, in counter circuits, in character-by-character code conversion circuits, in the synthesis of switching functions, and also in the theoretical derivations of complexity bounds for arbitrary switching networks. In particular, complete decoding network of n-input variables provides a distinct output for each of the 2^n possible combinations of the variables. For disjunctive contact networks, hence for networks of branch type elements, the familiar transfer trees using $2(2^n - 1)$ contacts are the only possible minimal n-variable decoding networks. In applications where disjunctivity of the outputs is not required, Lupanov has shown that the required number of contacts can be reduced considerably (by a factor of two as n grows arbitrarily large). In this chapter we shall present systematic procedures for the design of Lupanov trees. The Lupanov trees obtained have the best minimal costs known so far for all values of n. As an example, a 5-variable decoder has been designed which uses only 58 contacts compared to 60 contacts used by Lupanov.

I. INTRODUCTION

Decoding networks find applications in computer memories for the selection of a particular item of data addressed by a binary code, in counter circuits, in character-by-character code conversion circuits, in the synthesis of switching functions and also in the theoretical derivations of complexity bounds for

57

arbitrary switching networks (Harrison, 1965). In particular, complete decoding network of n-input variables provides a distinct output for each of the 2^n possible combinations of the variables. These networks have been studied extensively, both in terms of relay contact realizations (Shannon, 1949; Burks *et al.*, 1955) and in terms of gate-type element realizations (Burks *et al.*, 1954). Moore (1960) has shown that for disjunctive contact networks, hence for networks of branch-type elements, the familiar transfer trees using $C_M = 2(2^n - 1)$ contacts are the only possible minimal n-variable decoding networks.

In applications where disjunctivity of the outputs is not required, Lupanov (1958) has shown that the required number of contacts can be reduced considerably (by a factor of two as n grows arbitrarily large). Short (1963) has examined specific design procedures using Lupanov's approach. The main idea is based on the design of a r-variable partial decoder where $r = 2^k$, $k = 2, 3, \ldots$ whose $2^r/r$ outputs consist of mutually disjoint groups of r vertices, collectively exhausting all the vertices of the r-cube such that any vertex is distinguished from the rest of the vertices in the group by the value of a single variable.

In this chapter we shall present systematic procedures for the design of Lupanov trees. The Lupanov trees obtained have the best minimal costs known so far for all values of n. As an example, a 5-variable decoder has been designed, which uses only 58 contacts compared to 60 contacts used by Lupanov.

II. DISJUNCTIVE AND NONDISJUNCTIVE COMPLETE DECODING NETWORKS

Consider the 4-variable standard complete decoding tree using relay contacts shown in Fig. 1. It can be viewed as consisting of a partial decoder P (dotted lines) wihch generates as its outputs the set of functions $y_0 = \bar{x}_4 \bar{x}_3 \bar{x}_2$, $y_1 = \bar{x}_4 \bar{x}_3 x_2, \ldots, y_7 = x_4 x_3 x_2$ followed by a set of output " bundles " which performs the last step of decoding and which needs exactly one contact per output terminal. In general, for n variables, the 2^{n-1} outputs from the partial decoder can be produced by a complete tree in $n - 1$ variables and the network requires a total of $2(2^n - 1)$ contacts. Note that the network is *disjunctive*, i.e., it has the property that no two output terminals can ever be connected together.

In applications where disjunctivity of the outputs is not required, Lupanov (1958) suggested an alternative construction of a complete decoding tree in which the partial decoder generates a lesser number of more complex functions $y_0, y_1, \ldots$. As an example, consider the set of four functions $y_0, y_1, y_2,$

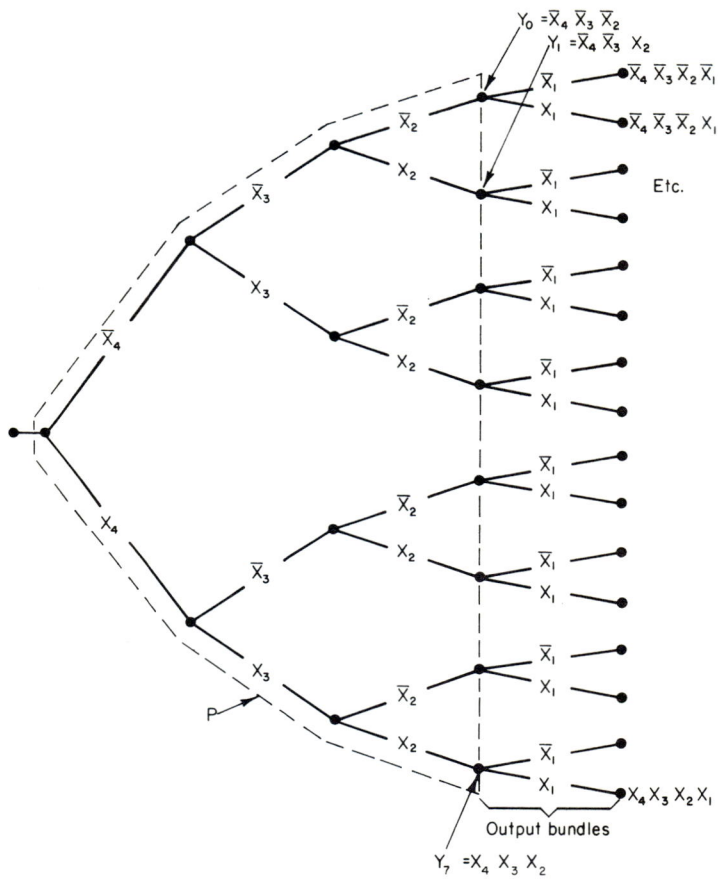

FIG. 1. Four-variable complete decoding tree.

and y_3 represented on the Karnaugh map in Fig. 2. Each vertex (fundamental product or minterm) is labeled with the number of the y-function to which it belongs. Each of these functions is like a "star" or a "shell" of four vertices that are unit distance from a center vertex, called the *shell center*. The shell centers of the four functions are

$$y_0 : \bar{x}_4 \bar{x}_3 \bar{x}_2 x_1 \quad (0001) = c_0$$

$$y_1 : \bar{x}_4 \bar{x}_3 \bar{x}_2 \bar{x}_1 \quad (0000) = c_1$$

$$y_2 : x_4 x_3 x_2 \bar{x}_1 \quad (1110) = c_2$$

$$y_3 : x_4 x_3 x_2 x_1 \quad (1111) = c_3$$

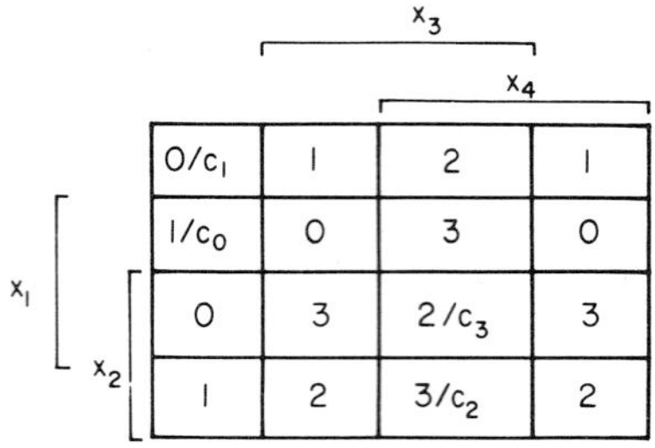

	x_3		
		x_4	
$0/c_1$	1	2	1
$1/c_0$	0	3	0
0	3	$2/c_3$	3
1	2	$3/c_2$	2

FIG. 2.   Representation of y_0, y_1, y_2, and y_3 in a Karnaugh map.

It is obvious, therefore, that any vertex of a given y-function can be distinguished from the rest of the vertices of the same y-function by the value of a single variable. Any function having this property will be called a *shell function*. The shell functions in our example can be written as

$$y_0 = S_0 + x_1 S_2, \qquad y_1 = S_1$$
$$y_2 = S_4 + \bar{x}_1 S_2, \qquad y_3 = S_3$$

where S_0, S_1, S_2, and S_3 are totally symmetric functions of variables x_1, x_2, x_3, and x_4 having Shannon's (1949) a-members 0, 1, 2, and 3, respectively. The partial decoder which generates these shell functions is derived from the four variable symmetric function network and needs 20 contacts. The complete 4-variable tree which is shown in Fig. 3, needs four 4-contact output bundles and a total of 36 contacts compared to 30 needed for the standard tree of Fig. 1. Apparently, therefore, we have not gained anything except making the network nondisjunctive! (For example, if $x_4 = x_3 = x_2 = x_1 = 1$, the output terminal 15 will be connected to the input, but the set of terminals 1, 2, 4, and 8 will be connected to each other and same is true for the set of terminals 3, 5, and 9, but none of them is connected to the input terminal.) However, let n the number of variables now be increased and let us adopt the structure of the decoding network as shown in Fig. 4. The partial decoder P produces the four outputs y_0, y_1, y_2, and y_3 each of which is connected to four complete decoding trees M in the remaining $n - 4$ variables. To each of the 2^{n-4} outputs from the M-tree connected to y_i ($0 \leq i \leq 3$), the original output bundle corresponding to y_i is attached. The entire network is, therefore, a valid realization of an n-variable decoding tree and needs a total number of contacts given by

$$C_1 = 20 + 4c(n - 4) + 2^n \qquad (1)$$

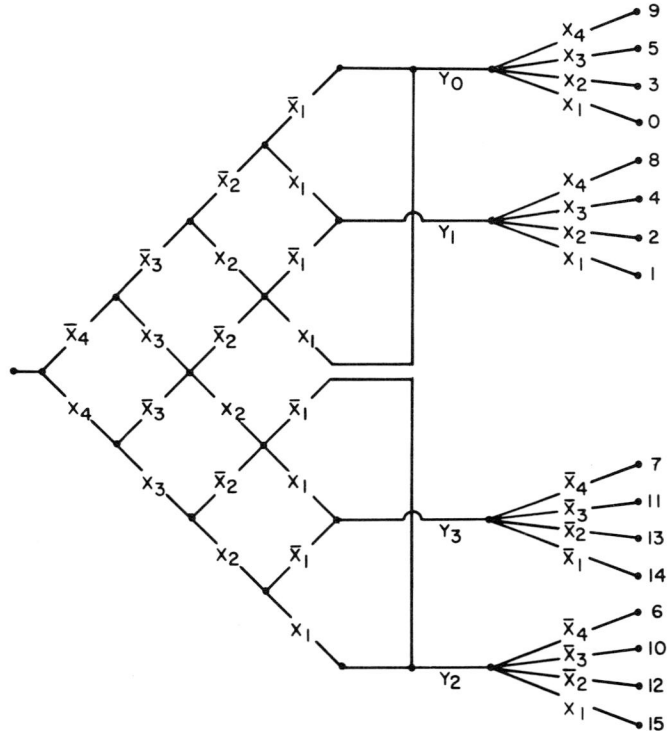

FIG. 3.   A 4-variable Lupanov tree.

where $c(n-4)$ is the cost of the M-tree. M could be any known cheapest realization of an $(n-4)$-variable tree. The values of C_1 for n up to 20 are given in Table I using Moore-bound C_M for $c(n-4)$. It is noted that for $n = 5$, the network requires 60 contacts compared to 62 given by Moore. For larger n, bigger savings are obtained, viz for $n = 9$, $C_1 = 772$ compared to 1022 given by Moore.

Actually we can obtain a still greater economy by generalizing the process used to cover the 4-cube with a set of four shell functions. The problem is to find a minimum number of shell centers in a r-cube such that every vertex of the r-cube is included in at least one shell function. To be most efficient, the shell functions should be able to cover mutually exclusive but collectively exhaustive sets of vertices of the r-cube. The solution to this problem is provided from the coding theory for the case when $r = 2^k$, $k = 2, 3, \ldots$.

We know from the theory of lossless single-error-correcting Hamming

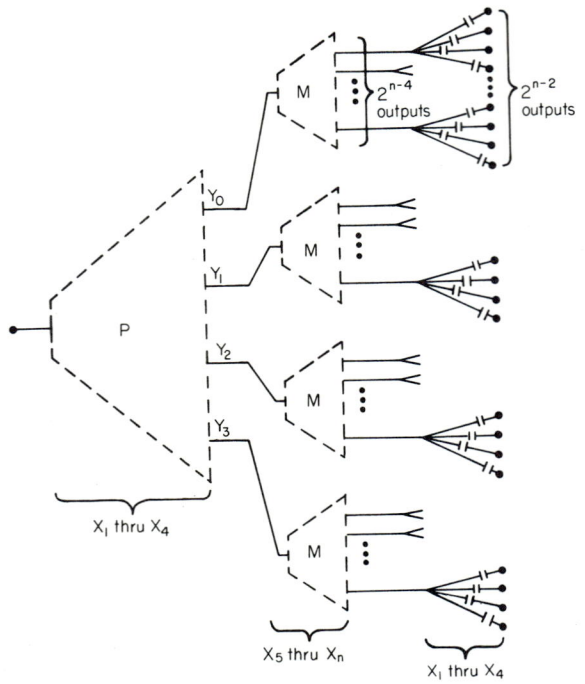

FIG. 4. The structure of an n-variable decoding tree.

codes (Peterson, 1961) that in a cube of dimension $r' = 2^k - 1$, exactly $2^{r'} - k$ nodes exist that are distance three apart. If we let the code words of such a code correspond to the shell centers, then it is clear that the unit spheres or the shells around these centers will partition the r'-cube completely and disjointly. This is not exactly what we want because the shell centers themselves are left out of the spheres. This difficulty can be overcome by just duplicating the entire shell center pattern with the use of one additional variable, x_r, say, where $r = r' + 1 = 2^k$, $k = 2, 3, \ldots$. One pattern corresponds to the value of $x_r = 0$ and the other for $x_r = 1$. This structure is shown in Fig. 5, which shows that each unit sphere of the pair now covers the center of the other. Thus, we shall have a total of $2 \cdot 2^{r'-k} = 2^r/r$ shell centers.

The general scheme for Lupanov's construction of an n-variable complete decoding network can now be described in three steps:

(i) Select a value of $r < n$ that is a power of two and design the partial decoder P whose $2^r/r$ outputs correspond to the shell functions, $y_0, y_1, \ldots, y_i, \ldots, y_{(2^r/r-1)}$. Let the total cost of P be $C(r)$.

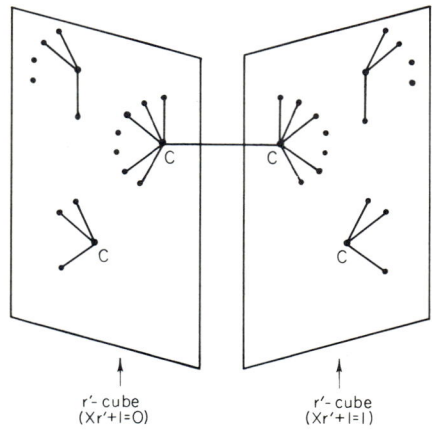

r'– cube
(Xr'+l=0)

r'– cube
(Xr'+l=l)

FIG. 5. Augmentation of Hamming code for covering the shell centers.

(ii) Each of the outputs of P is connected to a complete decoding network of the remaining $n - r$ variables. We can use any known cheapest realization for this network and let its cost be $c(n - r)$.

(iii) Each output of the $(n - r)$-variable tree connected to the y_i terminal of P is now connected to an output bundle of r contacts $\bar{x}_1, \bar{x}_2, \ldots, \bar{x}_r$, if y_i is defined with respect to the shell center $(\dot{x}_1 \dot{x}_2 \cdots \dot{x}_r)$ where the dot stands for complementation or no complementation.

This tree has a total number of contacts

$$C_L = C(r) + (2^r/r)c(n - r) + 2^n \tag{2}$$

If r is chosen to grow with n so that both r and $n - r$ increase with increasing n, then it can be shown that $C_L \sim 2^n$ as $n \to \infty$, which is half of the asymptotic cost required for the standard tree: $C_M \sim 2^{n+1}$.

Short (1963) has examined specific design procedures using Lupanov's idea and has obtained a 8-variable partial decoder whose cost $C(8) = 230$.[1] Thus the total cost using Short's decoder is

$$C_2 = 230 + 32\,c(n - 8) + 2^n \tag{3}$$

which is also tablulated in Table I using $c(n - 8)$ from C_1.[2]

---

[1] The figure 224 cited in Short's original paper is apparently in error (Short, 1968).
[2] This section is primarily based on a tutorial article by Kautz (1964).

TABLE I
The Costs for the Decoding Networks

n	Moore C_M	Lupanov C_1	Short C_2	Author C_1'	C_2'	C_3	C_4
1	2						
2	6						
3	14						
4	30	36		34			
5	62	60		58			
6	126	108		106		149	
7	254	204		202		237	283
8	510	396	486	394	454	413	451
9	1022	772	806	762	774	765	787
10	2046	1548	1446	1538	1414	1469	1459
11	4094	2884	2726	2874	2694	2829	2703
12	8190	5700	5286	5690	5254	5453	5411
13	16382	11300	10342	11258	10246	10701	10467
14	32766	22596	20070	22058	19974	21197	20579
15	65534	44324	39526	43562	39432	41997	40803
16	131070	88356	78438	86570	78342	82589	80931
17	262142	176292	156006	172074	155654	163485	159507
18	524286	372548	308646	342058	306590	325277	316179
19	1048576	701004	611750	682034	610694	647325	629523
20	2097150	1402012	1217958	1361962	1216902	1288349	1253651

III. THE CASE WHEN $r \neq 2^k$

Consider the design of a p-variable partial decoder when $p \neq 2^k$. We could, if we wish, arbitrarily define the set of output functions of the partial decoder to consist of a set of s functions whose union is the p-cube. If we insist that all these s functions have to be shell functions having vertices as shell centers, then these functions will have overlapping vertices since closely packed single-error-correcting Hamming codes do not exist for $p' \neq 2^k - 1$. We could then write s as

$$s = [2^p/p] + e$$

where e is a nonzero positive integer. Then the cost of the Lupanov tree can be written approximately as

$$C(p) + [2^p/p]c(n - p) + 2^n + ec(n - p) + ep2^{n-p}$$

We have now added an extra term $ep2^{n-p}$ which grows exponentially with n whereas the other terms have growth rates which are approximately the same as those of similar terms in C_L. Hence, we cannot get a cost function better than C_L for large n although the cost may be better than the Moore bound. For small values of n, if we take $2^k < p < n$ such that k is the largest integer satisfying $2^k < n$, although the cost of $c(n - p)$ is reduced compared with the case when $p = r = 2^k$, e is sufficiently large so that the contributions $ec(n - p)$ and $ep2^{n-p}$ taken together plus the increase in the cost of the partial decoder $C(p)$ lead to a less economic realization. On the other hand, if we decide that not all the s functions should be shell functions, then the output "bundles" associated with these nonshell functions become more complex (in fact, the bundles are now partially developed trees) and need more than one contact per output terminal. In this case the cost function depends very much on the actual selection of the s functions and can be written as

$$C(p) + s\,c(n - p) + 2^{n-p}(2^p + t)$$

where t is a nonzero positive integer. Here again we have added a term $t2^{n-p}$ which grows exponentially with n and therefore for large n the cost is larger than C_L. For smaller values of n, our experience in working with nonshell functions leads to similar conclusions in almost all cases.

Thus, when $p \neq 2^k$, we must more or less stick to the idea of shell functions. For this we have to generalize the notion of shell functions as described below. We shall illustrate the idea, taking $p = 6$.

Consider a k-subcube $\dot{x}_1 \dot{x}_2 \cdots \dot{x}_{p-k}$ in the p-cube $(k < p)$. Let G_i be the set of vertices which differ in the coordinate x_i $(1 \leq i \leq p - k)$ from each of the vertices in the k-subcube. There will be $p - k$ such sets $G_1, G_2, \ldots, G_{p-k}$ defined with respect to the coordinate $x_1, x_2, \ldots, x_{p-k}$, respectively, each having a set of 2^k vertices. Let us define a function y as follows: y contains the set of vertices $g_1, g_2, \ldots, g_{p-k}$ such that $g_i \in G_i$. Then y is obviously a shell function with respect to the coordinates $x_1, x_2, \ldots, x_{p-k}$ in that each vertex in the set $g_1, g_2, \ldots, g_{p-k}$ can be distinguished from all others in the set by the value of a single variable in $x_1, x_2, \ldots, x_{p-k}$. If we use y as a partial decoder output function, the associated output bundle should have $p - k$ contacts, one contact for each output terminal. The shell center associated with y is now a k-subcube and we can actually derive a total of

$$\rho = \overbrace{2^k \cdot 2^k \cdots}^{(p-k)\ \text{times}} = 2^{k(p-k)} \tag{4}$$

y functions which are shell functions with respect to a given k-subcube. When the shell center is a 0-subcube, i.e., a single vertex, there is only one shell function containing those vertices which are obtained by complementing one coordinate or the shell center at a time. Thus for $p = r = 4$ with respect to the shell center (0000), the shell function contains the set of vertices (0001, 0010, 0100, 1000). If the shell center is taken to be a 1-cube (000-), there could be $2^1 \cdot 2^1 \cdot 2^1 = 8$ different shell functions, a particular function being the collection of vertices (001∅, 010∅, 100∅) where ∅ is either 0 or 1.

Consider now the case when $p = 6$. With reference to the Karnaugh map shown in Fig. 6, we define a set of twelve partial decoder output functions

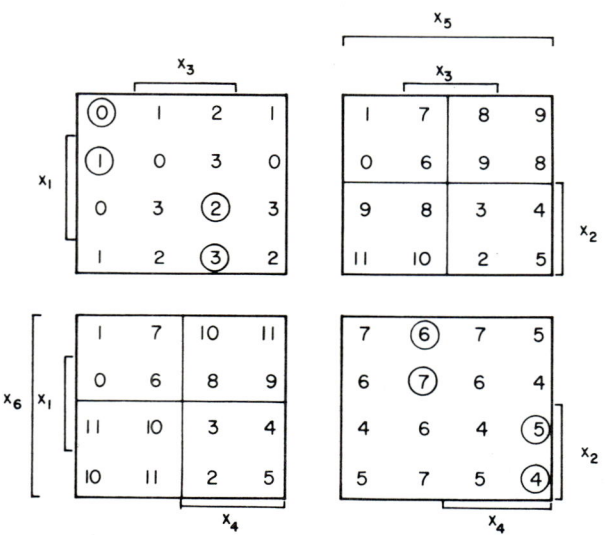

FIG. 6.   The function $y_0, \ldots, y_{11}$ plotted in the 6-variable Karnaugh map.

$y_0, y_1, \ldots, y_{11}$. Note that the first eight functions $y_0, \ldots, y_7$ are defined with respect to shell centers which are vertices or 0-cubes (marked by circles). The rest y_8, y_9, y_{10}, y_{11} are defined with respect to shell centers which are 2-cubes. The shell centers for these functions are

$$y_8 : \bar{x}_6 x_3 \bar{x}_2 x_1 \qquad (0\text{-}\text{-}101)$$
$$y_9 : \bar{x}_6 \bar{x}_3 x_2 \bar{x}_1 \qquad (0\text{-}\text{-}010)$$
$$y_{10} : x_6 \bar{x}_3 x_2 \bar{x}_1 \qquad (1\text{-}\text{-}010)$$
$$y_{11} : x_6 x_3 \bar{x}_2 x_1 \qquad (1\text{-}\text{-}101)$$

The geometrical representation of one of these functions y_8 is shown in Fig. 7. The vertices which belong to y_8 are marked by solid dots.

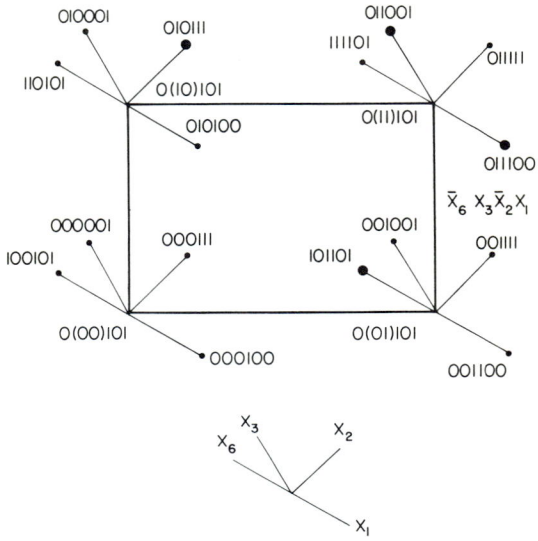

FIG. 7.   The geometrical representation of y_8.

While writing the Boolean expression for a shell function, the vertices which form the shell center can be regarded as "optional" terms for the shell function. Using such redundancies, the functions $y_8, \ldots, y_{11}$ become

$$y_8 = \bar{x}_6 \bar{x}_2 S_3 + (\bar{x}_6 \bar{x}_4 + \bar{x}_5 \bar{x}_3)S_4$$
$$y_9 = \bar{x}_6 x_5 x_2 S_3 + (x_6 x_2 + x_5 x_4)S_2$$
$$y_{10} = x_6 x_2 S_3 + (x_6 x_4 + x_5 x_2)S_2$$
$$y_{11} = x_6 \bar{x}_5 \bar{x}_2 S_3 + (\bar{x}_6 \bar{x}_2 + \bar{x}_5 \bar{x}_4)S_4$$

The other shell functions $y_0, \ldots, y_7$ are given by

$$y_0 = S_0 + x_1 S_2 \qquad\qquad y_4 = \bar{x}_3 S_4 + S_6$$
$$y_1 = S_1 \qquad\qquad y_5 = \bar{x}_1 \bar{x}_3 S_3 + (\bar{x}_1 + \bar{x}_3)S_5$$
$$y_2 = \bar{x}_6 \bar{x}_5 \bar{x}_1 S_2 + x_4 x_3 x_2 S_4 \qquad y_6 = \bar{x}_4 \bar{x}_2 S_3 + (\bar{x}_4 + \bar{x}_2)S_5$$
$$y_3 = \bar{x}_6 \bar{x}_5 S_3 + (\bar{x}_6 + \bar{x}_5)S_5 \qquad y_7 = x_6 x_5 x_3 S_4 + \bar{x}_4 \bar{x}_2 \bar{x}_1 S_2$$

The partial decoder is shown in Fig. 8 and needs 85 contacts. For a complete n-variable decoder, the cost function is given by

$$C_3 = 85 + 12c(n-6) + 2^{n-6}(8 \cdot 6 + 4 \cdot 4)$$
$$= 85 + 12c(n-6) + 2^n \qquad\qquad (5)$$

C_3 is tabulated in Table I for n up to 20 using the cheapest realization for $c(n-6)$ (see Section IV).

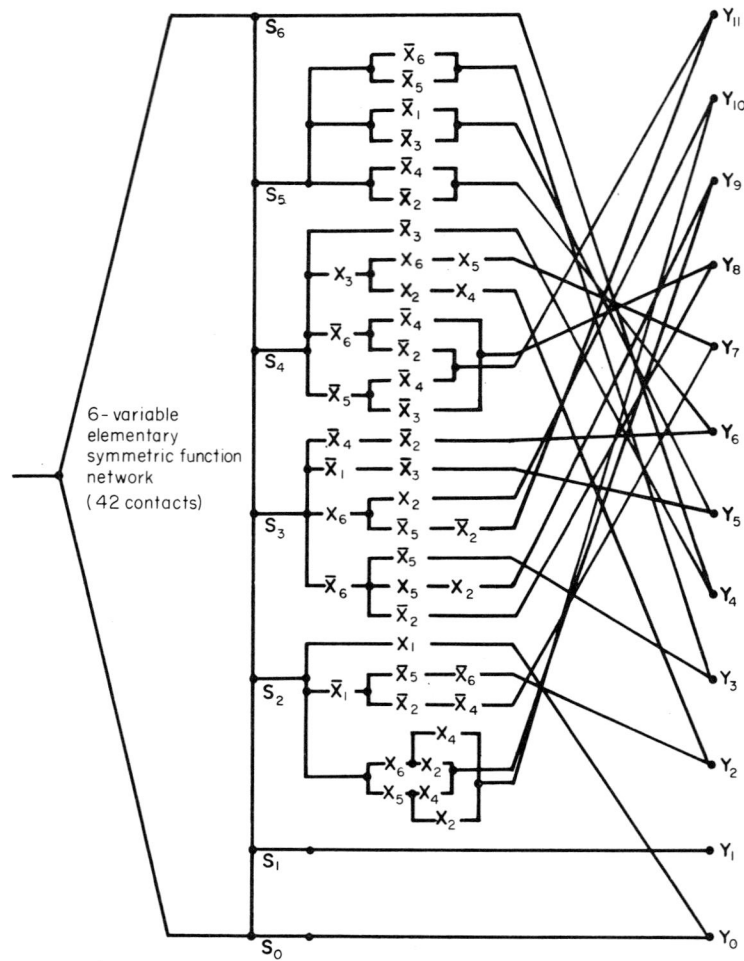

FIG. 8. The 6-variable partial decoder.

IV. THE OPTIONAL TERMS

We should like to emphasize the significance of the "optional terms" for simplification of the shell functions. In general, if the shell center is a k-cube $\dot{x}_1 \dot{x}_2 \cdots \dot{x}_{p-k}$, then, to any shell function associated with this shell center, it is always possible to add a set of 2^k optional terms $(\dot{x}_1 \dot{x}_2 \cdots \dot{x}_{p-k} \emptyset_1 \emptyset_2 \cdots \emptyset_k)$ where $\emptyset_j (1 \le j \le k)$ could be either 0 or 1. If the shell center is a single vertex $\dot{x}_1 \dot{x}_2 \cdots \dot{x}_p$, only one optional term $\dot{x}_1 \dot{x}_2 \cdots \dot{x}_p$ can be added to the shell function. If this is done to the four shell functions for the case $p = r = 4$, they are modified to

$$y_0 = S_0 + x_1 S_2, \qquad y_1 = S_0 + S_1$$
$$y_2 = S_4 + \bar{x}_1 S_2, \qquad y_3 = S_3 + S_4$$

Using these expressions, the partial decoder now needs only 18 contacts and C_1 is modified to C_1' as

$$C_1' = 18 + 4\,c(n-4) + 2^n \tag{6}$$

Thus for $n = 5$, we have a Lupanov complete decoder employing only 58 contacts instead of the 60 used by Lupanov. The 5-variable complete decoder is shown in Fig. 9. The value of C_1' for n up to 20 is also tabulated in Table I.

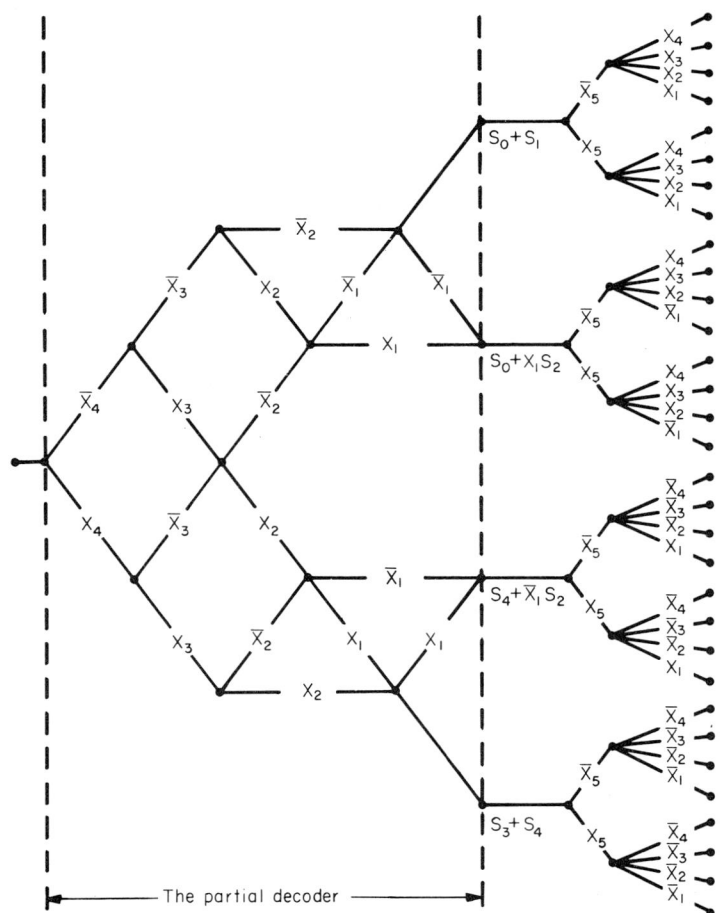

FIG. 9.   The 5-variable complete decoder.

The same technique can be applied to Short's 32 shell functions $s_1, \ldots, s_{32}$ for $r = 8$. The modified set of functions become

$$c_1 : \bar{s}\bar{t}\bar{u}\bar{v}\bar{w}\bar{x}\bar{y}\bar{z}; \qquad s_1 = S_0 + S_1\bar{z}$$

$$c_2 : \bar{s}\bar{t}\bar{u}v\bar{w}xy\bar{z}; \qquad s_2 = vxyS_3 + (\bar{s}\bar{t}\bar{u}\bar{w}S_2 + vxyS_4)\bar{z}$$

$$c_3 : \bar{s}\bar{t}u\bar{v}w\bar{x}y\bar{z}; \qquad s_3 = uwyS_3 + (\bar{s}\bar{t}\bar{v}\bar{x}S_2 + uwyS_4)\bar{z}$$

$$c_4 : \bar{s}\bar{t}uvwx\bar{y}\bar{z}; \qquad s_4 = \bar{s}\bar{t}\bar{y}S_4 + (styS_4 + uvwxS_5)\bar{z}$$

$$c_5 : \bar{s}\bar{t}\bar{u}\bar{v}wx\bar{y}\bar{z}; \qquad s_5 = twxS_3 + (\bar{s}\bar{u}\bar{v}\bar{y}S_2 + twxS_4)\bar{z}$$

$$c_6 : \bar{s}\bar{t}\bar{u}vw\bar{x}y\bar{z}; \qquad s_6 = \bar{s}\bar{u}\bar{x}S_4 + (\bar{s}\bar{u}\bar{x}S_3 + tvwyS_5)\bar{z}$$

$$c_7 : \bar{s}\bar{t}u\bar{v}\bar{w}xy\bar{z}; \qquad s_7 = \bar{s}\bar{v}\bar{w}S_4 + (\bar{s}\bar{v}\bar{w}S_3 + tuxyS_5)\bar{z}$$

$$c_8 : \bar{s}\bar{t}uv\bar{w}\bar{x}\bar{y}\bar{z}; \qquad s_8 = tuvS_3 + (\bar{s}\bar{w}\bar{x}\bar{y}S_2 + tuvS_4)\bar{z}$$

$$c_9 : s\bar{t}\bar{u}\bar{v}wxyz; \qquad s_9 = \bar{t}\bar{u}\bar{v}S_4 + (\bar{t}\bar{u}\bar{v}S_3 + swxyS_5)\bar{z}$$

$$c_{10} : s\bar{t}\bar{u}vw\bar{x}\bar{y}\bar{z}; \qquad s_{10} = svwS_3 + (\bar{t}\bar{u}\bar{x}\bar{y}S_2 + svwS_3)\bar{z}$$

$$c_{11} : s\bar{t}u\bar{v}\bar{w}x\bar{y}\bar{z}; \qquad s_{11} = suxS_3 + (\bar{t}\bar{v}\bar{w}\bar{y}S_2 + suxS_4)\bar{z}$$

$$c_{12} : s\bar{t}uv\bar{w}\bar{x}y\bar{z}; \qquad s_{12} = \bar{t}\bar{w}\bar{x}S_4 + (\bar{t}\bar{w}\bar{x}S_3 + suvyS_5)\bar{z}$$

$$c_{13} : st\bar{u}\bar{v}\bar{w}\bar{x}y\bar{z}; \qquad s_{13} = styS_3 + (\bar{u}\bar{v}\bar{w}\bar{x}S_2 + styS_4)\bar{z}$$

$$c_{14} : st\bar{u}\bar{u}vw\bar{x}y\bar{z}; \qquad s_{14} = \bar{u}\bar{w}\bar{y}S_4 + (\bar{u}\bar{w}\bar{y}S_3 + stvxS_5)\bar{z}$$

$$c_{15} : stu\bar{v}w\bar{x}\bar{y}\bar{z}; \qquad s_{15} = \bar{v}\bar{x}\bar{y}S_4 + (\bar{v}\bar{x}\bar{y}S_3 + stuwS_5)\bar{z}$$

$$c_{16} : stuvwxy\bar{z}; \qquad s_{16} = S_7 + S_6\bar{z}$$

$$c_{17} : \bar{s}\bar{t}\bar{u}\bar{v}\bar{w}\bar{x}\bar{y}z; \qquad s_{17} = S_0 + S_1z$$

$$c_{18} : \bar{s}\bar{t}\bar{u}v\bar{w}xyz; \qquad s_{18} = vxyS_3 + (\bar{s}\bar{t}\bar{u}\bar{w}S_2 + vxyS_4)z$$

$$c_{19} : \bar{s}\bar{t}u\bar{v}w\bar{x}yz; \qquad s_{19} = uwyS_3 + (\bar{s}\bar{t}\bar{v}\bar{x}S_2 + uwyS_4)z$$

$$c_{20} : \bar{s}\bar{t}uvwx\bar{y}z; \qquad s_{20} = \bar{s}\bar{t}\bar{y}S_4 + (\bar{s}\bar{t}\bar{y}S_3 + uvwxS_5)z$$

$$c_{21} : \bar{s}\bar{t}\bar{u}\bar{v}wx\bar{y}z; \qquad s_{21} = twxS_3 + (\bar{s}\bar{u}\bar{v}\bar{y}S_2 + twxS_4)z$$

$$c_{22} : \bar{s}\bar{t}\bar{u}vw\bar{x}yz; \qquad s_{22} = \bar{s}\bar{u}\bar{x}S_4 + (\bar{s}\bar{u}\bar{x}S_3 + tvwyS_5)z$$

$$c_{23} : \bar{s}\bar{t}u\bar{v}\bar{w}xyz; \qquad s_{23} = \bar{s}\bar{v}\bar{w}S_4 + (\bar{s}\bar{v}\bar{w}S_3 + tuxyS_5)z$$

$$c_{24} : \bar{s}tuv\bar{w}\bar{x}\bar{y}z; \qquad s_{24} = tuvS_3 + (\bar{s}\bar{w}\bar{x}\bar{y}S_2 + tuvS_4)z$$

$$c_{25} : s\bar{t}\bar{u}\bar{v}wxyz; \qquad s_{25} = \bar{t}\bar{u}\bar{v}S_4 + (\bar{t}\bar{u}\bar{v}S_3 + swxyS_5)z$$

$$c_{26} : s\bar{t}\bar{u}vw\bar{x}\bar{y}z; \qquad s_{26} = svwS_3 + (\bar{t}\bar{u}\bar{x}\bar{y}S_2 + svwS_4)z$$

$$c_{27} : s\bar{t}u\bar{v}\bar{w}x\bar{y}z; \qquad s_{27} = suxS_3 + (\bar{t}\bar{v}\bar{w}\bar{y}S_2 + suxS_4)z$$

$$c_{28} : s\bar{t}uv\bar{w}\bar{x}yz; \qquad s_{28} = \bar{t}\bar{w}\bar{x}S_4 + (\bar{t}\bar{w}\bar{x}S_3 + suvyS_5)z$$

$$c_{29} : st\bar{u}\bar{v}\bar{w}\bar{x}yz; \qquad s_{29} = styS_3 + (\bar{u}\bar{v}\bar{w}\bar{x}S_2 + styS_4)z$$

$$c_{30} : st\bar{u}\bar{u}vw\bar{x}\bar{y}z; \qquad s_{30} = \bar{u}\bar{w}\bar{y}S_4 + (\bar{u}\bar{w}\bar{y}S_3 + stvxS_5)z$$

$$c_{31} : stu\bar{v}w\bar{x}\bar{y}z; \qquad s_{31} = \bar{v}\bar{x}\bar{y}S_4 + (\bar{v}\bar{x}\bar{y}S_3 + stuwS_5)z$$

$$c_{32} : stuvwxyz; \qquad s_{32} = S_7 + S_6z$$

The symmetric functions are understood to be in terms of the first seven variables, so that in all cases $S_i = S_i(s, t, u, v, w, x, y)$. The structure of the partial decoder is shown in Fig. 10. The four functions $s_1, s_{16}, s_{17},$ and s_{32} are obtained at the edges of the symmetric network. The remaining functions can be grouped into 14 groups if advantage is taken of the similarity between pairs of functions like s_i and s_{i+16} ($2 \leq i \leq 15$). The intervening nets between

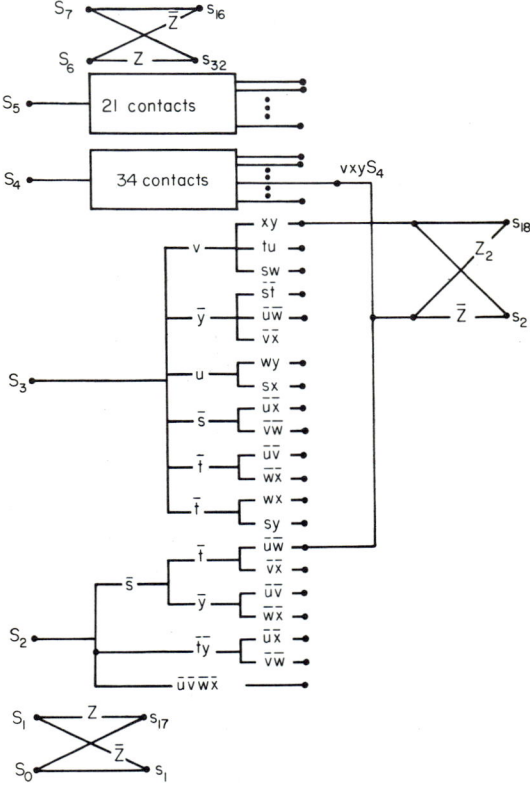

FIG. 10.   The 8-variable partial decoder.

the symmetric function net and the terminal z contacts on S_2 and S_3 are also shown; those on S_4 and S_5 are similar. Realization of a particular pair like s_2 and s_{18} is also shown. The network requires a total of 198 contacts (56 for symmetric network, 110 for intervening networks, and 32 for the terminal z contacts). The modified C_2, written C_2', is given by

$$C_2' = 198 + 32c(n - 8) + 2^n \qquad (7)$$

which is also tabulated in Table I. Note that the network of Fig. 10 was originally proposed by Short (1963); the improvement that has been incorporated is the elimination of 32 contacts from the output z terminals and this has been possible because of incorporation of shell centers as optimal terms.

V. TOWARD A GENERAL THEORY

The central problem in the design of Lupanov decoding networks seems to be the problem of finding a minimum set of shell functions which should form the outputs of the partial decoder. An equivalent way of stating the problem is to say that we wish to find a minimum number of shell centers, which are in general k-subcubes in the p-cube $(0 \leq k < p)$, such that every vertex of the p-cube is edge adjacent to at least one shell center. Lupanov's solution applies to the case when $p = r = 2^k$ and the shell centers are all vertices in the r-cube and represents a minimum solution.

Consider now the design of a p-variable partial decoder such that $r = 2^k < p < 2^{k+1}$. Our design procedure will be based on two steps. We believe that this procedure should lead to one of the minimal solutions although no proof is known for this conjecture. The steps are: (1) Derive a set of shell functions in the p-cube with respect to a set of vertices as shell centers such that the shell functions are disjoint and cover a maximum number of vertices in the p-cube. (2) The residual set of vertices are then covered by using a minimum number of shell functions whose centers are in general k-subcubes.

Step 1. Let H_0 be a set of single-error-correcting Hamming code words in the $r' = r - 1$ cube of variables $x_1, x_2, \ldots, x_{r'}$. Let H_i be the set of code words obtained from H_0 by complementing the ith coordinate x_i in H_0. It is well known that $H_0, H_1, \ldots, H_{r'}$ are disjoint and exhaust the entire r'-cube. Obviously, therefore, by the method of doubling using an extra variable x_r, as discussed in the previous sections, it is possible to define $2^r/r$ shell functions with respect to single vertices in r different ways. Denote these possible groups of shell centers as $P_1, P_2, \ldots, P_r$. Now let $p = r + m$. The p-cube can be viewed as consisting of a set of 2^m r-subcubes $x_{r+1}x_{r+2} \cdots x_{r+m}, \bar{x}_{r+1} x_{r+2} \cdots x_{r+m}, \ldots, \bar{x}_{r+1}\bar{x}_{r+2} \cdots \bar{x}_{r+m}$. Two groups of shell centers P_i and P_j $(1 \leq i, j \leq r, i \neq j)$ can be assigned to two of these subcubes if and only if the "distance" between the subcubes is greater than or equal to 2. This insures that the shell functions defined with respect to vertices in P_i and P_j will cover disjoint sets of vertices in the p-cube. Since the m-cube of variables $x_{r+1}, x_{r+2}, \ldots, x_{r+m}$ (for that matter any arbitrary n-cube) is a bipartite graph (Berge, 1962), we can find a maximum of 2^{m-1} vertices in the m-cube (that is, 2^{m-1} r-subcubes in the p-cube) such that the distance between any pair of vertices in the m-cube (i.e., the "distance" between any pair of r-subcubes in the p-cube) is greater than or equal to 2. If $r \geq 2^{m-1}$, we can arbitrarily pick up a set of

2^{m-1} shell center patterns from $P_1, P_2, \ldots, P_r$ and can define a total number of $(2^r/r)2^{m-1}$ shell functions which will cover a total of V_1 vertices of the p-cube given by

$$V_1 = p(2^r/r)2^{m-1} \qquad (8)$$

and the number of vertices left uncovered is

$$N_1 = 2^p - V_1 = 2^p - p(2^r/r)2^{m-1} \qquad (9)$$

There are a total of $\binom{r}{2^{m-1}}$ choices for the shell center patterns, and a total of $2^{m-1}!$ ways of distributing the shell pattern in the different subcubes and 2 ways to pick up a set of 2^{m-1} vertices of distance greater or equal to 2 in the m-cube. Therefore, the number of possible ways that a set of V_1 vertices can be covered is

$$W_1 = 2\binom{r}{2^{m-1}}2^{m-1}! \qquad (10)$$

On the other hand, if $2^{m-1} > r$, we can utilize all the shell center patterns $P_1, P_2, \ldots, P_r$ and the number of vertices that can be covered in the p-cube V_2 is given by

$$V_2 = p(2^r/r)r = p2^r \qquad (11)$$

and the number of vertices left uncovered will be

$$N_2 = 2^p - V_2 = 2^p - p2^r \qquad (12)$$

and the number of possible ways W_2 that the V_2 vertices can be covered is

$$W_2 = 2\binom{2^{m-1}}{r}r! \qquad (13)$$

Let us take two examples to illustrate the above ideas

Example 1. Let $r' = 3$, $r = 4$, $p = 5$, and therefore $m = 2$. The set of Hamming codes H_0, H_1, H_2, and H_3 are given by

$$H_0 = \begin{bmatrix} 0 & 0 & 0 \\ 1 & 1 & 1 \end{bmatrix}, \quad H_1 = \begin{bmatrix} 1 & 0 & 0 \\ 0 & 1 & 1 \end{bmatrix}, \quad H_2 = \begin{bmatrix} 0 & 1 & 0 \\ 1 & 0 & 1 \end{bmatrix}, \quad H_3 = \begin{bmatrix} 0 & 0 & 1 \\ 1 & 1 & 0 \end{bmatrix}$$

The shell center patterns P_1, P_2, P_3, and P_4 are given by

$$P_1 = \begin{bmatrix} 0 & 0 & 0 & 0 = a_1 \\ 0 & 0 & 0 & 1 = a_2 \\ 1 & 1 & 1 & 0 = a_3 \\ 1 & 1 & 1 & 1 = a_4 \end{bmatrix}, \quad P_2 = \begin{bmatrix} 1 & 0 & 0 & 0 = b_1 \\ 1 & 0 & 0 & 1 = b_2 \\ 0 & 1 & 1 & 0 = b_3 \\ 0 & 1 & 1 & 1 = b_4 \end{bmatrix}$$

$$P_3 = \begin{bmatrix} 0 & 1 & 0 & 0 = c_1 \\ 0 & 1 & 0 & 1 = c_2 \\ 1 & 0 & 1 & 0 = c_3 \\ 1 & 0 & 1 & 1 = c_4 \end{bmatrix}, \quad P_4 = \begin{bmatrix} 0 & 0 & 1 & 0 = d_1 \\ 0 & 0 & 1 & 1 = d_2 \\ 1 & 1 & 0 & 0 = d_3 \\ 1 & 1 & 0 & 1 = d_4 \end{bmatrix}$$

The shell center patterns are depicted in a Karnaugh map shown in Fig. 11. One assignment of shell centers in the 6-cube has already been shown in Fig. 6 where P_1 and P_3 have been assigned in subcubes $\bar{x}_6 \bar{x}_5$ and $x_6 x_5$, respectively. We could have interchanged P_1 and P_3 or selected any other pair like P_1 and P_2 from P_1, P_2, P_3, and P_4 or could have taken the subcube $x_6 \bar{x}_5$ and $\bar{x}_6 x_5$ instead, so that there are actually $2 \cdot \binom{4}{2} \cdot 2! = 24$ valid assignments of shell centers. The number of vertices covered by the shell functions is $V_1 = 6 \cdot 2^4/4 \cdot 2 = 48$, and the algebraic expressions for the shell functions have been given in Section III.

Example 2. Let $r' = 3$, $r = 4$, $p = 7$, and hence $m = 3$. In this case, $r = 2^{m-1}$ and we shall use all the shell center patterns P_1, P_2, P_3, and P_4. The assignment is shown in Fig. 12. The patterns P_1, P_2, P_3, and P_4 have been assigned to subcubes $\bar{x}_7 \bar{x}_6 \bar{x}_5$, $x_7 x_6 \bar{x}_5$, $\bar{x}_7 x_6 x_5$, and $x_7 x_6 x_5$, respectively.

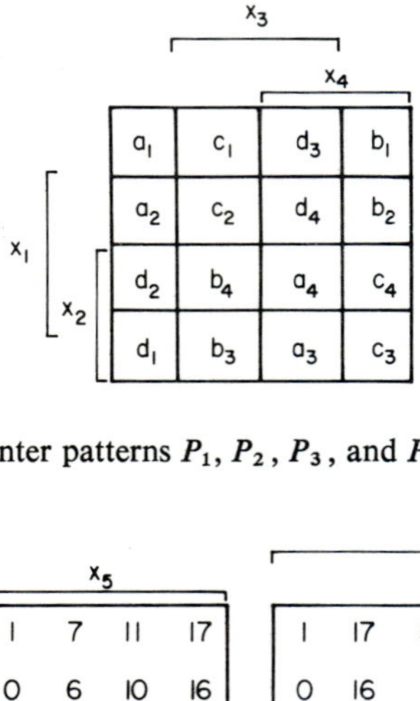

FIG. 11.   The shell center patterns P_1, P_2, P_3, and P_4 on a Karnaugh map.

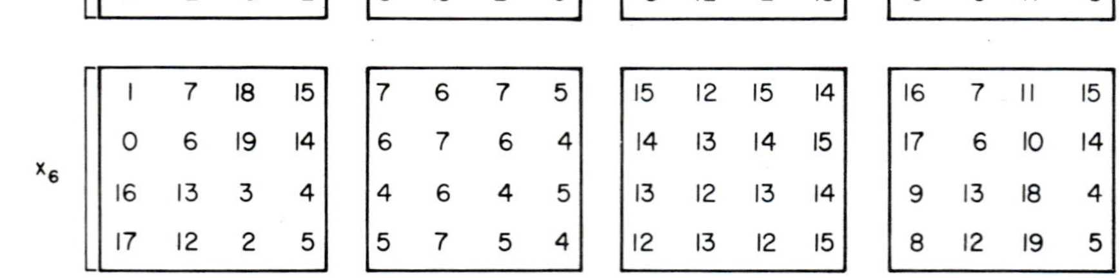

FIG. 12.   The functions $z_0, \ldots, z_{19}$ plotted in the 7-variable Karnaugh map.

In this case, there are actually $2 \cdot \binom{4}{4} \cdot 4! = 48$ different assignments of shell patterns and the number of vertices covered by the shell functions is $7 \cdot 2^4/4 \cdot 2^2 = 112$. The shell functions z_i have been simplified, taking the shell centers as optional terms and are given below. Here S_i denotes a symmetric function $S_i(x_2, x_3, x_4, x_5, x_6, x_7)$

$c_0 : \bar{x}_1 \bar{x}_2 \bar{x}_3 \bar{x}_4 \bar{x}_5 \bar{x}_6 \bar{x}_7 ;$ $\qquad z_0 = S_0 + \bar{x}_1 S_1$

$c_1 : x_1 \bar{x}_2 \bar{x}_3 \bar{x}_4 \bar{x}_5 \bar{x}_6 \bar{x}_7 ;$ $\qquad z_1 = S_0 + x_1 S_1$

$c_2 : \bar{x}_1 x_2 x_3 x_4 \bar{x}_5 \bar{x}_6 \bar{x}_7 ;$ $\qquad z_2 = x_2 x_3 x_4 S_3 + (x_2 x_3 x_4 S_4 + \bar{x}_5 \bar{x}_6 \bar{x}_7 S_2)\bar{x}_1$

$c_3 : x_1 x_2 x_3 x_4 \bar{x}_5 \bar{x}_6 \bar{x}_7 ;$ $\qquad z_3 = x_2 x_3 x_4 S_3 + (x_2 x_3 x_4 S_4 + \bar{x}_5 \bar{x}_6 \bar{x}_7 S_2)x_1$

$c_4 : \bar{x}_1 x_2 \bar{x}_3 x_4 x_5 x_6 x_7 ;$ $\qquad z_4 = \bar{x}_3 \bar{x}_7 S_4 + [\bar{x}_3 \bar{x}_7 S_3 + (\bar{x}_3 + \bar{x}_7)S_5]\bar{x}_1$

$c_5 : x_1 x_2 \bar{x}_3 x_4 x_5 x_6 \bar{x}_7 ;$ $\qquad z_5 = \bar{x}_3 \bar{x}_7 S_4 + [x_3 \bar{x}_7 S_3 + (\bar{x}_3 + \bar{x}_7)S_5]x_1$

$c_6 : \bar{x}_1 \bar{x}_2 \bar{x}_3 \bar{x}_4 x_5 x_6 \bar{x}_7 ;$ $\qquad z_6 = x_3 x_5 x_6 S_3 + (x_3 x_5 x_6 S_4 + \bar{x}_2 \bar{x}_4 \bar{x}_7 S_2)\bar{x}_1$

$c_7 : x_1 \bar{x}_2 x_3 \bar{x}_4 x_5 x_6 \bar{x}_7 ;$ $\qquad z_7 = x_3 x_5 x_6 S_3 + (x_3 x_5 x_6 S_4 + \bar{x}_2 \bar{x}_4 \bar{x}_7 S_2)x_1$

$c_8 : \bar{x}_1 x_2 \bar{x}_3 \bar{x}_4 x_5 \bar{x}_6 x_7 ;$ $\qquad z_8 = x_2 x_5 x_7 S_3 + (x_2 x_5 x_7 S_4 + \bar{x}_3 \bar{x}_4 \bar{x}_6 S_2)\bar{x}_1$

$c_9 : x_1 x_2 \bar{x}_3 \bar{x}_4 x_5 \bar{x}_6 \bar{x}_7 ;$ $\qquad z_9 = x_2 x_5 x_7 S_3 + (x_2 x_5 x_7 S_4 + \bar{x}_3 \bar{x}_4 \bar{x}_6 S_2)x_1$

$c_{10} : \bar{x}_1 \bar{x}_2 x_3 x_4 x_5 \bar{x}_6 x_7 ;$ $\qquad z_{10} = \bar{x}_2 \bar{x}_6 S_4 + [\bar{x}_2 \bar{x}_6 S_3 + (\bar{x}_2 + \bar{x}_6)S_5]\bar{x}_1$

$c_{11} : x_1 \bar{x}_2 x_3 x_4 x_5 \bar{x}_6 x_7 ;$ $\qquad z_{11} = \bar{x}_2 x_6 S_4 + [\bar{x}_2 \bar{x}_6 S_3 + (\bar{x}_2 \bar{x}_6)S_5]x_1$

$c_{12} : \bar{x}_1 x_2 x_3 \bar{x}_4 \bar{x}_5 x_6 x_7 ;$ $\qquad z_{12} = \bar{x}_4 \bar{x}_5 S_4 + [x_4 \bar{x}_5 S_3 + (\bar{x}_4 + \bar{x}_5)S_5]\bar{x}_1$

$c_{13} : x_1 x_2 x_3 \bar{x}_4 \bar{x}_5 x_6 x_7 ;$ $\qquad z_{13} = \bar{x}_4 \bar{x}_5 S_4 + [\bar{x}_4 \bar{x}_5 S_3 + (\bar{x}_4 + \bar{x}_5)S_5]x_1$

$c_{14} : \bar{x}_1 \bar{x}_2 \bar{x}_3 x_4 \bar{x}_5 x_6 x_7 ;$ $\qquad z_{14} = x_4 x_6 x_7 S_3 + (x_4 x_6 x_7 S_4 + \bar{x}_2 \bar{x}_3 \bar{x}_5 S_2)\bar{x}_1$

$c_{15} : x_1 \bar{x}_2 \bar{x}_3 x_4 \bar{x}_5 x_6 x_7 ;$ $\qquad z_{15} = x_4 x_6 x_7 S_3 + (x_4 x_6 x_7 S_4 + \bar{x}_2 \bar{x}_3 \bar{x}_5 S_2)x_1$

Step 2. In this part we are concerned with the problem of covering the residual set of vertices N_1 or N_2 using shell functions defined with respect to k-subcubes in general. The solution to this problem seems to be complex. Our approach will be intuitive and we hope to bring out certain notions which might stimulate further work. We need a few ideas defined.

DEFINITION (The Residual Function). Let us define a switching function $R_p(x_1, \ldots, x_p)$ called the residual function whose true vertices in the p-cube are the residual vertices N_1 or N_2 of step 1. An algebraic expression for R_p can be easily written down: let $X = \{(\dot{x}_1, \dot{x}_2, \ldots, \dot{x}_p)\}$ denote the set of shell centers with respect to which the shell functions in step 1 were defined. The $\bar{R}_p$, corresponding to vertices being covered, can be written as $\bar{R}_p = \sum S_{0,1}(\bar{x}_1, \bar{x}_2, \ldots, \bar{x}_p)$ for all shell centers in X. Therefore,

$$R_p = \prod S_{2, 3, \ldots, p}(\bar{x}_1, \bar{x}_2, \ldots, \bar{x}_p)$$
$$= \prod (\bar{x}_1 \bar{x}_2 + \bar{x}_1 \bar{x}_3 + \cdots + \bar{x}_{p-1} \bar{x}_p) \qquad (14)$$

Simplification of an expression like Eq. (18) is very cumbersome. An alternative way of obtaining R_p is given below. (The justification for this is left to the reader.)

$\bar{R}_p$ = (algebraic expressions of the r-cubes totally covered in step 1)

$+ \sum$ (algebraic expressions of the shell centers employed in the ith totally covered r-cube)

(algebraic expression for the r-cubes adjacent to the ith totally covered r-cube) (15)

Thus for Fig. 6, the vertices for $\bar{R}_6$ are marked by integer $0, 1, \ldots, 7$ and is given by

$$\bar{R}_6 = (x_6 x_5 + \bar{x}_6 \bar{x}_5) + (\bar{x}_4 \bar{x}_3 \bar{x}_2 + x_4 x_3 x_2)(x_6 \bar{x}_5 + \bar{x}_6 x_5)$$
$$+ (x_4 \bar{x}_3 x_2 + \bar{x}_4 x_3 x_2)(x_6 \bar{x}_5 + \bar{x}_6 x_5) \qquad (16)$$

Therefore,

$$R_6 = (x_6 \bar{x}_5 + \bar{x}_6 x_5)[x_6 x_5 + \bar{x}_6 \bar{x}_5$$
$$+ (x_4 + x_3 + x_2)(\bar{x}_4 + \bar{x}_3 + \bar{x}_2)(\bar{x}_4 + x_3 + \bar{x}_2)(x_4 + \bar{x}_3 + x_2)]$$
$$= (x_6 \bar{x}_5 + \bar{x}_6 x_5)(x_4 \bar{x}_2 + \bar{x}_4 x_2) \qquad (17)$$

Similarly, for Fig. 12 the vertices for $\bar{R}_7$ are marked by integers $0, 1, \ldots, 15$ and $\bar{R}_7$ is given by

$$\bar{R}_7 = \bar{x}_7 \bar{x}_6 \bar{x}_5 + \bar{x}_7 x_6 x_5 + x_7 x_6 \bar{x}_5 + x_7 \bar{x}_6 x_5$$
$$+ (\bar{x}_4 \bar{x}_3 \bar{x}_2 + x_4 x_3 x_2)(\bar{x}_7 \bar{x}_6 x_5 + \bar{x}_7 x_6 \bar{x}_5 + x_7 \bar{x}_6 \bar{x}_5)$$
$$+ (\bar{x}_4 x_3 \bar{x}_2 + x_4 \bar{x}_3 x_2)(\bar{x}_7 x_6 \bar{x}_5 + \bar{x}_7 \bar{x}_6 x_5 + x_7 x_6 x_5)$$
$$+ (x_4 \bar{x}_3 \bar{x}_2 + \bar{x}_4 x_3 x_2)(x_7 \bar{x}_6 \bar{x}_5 + x_7 x_6 x_5 + \bar{x}_7 x_6 \bar{x}_5)$$
$$+ (x_4 x_3 x_2 + \bar{x}_4 \bar{x}_3 x_2)(x_7 \bar{x}_6 \bar{x}_5 + x_7 x_6 x_5 + \bar{x}_7 \bar{x}_6 x_5) \qquad (18)$$

and R_7 after simplification is given by

$$R_7 = \bar{x}_7 x_6 \bar{x}_5(x_4 x_3 \bar{x}_2 + \bar{x}_4 \bar{x}_3 x_2) + \bar{x}_7 \bar{x}_6 x_5(x_4 \bar{x}_3 \bar{x}_2 + \bar{x}_4 x_3 x_2)$$
$$+ x_7 \bar{x}_6 \bar{x}_5(\bar{x}_4 x_3 \bar{x}_2 + x_4 \bar{x}_3 x_2) + x_7 x_6 x_5(\bar{x}_4 \bar{x}_3 \bar{x}_2 + x_4 x_3 x_2) \qquad (19)$$

DEFINITION (Projection of a Switching Function). Let $f(x_1, \ldots, x_n)$ be a switching function. Consider a s-subcube of the n-cube in a set of variables which is a proper subset of $(x_1, \ldots, x_n)$, say, without loss of generality, $x_1, x_2, \ldots, x_s$. By the projection of f on the s-subcube we mean the transformation

$$f(x_1, x_2, \ldots, x_n) \to g(x_1, x_2, \ldots, x_s)$$

where $\dot{x}_1 \dot{x}_2 \cdots \dot{x}_s$ is a true vertex of g in the s-subcube if and only if there exists at least one vertex $\dot{x}_1 \dot{x}_2 \cdots \dot{x}_s \dot{x}_{s+1} \cdots \dot{x}_n$ in the n-cube which is a true vertex of f ($\dot{x}_i$ is either x_i or $\bar{x}_i$). If every true vertex of g is a projection of exactly one true vertex of f, we call the projection one-to-one; otherwise the projection is many-to-one. If g is an identity function, we say that the projection of f on the s-subcube is an *identity projection*. Furthermore, if the projection is one-to-one and identity, we call the projection a *true identity projection*. An algebraic test for deciding whether or not any particular projection of f on a subcube is identity or true identity can be stated utilizing expansion theorem

$$f(x_1, x_2, \ldots, x_n) = \sum_{i=0}^{2^{n-s}} \phi_i(x_{s+1}, \ldots, x_n)\sigma_i(x_1, \ldots, x_s) \qquad (20)$$

where ϕ_i is a "fundamental product" associated with the ith configuration of $(x_{s+1}, \ldots, x_n)$ and σ_i is the ith residue of f. Obviously, then f has an identity projection on the s-cube in variables $x_1, \ldots, x_s$ if and only if $\sum \sigma_i = 1$. The projection is a true identity projection if, furthermore, $\sigma_i \cdot \sigma_j = 0$ for any pair i, j such that $i \neq j$.

We are now in a position to describe our procedure of step 2 as follows: Compute the residual function $R_p(x_1, \ldots, x_p)$ and check whether R_p has identity projection on some s-cube ($s < p$). If there are more than one, pick up one for which s has a maximum value. (If f has more than one true terms, a trivial projection always exists for $s = 1$.) We have to consider now two cases:

Case I. The projection is a true identity projection. Here we have to consider two subcubes:

Case a: s *is a power of two.* In this case we have a neat solution. As discussed in the earlier sections, we can derive a set of $2^s/s$ shell functions defined with respect to vertices in the s-cube (actually the shell centers are $(p - s)$-cubes in the p-cube).

Example. For $p = 6$, as is seen from Eq. (17), the residual function R_6 has a true identity projection on the 4-cube of variables $x_6, x_3, x_2,$ and x_1. We are, therefore, able to define a set of four shell functions $y_8, y_9, y_{10},$ and y_{11} which have already been discussed. The situation is similar for $p = 7$, as can be verified from Eq. (19); R_7 has a true identity projection on the 4-cube of variables $x_4, x_3, x_2,$ and x_1. We can, therefore, define a set of four shell functions $z_{16}, z_{17}, z_{18},$ and z_{19} as shown in Fig. 12 (the sets of vertices marked 16–19) having shell centers

$$\begin{aligned} z_{16} &: \bar{x}_4 \bar{x}_3 \bar{x}_2 x_1 && (\text{- - -}0001) \\ z_{17} &: \bar{x}_4 \bar{x}_3 \bar{x}_2 \bar{x}_1 && (\text{- - -}0000) \\ z_{18} &: x_4 x_3 x_2 \bar{x}_1 && (\text{- - -}1110) \\ z_{19} &: x_4 x_3 x_2 x_1 && (\text{- - -}1111) \end{aligned}$$

and the functions $z_{16}, \ldots, z_{19}$ are given by the following matrices of true fundamental product terms:

$$z_{16} = \begin{array}{ccccccc} x_7 & x_6 & x_5 & x_4 & x_3 & x_2 & x_1 \\ 1 & 1 & 1 & 0 & 0 & 0 & 0 \\ 0 & 1 & 0 & 0 & 0 & 1 & 1 \\ 1 & 0 & 0 & 0 & 1 & 0 & 1 \\ 0 & 0 & 1 & 1 & 0 & 0 & 1 \end{array},
\qquad
z_{17} = \begin{array}{ccccccc} x_7 & x_6 & x_5 & x_4 & x_3 & x_2 & x_1 \\ 1 & 1 & 1 & 0 & 0 & 0 & 1 \\ 0 & 1 & 0 & 0 & 0 & 1 & 0 \\ 1 & 0 & 0 & 0 & 1 & 0 & 0 \\ 0 & 0 & 1 & 1 & 0 & 0 & 0 \end{array}$$

$$z_{18} = \begin{array}{ccccccc} 1 & 1 & 1 & 1 & 1 & 1 & 1 \\ 0 & 1 & 0 & 1 & 1 & 0 & 0 \\ 1 & 0 & 0 & 1 & 0 & 1 & 0 \\ 0 & 0 & 1 & 0 & 1 & 1 & 0 \end{array},
\qquad
z_{19} = \begin{array}{ccccccc} 1 & 1 & 1 & 1 & 1 & 1 & 0 \\ 0 & 1 & 0 & 1 & 1 & 0 & 1 \\ 1 & 0 & 0 & 1 & 0 & 1 & 1 \\ 0 & 0 & 1 & 0 & 1 & 1 & 1 \end{array}$$

The functions $z_0, \ldots, z_{15}$ have already been given. The network which produces $z_0, \ldots, z_{15}$ is shown in Fig. 13 and needs 105 contacts. The network of Fig. 14 produces z_{16}, z_{17}, z_{18}, and z_{19} and needs 50 contacts. The partial

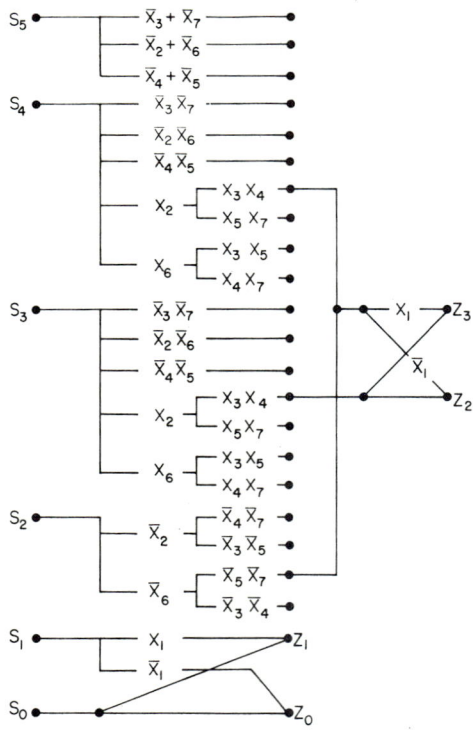

FIG. 13.   The contact network for $z_0, \ldots, z_{15}$.

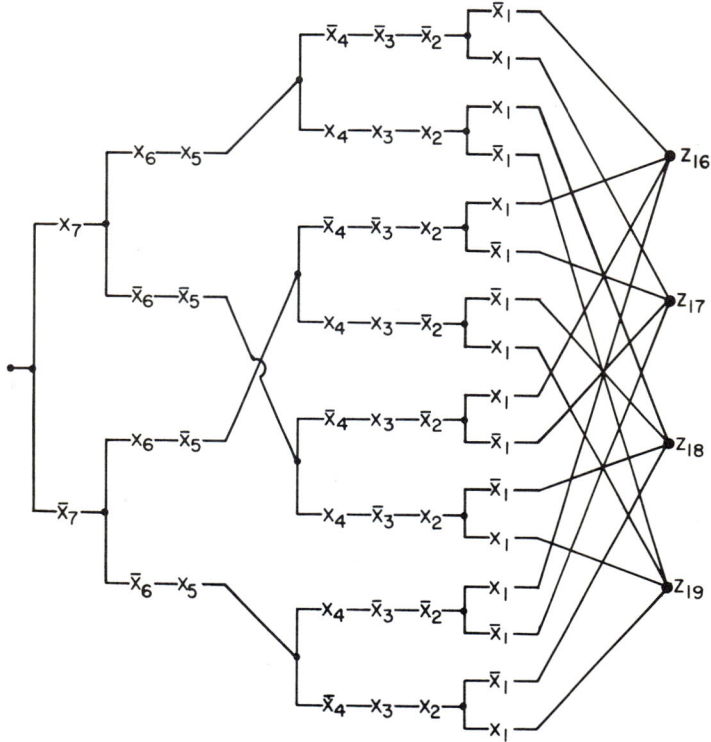

FIG. 14.   The contact network for z_{16}, z_{17}, z_{18}, and z_{19}.

decoder for $p = 7$, therefore, needs a total of 155 contacts. Thus for a n-variable complete decoder we can write

$$C_4 = 155 + 20c(n - 7) + 2^n. \tag{21}$$

The value of C_4 for n up to 20 is tabulated in Table I.

Case b: *s is not a power of two.* In this case we go back to step 1 and repeat the entire procedure with respect to s variables, i.e., taking $p = s$. The shell functions will be determined with respect to shell centers in the s-cube, but their relationship with respect to the original p-cube is a matter of dimension only. At some stage, s must become a power of two so that this case will degenerate.

A theoretical question that can be raised in this context is to inquire under what conditions the projection of a residual function R_p is a true identity projection. A necessary condition is that the number of the residual vertices N_1 or N_2 must be a power of two. Taking $r \geq 2^{m-1}$ means that

$$2^p - p(2^r/r)2^{m-1} = 2^{p-1}(1 - (m/r))$$

must be a power of two, that is, m/r has values $\frac{1}{2}, \frac{3}{4}, \frac{7}{8}, \frac{15}{16}, \frac{31}{32}, \ldots$. This gives Table II for the possible values of p for which R_p has a true identity projection.

TABLE II
Possible Values of p for Which
R_p Has True Identity Projection

r	m	p
2	1	3
4	2	6
	3	7
8	4	12
	6	14
	7	15
16	8	24
	9	25
	14	30
	15	31

We have already noted that for $p = 3, 6,$ and 7 the residual functions do actually have true identity projections. The other cases have not been resolved and it is not known whether the above condition is also a sufficient condition for true projection.

Case II. The projection is an identity but not a true identity projection: In this case there are several alternatives that can be followed and it is not known which one of them should lead to an optimal solution.

A. If R_p does not have a true identity projection, then there must exist some function Q_p where $R_p > Q_p$ (i.e., $R_p Q_p = Q_p$), whose projection is a true identity projection. We will treat Q_p as in case I and the remainder function $R_p \bar{Q}_p$ is then considered as a residual function for which the entire procedure for step 2 is repeated.

B. The projection of R_p can be expressed as a union of a number of true identity projections. Each true identity projection is then treated separately as in Case I.

C. We can augment R_p by adding to R_p a function $H_p(x_1, \ldots, x_p)$ such that $R_p + H_p$ has a true identity projection where H_p is contained in $\bar{R}_p$.

We shall illustrate Case II by an example. Take $p = 5$. In step 1, we would have defined a set of four shell functions as shown in the Karnaugh map of Fig. 15 and have the residual function as

$$R_5 = x_5[x_4(\bar{x}_3 + \bar{x}_2) + x_3(\bar{x}_4 + \bar{x}_2) + x_2(\bar{x}_4 + \bar{x}_3)]$$

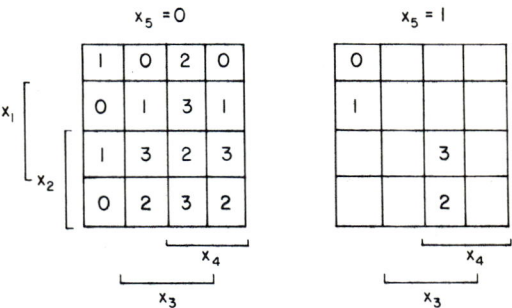

FIG. 15.   The first step for $p = 5$.

which can be written as a matrix as

$$
\begin{array}{c@{\quad}ccccc}
 & x_5 & x_4 & x_3 & x_2 & x_1 \\
(2) & 1 & 0 & 0 & 1 & 0 \\
(2) & 1 & 0 & 0 & 1 & 1 \\
(2) & 1 & 1 & 1 & 0 & 0 \\
(2) & 1 & 1 & 1 & 0 & 1 \\[4pt]
(1) & 1 & 0 & 1 & 0 & 0 \\
(1) & 1 & 0 & 1 & 0 & 1 \\
(1,2) & 1 & 0 & 1 & 1 & 0 \\
(1,2) & 1 & 0 & 1 & 1 & 1 \\
(1,2) & 1 & 1 & 0 & 0 & 0 \\
(1,2) & 1 & 1 & 0 & 0 & 1 \\
(1) & 1 & 1 & 0 & 1 & 0 \\
(1) & 1 & 1 & 0 & 1 & 1 \\
\end{array}
\qquad
\begin{array}{l}
R_p \bar{Q}_p \\[60pt]
Q_p
\end{array}
$$

Note that the portion of the matrix marked Q_p (a subfunction of R_p) has true identity projection on x_3, x_2, and x_1. The remainder $R_p \bar{Q}_p$ in this case also projects as a true identity projection on x_2 and x_1. Note also that R_5 can be expressed also as union of two identity projections on variables x_3, x_2, and x_1 if we let $R_5 = R^1 + R^2$ where terms of R^1 and R^2 are marked by integer 1 and 2, respectively, at the left hand side of the matrix. And if we let $H_5 = x_5(x_4 x_3 x_2 + \bar{x}_4 \bar{x}_3 \bar{x}_2)$, it is easily verified that $R_5 + H_5 = x_5$, i.e., if R_5 is augmented with H_5 we get a true identity projection on x_4, x_3, x_2, and x_1. If we take the last possibility, we will need four more shell functions. For all other cases, the number of shell functions needed is greater than 4. This example illustrates that when R_p does not have a true identity projection, we may have to consider several possibilities.

VI. CONCLUSIONS

The results of our investigations are summarized in Table I. It is seen that the Lupanov decoding networks obtained have the best minimal costs (underlined in the table) known so far for all values of n. Recognition of the fact that shell centers can be treated as optional terms, has yielded improved contact counts for the partial decoders for $r = 4$ and 8, resulting, as an example, in a 58-contact 5-variable complete decoder. For smaller values of n, the cost of the partial decoder forms a significant fraction of the total cost and therefore they have to be carefully designed. We have employed the basic symmetric network as a nucleus in all the cases, but alternative approaches appropriate for the synthesis of multioutput networks could be taken.

It should be noted that, although the designs presented here have all followed the approach that in the first step a maximal number of nonoverlapping shell functions with vertices as shell centers should be obtained, other approaches are possible. For example, Lupanov's original 5-variable decoder is based on a set of 12 shell functions, as shown in Fig. 16, none of whose

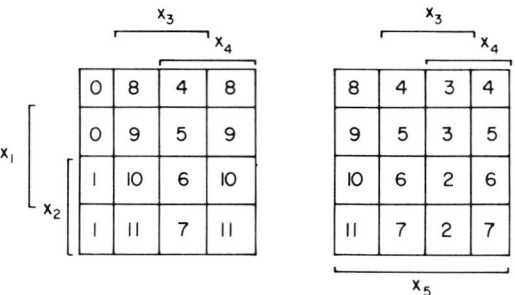

FIG. 16.   The partial decoder output functions of Lupanov's 5-variable decoder.

shell centers is a single vertex and the number of shell centers is also not minimal. In fact, an exact method of obtaining the best shell function covering of the n-cube can be stated as: (a) obtain a list of all shell functions derived with respect to vertices, 1-cubes, 2-cubes, ..., $(n - 1)$-cubes; (b) pick up a set of shell functions which will cover the n-cube such that the network which realizes them is least costly of all such possible covers of the n-cube. But, the total number of shell functions is very large and is given by

$$\sum_{k=0}^{n-1} \binom{n}{k} \cdot 2^{n-k} \cdot 2^{k(n-k)} = \sum_{k=0}^{n-1} (n!/(n-k)!\,k!)2^{(n-k)(1+k)} \tag{22}$$

so that for $n > 3$, an enumerative solution to the problem becomes practically unmanageable. Hence one needs to develop systematic procedures for obtaining the cover from some other approach. We have explored the linear coding theory approach to the problem. It might be worthwhile to investigate whether the theory of nonlinear codes, particularly the distance-three, closely-packed binary codes discovered by Vasilev (1962), could afford a better solution.

ACKNOWLEDGMENT

Presénted at the Ninth Annual Symposium on Switching and Automata Theory, October 15–18, 1968, Schenectady, New York. Reproduced with permission from the IEEE. The research reported in this chapter was partially supported by NSF GJ-158. Part of this work was done while the author was with the Tata Institute of Fundamental Research, India (Mukhopadhyay, 1967).

REFERENCES

BERGE, C. (1962). "Theory of Graphs and Applications." Methuen, London.

BURKS, A. W., McNAUGHTON, R., POLLMER, C. H., WARREN, D. W., and WRIGHT, J. B. (1954). Complete decoding nets: general theory and minimality. *SIAM (Soc. Ind. Appl. Math.) J. Appl. Math.* 2, 201–243.

BURKS, A. W., McNAUGHTON, R., POLLMER, C. H., WARREN, D. W., and WRIGHT, J. B. (1955). The folded tree. *J. Franklin Inst.* 260, 9–24, 115–126.

HARRISON, M. A. (1965). "Introduction to Switching and Automata Theory." McGraw-Hill, New York.

KAUTZ, W. H. (1964). Lupanov decoding tree. Tutorial article, private communication.

LUPANOV, O. B. (1958). O sinteze kontaktnykh skhem. *Dokl. Akad. Nauk. SSSR* 119, 23–26. [English transl.: On the synthesis of contact networks. *Automat. Express* 1, 7–8].

MOORE, E. F. (1960). Minimal complete relay decoding networks. *IBM J. Res. Develop.* 4, 525–531.

MUKHOPADHYAY, A. (1967). Lupanov Decoding Trees. Tech. Rep. No. 29, July, Tata Institute of Fundamental Research, Computer Group, Bombay, India.

PETERSON, W. W. (1961). "Error Correcting Codes." Wiley, New York.

SHANNON, C. E. (1949). The synthesis of two-terminal switching circuits. *Bell. Syst. Tech. J.* 28, 59–98.

SHORT, R. A. (1963). Contact decoding networks. Presented at the 10th Annual Symp. on Computers and Data Processing, Logic Design Session, Denver Research Institute, Denver, Colorado, June 26–27.

SHORT, R. A. (1968). Private communication.

VASILEV, Yu. L. (1962). Non-group closely-packed codes. *Probl. Kibern.* 8, 337–339 [Abstract: *IEEE Trans. Inform. Theory*, IT-10, 261.]

Chapter IV

COUNTING THEOREMS AND THEIR APPLICATIONS TO CLASSIFICATION OF SWITCHING FUNCTIONS

MICHAEL A. HARRISON

I. INTRODUCTION TO BOOLEAN FUNCTIONS AND CLASSIFICATION PROBLEMS 86

II. GROUP THEORY AND PÓLYA'S THEOREM 87

III. SOME APPLICATIONS OF PÓLYA'S THEOREM TO SWITCHING FUNCTIONS 98

IV. STRUCTURE THEOREMS FOR PERMUTATION GROUPS AND THE DETERMINATION OF CYCLE INDICES 101

V. OPERATIONS ON THE RANGE, GENERA, AND A LOWER BOUND 113

APPENDIXES
1. Cycle Index Polynomials for S_n. 116
2. Cycle Index Polynomials for G_n. 117
3. Cycle Index Polynomials for $GL_n(\mathbb{Z}_2)$. 118
4. Cycle Index Polynomials for $A_n(\mathbb{Z}_2)$. 119

REFERENCES 120

ABSTRACT. In this chapter we begin by introducing a powerful combinatorial theorem due to Pólya. This result is applied to switching theory to produce a number of different classifications. Specifically, we count the number of equivalence classes under the following notions of equivalence: (1) complementation of some of the variables, (2) permutation of some of the variables, (3) complementations and/or permutations of the variables, (4) any linear function of the variables, (5) any affine function of the variables. We obtain cycle index polynomials in all cases above. This leads to the desired numbers which are tabulated. A generalization of the Pólya result by DeBruijn is considered next. Applications are made which allow us to include negation of functions as part of the definition of equivalence. In this manner we obtain the number of genera and the number of equivalence classes of networks.

85

I. INTRODUCTION TO BOOLEAN FUNCTIONS AND CLASSIFICATION PROBLEMS

Many of the problems in switching theory are compounded because of the large number of switching functions. If we let $\mathbb{Z}_2 = \{0, 1\}$ and let $\mathbb{Z}_2^n = \{(a_1, \ldots, a_n) \mid a_i \in \mathbb{Z}_2\}$, then the class F_n of *boolean functions* is the set of mappings from $\mathbb{Z}_2^n$ into $\mathbb{Z}_2$. Consequently there are 2^{2^n} such functions. Techniques which involve enumeration of functions can only be used if n is trivially small. A common technique for extending the scope of such enumerative methods is to classify the set of functions into equivalence classes under some natural equivalence relation.

As an example, let us consider two functions of n variables to be "equivalent under C_2^{n}"[1] if one can be transformed into the other by complementing some of the variables.

For example if $f(x_1, x_2) = x_1 + \bar{x}_2$, then $g(x_1, x_2) = f(\bar{x}_1, \bar{x}_2) = \bar{x}_1 + x_2$ is equivalent to f under C_2^n. This is a "natural" equivalence relation if we are dealing with a digital technology in which both signals and their complements are available. Thus, both f and g can be realized by the same circuit if we replace some of the input signals by their complements.

Another natural equivalence relation occurs if we say that functions are "equivalent under S_n" when one can be obtained from the other by permuting some of the variables. For instance

$$f(x_1, x_2, x_3) = x_1 + x_2 \bar{x}_3 \tag{1}$$

is equivalent under S_3 to

$$g(x_1, x_2, x_3) = f(x_2, x_3, x_1) = x_2 + \bar{x}_1 x_3 \tag{2}$$

This type of equivalence appears to be natural in most technologies.

Whenever we have found two types of operations which naturally partition the boolean functions into equivalence classes, it is of interest to consider an equivalence which allows both types of operations. For example, we say that two switching functions are "equivalent under G_n" if one can be obtained from the other by a complementation and/or permutation of the variables. That is, if

$$f(x_1, x_2, x_3) = x_1 + x_2 \bar{x}_3$$

then

$$g(x_1, x_2, x_3) = f(\bar{x}_2, \bar{x}_3, x_1) = \bar{x}_2 + \bar{x}_1 \bar{x}_3$$

is equivalent to $f(x_1, x_2, x_3)$ under G_3.

---

[1] The reason for this terminology will become clear in a later section.

We shall introduce other equivalence relations later but the general idea should now be clear. Since we want to classify switching functions into classes, two natural problems occur. First, we ask for the number of such equivalence classes. This problem will be treated in detail in this chapter for the family of equivalence relations which are "induced by groups."

Another problem in such a study is to give a method which will decide the class to which an arbitrary function belongs. This problem will not be treated in the present chapter but relevant information can be found in Chapter V, in Lechner (1968), and in Harrison (1965).

II. GROUP THEORY AND PÓLYA'S THEOREM

The types of problems discussed in Section I are all special cases of a general and important result in combinatorial analysis. In the present section we shall develop this theory and prove a fundamental result due to Pólya (1937).

Let us recall that a group is an algebraic system with a nonempty domain G and a binary operation $\circ$ defined on G which satisfies the following axioms:

1. ASSOCIATIVITY.   For each a, b, and c in G,

$$a \circ (b \circ c) = (a \circ b) \circ c$$

2. IDENTITY.   There is an element 1 in G so that for each a in G

$$a \circ 1 = 1 \circ a = a \tag{3}$$

3. INVERSES.   For each a in G there is some element a^{-1} in G so that

$$a \circ a^{-1} = a^{-1} \circ a = 1 \tag{4}$$

We say that 1 is the *identity* while a^{-1} is called the *inverse* of a. Let us agree to write $a \circ b$ as ab.

As an example, notice that the set $\{1\}$ is a group under the operation $11 = 1$. A more interesting example comes from considering $\mathbb{Z}_m = \{0, 1, \ldots, m-1\}$ and the group operation is addition modulo m. Another example is S_n, the *symmetric group* on $\{1, \ldots, n\}$ which consists of all permutations[2] of $\{1, \ldots, n\}$ under the operation of composition.[3]

---

[2] A permutation of a set S is a one-to-one map from S onto S.

[3] If f maps Y into Z and g maps X into Y then the *composition fg* of f and g maps X into Z by the rule $fg(x) = f(g(x))$ for each $x \in X$.

DEFINITION.   If G is a group, then the *order* of G is its cardinality and is written $|G|$. If G is a permutation group acting on a set S, then the *degree* of G is the cardinality of S.

For example, S_n has order $n!$ and degree n.

If G is a group and H is a subset of G which is also a group, then we say that H is a *subgroup* of G. If G is finite, it is interesting to observe that a subset H is a subgroup if and only if it is closed under the group operation and it satisfies Axiom 2.

DEFINITION.   Let H be a subgroup of a group G and suppose that β is in G. Then the set

$$H\beta = \{\alpha\beta \,|\, \alpha \in H\}$$

is called a *right coset* of H.

Similarly, the set $\beta H = \{\beta\alpha \,|\, \alpha \in H\}$ is called a left coset. Note that $\beta \in H$ implies that $H\beta = H$.

FACT 2.1.   Let H be a subgroup of a group G. Two right (left) cosets of H in G are either disjoint or identical. The number of elements in a right (left) coset is the same as the order of H.

Proof: Suppose that

$$H\beta_1 \cap H\beta_2 \neq \varnothing \tag{5}$$

Then, there exist α_1 and α_2 in H so that

$$\alpha_1\beta_1 = \alpha_2\beta_2$$

which implies that $\beta_1 = \alpha_1^{-1}\alpha_2\beta_2$. Thus every element of the form $\alpha\beta_1$ (i.e., in $H\beta_1$) is in $H\beta_2$. By symmetry, we have that $H\beta_2 \subseteq H\beta_1$ which implies that $H\beta_1 = H\beta_2$.

To see that[4] $|H| = |H\beta|$, simply note that the map $\varphi_\beta : \alpha \to \alpha\beta$ is one-to-one and onto.

We have seen that distinct right cosets are disjoint. Also note that each element is in at least one coset since if $\beta \in G$, then $\beta \in H\beta$ because $1 \in H$.

If H is a subgroup of a group G then the number of left (or right) cosets of H in G is called the *index* of H in G and is written $[G : H]$.

---

[4] For any set E, the cardinality of E is written $|E|$.

The next result, while trivial, is nonetheless fundamental.

FACT 2.2.   Let G be a group with a subgroup H. We have

$$|G| = |H|[G : H]$$

Proof: G is partitioned into $[G : H]$ cosets each of which has $|H|$ elements.
Thus the order and the index of any subgroup divide[5] the order of the
group.

Next, we return to considering permutation groups.

FACT 2.3.   Let G be a permutation group on S and fix $s \in S$. Then the set
$J_s = \{\alpha \in G \mid \alpha(s) = s\}$ is a subgroup of G.

Proof: If $\alpha(s) = s$ and $\beta(s) = s$ then $\alpha\beta(s) = s$.

Now we begin to relate groups to counting problems through the following important definition:

DEFINITION.   Let G be a permutation group acting on a set S and suppose
that s and s' are in S. We say that s is *equivalent* to s' (written $s \sim s'$) if there
exists $\alpha \in G$ so that $s' = \alpha(s)$.

FACT 2.4.   The relation $\sim$ of equivalence under a group G is an equivalence
relation.

Proof: Since $s = 1(s)$, we have $s \sim s$. If $s \sim s'$, then $s' = \alpha(s)$. But then
$s = \alpha^{-1}(s')$ and we have shown that $s' \sim s$. Lastly, if $s \sim s'$ and $s' \sim s''$, then
we have $s' = \beta(s)$ and $s'' = \alpha(s')$. This shows that $s \sim s''$ since $\alpha\beta(s) = s''$.

For future applications, we want to relate the size of the equivalence class
containing an element s of S (the class containing s will be written $[s]$), to the
subgroup J_s.

---

[5] We say that *a divides b*, written $a \mid b$, if $b = ac$ for some c.

LEMMA 2.1. Let G be a permutation group acting on S and let s be in S. The number of elements in the equivalence class containing s is $[G:J_s]$ where $J_s = \{\alpha \in G \,|\, \alpha(s) = s\}$.

Proof: Consider the map φ which takes, for any $\alpha \in G$, the element $\alpha(s)$ into αJ_s. Since J_s is a subgroup by Fact 2.3, φ is a map from $[s] = \{\alpha(s) \,|\, \alpha \in G\}$ into left cosets of J_s. Note that φ is onto since given any βJ_s, surely $\beta(s)$ maps onto it. Moreover, φ is one-to-one since $\alpha J_s = \beta J_s$ implies that there exist γ_1, γ_2, in J_s so that for each x in S

$$\alpha\gamma_1(x) = \beta\gamma_2(x) \tag{6}$$

By setting $x = s$, we see that

$$\alpha(s) = \beta(s) \tag{7}$$

which shows that φ is one-to-one.

We can now prove an interesting theorem due to Froebenius which is useful in obtaining Pólya's theorem. Moreover, this result has a number of direct applications.

THEOREM 2.1 (Froebenius). Let G be a permutation group of finite order g acting on a set S. The number of equivalence classes induced on S by G is

$$\sum_{[s] \subseteq S} 1 = (1/g) \sum_{\alpha \in G} I(\alpha) \tag{8}$$

where $I(\alpha)$ is the number of fixed points of S under α.

Proof: If we let

$$\delta_{\alpha(s),\,s} = \begin{cases} 1 & \text{if } \alpha(s) = s \\ 0 & \text{otherwise} \end{cases}$$

then we have

$$I(\alpha) = \sum_{s \in S} \delta_{\alpha(s),\,s} \tag{9}$$

We can compute as follows

$$(1/g) \sum_{\alpha \in G} I(\alpha) = (1/g) \sum_{\alpha \in G} \sum_{s \in S} \delta_{\alpha(s),\,s} \tag{10}$$

$$= (1/g) \sum_{s \in S} \sum_{\alpha \in G,\, \alpha(s) = s} 1 \tag{11}$$

If we let $h_s = |J_s|$ where $J_s = \{\alpha \in G \mid \alpha(s) = s\}$, then we have that

$$(1/g) \sum_{\alpha \in G} I(\alpha) = (1/g) \sum_{s \in S} h_s \tag{12}$$

$$= (1/g) \sum_{[s] \leq S} \sum_{t \in [s]} h_s \tag{13}$$

$$= (1/g) \sum_{[s] \leq S} h_s \sum_{t \in [s]} 1 \tag{14}$$

The sum in the last line is well defined because if $s \sim s'$ then J_s is *isomorphic*[6] to $J_{s'}$ [under a map φ which takes α into $\gamma\alpha\gamma^{-1}$ where $s' = \gamma(s)$] which implies that $h_s = h_{s'}$. By Lemma 2.1, we have that

$$(1/g) \sum_{\alpha \in G} I(\alpha) = (1/g) \sum_{[s] \leq S} h_s [G : J_s] \tag{15}$$

$$= \sum_{[s] \leq S} (h_s/g)(g/h_s) \tag{16}$$

$$= \sum_{[s] \leq S} 1 \tag{17}$$

As an interesting example of this theorem, let F_n denote the class of all switching functions of n variables. Let L_n be the class of all linear boolean functions of n variables. Thus L_n consists of all functions of the form

$$c_0 \oplus c_1 x_1 \oplus \cdots \oplus c_n x_n \tag{18}$$

where $c_i \in \{0, 1\}$. Note that L_n is in fact a (commutative) group under the operation of "ring sum." Two boolean functions f and g are said to be *equivalent* if there is some $l(x_1, \ldots, x_n)$ in L_n so that

$$g(x_1, \ldots, x_n) = f(x_1, \ldots, x_n) \oplus l(x_1, \ldots, x_n) \tag{19}$$

It is easy to verify that the map $l' : f \to f \oplus l$ is in fact a permutation of F_n so that Theorem 2.1 applies. In order to obtain the number of classes, we need only compute $I(l')$ for each l in L_n. It is easy to verify that $f \oplus l = f$ if and only if $l = 0$. Therefore, the number of classes is

$$(1/g) \sum_{[f] \leq F_n} 1 = (1/2^{n+1}) \sum_{\alpha \in L_n} I(\alpha) = (1/2^{n+1})2^{2^n} = 2^{2^n - (n+1)} \tag{20}$$

In order to derive our main theorem we shall need a generating function which exhibits the cycle structure of the group G acting on set S. It turns out this generating function is fundamental and will appear in many of our results.

---

[6] Let G and G' be groups. A map φ from G into G' is said to be a *homomorphism* if $\varphi(ab) = \varphi(a)\varphi(b)$ for each a, b in G. G' is said to be a *homomorphic image* of G if there is a homomorphism φ so that $G' = \varphi G$. G is *isomorphic* to G' if there is a one-to-one homomorphism from G onto G'.

DEFINITION.   Let G be a permutation group of finite order g which acts on a finite set S with s elements. Let $f_1, \ldots, f_s$ be s indeterminates and let $g_{(j_1,\cdots,j_s)} = g_{(j)}$ be the number of permutations in G with j_i cycles of length i for $1 \leq i \leq s$. The *cycle index polynomial* of G acting on S is

$$Z_G(f_1, \ldots, f_s) = (1/g) \sum_{(j)} g_{(j_1, \ldots, j_s)} f_1^{j_1} \cdots f_s^{j_s} \tag{21}$$

where the sum is over all nonnegative integers $j_1, \ldots, j_s$ such that

$$\sum_{i=1}^{s} ij_i = s \tag{22}$$

Since every element of S is in one and only one cycle of a permutation, the cycle structure $(j_1, \ldots, j_s)$ must satisfy Eq. (22). Nonnegative integers which satisfy Eq. (22) are called *partitions* of s. There may be partitions (j) of s which do not occur in G and in this case $g_{(j)} = 0$. Thus, the cycle index is really a sum over the entire group G. We can therefore conclude that

$$Z_G(1, \ldots, 1) = 1 \tag{23}$$

Let us consider the identity group I acting on a set of s elements. Clearly

$$Z_I(f_1, \ldots, f_s) = f_1^{s} \tag{24}$$

since each element in S is fixed under the identity.

Another example can be obtained by letting $S = \{0, 1, \ldots, m-1\}$ and letting C_m denote the cyclic group generated by the cycle $(0, 1, \ldots, m-1)$. It is easy to see that the cycle index of C_m is

$$Z_{C_m}(f_1, \ldots, f_m) = (1/m) \sum_{d|m} \varphi(d) f_d^{m/d} \tag{25}$$

where $\varphi(d)$ is the Euler φ function, i.e., the number of positive integers not exceeding d and relatively prime to it. Another useful example can be obtained by choosing $S = \{1, \ldots, n\}$ and letting S_n be the group of all $n!$ permutations of S. In order to compute Z_{S_n}, we must ask for the number of permutations which have cycle structure $(j_1, \ldots, j_n)$. The set of all permutations with cycle structure $(j_1, \ldots, j_n)$ has redundancies of two types. It does not matter which element comes first in a cycle (e.g., $(123) = (231)$) nor does the order of disjoint cycles matter (e.g., $(12)(34) = (34)(12)$). Thus the number of such cycles is easily seen to be

$$C_{(j)} = n! \left/ \prod_{i=1}^{n} j_i! \, i^{j_i} \right. \tag{26}$$

because in each cycle of length i, there are i choices for the first element.

Moreover there are $j_i!$ ways to permute the cycles of length i. Therefore we have that

$$Z_{S_n}(f_1, \ldots, f_n) = (1/n!) \sum_{(j)} \left(n! \Big/ \prod_{i=1}^{n} j_i! \, i^{j_i} \right) f_1^{j_1} \cdots f_n^{j_n} \qquad (27)$$

where the sum is over all partitions of n.

It should be noted that the cycle index does not give much information concerning the structure of a group. For instance, one can find two non-isomorphic groups G_1 and G_2 of order p^3 where p is an odd prime and both groups have cycle index

$$(1/p^3)(f_1^{p^3} + (p^3 - 1)f_p^{p^2}) \qquad (28)$$

This can be done with G_1 abelian[7] while G_2 is not. Further details can be found in Pólya (1937).

We can now begin our development of Pólya's Theorem. Let D and R be finite sets with $|D| = s$ and $|R| = q$. For each $r \in R$, there is an associated weight $w(r)$ which is a t-tuple of natural numbers, i.e., w is a map from R into $\mathbb{N}^t$ where $\mathbb{N}$ denotes the set of natural numbers. We shall need a generating function which enumerates the range. Define

$$\Psi(x_1, \ldots, x_t) = \sum_{k_1, \ldots, k_t \geq 0} \Psi_{(k_1, \ldots, k_t)} x_1^{k_1} \cdots x_t^{k_t} \qquad (29)$$

as the *range enumerating function* where $\Psi_{(k_1, \ldots, k_t)}$ is the number of elements of R with weight $(k_1, \ldots, k_t)$.

Our main interest is in functions from D into R and we write R^D for the class of all such maps.

To extend the notion of weight to such functions, suppose that

$$(k_1, \ldots, k_t) = \sum_{d \in D} w(f(d)) \qquad (30)$$

and define

$$w(f) = x_1^{k_1} \cdots x_t^{k_t} \qquad (31)$$

If $f, g \in R^D$ and G is a permutation group on D, define f to be *equivalent* to g if there exists $\alpha \in G$ so that $g(d) = f(\alpha(d))$ for each $d \in D$. It is easily seen that this is an equivalence relation on R^D.

The following fact is important though trivial to verify.

FACT 2.5.   If f is equivalent to g, then $w(f) = w(g)$.

---

[7] A group G is *abelian* if $ab = ba$ for each a, b in G.

Proof: If $g(d) = f(\alpha(d))$ for all $d \in D$, then

$$\sum_{d \in D} w(g(d)) = \sum_{d \in D} w(f(\alpha(d))) = \sum_{d \in D} w(f(d)) \qquad (32)$$

because α is a permutation and the last expression is merely the previous sum written in a different order.

Next we show how the permutation group G acting on D induces a group which permutes the elements of R^D.

LEMMA 2.2.   Let G be a permutation group acting on D. G induces a permutation group G' acting on R^D which is a homomorphic image of G. Moreover, there is $\alpha \in G$ so that $f = g\,\alpha$ if and only if there is a permutation π_α in G' so that $\pi_\alpha(f) = g$.

Proof: Let α be a permutation in G and define the map π_α by $\pi_\alpha(f) = f\alpha^{-1}$. Clearly π_α is a permutation on R^D [π_α is onto since given any function f, the function $f\alpha$ maps onto it under π_α. Moreover, if $\pi_\alpha(f) = \pi_\alpha(g)$ then $f(\alpha^{-1}(d)) = g(\alpha^{-1}(d))$ for all $d \in D$. Since α^{-1} is a permutation, the last condition implies $f(d') = g(d')$ for all $d' \in D$. That is $f = g$ and π_α is one-to-one].

Let us define a mapping φ which carries α in G into π_α. Define $G' = \varphi G$. First we claim that φ is a homomorphism, i.e., $\varphi(\alpha\beta) = \varphi(\alpha)\varphi(\beta)$. We compute that $\varphi(\alpha\beta) = \pi_{\alpha\beta}$ and $(\varphi\alpha)(\varphi\beta) = (\varphi\alpha)\pi_\beta = \pi_\alpha\pi_\beta$. Let f be in R^D and then we find that

$$\begin{aligned}
\varphi(\alpha\beta)(f) &= \pi_{\alpha\beta}(f) = f(\alpha\beta)^{-1} = f\beta^{-1}\alpha^{-1} \\
&= \pi_\alpha(f\beta^{-1}) = \pi_\alpha\pi_\beta f = \varphi(\alpha)\varphi(\beta)(f) \qquad (33)
\end{aligned}$$

Therefore

$$\varphi(\alpha\beta) = \varphi(\alpha)\varphi(\beta) \qquad (34)$$

and φ is a homomorphism.

Finally by the definition of π_α we see that there is α in G so that $f = g\alpha$ if and only if $\pi_\alpha f = g$.

Our next lemma is a generalization of the result by Froebenius.

LEMMA 2.3.   If G is a permutation group acting on D, then

$$\sum_{[f] \subseteq R^D} w([f]) = (1/g) \sum_{\alpha \in G} \sum_{f \in R^D,\, f\alpha = f} w(f) \qquad (35)$$

The last summation is over all functions from D into R for which $f\alpha = f$.

Proof: Note that the first sum in Eq. (35) is well defined by Fact 2.5. Let $(k_1, \ldots, k_t)$ be a weight and let $x_1^{k_1} \cdots x_t^{k_t}$ be the corresponding weight of a function. Since $f\alpha = g$ if and only if $\pi_\alpha(g) = f$ and since $f\alpha = g$ implies $w(f) = w(g)$, we use Theorem 2.1 to conclude that

$$\sum_{[f] \subseteq R_{(k)}^D} 1 = (1/g) \sum_{\alpha \in G} I_{(k)}(\alpha) \tag{36}$$

where $R_{(k)}^D = \{f \in R^D \mid w(f) = x_1^{k_1} \cdots x_t^{k_t}, \ (k) = (k_1, \ldots, k_t)\}$ and $I_{(k)}(\alpha)$ is the number of functions of weight $x_1^{k_1} \cdots x_t^{k_t}$ which are fixed under α. We multiply (36) by $w(f) = w([f])$ and sum over all such weights to get

$$\sum_{k_1} \cdots \sum_{k_t} \sum_{[f] \subseteq R_{(k)}^D} w(f) = (1/g) \sum_{k_1} \cdots \sum_{k_t} \sum_{\alpha \in G} I_{(k)}(\alpha) w(f) \tag{37}$$

But the left-hand side is simply the sum over all equivalence classes and we have that

$$\sum_{[f] \subseteq R^D} w(f) = (1/g) \sum_{\alpha \in G} \sum_{f \in R^D, f\alpha = f} w(f) \tag{38}$$

since the right-hand side is the sum over all weights when f is a fixed point of α.

The previous proof works for any weight function which has the property that equivalent functions have the same weight.

In our next results we have to consider the relationship between the cycles of a permutation α on D and values that f can take and still have the property that $f = f\alpha$.

FACT 2.6. Let $f \in R^D$ and let α be a permutation on D. Suppose that $\alpha = \sigma_1 \cdots \sigma_k$ where the σ_i are (disjoint) cycles and $D = D_1 \cup \cdots \cup D_k$ where $D_i \cap D_j = \emptyset$ if $i \neq j$ and the D_i are the elements in the σ_i for $1 \leq i \leq k$. Then $f\alpha = f$ if and only if for each i, $1 \leq i \leq k$, we have that $f(d) = f(d')$ for all d, d' in D_i (i.e., f is constant on each block D_i).

Proof: If $f\alpha = f$, we have that for each $d \in D$

$$f(d) = f(\alpha d) = f(\alpha^2 d) = \cdots = f(\alpha^s d) \tag{39}$$

Therefore, f is constant on each cycle. The converse is also immediate.

We need only one more lemma to complete the proof of Pólya's Theorem.

We must still evaluate $\sum_{f \in R^D, f\alpha = f} w(f)$ where we sum over all functions which are fixed by α.

LEMMA 2.4. If α be a permutation of D which has j_i cycles of length i where $1 \leq i \leq s$, then

$$\sum_{f \in R^D, f\alpha = f} w(f) = \prod_{i=1}^{s} \Psi^{j_i}(x_1^{i}, \ldots, x_t^{i}) \tag{40}$$

Proof: Suppose that σ is a cycle of length i in α and that $f\alpha = f$. We know that $f(d) = f(\alpha d) = \cdots = f(\alpha^{s-1}d)$. By Fact 2.6 all of the functions fixed by α can be obtained by picking a range element $r(\sigma)$ for each cycle σ in α and requiring that $f(d) = r(\sigma)$ for all d in D_i, the orbit of σ. Clearly, each such cycle σ contributes the following amount to the desired sum

$$\sum_{f \in R^D, f\sigma = f} w(f) = \sum_{r \in R, w(r)=(k_1, \ldots, k_r)} (x_1^{k_1} \cdots x_t^{k_t})^i$$

$$= \sum_{r \in R} (x_1^{i})^{k_1} \cdots (x_t^{i})^{k_t} = \Psi(x_1^{i}, \ldots, x_t^{i}) \tag{41}$$

since we need only assign range elements to all the elements of D_i and there are i elements to assign.

To complete the proof, note that since the cycles in α are disjoint, the choices are independent and the total contribution of α to the sum is the product of the individual choices, namely

$$\prod_{i=1}^{s} \Psi^{j_i}(x_1^{i}, \ldots, x_r^{i})$$

Now we can state and prove Pólya's Theorem. We let G be a permutation group on D and consider equivalence classes of functions from D into R. Each element r in R has a weight $w(r)$ in $\mathbb{N}^t$ and hence each function $f \in R^D$ has a weight $w(f) = x_1^{k_1} \cdots x_t^{k_t}$ where $(k_1, \ldots, k_t) = \sum_{d \in D} w(f(d))$. We denote the function counting series by

$$P(x_1, \ldots, x_t) = \sum_{k_1} \cdots \sum_{k_t} p_{k_1, \ldots, k_t} x_1^{k_1} \cdots x_t^{k_t} \tag{42}$$

where $p_{k_1, \ldots, k_t}$ is the number of equivalence classes of weight $x_1^{k_1} \cdots x_t^{k_t}$.

Pólya's Theorem allows us to determine $P(x_1, \ldots, x_t)$ in terms of the cycle index polynomial $Z_G(f_1, \ldots, f_s)$ and the range enumerating series $\Psi(x_1, \ldots, x_t)$.

THEOREM 2.2 (Pólya, 1937).

$$P(x_1, \ldots, x_t) = Z_G(\Psi(x_1, \ldots, x_t), \Psi(x_1^2, \ldots, x_t^2), \ldots, \Psi(x_1^s, \ldots, x_t^s)) \quad (43)$$

Proof:

$$P(x_1, \ldots, x_t) = \sum_{k_1} \cdots \sum_{k_t} p_{k_1, \ldots, k_t} x_1^{k_1} \cdots x_t^{k_t} \quad (44)$$

$$= \sum_{[f] \subseteq R^D} w([f]) \quad (45)$$

By Lemma 2.3 applied to (45),

$$P(x_1, \ldots, x_t) = (1/g) \sum_{\alpha \in G} \sum_{f \in R^D, \, f\alpha = f} w(f) \quad (46)$$

By Lemma 2.4 applied to (46), we obtain

$$P(x_1, \ldots, x_t) = (1/g) \sum_{\alpha \in G} \prod_{i=1}^{s} \Psi^{j_i}(x_1^i, \ldots, x_t^i) \quad (47)$$

$$= Z_G(\Psi(x_1, \ldots, x_t), \ldots, \Psi(x_1^s, \ldots, x_t^s)) \quad (48)$$

COROLLARY 2.1. The total number of equivalence classes of functions from D into R with G acting on D is

$$\sum_{[f] \subseteq R^D} 1 = Z_G(q, \ldots, q) \quad (49)$$

where $|R| = q$.

Proof: For each $r \in R$, take $w(r) = 0$. Then $\Psi(x_1, \ldots, x_t) = q$ and $w([f]) = 1$ for each class $[f]$.

EXERCISES

1. Let G be a finite group and let H be a subset of G which contains the identity and is closed under the group operation. Show that H is a subgroup of G.

2. Let a and b be elements of a group G. We say that a and b are *strongly enumeratively equivalent* if a and b have the same set of fixed points for every permutation representation of G. Show that a and b are strongly enumeratively equivalent if and only if a and b are powers of each other (i.e., $a = b^p$ and $b = a^q$ for some p and q).

III. SOME APPLICATIONS OF PÓLYA'S THEOREM TO SWITCHING FUNCTIONS

In Section I, we indicated a general plan for classifying Boolean functions. We now use Pólya's theorem to enumerate the number of equivalence classes.

As the most trivial kind of equivalence relation, let us say that two functions are equivalent if and only if they are identical. In order to use Pólya's theorem, let I_n be the identity group on $\mathbb{Z}_2^n$. The range enumerating series is $\Psi(x) = 1 + x$ where $w(0) = 0$ and $w(1) = 1$. The cycle index of I_n is clearly

$$Z_{I_n}(f_1, \ldots, f_{2^n}) = f_1^{2^n} \tag{50}$$

Thus, the number of equivalence classes is

$$Z_{I_n}(2, \ldots, 2) = 2^{2^n} \tag{51}$$

as it should be. The generating function is

$$P(x) = Z_I(1 + x, \ldots, 1 + x^{2^n}) = (1 + x)^{2^n} \tag{52}$$

$$= \sum_{k=0}^{2^n} \binom{2^n}{k} x^k \tag{53}$$

The weight of a switching function, according to our definition is x^k where $k = |f^{-1}(1)|$. Note that this is essentially the same definition as in ordinary switching theory since k is the number of ones in the truth table.

Although, our first example was trivial we can abstract from it. We shall deal with "natural" equivalence relations which involve the variables of some boolean function. Such transformations induce permutations on the domain of the boolean functions, namely $\{0, 1\}^n$. These mappings form a group whose cycle index must be computed and from which the number of equivalence classes can be determined.

As an example, consider Boolean functions of n variables and define a group which we shall call C_2^n as follows:

$$C_2^n = \{(i_1, \ldots, i_n) \mid i_j \in \{0, 1\} \quad \text{for } 1 \leq j \leq n\} \tag{54}$$

Intuitively, C_2^n will complement some of the variables of a switching function. If $i = (i_1, \ldots, i_n)$ is in C_2^n, define

$$i(x_1, \ldots, x_n) = (x_1^{i_1}, \ldots, x_n^{i_n}) \tag{55}$$

where

$$x_j^{i_j} = \begin{cases} x_j & \text{if } i_j = 0 \\ \bar{x}_j & \text{if } i_j = 1 \end{cases}$$

The group operation is sum mod 2 and written $\oplus$. For instance if $i = (1, 0)$ then $(1, 0)(x_1, x_2) = (\bar{x}_1, x_2)$ and i induces a map on $\{0, 1\}^2$ which is

$$(0, 0) \leftrightarrow (1, 0), \qquad (0, 1) \leftrightarrow (1, 1)$$

If we agree to regard each n-tuple as a binary number, then the permutation can be written in cyclic notation as $(0, 2)(1, 3)$.

Again these permutations act on functions by their behavior of the domain. Thus the boolean functions of two variables are grouped into seven classes under $C_2{}^2$. These classes are

$$[0]$$
$$[\bar{x}_1 \bar{x}_2, \bar{x}_1 x_2, x_1 \bar{x}_2, x_1 x_2]$$
$$[x_1, \bar{x}_1], \quad [x_2, \bar{x}_2]$$
$$[x_1 \oplus x_2, x \equiv x_2]$$
$$[\bar{x}_1 + \bar{x}_2, \bar{x}_1 + x_2, x_1 + \bar{x}_2, x_1 + x_2]$$
$$[1]$$

Another important classification occurs if variables are permuted. If π is any permutation on the indices $\{1, \ldots, n\}$, then π acts on variables by

$$\pi(x_1, \ldots, x_n) = (x_{\pi^{-1}(1)}, \ldots, x_{\pi^{-1}(n)})$$

Such permutations induce maps on the domain $\{0, 1\}^n$. For instance $\sigma = (12)$ induces the permutation h_σ on $\{0, 1\}^3$, where $h_\sigma = (2, 4)(3, 5)$ and we again use the decimal equivalents of the domain elements. We shall denote by S_n' the representation of S_n which is induced on the domain by permutations of the variables. For example, the functions of two variables are classified into 12 classes as follows:

$$[0]$$
$$[\bar{x}_1 \bar{x}_2], \quad [\bar{x}_1 x_2, x_1 \bar{x}_2], \quad [x_1 x_2]$$
$$[x_1, x_2], \quad [x_1 \oplus x_2], \quad [x_1 \equiv x_2], \quad [\bar{x}_1, \bar{x}_2]$$
$$[\bar{x}_1 + \bar{x}_2], \quad [\bar{x}_1 + x_2, x_1 + \bar{x}_2], \quad [x_1 + x_2]$$
$$[1]$$

Note that a function is in a singleton equivalence class under S_n if and only if it is a symmetric function.

If we allow both complementations and permutations of the variables, then another group, called G_n, is induced. The group action on variables is represented by

$$((i_1, \ldots, i_n), \pi)(x_1, \ldots, x_n) = (x_{\pi^{-1}(1)}^{i_1}, \ldots, x_{\pi^{-1}(n)}^{i_n})$$

where $i_j \in \{0, 1\}$ for $1 \le j \le n$ and $\pi \in S_n$. G_n is especially important in switching theory and other areas of discrete mathematics since it is the symmetry group of the n-cube. The classification of Boolean functions of two variables into six classes is shown below.

$$[0]$$
$$[\bar{x}_1\bar{x}_2, \bar{x}_1x_2, x_1\bar{x}_2, x_1x_2]$$
$$[x_1, x_2, \bar{x}_1, \bar{x}_2], [x_1 \oplus x_2, x_1 \equiv x_2]$$
$$[\bar{x}_1 + \bar{x}_2, \bar{x}_1 + x_2, x_1 + \bar{x}_2, x_1 + x_2]$$
$$[1]$$

There are a variety of applications such as coding theory in which linear operations are allowed on the variables. In order to preserve the group property we shall consider invertible linear transformations. We shall say that two Boolean functions f and g are equivalent under $GL_n(\mathbb{Z}_2)$ (which stands for the general linear group on an n-dimensional vector space over the field $\mathbb{Z}_2$) if there is a nonsingular matrix $A = (a_{ij})$ so that

$$g(x_1, \ldots, x_n) = f((x_1, \ldots, x_n)A)$$

$$= f\left(\sum_{k=1}^{n} a_{k1}x_k, \ldots, \sum_{k=1}^{n} a_{kn}x_k\right)$$

where $\sum$ denotes extended summation modulo 2. One of the motivations for considering this group is that there are technologies in which linear operations are inexpensive and the number of classes is much smaller than with (say) G_n because the order of $GL_n(\mathbb{Z}_2)$ is so large (see Chapter V) and Lechner, 1968. The order of G_n is $n! \, 2^n$ while for $GL_n(\mathbb{Z}_2)$ the corresponding order is

$$2^{n(n-1)/2} \prod_{i=1}^{n} (2^i - 1) \tag{56}$$

For $n = 2$, there are eight classes of functions,

$$[0]$$
$$[\bar{x}_1\bar{x}_2]$$
$$[\bar{x}_1x_2, x_1\bar{x}_2, x_1x_2]$$
$$[x_1, x_2, x_1 \oplus x_2], \quad [\bar{x}_1, \bar{x}_2, x_1 \equiv x_2]$$
$$[x_1 + x_2], [\bar{x}_1 + x_2, \bar{x}_1 + x_2, x_1 + \bar{x}_2]$$
$$[1]$$

The last group that we shall consider here is the group which allows linear mappings as in $GL_n(\mathbb{Z}_2)$ and complementations of the variables. This group denoted by $A_n(\mathbb{Z}_2)$ (for the affine group) has elements (b, A) where $b = (b_1, \ldots, b_n) \in C_2{}^n$ and $A \in GL_n(\mathbb{Z}_2)$. For example, when $n = 2$, $b = (1, 0)$ and $A = \begin{pmatrix} 1 & 0 \\ 1 & 1 \end{pmatrix}$, we have

$$(b, A)(x_1, x_2) = b((x_1, x_2)A) \tag{57}$$

$$= (1, 0)\left((x_1, x_2)\begin{pmatrix} 1 & 0 \\ 1 & 1 \end{pmatrix}\right) \tag{58}$$

$$= (1, 0)(x_1 \oplus x_2, x_2) = (x_1 \equiv x_2, x_2) \tag{59}$$

This group leads to a still smaller number of classes. If $n = 2$, we find five classes which are

$$[0]$$
$$[\bar{x}_1\bar{x}_2, \bar{x}_1 x_2, x_1\bar{x}_2, x_1 x_2]$$
$$[x_1, x_2, \bar{x}_1, \bar{x}_2, x \oplus y, x \equiv y]$$
$$[\bar{x}_1 + \bar{x}_2, \bar{x}_1 + x_2, x_1 + \bar{x}_2, x_1 + x_2]$$
$$[1]$$

which is the least number of classes that one could have under any "weight preserving" definition of equivalence since there are $2^n + 1$ possible weights for Boolean functions.

Now that all the groups have been introduced, we can begin to solve the associated enumeration problems. In order to calculate the desired numbers, it suffices to construct the appropriate cycle indices. In the next section we shall state some general theorems, and construct and tabulate our results.

IV. STRUCTURE THEOREMS FOR PERMUTATION GROUPS AND THE DETERMINATION OF CYCLE INDICES

In order to determine the cycle indices we shall exploit the structure theory of permutation groups. We shall need a concept of equivalence for these groups which is more restrictive than mere isomorphism.

DEFINITION. Let A and B be permutation groups acting on object sets X and Y, respectively. A is said to be *permutationally equivalent* to B if (1) A is isomorphic to B under a map φ, (2) there is a one-to-one correspondence h from X onto Y, and (3) for each $x \in X$, $\alpha \in A$, we have

$$h(\alpha x) = (\varphi\alpha)h(x) \tag{60}$$

This condition can be represented by the following commutative diagram

$$
\begin{array}{ccc}
X & \xrightarrow{\;h\;} & Y \\
\downarrow{\scriptstyle \alpha} & & \downarrow{\scriptstyle \varphi\alpha} \\
X & \xrightarrow[\;h\;]{} & Y
\end{array}
$$

As an example of two groups which are isomorphic but not permutationally equivalent, consider the identity group I which acts on the n variables of the Boolean functions. This group has order 1 and degree n. The group I_n of the previous section was the identity group on the *domain* of the Boolean functions of n variables and hence has degree 2^n. Consequently, I_n and I are not permutationally equivalent. The same argument applies to S_n and S_n'.

The following simple proposition is important in our later work:

THEOREM 4.1. If A and B are permutationally equivalent groups, then $Z_A = Z_B$ where Z_A denotes the cycle index of A.

Proof: The trivial proof is left as problem 3.

In dealing with ordinary or generalized switching functions, we always find ourselves dealing with the cartesian product of sets, e.g., $\mathbb{Z}_2^n$. We now introduce several products of permutation groups which are closely related to cartesian products of their object sets. Our first goal will be to consider the "direct product" of permutation groups. In order to do so, we must first compute the effect of the direct product of two permutations.

LEMMA 4.1. If α is a permutation on a set X with $|X| = a$ and α has cycle structure denoted by $f_1^{j_1} \cdots f_a^{j_a}$, and β is a permutation on Y with $|Y| = b$ and β has cycle structure $f_1^{k_1} \cdots f_b^{k_b}$, then the permutation (α, β) acting on $X \times Y$ by the rule

$$(\alpha, \beta)(x, y) = (\alpha x, \beta y) \tag{61}$$

has cycle structure given by

$$\left(\prod_{p=1}^{a} f_p^{j_p}\right) \times \left(\prod_{q=1}^{b} f_q^{k_q}\right) = \prod_{p=1}^{a} \prod_{q=1}^{b} (f_p^{j_p} \times f_q^{k_q}) \tag{62}$$

$$= \prod_{p=1}^{a} \prod_{q=1}^{b} f_{\langle p, q \rangle}^{j_p k_q (p, q)} \tag{63}$$

where $\langle p, q \rangle$ is the least common multiple of p and q while (p, q) is the greatest common divisor of p and q.

Before giving the proof, we do two simple examples of the cross operation

$$f_1^2 \times f_3 = f_3^2 \tag{64}$$

$$f_2 \times (f_1 f_2) = (f_2 \times f_1)(f_2 \times f_2) = (f_2)(f_2^2) = f_2^3 \tag{65}$$

Proof: It clearly suffices to let α and β be cycles of length p and q, respectively. The result is immediate if $p = q$ so suppose $p < q$. We write $\alpha = (a_1, \ldots, a_p)$, $\beta = (b_1, \ldots, b_q)$. Consider the element (a_1, b_1). Clearly (a_1, b_1) goes to $(a_2, b_2), \ldots,$ to (a_p, b_p), to $(a_1, b_{p+1}), \ldots,$ and finally to (a_1, b_1) after $\langle p, q \rangle$ steps. There will be (p, q) cycles of length $\langle p, q \rangle$ since we are permuting pq objects and we have $pq = \langle p, q \rangle (p, q)$. Note that the choice of a_1 and b_1 was convenient but inessential.

We shall expand the cross operation to be distributive over addition and now we may define this operation on polynomials.

Example. Let $Z_I = f_1^2$ and $Z_{C_2} = \frac{1}{2}(f_1^2 + f_2)$. Then we compute

$$Z_I \times Z_{C_2} = f_1^2 \times \frac{1}{2}(f_1^2 + f_2) \tag{66}$$

$$= \frac{1}{2}((f_1^2 \times f_1^2) + (f_1^2 \times f_2)) \tag{67}$$

$$= \frac{1}{2}(f_1^4 + f_2^2) \tag{68}$$

The following useful definition and theorem justifies this notion.

DEFINITION. Let A be a permutation group acting on X and B a permutation group acting on Y. Define $A \times B = \{(\alpha, \beta) \mid \alpha \in A, \beta \in B\}$ and we define $A \times B$ to act on $X \times Y$ as follows. For each $(\alpha, \beta) \in A \times B$ and each $(x, y) \in X \times Y$

$$(\alpha, \beta)(x, y) = (\alpha x, \beta y) \tag{69}$$

THEOREM 4.2. $A \times B$ is a permutation group of order ab and degree xy on $X \times Y$. If the cycle index of A is Z_A and of B is Z_B, then $Z_{A \times B} = Z_A \times Z_B$.

Proof: Immediate from Lemma 4.1.

We can now give an application of Theorem 4.2 to switching theory. Let us consider the counting problem for switching functions under the complementing group C_2^n. Recall that the elements of C_2^n are n-tuples $(i_1, \ldots, i_n)$ with each $i_j \in \{0,1\}$. The following simple fact establishes the link between Boolean functions and the abstract theory.

LEMMA 4.2. C_2^n is permutationally equivalent to $X_{i=1}^n C_2$.

Proof: The result is immediately by virtue of the action of C_2^n on $\mathbb{Z}_2^n$.

We can now settle all the counting problems for C_2^n.

THEOREM 4.3. The cycle index of C_2^n is given by

$$Z_{C_2^n}(f_1, f_2) = \frac{1}{2^n}(f_1^{2^n} + (2^n - 1)f_2^{2^{n-1}}) \tag{70}$$

Proof: The argument is by induction on n and follows immediately from the following computation

$$\left(\frac{1}{2^{n-1}}(f_1^{2^{n-1}} + (2^{n-1} - 1)f_2^{2^{n-2}})\right) \times (\tfrac{1}{2}(f_1^2 + f_2)) \tag{71}$$

$$= \frac{1}{2^n}(f_1^{2^n} + (2^{n-1} - 1)f_2^{2^{n-1}} + f_2^{2^{n-1}} + (2^{n-1} - 1)f_2^{2^{n-1}}) \tag{72}$$

$$= \frac{1}{2^n}(f_1^{2^n} + (2^n - 1)f_2^{2^{n-1}}) \tag{73}$$

COROLLARY 4.1. The number of equivalence classes of Boolean functions under C_2^n is

$$\frac{1}{2^n}(2^{2^n} + (2^n - 1)2^{2^{n-1}}) \tag{74}$$

COROLLARY 4.2. The number of equivalence classes of Boolean functions with k ones in their truth tables is

$$\frac{1}{2^n}\binom{2^n}{k} \qquad\qquad \text{if } k \text{ is odd} \tag{75}$$

$$\frac{1}{2^n}\left[\binom{2^n}{k} + (2^n - 1)\binom{2^{n-1}}{k/2}\right] \qquad \text{if } k \text{ is even} \tag{76}$$

Proof: By Pólya's theorem

$$P(x) = Z_{C_{2^n}}(1 + x, 1 + x^2) = \frac{1}{2^n}((1 + x)^{2^n} + (2^n - 1)(1 + x^2)^{2^{n-1}}) \quad (77)$$

Two applications of the binomial theorem and collecting the coefficients of x^k in $P(x)$ will finish the proof.

Next, we turn our attention to finding the cycle indices for S_n' and for G_n. It is possible to directly derive the cycle indices [see Pólya (1940), Slepian (1953), and Harrison (1964c)] but we shall briefly study the group structure of G_n. We shall obtain the desired polynomials as instances of a useful general theorem which is proven in Harrison and High (1968). This theorem can also be applied to yield related cycle indices which occur in the theory of many valued logical functions.

In general, let X be a finite set of m elements and let B be a permutation group of order b which acts on X. Also let S_n be the symmetric group acting on $\{1, \ldots, n\}$. We define a group $S_n \otimes B$ as

$$S_n \otimes B = \{(\beta, \sigma) \mid \beta \in B^n, \sigma \in S_n\}$$

which operates on n tuples of elements from X by the following rule: for each $\beta = (\beta_1, \ldots, \beta_n) \in B^n$, $\sigma \in S_n$ and each $(x_1, \ldots, x_n) \in X^n$, we have

$$(\beta, \sigma) = (\beta_1 x_{\sigma^{-1}(1)}, \ldots, \beta_n x_{\sigma^{-1}(n)}) \quad (78)$$

If we define $\sigma(\beta) = (\beta_{\sigma^{-1}(1)}, \ldots, \beta_{\sigma^{-1}(n)})$, then we have

$$(\beta_2, \sigma_2)(\beta_1, \sigma_1) = (\beta_2 \sigma_2(\beta_1), \sigma_2 \sigma_1) \quad (79)$$

Example.   Let $B = C_2$ and suppose that $n = 3$ and

$$(\beta_1, \sigma_1) = ((100), (123))$$
$$(\beta_2, \sigma_2) = ((101), (13))$$

Then $\sigma_2(\beta_1) = (001)$ and we have that

$$(\beta_2, \sigma_2)(\beta_1, \sigma_1) = ((101) \oplus (001), \sigma_2 \sigma_1)$$
$$= (100, (12))$$

The structure of $S_n \otimes B$ is explored in detail by Harrison and High (1968) and the cycle index is derived there. Before stating the main result, we need some preliminary definitions.

DEFINITION. Let $\beta \in B$ have k_i cycles of length i for $1 \le i \le m$ and let $K = \{j \mid k_j \ne 0\}$. For any positive integer i' and any set $S \subseteq K$ define

$$G(S, i') = \sum_{T \subseteq S} (-1)^{|S-T|} \left(\sum_{t \in T} t k_t \right)^{i'} \tag{80}$$

If i is a positive integer and $S \subseteq K$, write $i = i_{S_1} i_{S_2}$ where i_{S_2} is the largest factor of i such that[8] $(i_{S_2}, \langle S \rangle) = 1$.

DEFINITION. Let i be given with $(k) = (k_1, \ldots, k_m)$. Fixing j and $S \subseteq K$, define

$$D_{ji(k)S} = \{t : t \mid ji_{S_1}\langle S \rangle, \ t \nmid ji_{S_1} l \text{ for any } l < \langle S \rangle\} \tag{81}$$

and $g_{ji(k)S}(d) = 0$ if $d \ne ji_{S_1}\langle S \rangle$ with $j \mid i_{S_2}$. If $d = ji_{S_1}\langle S \rangle$ where $j \mid i_{S_2}$

$$g_{ji(k)S}(ji_{S_1}\langle S \rangle) = (1/ji_{S_1}\langle S \rangle) \sum_{t \in D_{ji(k)S}} G(S, t/\langle S \rangle)\mu(ji_{S_1}\langle S \rangle/t) \tag{82}$$

where μ is the Möbius function.[9] Finally, define

$$G_{ji(k)}(d) = \sum_{S \subseteq K} g_{ji(k)S}(d) \tag{83}$$

This definition is complicated and a few examples are in order.

Example 1. Suppose that $m = 2$ and that $(k_1, k_2) = (2, 0)$. Then $K = \{1\}$ and

$$D_{ji(k)S} = \{t : t \mid j\} \tag{84}$$

$$G(\{1\}, i') = 2^{i'} \tag{85}$$

and

$$G_{ji(k)}(d) = (1/j) \sum_{d \mid j} 2^d \mu(j/d) \tag{86}$$

Example 2. Suppose that $m = 2$ and that $(k_1, k_2) = (0, 1)$. Then $K = \{2\}$ and

$$G(\{2\}, i') = 2^{i'} \tag{87}$$

---

[8] $\langle S \rangle$ denotes the least common multiple of the set of integers in S.
[9] The Möbius function is defined as follows: $\mu(1) = 1$, $\mu(a) = 0$ if $p^2 \mid a$ for any prime p, and $\mu(a) = (-1)^k$ if $a = p_1 \cdots p_k$ where the p_i are distinct primes.

A short computation yields that

$$D_{ji(k)S} = \{t : t\,|\,2j \quad \text{and} \quad t \nmid j\} \tag{88}$$

and

$$G_{ji(k)}(2t) = (1/2t) \sum_{d\,|\,2t,\; d \nmid t} 2^{d/2}\mu(2t/d) \tag{89}$$

We can now state an important theorem which gives us the cycle index of $S_n \otimes B$ in terms of the cycle index of B.

THEOREM 4.4 (Harrison and High, 1968).   Let B be a permutation group with cycle index

$$Z_B(f_1, \ldots, f_m) = \sum_{(k)} b_{(k)} f_1^{k_1} \cdots f_m^{k_m}$$

then

$$Z_{S_n \otimes B} = \frac{1}{n!\,b^n} \sum_{(j)} \frac{n!\,b^n}{(\prod_{i=1}^n j_i!\,(ib)^{j_i})} \overset{n}{\underset{i=1}{\times}} \left(\sum_k b_{(k)} \prod_{d \in D} f_d^{G(d)} \right)^{\times j_i}$$

where the sum is over all partitions of n, $D = D_{ji(k)}$ and $G(d) = G_{ji(k)}(d)$. We have written the exponent as $\times j_i$ to indicate repeated applications of the cross product.

It is now a routine matter to compute the cycle index of S_n' since we can use the previous result.

THEOREM 4.5.   The cycle index of S_n' is

$$Z_{S_n'} = \frac{1}{n!} \sum_{(j)} \frac{n!}{(\prod_{i=1}^n j_i!\,i^{j_i})} \overset{n}{\underset{i=1}{\times}} \left(\prod_{d\,|\,i} f_d^{g_i(d)} \right)^{\times j_i}$$

where the sum is over all partitions of n and

$$g_i(s) = (1/s) \sum_{d\,|\,s} 2^d \mu(s/d) \qquad \text{if} \quad s\,|\,i$$

$$g_i(s) = 0 \qquad\qquad\qquad \text{otherwise}$$

Proof.   Let I be the identity group on $\mathbb{Z}_2$ which has cycle index f_1^2. Thus S_n' is permutationally equivalent to $S_n \otimes I$. The result follows from Theorem 4.4 and Example 1.

As an example of the computation, we work out the formula for $n = 2$ in order to compare with the classes which were displayed earlier. The formula yields for $n = 2$

$$Z_{S_{2'}} = \tfrac{1}{2}((f_1^2)^{\times 2} + f_1^2 f_2)$$
$$= \tfrac{1}{2}(f_1^4 + f_1^2 f_2)$$

Then $P(x) = Z_{S_{2'}}(1 + x, 1 + x^2) = 1 + 3x + 4x^2 + 3x^3 + x^4$ which agrees with the earlier result. A table of values is given for S_n' and the first few cycle indices are in Appendix 1.

TABLE I
The Total Number of Classes under S_n'

n	T_n
1	4
2	12
3	80
4	3984
5	37,333,248
6	25,626,412,338,274,304

Next, we shall turn our attention to G_n. Calculation of the cycle index is now easy to view of Theorem 4.4.

THEOREM 4.6.   The cycle index of G_n is given by

$$Z_{G_n} = \frac{1}{n! \, 2^n} \sum_{(j)} \frac{n! \, 2^n}{\left(\prod_{i=1}^{n} j_i ! (2i)^{j_i}\right)} \overset{n}{\underset{i=1}{\times}} \left(\prod_{d|i} f_d^{g_i(d)} + \prod_{d|2i, \, d \nmid i} f_d^{h_i(d)} \right)^{\times j_i}$$

where the sum is over all partitions of n, $g_i(d)$ is defined as in Theorem 4.5 and

$$h_i(2s) = (1/2s) \sum_{d|2s, \, d \nmid s} 2^{d/2} \mu(2s/d)$$

Proof: Since G_n is permutationally equivalent to $S_n \otimes C_2$ where $Z_{C_2} = \tfrac{1}{2}(f_1^2 + f_2)$, use Theorem 4.4 and the examples on page 106 to yield the result.

To check our earlier result, we compute the case when $n = 2$.

$$Z_{G_2} = \tfrac{1}{8}((f_1^2 + f_2)^{\times 2} + 2(f_1^2 f_2 + f_4)) \qquad (90)$$
$$= \tfrac{1}{8}(f_1^4 + 3f_2^4 + 2f_1^2 f_2 + 2f_4) \qquad (91)$$

Then we find that

$$P(x) = Z_{G_2}(1 + x, \ldots) = 1 + x + 2x^2 + x^3 + x^4 \qquad (92)$$

The following table contains some of the relevant numbers for G_n while Appendix 2 has a table of cycle indices for G_n for small values of n.

TABLE II
The Total Number of Classes Under G_n

n	T_n
1	3
2	6
3	22
4	402
5	1,228,158
6	400,507,806,843,728

In order to derive the cycle indices of $GL_n(\mathbb{Z}_2)$ and $A_n(\mathbb{Z}_2)$ we will have to introduce some algebraic machinery. While a complete derivation of these cycle indices is quite lengthy, we can give a sketch of the techniques. More details can be found in Harrison (1964) and the references cited there.

The properties of $GL_n(\mathbb{Z}_2)$ are intimately related to the theory of irreducible polynomials over $\mathbb{Z}_2$. For each positive integer m, there are only a finite number of irreducible polynomials of degree m. Assume that we have the irreducible polynomials over $\mathbb{Z}_2$ sorted by degree and enumerated. We exclude $p(x) = x$ from the list. Let d_i denote the degree of the ith polynomial and let e_i be its exponent (the least positive integer k_i such that $p_i(x) \mid x^{k_i} - 1$).

The irreducible factors of $x^{2^n} - x$ are exactly the irreducible polynomials (over $\mathbb{Z}_2$) of all degrees which divide n. If we let $I(m)$ be the number of irreducible polynomials of degree m over $\mathbb{Z}_2$, then by equating degrees it follows that

$$\sum_{d \mid n} d I(d) = 2^n \qquad (93)$$

By the Möbius inversion formula

$$I(n) = (1/n) \sum_{d \mid n} 2^d \mu(n/d) \qquad (94)$$

Letting t_n be the number of irreducible polynomials in $\mathbb{Z}_2[x]$ of degree at most n (and excluding $p(x) = x$), we have that

$$t_n = \sum_{m=1}^{n} I(m) - 1 \qquad (95)$$

We now have the vocabulary to describe the cycle indices of the linear groups on $\{0, 1\}^n$. First one finds all possible nonnegative integer solutions $(a_1, \ldots, a_{t_n})$ of the equation

$$\sum_{i=1}^{t_n} a_i d_i = n \tag{96}$$

For each element a_i, we write all partitions of a_i in the form

$$a_i = \sum_{j=1}^{a_i} j\alpha_{ij} \tag{97}$$

Next, let b_j be the least integer such that $2^{b_j} \geq j$ and then define $q_{ij} = e_i 2^{b_j}$ where e_i is the exponent of the ith irreducible polynomial in our enumeration. Also define

$$h_{ij} = 2^{d_i(j-1)}(2^{d_i} - 1)/q_{ij} \tag{98}$$

Using these parameters, we can now give the result for $GL_n(\mathbb{Z}_2)$.

THEOREM 4.7.   The cycle index of $GL_n(\mathbb{Z}_2)$ as a permutation group on $\{0, 1\}^n$ is

$$Z_{GL_n(\mathbb{Z}_2)} = \left[2^{n(n-1)/2} \prod_{i=1}^{n} (2^i - 1) \right]^{-1}$$

$$\cdot \sum \left\{ 2^{n(n-1)/2} \prod_{i=1}^{n} (2^i - 1) \underset{i=1}{\overset{t_n}{\mathsf{X}}} \underset{j=1}{\overset{a_i}{\mathsf{X}}} \left(\prod_{k=1}^{j} f_1 f_{q_{ik}}^{h_{ik}} \right)^{\times \alpha_{ij}} \right.$$

$$\cdot \left[\prod_{j=1}^{t_n} \left(\prod_{k=1}^{a_j} 2^{a_{jk}^2(k-1)} \prod_{k=1}^{a_j-1} \prod_{l=k+1}^{a_j} 2^{2k\alpha_{jk}\alpha_{jl}} \right)^{d_j} \right.$$

$$\left. \left. \cdot \prod_{p=1}^{a_j} 2^{d_j(\alpha_{jp}-1)\alpha_{jp}/2} \prod_{q=1}^{\alpha_{jp}} (2^{qd_j} - 1) \right]^{-1} \right\} \tag{99}$$

where the sum is over all partitions of the form (97) associated with solutions of (98).

A detailed proof may be found in Harrison (1964).

Example.   If $n = 2$, then $t_n = 2$ and there are two solutions of $a_1 + 2a_2 = 2$, namely, $(2, 0)$ and $(0, 1)$. Thus there are a total of three solutions to (97). These are

$$(\alpha_{11} = 2, 0), \qquad (\alpha_{12} = 1, 0), \qquad (0, \alpha_{21} = 1)$$

Using these values in Eq. (99) we obtain

$$Z_{GL_2(\mathbb{Z}_2)} = \tfrac{1}{6}(f_1{}^4 + 3f_1{}^2 f_2 + 2f_1 f_3)$$

from which it follows that

$$P(x) = 1 + 2x + 2x^2 + 2x^3 + x^4$$

which agrees with our earlier tabulation.

It is not too difficult to derive the cycle index of $A_n(\mathbb{Z}_2)$ from that of $GL_n(\mathbb{Z}_2)$. The result hinges on the following lemma which is stated without proof:

LEMMA 4.3.   For each $1 \leq i \leq n$ and $1 \leq j \leq a_i$ the contribution to the cycle structure of $A_n(\mathbb{Z}_2)$ is given by

$$u_{ij} = \begin{cases} 2^{j-1} \prod_{k=1}^{j} f_1 f_{q1k}^{h1k} + 2^{j-1} f_{q1\ j+1}^{2^j/q_1\ j+1} \\ 2^{d_{ij}} \prod_{k=1}^{j} f_1 f_{qik}^{hik}, & \text{if } i > 1 \end{cases} \tag{100}$$

We can now state the main result for $A_n(\mathbb{Z}_2)$.

THEOREM 4.8.   The cycle index for $A_n(\mathbb{Z}_2)$ acting as a permutation group on $\{0, 1\}^n$ is given by

$$Z_{A_n(\mathbb{Z}_2)} = \left[2^{n(n+1)/2} \prod_{i=1}^{n} (2^i - 1) \right]^{-1}$$

$$\cdot \sum \left\{ 2^{n(n-1)/2} \prod_{i=1}^{n} (2^i - 1) \underset{i=1}{\overset{n}{X}} \underset{j=1}{\overset{a_i}{X}} u_{ij}^{\times \alpha_{ij}} \right.$$

$$\cdot \left[\prod_{j=1}^{t_n} \left(\prod_{k=1}^{a_j} 2^{\alpha_{jk}^2(k-1)} \prod_{k=1}^{a_j-1} \prod_{l=k+1}^{a_j} 2^{2k\alpha_{jk}\alpha_{jl}} \right)^{d_j} \right.$$

$$\left. \cdot \prod_{p=1}^{a_j} 2^{d_j \alpha_{jp}(\alpha_{jp}-1)/2} \prod_{q=1}^{\alpha_{jp}} (2^{qd_j} - 1) \right]^{-1} \right\} \tag{101}$$

where the sum is as in Theorem 4.7.

Example.   For $n = 2$, we find that

$$Z_{A_n(\mathbb{Z}_2)} = \tfrac{1}{24}(f_1^4 + 3f_2^2 + 6f_1^2 f_2 + 6f_4 + 8f_1 f_3)$$

and hence

$$P(x) = 1 + x + x^2 + x^3 + x^4$$

which is in accord with an earlier tabulation.

We conclude the present section by giving the number of classes for both linear groups when n is small. The reader who is interested in further details concerning these groups and other generalizations should consult Harrison (1964), Lechner (1968), and Chapter V.

TABLE III
The Number of Classes Under $GL_n(\mathbb{Z}_2)$

n	T_n
1	4
2	8
3	20
4	92
5	2744
6	950,998,216

TABLE IV
The Number of Classes Under $A_n(\mathbb{Z}_2)$

n	T_n
1	3
2	5
3	10
4	32
5	382
6	15,768,919

EXERCISES

3. Prove Theorem 4.1 by showing that if a group A is permutationally equivalent to a group B, then they have the same cycle index.

4. Let G be a permutation group on $\{0, 1\}^n$. Show that the number of equivalence classes of Boolean functions of n variables is always greater than 2^n.

5. Prove the following identities in the group $S_n \otimes B$ where 1 denotes the identity of B, S_n, and B^n.

   (a) $(\sigma(\beta), \sigma) = (1, \sigma)(\beta, 1)$
   (b) $(1, \sigma)(\beta, 1)(1, \sigma^{-1}) = (\sigma(\beta), 1)$
   (c) $(\beta, \sigma)(\beta, 1)(1, \sigma^{-1}) = (\sigma(\beta), 1)$

   where $\sigma \in S_n$ and $\beta \in B^n$.

V. OPERATIONS ON THE RANGE, GENERA, AND A LOWER BOUND

Suppose that we examine the six equivalence classes of Boolean functions of two variables under G_2. There are as follows:

[0],   [1]

$[\bar{x}_1\bar{x}_2, \bar{x}_1 x_2, x_1\bar{x}_2, x_1 x_2]$, $[\bar{x}_1 + \bar{x}_2, \bar{x}_1 + x_2, x_1 + \bar{x}_2, x_1 + x_2]$

$[x_1, \bar{x}_1, x_2, \bar{x}_2]$, $[x_1 \oplus x_2, x_1 \equiv x_2]$

Note that the classes on each line are "complements" of one another. If we consider a function to be equivalent to its complement as well as using normal equivalence under G_n, then there are only four of the new classes which we shall call genera of G_2.

Formally, we say that f and g are in the same G-*genus* if f is equivalent to g or to $\bar{g}$ under G. This is a special case of a more general situation in which a permutation group G acts on a domain D and a permutation group H acts on the range R of a set of functions. We say that F is *equivalent* to g under G and H if $g = \beta f\alpha$ for some $\alpha \in G$ and $\beta \in H$. There is an important theorem due to DeBruijn (1959) which counts the number of classes under G and H. We will neither state nor prove this result here but mention that it suffices to know the cycle index of G and H in order to compute the desired numbers. The interested reader should consult DeBruijn (1959) and Harrison (1965).

In Harrison (1963d) the theorem of DeBruijn is used to prove the following result which deals with genera:

THEOREM 5.1. Let G be any group acting on $\mathbb{Z}_2{}^n$. The number of G-genera is

$$\tfrac{1}{2}(Z_G(2, \ldots, 2) + Z_G(0, 2, 0, 2, \ldots, 0, 2))$$

where Z_G is the cycle index of G.

COROLLARY 5.1. The number of $C_2{}^n$ genera is

$$\frac{1}{2^{n+1}}(2^{2^n} + (2^n - 1)2^{2^{n-1}+1})$$

COROLLARY 5.2. If T_n denotes the number of classes of functions under S_n' then the number of S_n'-genera is $T_n/2$.

Proof: This follows from the fact that each term of Z_{S_n} is of the form $f_1^a x$ where $a \geq 2$.

We may mention in passing that any equivalence class S of functions is *self-complementary* if $f \in S$ implies $\bar{f} \in S$. It is easy to determine the number of self-complementary classes.

THEOREM 5.2.   The number of self-complementary classes of Boolean functions under G is

$$Z_G(0, 2, \ldots, 0, 2)$$

where Z_G is the cycle index of G.

Proof: Let T_n be total number of classes and N_{SC} be the number of self-complementary classes. Thus the number of genera is

$$\tfrac{1}{2}(T_n - N_{SC}) + N_{SC} = \tfrac{1}{2}(Z_G(2, \ldots, 2) + Z_G(0, 2, \ldots, 0, 2)) \qquad (102)$$

Using the corollary to Pólya's theorem, we have that $N_{SC} = Z_G(0, 2, \ldots, 0, 2)$.

Our methods have yielded solutions to a number of counting problems in terms of the cycle index polynomials. These results are exact in that they determine the correct number of classes. It is often possible to approximate the desired numbers by a trivial calculation. Our next result gives a method for doing this.

THEOREM 5.3.   Let G be any group defined on the n-variables of the Boolean functions. Let $\varepsilon > 0$ be fixed. If $|G| < 2^{(2^{n-1} - \varepsilon \log_2 n)}$, then the number of classes of boolean functions under G is asymptotic[9] to

$$2^{2^n}/|G|$$

In any event, $2^{2^n}/|G|$ is a lower bound on the number of classes.

Proof: The number of classes is

$$Z_G(2, \ldots, 2) = (1/|G|)(2^{2^n} + \theta) \qquad (103)$$

from which the lower bound follows.

---

[9] We write $f(x) = o(g(x))$ if $\lim_{x \to \infty}(f(x)/g(x)) = 0$. Also, define, $f(x) \sim g(x)$ if $\lim_{x \to \infty}(f(x)/g(x)) = 1$.

Any group which is defined on the variables may be taken as a sum on the number of partitions of n. We bound θ from above by taking the largest nonidentity term of G. This occurs with the partition $j_1 = n - 2, j_2 = 1$ which produces $f_1^{2^{n-2}} \times f_1 = f_1^{2^{n-1}}$ which will contribute $2^{2^{n-1}}$ to θ. One may compute that

$$(1/|G|)(2^{2^n} + \theta) = (2^{2^n}/|G|) + \theta/|G| \tag{104}$$

where

$$(\theta/|G|)/(2^{2^n}/|G|) \leq [(|G| - 1)2^{2^{n-1}}/|G|]/(2^{2^n}/|G|) \tag{105}$$

$$= (|G| - 1)/2^{2^{n-1}} < |G|/2^{2^{n-1}} < 1/n^{\varepsilon} = o(1) \tag{106}$$

Thus the number of classes is

$$(1/|G|)(2^{2^n} + \theta) < 2^{2^n}/|G| + o(2^{2^n}/|G|) \sim 2^{2^n}/|G| \tag{107}$$

which completes the proof.

As an example of this method, we compare the exact number for G_n with the approximation. Further details can be found in Harrison (1966).

TABLE V
The Exact and Approximate Number of Classes under G_n

n	T_n	Approximation
1	3	2
2	6	2
3	22	6
4	402	171
5	1,228,158	1,118,482
6	400,507,806,843,728	400,319,966,877,378

EXERCISES

6. Show that the number of self-complementary equivalence classes under $GL_n(\mathbb{Z}_2)$ is zero. Conclude that the number of $GL_n(\mathbb{Z}_2)$ genera is one-half the total number of classes.

7. Show that the number of self-complementary equivalence classes under C_{2^n} is $(2^n - 1)2^{2^{n-1}-n}$

8. Let G be any group defined on $X_n = \{x_1, \ldots, x_n\}$. Fix $\varepsilon > 0$ and call a function of n variables *neutral* if it has the same number of zeroes and ones in its truth table. Show that if the order of G does not exceed $(2/\pi)^{1/2} 2^{2^{n-1} - n/2 - \varepsilon \log_2 n}$ then the number of self-complementary equivalence classes of functions tends to zero with the number of classes of neutral functions for increasing n.

APPENDIX 1. CYCLE INDEX POLYNOMIALS FOR S_n

n	Z_{S_n}
1	f_1^2
2	$\dfrac{1}{2}(f_1^4 + f_1^2 f_2)$
3	$\dfrac{1}{6}(f_1^8 + 3f_1^4 f_2^2 + 2f_1^2 f_3^2)$
4	$\dfrac{1}{24}(f_1^{16} + 6f_1^8 f_2^4 + 3f_1^4 f_2^6 + 8f_1^4 f_3^4 + 6f_1^2 f_2 f_4^3)$
5	$\dfrac{1}{120}(f_1^{32} + 10f_1^{16} f_2^8 + 15f_1^8 f_2^{12} + 20f_1^8 f_3^8 + 20f_1^4 f_2^2 f_3^4 f_6^2$ $+ 30f_1^4 f_2^2 f_4^6 + 24f_1^2 f_5^6)$
6	$\dfrac{1}{720}(f_1^{64} + 15f_1^{32} f_2^{16} + 45f_2^{16} f_2^{24} + 40f_1^{16} f_3^{16} + 15f_1^8 f_2^{28}$ $+ 120f_1^8 f_2^4 f_3^8 f_6^4 + 90f_1^8 f_2^4 f_4^{12} + 40f_1^4 f_3^{20} + 90f_1^4 f_2^6 f_4^{12}$ $+ 144f_1^4 f_5^{12} + 120f_1^2 f_2 f_3^2 f_6^9)$

APPENDIX 2. CYCLE INDEX POLYNOMIALS FOR G_n

n	Z_{G_n}
1	$\dfrac{1}{2}(f_1{}^2 + f_2)$
2	$\dfrac{1}{8}(f_1{}^4 + 3f_2{}^2 + 2f_1{}^2f_2 + 2f_4)$
3	$\dfrac{1}{48}(f_1{}^8 + 13f_2{}^4 + 8f_1{}^2f_3{}^2 + 8f_2f_6 + 6f_1{}^4f_2{}^2 + 12f_4{}^2)$
4	$\dfrac{1}{384}(f_1^{16} + 51f_2{}^8 + 48f_1{}^2f_2f_4{}^3 + 48f_8{}^2 + 12f_1{}^8f_2{}^4 + 84f_4{}^4$ $\qquad + 12f_1{}^4f_2{}^6 + 32f_1{}^4f_3{}^4 + 96f_2{}^2f_6{}^2)$
5	$\dfrac{1}{3840}(f_1^{32} + 231f_2^{16} + 20f_1^{16}f_2{}^8 + 520f_4{}^8 + 80f_1{}^8f_3{}^8 + 720f_2{}^4f_6{}^4$ $\qquad + 160f_1{}^4f_2{}^2f_3{}^4f_6{}^2 + 320f_4{}^2f_{12}^2 + 240f_1{}^4f_2{}^2f_4{}^6 + 480f_8{}^4$ $\qquad + 240f_2{}^4f_4{}^6 + 60f_1{}^8f_2^{12} + 384f_1{}^2f_5{}^6 + 384f_2f_{10}^3)$
6	$\dfrac{1}{46080}(f_1^{64} + 1053f_2^{32} + 30f_1^{32}f_2^{16} + 4920f_4^{16} + 180f_1^{16}f_2^{24}$ $\qquad + 160f_1^{16}f_3^{16} + 5280f_2{}^8f_6{}^8 + 120f_1{}^8f_2^{28} + 960f_1{}^8f_2{}^4f_3{}^8f_6{}^4$ $\qquad + 3840f_4{}^4f_{12}^4 + 720f_1{}^8f_2{}^4f_4^{12} + 5760f_8{}^8 + 2160f_2{}^8f_4^{12}$ $\qquad + 640f_1{}^4f_3^{20} + 1920f_2{}^2f_6^{10} + 1440f_1{}^4f_2{}^6f_4^{12} + 2304f_1{}^4f_5^{12}$ $\qquad + 6912f_2{}^2f_{10}^6 + 3840f_1{}^2f_2f_3{}^2f_6{}^9 + 3840f_4f_{12}^5)$

APPENDIX 3. CYCLE INDEX POLYNOMIALS FOR $GL_n(\mathbb{Z}_2)$

n	$Z_{GL_n(\mathbb{Z}_2)}$
1	f_1^2
2	$\dfrac{1}{6}(f_1^4 + 3f_1^2 f_2 + 2f_1 f_3)$
3	$\dfrac{1}{168}(f_1^8 + 21f_1^4 f_2^2 + 42f_1^2 f_2 f_4 + 56f_1^2 f_3^2 + 48f_1 f_7)$
4	$\dfrac{1}{20160}(f_1^{16} + 105f_1^8 f_2^4 + 210f_1^4 f_2^6 + 1260f_1^4 f_2^2 f_4^2 + 2520f_1^2 f_2 f_4^3$
	$\quad + 112f_1 f_3^5 + 1680f_1 f_3 f_6^2 + 1120f_1^4 f_3^4 + 3360f_1^2 f_2 f_3^2 f_6$
	$\quad + 5760f_1^2 f_7^2 + 1344f_1 f_5^3 + 2688f_1 f_{15})$
5	$\dfrac{1}{9,999,360}(f_1^{32} + 465f_1^{16} f_2^8 + 26{,}040f_1^8 f_2^4 f_4^4 + 312{,}480f_1^4 f_2^2 f_4^6$
	$\quad + 624{,}960f_1^2 f_2 f_4^3 f_8^2 + 6510f_1^8 f_2^{12} + 78{,}120f_1^4 f_2^6 f_4^4$
	$\quad + 19{,}840f_1^8 f_3^8 + 416{,}640f_1^4 f_2^2 f_3^4 f_6^2 + 833{,}280f_1^2 f_2 f_3^2 f_4 f_6 f_{12}$
	$\quad + 55{,}552f_1^2 f_3^{10} + 833{,}280f_1^2 f_3^2 f_6^4 + 476{,}160f_1^4 f_7^4$
	$\quad + 1{,}428{,}480f_1^2 f_2 f_7^2 f_{14} + 952{,}320f_1 f_3 f_7 f_{21} + 666{,}624f_1^2 f_5^6$
	$\quad + 1{,}333{,}248f_1^2 f_{15}^2 + 1{,}935{,}360f_1 f_{31})$

APPENDIX 4. CYCLE INDEX POLYNOMIALS FOR $A_n(\mathbb{Z}_2)$

n	$Z_{A_n(\mathbb{Z}_2)}$

$1 \quad \dfrac{1}{2}(f_1{}^2 + f_2)$

$2 \quad \dfrac{1}{24}(f_1{}^4 + 3f_2{}^2 + 6f_1{}^2 f_2 + 6f_4 + 8f_1 f_3)$

$3 \quad \dfrac{1}{1344}(f_1{}^8 + 49f_2{}^4 + 42f_1{}^4 f_2{}^2 + 252f_4{}^2 + 168f_1{}^2 f_2 f_4$

$\qquad + 384 f_1 f_7 + 224 f_1{}^2 f_3{}^2 + 224 f_2 f_6)$

$4 \quad \dfrac{1}{322,560}(f_1^{16} + 645 f_2{}^8 + 210 f_1{}^8 f_2{}^4 + 13{,}440 f_4{}^4 + 840 f_1{}^4 f_2{}^6$

$\qquad + 5{,}040 f_1{}^4 f_2{}^2 f_4{}^2 + 5{,}040 f_2{}^4 f_4{}^2 + 20{,}160 f_1{}^2 f_2 f_4{}^3 + 20{,}160 f_8{}^2$

$\qquad + 1792 f_1 f_3{}^5 + 26{,}880 f_1 f_3 f_6{}^2 + 26{,}880 f_1{}^2 f_2 f_3{}^2 f_6 + 26{,}880 f_4 f_{12}$

$\qquad + 4480 f_1{}^4 f_3{}^4 + 13{,}440 f_2{}^2 f_6{}^2 + 46{,}080 f_1{}^2 f_7{}^2 + 46{,}080 f_2 f_{14}$

$\qquad + 43{,}008 f_1 f_{15} + 21{,}504 f_1 f_5{}^3)$

$5 \quad \dfrac{1}{319,979,520}(f_1^{32} + 32{,}581 f_2^{16} + 930 f_1^{16} f_2{}^8 + 2{,}455{,}200 f_4{}^8$

$\qquad + 104{,}160 f_1{}^8 f_2{}^4 f_4{}^4 + 312{,}480 f_2{}^8 f_4{}^4 + 2{,}499{,}840 f_1{}^4 f_2{}^2 f_4{}^6$

$\qquad + 3{,}888{,}640 f_2{}^4 f_6{}^4 + 14{,}999{,}040 f_8{}^4 + 9{,}999{,}360 f_1{}^2 f_2 f_4{}^3 f_8{}^2$

$\qquad + 26{,}040 f_1{}^8 f_2^{12} + 624{,}960 f_1{}^4 f_2{}^6 f_4{}^4 + 79{,}360 f_1{}^8 f_3{}^8$

$\qquad + 3{,}333{,}120 f_1{}^4 f_2{}^2 f_3{}^4 f_6{}^2 + 19{,}998{,}720 f_4{}^2 f_{12}^2$

$\qquad + 13{,}332{,}480 f_1{}^2 f_2 f_3{}^2 f_4 f_6 f_{12} + 888{,}832 f_1{}^2 f_3^{10}$

$\qquad + 14{,}221{,}312 f_2 f_6{}^5 + 13{,}332{,}480 f_1{}^2 f_3{}^2 f_6{}^4 + 3{,}809{,}280 f_1{}^4 f_7{}^4$

$\qquad + 11{,}427{,}840 f_2{}^2 f_{14}^2 + 22{,}855{,}680 f_1{}^2 f_2 f_7{}^2 f_{14} + 22{,}855{,}680 f_4 f_{28}$

$\qquad + 30{,}474{,}240 f_1 f_3 f_7 f_{21} + 10{,}665{,}984 f_1{}^2 f_5{}^6 + 10{,}665{,}984 f_2 f_{10}^3$

$\qquad + 21{,}331{,}968 f_1{}^2 f_{15}^2 + 21{,}331{,}968 f_2 f_{30} + 61{,}931{,}520 f_1 f_{31} + 2{,}499{,}840 f_2{}^4 f_4{}^6)$

ACKNOWLEDGMENTS

The writing of this chapter was supported in part by the John Simon Guggenheim Foundation, the National Science Foundation under Grant GJ-474, and the Air Force Office of Scientific Research under Grant AF-AFOSR-70-1845. Special thanks are also due to Project MAC at MIT where most of this work was accomplished.

REFERENCES

BECKENBACH, E. F. (ed.) (1964). "Applied Combinatorial Mathematics." Wiley, New York.

DEBRUIJN, N. G. (1959). Generalization of Pólya's fundamental theorem in enumerative combinatorial analysis. *Kon. Ned. Akad. Wetensch. Ser. A* **62**, 56–79.

GOLOMB, S. W. (1961). A mathematical theory of discrete classification, *in* " Proceedings of the Fourth London Symposium on Information theory." Butterworths, London.

GOLOMB, S. W., and HALES, A. W. (1968). On enumerative equivalence of group elements. *J. Combinatorial Theory* **5**, 308–312.

HALL, M. (1959). "The Theory of Groups." Macmillan, New York.

HARRISON, M. A. (1963a). Combinatorial Problems in Boolean Algebras and Applications to the Theory of Switching. Ph.D. Thesis, University of Michigan, Ann Arbor.

HARRISON, M. A. (1963b). On the number of classes of (n, k) switching networks. *J. Franklin Inst.* **276**, 313–327.

HARRISON, M. A. (1963c). The number of transitivity sets of Boolean functions. *J. Soc. Ind. Appl. Math.* **11**, 808–828.

HARRISON, M. A. (1963d). The number of equivalence classes of Boolean functions under groups containing negation. *IEEE Trans. Electron. Comput.* **EC-12**, 559–561.

HARRISON, M. A. (1964). On the classification of Boolean functions by the general linear and affine groups. *J. Soc. Appl. Ind. Math.* **12**, 285–299.

HARRISON, M. A. (1965). "Introduction to Switching and Automata Theory." McGraw-Hill, New York.

HARRISON, M. A. (1966). On asymptotic estimates in switching and automata theory. *J. Assoc. Comput. Mach.* **13**, 151–157.

HARRISON, M. A. and HIGH, R. G. (1968). On the cycle index of a product of permutation groups. *J. Combinatorial Theory* **4**, 277–299.

LECHNER, R. J. (1968). A transform approach to logic design. *Proc. 1968 9th Symp. on Switching & Automata Theory*, pp. 213–234 (to appear in IEEE *Trans. Comput.* Vol. **C-19**).

LIU, C. L. (1968). "Introduction to Combinatorial Mathematics." McGraw-Hill, New York.

PÓLYA, G. (1937). Kombinatorische Anzahlbestinnumgen für Gruppen, Graphen, und Chemische Verbindungen. *Acta Math.* **68**, 145–253.

PÓLYA, G. (1940). Sur les types des propositions composées. *J. Symbolic Logic* **5**, 98–103.

SLEPIAN, D. (1953). On the number of symmetry types of Boolean functions of n variables. *Can. J. Math.* **5**, 185–193.

Chapter V

HARMONIC ANALYSIS OF SWITCHING FUNCTIONS

ROBERT J. LECHNER

I.	**SUMMARY**	122
II.	**SURVEY OF ABSTRACT HARMONIC ANALYSIS**	125
	A. The Fourier Transform of a Function on $\mathbb{Z}^n$	125
	B. Survey of Classical Properties	139
	C. Fundamental Theorem on Invariance	148
III.	**COMBINATORIAL APPLICATIONS**	156
	A. A Test for Implicants of f or $\bar{f}$	156
	B. Algorithm to Extract Prime Implicants	163
	C. Test for Disjunctive Decomposition	171
	D. Criteria for Factoring Variables	172
IV.	**ANALYSIS OF THE PROTOTYPE EQUIVALENCE RELATION**	173
	A. Definition, Interpretation, and Context	174
	B. Enumeration of Prototype Equivalence Classes	183
	C. Explicit Definition of 5-Variable Prototypes	191
V.	**SYNTHESIS OF ENCODED INPUT LOGIC**	196
	A. Review of LSI Logic Design Approaches	197
	B. Synthesis Techniques for Encoded Input Logic	204
	C. A 6-Variable Example	210
	D. Fundamental Problems and Extensions	218
	REFERENCES	225

ABSTRACT. This chapter is a self-contained development of abstract harmonic analysis applied to a single-output combinational logic functions; linear algebra and elementary group theory are the only mathematical prerequisites. New analysis and synthesis techniques are developed, and the groundwork is laid for future extensions to multiple output combinational logic, sequential machines, and real- or complex-valued functions of binary arguments. Harmonic analysis adds novel conceptual insights and unifying principles, improved computational techniques, and new measures of complexity to the traditional approach to switching theory.

The first section is a summary of the chapter. Section II surveys classical Fourier transform properties and introduces the canonical expansion of a switching function as an

n-dimensional abstract Fourier transform over the finite two-element field. The two most important transform properties are the convolution theorem, which leads to tests for prime implicants and disjunctive decompositions, and spectrum invariance which is basic to further theoretical developments and to a new synthesis technique called encoded input logic.

Section III develops a new algorithm which concurrently extracts prime implicants and detects disjunctive decompositions of a switching function. Implicants of both the function and its complement are detected simultaneously, and "core" implicants can be identified. The algorithm which is not sensitive to functional complexity has been programmed for a commercial time-sharing system.

Section IV introduces the restricted affine group (RAG) whose elements, called prototype transformations, encode the arguments and outputs of combinational logic functions. This group partitions the space $\mathscr{F}$ of all two-valued functions on $\mathbb{Z}^n$ into 3, 8, and 48 equivalence classes respectively for $n = 3, 4$, and 5. Unique representatives are identified for each class when $n = 3$ and 4 and for 46 of the 48 classes of 5-argument functions.

Section V applies the tools of abstract harmonic analysis to the synthesis problem for large truth tables (many-input combinational logic). A general multilevel synthesis approach, called encoded input logic, is introduced which is compatible with large-scale-integrated circuit technology. Both the conventional macrocellular and the newer microcellular array approach are included as special cases. Prototype encoding transformations are used to reduce the complexity of an imbedded normal form realization. Practical synthesis algorithms are based on Fourier analysis. A realistic 6-argument example is treated in detail. Section V concludes with a list of fundamental problems whose solutions would extend the research presented herein.

I. SUMMARY

Harmonic analysis is a new method for both theoretical analysis and practical synthesis of digital logic circuits. This chapter is a simplified exposition which treats fully the single-output combinational logic case. It thereby lays the groundwork for future extensions, first to multiple-output combinational logic and then to sequential machines or finite state automata.

The organization of this chapter was dictated by its two-fold purpose. One goal is to assist the engineering-oriented graduate student or logic designer in applying harmonic analysis techniques to logic design for the technological environment of large-scale-integrated semiconductor electronics. Another goal is to develop the theoretical properties of abstract Fourier transforms for the mathematically oriented graduate student or logic theoretician as a point of departure for further research.

These two goals are somewhat inconsistent. The first goal demands attention to those computational properties of Fourier transforms which are directly applicable to combinational logic design. These should be introduced with a minimum of theoretical prerequisites. The second goal requires an abstract development which introduces all the known theoretical properties of Fourier transforms including those which appear to have no direct application, to stimulate new directions of research.

This chapter treats both theoretical and practical aspects of abstract harmonic analysis, with particular emphasis on discrete-valued functions of binary arguments. Of course, extensions to continuous real or complex-valued functions are possible and have considerable practical interest. Typical applications include reliability analysis for discrete systems and modulation theory for digital communications.

Basic material on harmonic analysis is introduced first, in Section II. An introductory section (II,A) is followed by a survey of classical Fourier transform properties (II,B). The representation of a switching function as an n-dimensional abstract Fourier transform over the finite two-element field is unique, therefore canonical, and has many valuable properties. These properties have inspired new algorithms for some classical problems of combinational logic synthesis to be described in Section III.

The most important theoretical contributions of harmonic analysis to switching theory depend on the intimate connection between Fourier transforms and linear or affine operators (encoding transformations) on the domain and range of switching functions. The fundamental theorem on invariance of Fourier transforms under linear and affine operators is introduced in Section II,C. Although this section is not a prerequisite to the engineering application of Section III, it is basic to the analysis of equivalence classes under transformation groups in Section IV and to a new synthesis approach based on these groups in Section V.

Section III develops a new and unified algorithm to extract prime implicants which, as a by-product, also detects disjunctive decomposition of a switching function. Implicants of both the function and its complement are detected simultaneously, and the algorithm is not sensitive to functional complexity. Its computational burden grows as $n3^n$, where n is the number of function arguments, and its storage requirements grow as 2^n rather than 3^n (prime implicants need not be stored).

Section IV introduces a new class of transformation groups called the restricted affine groups (RAG). The elements of these groups are called prototype transformations, and they operate on the Cartesian product (or direct sum) of the domain $\mathbb{Z}^n$ and range $\mathbb{Z}$ of a switching function, where $\mathbb{Z} = \{0, 1\}$, the two-element Galois field. They comprise the largest subgroup of the $(n + 1)$-dimensional affine group which does not introduce feedback from the function output to its input. The RAG includes as subgroups all the transformation groups previously considered in the literature of switching theory. The prototype equivalence classes into which RAG partitions the space $\mathscr{F}$ of all 2-valued functions on $\mathbb{Z}^n$ are unions of the equivalence classes induced by its subgroups.

Section IV,A places RAG within the context of its classical subgroups, describes prototype transformations in mathematical and engineering terms, and establishes a useful connection between linear and affine representations of group elements. By means of this connection, rational canonical forms for prototype transformations and methods for computing the parameters required in Polya's counting theorem were developed and applied by Lechner (1963) to show that the set of 2^{32} truth tables of 5-argument functions are grouped into exactly 48 prototype equivalence classes by the RAG.

Section IV,B explains how this method differs from the method used by Harrison (1964) for direct product groups. This section also presents data on prototype classes for the 3- and 4-argument cases and class counts for functions of a given weight (number of points x such that $f(x) = 1$) under linear and affine groups on the *domain* of f. Ninomiya (1958) first applied the invariant properties of Fourier transforms to the classification of switching functions under prototype transformations for $n \leq 4$ arguments.

Section IV,C gives results of a sample and search technique which identified unique representatives for 46 of the 48 prototype equivalence classes for 5-argument functions. Parameters which separate (uniquely characterize) the 46 known classes are also tabulated; these can easily be computed from the Fourier transform of an arbitrary 5-argument function. Resolution of the remaining ambiguity (identification of the last two classes) remains an open problem.

Section V returns to the problem of combinational logic synthesis armed now with the tools of abstract harmonic analysis. Section V,A motivates the new approach by considering the overall context of large-scale-integrated circuit technology.

Section V,B proposes a new structural model or framework for combinational logic synthesis which reduces to either the conventional macrocellular or the newer microcellular array approach for extreme cases of particular functions. This approach, called encoded input logic, applies the prototype transformation of Section IV to reduce the complexity of normal form realization. The invariant properties of Fourier transforms in Section II,C yield practical synthesis algorithms which select the encoding transformations.

Section V,C presents a realistic example, a 6-input function taken from a universal logic module (ULM) example of Elspas *et al.* (1967). This example made use of a preliminary version of the prime implicant extraction algorithm written for an interactive time-sharing terminal. A minimal 2-level normal form of this function required eleven 7- or 8-input AND gates with a total of 104 gate inputs. By imbedding another function from the same prototype equivalence class within a network of eight EXCLUSIVE OR gates, an encoded-input logic realization was achieved with only 53 gate inputs. Eight AND gates, each with four or five inputs, were required in addition to the eight 2-input EXCLUSIVE OR gates. Of course, there are many ways to obtain an implementation which is less costly than a minimal 2-level normal form as soon as the 2-level (speed-related) constraint is relaxed. Besides its compatibility with macrocellular and microcellular array technology, the principal advantage of the encoded-input logic is that the new, apparently powerful, computationally effective, and easily understood techniques of abstract harmonic analysis can be brought to bear on the problem.

Section V,D considers fundamental problems and extensions that have been motivated by the research presented in this chapter. These include synthesis of encoded-input threshold logic, probability distributions for spectral coefficients, canonical forms of prototype transformations on the direct sum $\mathbb{Z}^n + \mathbb{Z}^m$ and their application to multiple-output and many-valued functions, state assignments for sequential circuits, and further development of harmonic analysis along lines suggested by topological dynamics.

In conclusion, the many-faceted literature of combinational switching theory justifies a continuing search for new conceptual insights, unifying principles, and quantitative properties for combinational logic functions. In this respect, abstract harmonic analysis appears to make a unique contribution.

II. SURVEY OF ABSTRACT HARMONIC ANALYSIS

The plan of this section is as follows: Section II,A formally defines the abstract Fourier transform of a combinational logic function, then develops generalizations and computational techniques. It also provides historical background on applications to both switching and coding theory and analogies to other engineering applications of related transform techniques. Section II,B surveys the classical properties of the Fourier transform representation including the convolution theorem which is basic to the combinatorial applications of Section III. Section II,C develops the invariant properties of Fourier transforms under affine operators. These will be used to analyze the prototype equivalence relation in Section IV and to synthesize logic with linearly encoded inputs in Section V. Since the Fourier transform possesses unique advantages as an analytical tool for further research on the structure of switching functions, notice is given to computational techniques that might be used in each section.

A. THE FOURIER TRANSFORM OF A FUNCTION ON $\mathbb{Z}^n$

1. Notation

In this chapter $\mathbb{Z}$ will denote the 2-element field $\{0, 1\}$, $\mathbb{Z}^n$ will denote the n-dimensional vector space over the field $\mathbb{Z}$, and $\mathscr{F}$ will denote the set of all n-input, single-output functions from $\mathbb{Z}^n$ into $\mathbb{Z}$. For $N > 2$ (and not necessarily a prime power), $\mathbb{Z}_N$ will identify the ring of residue classes mod N; i.e.,

$\mathbb{Z} = I/2I$, and $\mathbb{Z}_N = I/NI$ where I is the ring of all integers. For clarity, another symbol X will also be used to identify the set $\mathbb{Z}^n$ of all binary n-tuples when it is the domain for the space $\mathscr{F}$. The general element x of X will be represented by the row vector $(x_1, \ldots, x_n)$; the symbol x^t will denote the column vector which is the transpose of x. To index elements of X, we will use the natural ordering of n-tuples defined by the correspondence between integers $0 \le i < 2^n$ and their radix-two expansions or binary n-tuples $x(i)$. (In other words, if $x(i) = (x_1, x_2, \ldots, x_n)$, then $i = x_1 2^{n-1} + x_2 2^{n-2} + \cdots + x_n$.)

Elements of $\mathscr{F}$ will be denoted by letters f, g, h. To define f more explicitly, we use the graph or truth table $\{(x, f(x)) : x \in X, f(x) \in \mathbb{Z}\}$ (a subset of the Cartesian product or direct sum $X + \mathbb{Z}$), and its 2^n-tuple vector equivalent $\{f_i = f(x(i)), 0 \le i < 2^n\}$; we also use the standard sum or level set representation $f^{-1}(1) = \{x : f(x) = 1\}$ or its integer subset equivalent $\{i : f(x(i)) = 1\}$. The logical complement of a binary variable x_i will be denoted $\bar{x}_i$. We will also use the overbar symbol to denote the functional complement of a binary-valued function $(\bar{f} = 1 - f)$. The terms argument, input, and variable will be used interchangeably to mean one of the x_i.

The residue class, modulo two, of the integer i will be written $|i|_2$; $|b|$ also denotes the weight or real sum $b_1 + b_2 + \cdots + b_n$ of a *binary n-tuple*. The symbol $\oplus$ will denote vector addition mod 2.

The 2^n-tuple $\{f_i, 0 \le i < 2^n\}$ is in one : one correspondence with the disjunctive canonical form for $f(x)$, an expansion of f with respect to the fundamental product, or minterm, basis of the space $\mathscr{F}$. The minterm basis functions are defined for $0 \le i < 2^n$ by

$$p_i(x) = (x_1)^{i_1}(x_2)^{i_2} \cdots (x_n)^{i_n}$$

where $(i_1, i_2, \ldots, i_n)$ is the radix-two expansion of i, and $x^k = x$ or $\bar{x}$ if $k = 1$ or 0, respectively. The vector representation of p_i is

$$p_i(j) = p_{ij} = \begin{cases} 1 & \text{iff} \quad i = j \\ 0 & \text{otherwise} \end{cases}$$

In other words, p_i is the ith unit vector $0 \le i < 2^n$. Thus, instead of $f(x) = \sum f_i p_i(x) \in \mathscr{F}$, we may write $f = \sum f_i p_i \in \mathbb{Z}^{2^n}$. A similar correspondence between functions and vectors will be used when f is represented with respect to the Fourier transform basis.

2. Definition of the Fourier Transform of a Function on $\mathbb{Z}^n$

The representation of f by its truth table, binary 2^n-tuple, or disjunctive canonical form is of value for three reasons: (1) it is a unique representation, (2) it explicitly identifies the value of $f(x)$ for each state or configuration x

of its arguments, and (3) its tabular form provides a convenient starting point from which to derive simplified logic formulas for use in computer programs or logic circuit design. The abstract Fourier transform of f is also a unique representation. Although it does not have advantage (2) this is more than compensated by its other unique advantages that will be demonstrated in this chapter.

The space $\mathscr{F}$ of all two-valued functions on $\mathbb{Z}^n$ has a locally compact Abelian or commutative group for its domain, and its range elements 0 and 1 can be added and multiplied as complex numbers. These two simple requirements are both necessary and sufficient to bring the methods of abstract harmonic analysis to bear on the study of this function space. In other words, an orthogonal basis set of Fourier transform kernel functions can be constructed for $\mathscr{F}$. A real (integer) valued function on $\mathbb{Z}^n$ will have a real (integer) valued transform.

The kernel or basis functions of the Fourier transform pair are defined in terms of a particular type of mapping from the n-dimensional vector space $\mathbb{Z}^n$ to the direct product of n copies of the multiplicative subgroup $\{\pm 1\}$ on the unit circle of the complex plane. This mapping is such that the group sum of any two domain elements is mapped into the complex product of the images (i.e., the mapping is a group homomorphism, and group addition becomes complex multiplication). To motivate this definition, consider for example the cyclic group of integers modulo N. If the integers $x = 0, 1, 2, \ldots,$ $N - 1$ are mapped into the points $\exp(2\pi i x/N)$ on the unit circle, then this mapping preserves the group operation; that is, the image of $x \oplus y$ is $\exp[2\pi i(x + y)/N]$ which is identical to the product $\exp(2\pi i x/N) \cdot \exp(2\pi i y/N)$. In other words, multiplication of points on the unit circle can be carried out by adding their "angles" or exponents.

There are actually N group homomorphisms (rather than one) from the ring of integers modulo N onto the unit circle; they are defined by

$$Q_k(x) = \exp(2\pi i k x/N) \tag{1}$$

In the case of functions on $\mathbb{Z}$, N becomes 2 and the N equally spaced points on the unit circle (one of which must be the multiplicative identity) become two real integers: $Q_k(x) = \exp(2\pi i k x/2) = (-1)^{kx} = \pm 1 (k = 0 \text{ or } 1)$. This mapping must be extended to the direct sum of n 1-dimensional subspaces over $\mathbb{Z}$ (more generally, to the direct sum of n cyclic Abelian groups). Define the image of $(x_1, x_2, \ldots, x_n)$ under the homomorphism $Q_{k_1 k_2 \cdots k_n}(x) = Q_k(x)$ to be $(-1)^{k_1 x_1}(-1)^{k_2 x_2} \cdots (-1)^{k_n x_n}$ or $(-1)^{kx^t}$ where k is the vector $(k_1, k_2, \ldots, k_n)$ which indexes the homomorphism. These functions are called "group characters" or Fourier transform kernel functions (Littlewood, 1940). From now on we will use w instead of k to index these functions (by analogy with engineering use of the Greek letter ω for complex frequency).

Example 1. For the case $n = 2$, a geometric as well as numerical derivation of the transform pair can be illustrated. (Arguments x_0, x_1 will be used instead of x_1, x_2 in this example.) The space $\mathcal{F}$ consists of $2^4 = 16$ different functions of 2 arguments (x_0, x_1). To each integer between 0 and 15 (used as a function index), there corresponds a unique binary 4-tuple (the truth table for the function). The components of this 4-tuple are $f_0 = f(0, 0), f_1 = f(0, 1)$, $f_2 = f(1, 0)$, and $f_3 = f(1, 1)$, respectively. For example, the function $f^7 = \bar{x}_0 \vee \bar{x}_1$ has the truth table (1, 1, 1, 0). When its components f_i are written in ascending order from right to left (f_3, f_2, f_1, f_0), this 4-tuple is also the binary code for its index, 7.

In Fig. 1a, these sixteen 4-tuples are represented as vertices of a 4-dimen-

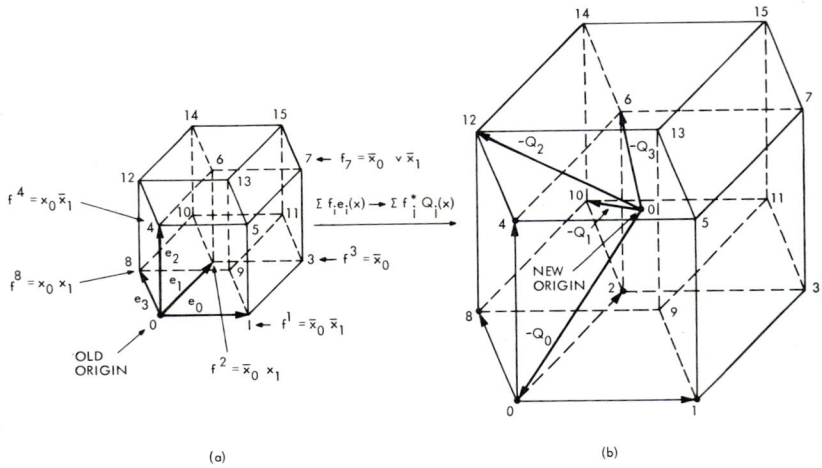

FIG. 1.   The Q-basis for functions of two variables. (a) Representation as vertices on a 4-dimensional unit cube. (b) Negatives of the new basis vectors after the transformation $y \to Q_w(x)$.

sional unit cube, with coordinate values 0 or 1. Formulas for some of the indexed functions are shown next to the corresponding vertices. Four unit basis vectors are also shown on Fig. 1a. These vectors are associated with the "minterm" functions $f(x) = x_1 x_2, x_1 \bar{x}_2, \bar{x}_1 \bar{x}_2$, and $\bar{x}_1 \bar{x}_2$, respectively. The expansion of any function (vector) with respect to this basis is called its disjunctive canonical form. For example, the function $\bar{x}_0 \vee \bar{x}_1 = \bar{x}_1 x_2 + x_1 \bar{x}_2 + \bar{x}_1 \bar{x}_2$ will be evaluated in two steps to illustrate the geometric correspondence shown on the figure: (1) For each w, evaluate $y = xw^t$ (mod 2) for each x. (2) Apply the transformation $y \to (-1)^y = 1 - 2y$ to each result of step (1). The following table shows y_w and Q_w for each w and x:

	$y_w(x) = xw^t \pmod 2$				$Q_w(x) = 1 - 2y_w(x)$			
$x =$	$0\,0$	$0\,1$	$1\,0$	$1\,1$	$0\,0$	$0\,1$	$1\,0$	$1\,1$
$w =$								
$0\ 0$	0	0	0	0	1	1	1	$1 = Q_0$
$0\ 0$	0	1	0	1	1	-1	1	$-1 = Q_1$
$1\ 0$	0	0	1	1	1	1	-1	$-1 = Q_2$
$1\ 1$	0	1	1	0	1	-1	-1	$1 = Q_3$

The rows defining $y_w(x)$ are mod 2 linear functions of x and correspond to vertices numbered 0, 10, 12, and 6 on the left side of the figure. The transformation $\dot{y} \to Q_w(x)$ represents a doubling in size and change of sign in all coordinates of the unit cube, followed by a translation of origin from one vertex to the center of the cube. Negatives of the new basis vectors are shown on Fig. 1b so their endpoints terminate on vertices of the corresponding linear functions xw^t. The new basis vectors are also orthogonal (as will be shown rigorously later).

Example 2. Representation of the function $\bar{x}_0 \vee \bar{x}_1$: Using ordinary Euclidean geometry, we can obtain the projections of the vector $f = (1, 1, 1, 0)$, which represents $f_7(x) = \bar{x}_0 \vee \bar{x}_1$ onto each row of the **Q** matrix (see Section II,A,4):

$$f^* = f \cdot Q = (1, 1, 1, 0)\begin{bmatrix} 1 & 1 & 1 & 1 \\ 1 & -1 & 1 & -1 \\ 1 & 1 & -1 & -1 \\ 1 & -1 & -1 & 1 \end{bmatrix} = (3, 1, 1, -1) \qquad (2)$$

Since $Q^{-1} = 2^{-n}Q$ (as will be shown later), $f = f^*Q^{-1} = 2^{-n}f^*Q$. This means that $f(x)$ has a unique expansion of the type $f(x) = \alpha \sum f_i^* Q_i(x)$; in this example, $f(x) = \alpha[3Q_0(x) + Q_1(x) + Q_2(x) - Q_3(x)]$ where $\alpha = \frac{1}{4}$ is a normalizing factor so that $f = 0$ or 1.

THEOREM 2.1.   The set of Fourier kernel or basis functions $\{Q_w(x) = (-1)^{wx^t}, w \in Z^n\}$ (also known as the character group of Z^n) is an orthogonal basis for the space of all complex-valued functions on Z^n.

Proof: The Q_w are all distinct; for suppose $Q_u = Q_v$, although u and v differ in (say) the ith coordinate. Then $u_i \neq v_i$ and $Q_u(e_i) \neq Q_v(e_i)$ when x is the ith unit vector e_i. Any set of 2^n distinct kernel functions $Q_w(x)$ will span

$\mathscr{F}$ (be a basis) if they are orthogonal. To prove orthogonality, expand the inner product:

$$(Q_w, Q_y) = \sum_x (-1)^{wx^t}(-1)^{yx^t} = \sum_x \prod_j (-1)^{(w_j+y_j)x_j} = \sum_x (-1)^{(w+y)x^t} \quad (3)$$

Now $(w_j + y_j)$ has the range 0, 1, and 2 in each exponent; however, because $(-1)^{2x_j} = (-1)^0 = 1$ regardless of the value of x, $w_j + y_j$ can be replaced by a modulo two sum for each j. Defining $u = w + y \pmod{2} = w \oplus y$,

$$(Q_w, Q_y) = \sum_x (-1)^{ux^t} \quad (4)$$

where the sum is over all of X.

If $w = y$, then $u = 0$ and every term of this sum is 1, which implies $(Q_w, Q_w) = 2^n$. If $w \neq y$, then $u \neq 0$ and the inner product ux^t is a projection of X into a 1-dimensional subspace, or 2-element subgroup. Its kernel, the set $K = \{x \in X : ux^t = 0\}$, is a subspace of dimension $n - 1$, and $X - K$ is a single coset $C = \{x \in X : ux^t = 1\}$. Splitting the sum into two parts, since K and C are both 2^{n-1}-element subset of X, we have

$$(Q_w, Q_y) = \sum_{x \in K} (-1)^{ux^t} + \sum_{x \in C} (-1)^{ux^t} = \sum_{x \in K} (-1)^0 + \sum_{x \in C} (-1)^1 = 0 \quad (5)$$

which proves orthogonality.

DEFINITION.   The *Abstract Fourier Transform* of f is the integer-valued function f^* which defines (up to a scale factor) the expansion coefficients or coordinates of f with respect to the basis $\{Q_w(x) = (-1)^{xw^t}; w \in Z^n\}$. The domain of definition of f^* is another vector space W of binary n-tuples isomorphic to X. For notational economy, the jth element $w(j)$ of W will be defined as the binary code for j, $f^*(w(j))$ will be denoted f_j^*, and $Q_j(x(i))$ or merely Q_{ji} will denote $Q_{w(j)}(x(i))$. Then the Fourier expansion of f can be represented two ways:

$$f(x) = 2^{-n} \sum_w f^*(w)(-1)^{xw^t} = 2^{-n} \sum_w f^*(w)Q_w(x) \quad (6)$$

or

$$f_i = f(x(i)) = 2^{-n} \sum_j f_j^*Q_j(x(i)) = 2^{-n} \sum_j f_j^*Q_{ji} \quad (7)$$

Example 3.   Take $n = 2$, f^* and Q from the preceding example. Then $f = 2^{-n}f^*Q$ becomes $(f_0, f_1, f_2, f_3) = 2^{-2}(3, 1, 1, -1)Q = (1, 1, 1, 0)$ which is the truth table for the function $f(x) = \bar{x}_0 \vee \bar{x}_1$.

3. Derivation of the Transform Pair

To derive the transform coefficients f_j^*, simply assume the existence of the Fourier series expansion $f(x) = 2^{-n} \sum_j f_j^* Q_j(x)$, multiply both sides by $Q_k(x)$, sum over X, and make use of the orthogonality property $(Q_j, Q_k) = 2^n \delta_{jk}$ of the basis functions $(\delta_{jj} = 1; \delta_{jk} = 0 \text{ for } j \neq k)$.

$$\sum_x f(x)Q_k(x) = 2^{-n} \sum_{j=0}^{2^n-1} f_j^* \sum_x Q_j(x)Q_k(x)$$

$$= 2^{-n} \sum_j f_j^*(2^n \delta_{jk}) = f_k^* \tag{8}$$

DEFINITION. Using the above notation, the abstract Fourier transform pair is defined for any (real-valued) function f in $\mathscr{F}$ as follows (using real arithmetic):

$$f_j^* = f^*(w(j)) = \sum_x f(x)Q_j(x) = \sum_i f_i Q_{ij} \tag{9}$$

$$f_i = f(x(i)) = 2^{-n} \sum_j f_j^* Q_j(x(i)) = 2^{-n} \sum_j f_j^* Q_{ji} \tag{10}$$

Example 4. Collecting the preceding examples together produces the following transform pair for the function $f^7 = \bar{x}_0 \vee \bar{x}_1$:

$$(f_0^*, f_1^*, f_2^*, f_3^*) = (1, 1, 1, 0)Q = (3, 1, 1, -1)$$
$$(f_0, f_1, f_2, f_3) = 2^{-2}(3, 1, 1, -1)Q = (1, 1, 1, 0)$$

where Q is the matrix of Example 2.

Since the domains of f and f^* are isomorphic, and the transform is symmetric, the same * superscript may be used to denote the *inverse transform* which maps f^* back into $f: f = 2^{-n}(f^*)^*$. The scale factor 2^{-n} which is required to normalize the transform pair is included in the inverse transform to make f^* an integer-valued function. It is explicitly written so that no ambiguity exists as to whether the * superscript denotes a forward or reverse transform.

The matrix formulation of this transform pair is convenient for computation. The integer-valued 2^n-tuples (row vectors) $f = \{f_j, 0 \leq j < 2^n\}$ and $f^* = \{f_i^*, 0 \leq i < 2^n\}$ are related by the symmetric transform matrix $Q = \{q_{ij}\}$ as follows:

$$f^* = fQ; \quad f = f^*Q^{-1} = 2^{-n}f^*Q \tag{11}$$

The direct analogy between this transform pair and the classical n-dimensional Fourier series representation of a function of period p in each of its coordinates is apparent if we use p instead of 2 for the characteristic of the *finite* field over which X and W are defined. In this case, the abstract transform pair becomes

$$f^*(w) = \sum_x f(x) \exp(2\pi i(wx^t)/p) \tag{12}$$

$$\cdot f(x) = p^{-n} \sum_w f^*(w) \exp(-2\pi i(wx^t)/p) \tag{13}$$

Note that the basis functions take p equally spaced values on the unit circle. For $p = 2$, these values are ± 1. For all other p, they become complex-valued. For a finite domain, both the function and its transform have finite and discrete domain of definition. This is a major difference from the conventional Fourier transform pair for a function of n variables, each with period p. The latter has a continuous finite domain X and a discrete countably infinite transform domain W. In this case W includes *all* n-tuples of integers.

$$f^*(w) = \int_0^p \cdots \int_0^p f(x) \exp(2\pi i(wx^t)/p)\, dx_1, \ldots, dx_n, \qquad w \in Z^n \tag{14}$$

$$f(x) = p^{-n} \sum_{w \in W^n} f^*(w) \exp(-2\pi i(wx^t)/p), \qquad x \in [0, p]^n \tag{15}$$

4.   A Fast Fourier Transform Algorithm

From Eq. (11) computation of f^* from f (or vice versa) apparently requires 2^n dot products, each involving $2^n - 1$ additions or subtractions. Ninomiya (1958) computed a slight variation of f^* by summing m basis vectors, where m is the number of nonzero components of f. Golomb (1959) overlaid a set of 2^n templates on an array representing f to compute a set of 2^n invariants completely equivalent to (but not as symmetrically defined as) f^*. Each template selected 2^{n-1} components of f to be summed. The following computational algorithm evaluates f^* or its inverse in no more than $n2^{n+1}$ integer-valued addition or subtraction operations (Lechner, 1963a,b). This algorithm has been programmed for a digital computer and requires only 2^n cells of working storage. The transform f^* can be overlaid within the 2^n cells of storage used for the truth table of f itself. This algorithm is a special case of the fast Fourier transform algorithm now in widespread use on digital computers (Good, 1958; Bergland and Hale, 1967).

The algorithm depends on the recursive definition of the matrix Q, which is identical to the Hadamard transform matrix (Golomb *et al.*, 1964). For example, if Q_n denotes the matrix Q of order 2^n, then

$$
Q_1 = \begin{bmatrix} 1 & 1 \\ 1 & -1 \end{bmatrix}, \qquad
Q_2 = \begin{bmatrix} 1 & 1 & 1 & 1 \\ 1 & -1 & 1 & -1 \\ 1 & 1 & -1 & -1 \\ 1 & -1 & -1 & 1 \end{bmatrix}
$$

and in general, if $Q^{[n]}$ denotes the nth Kronecker power of Q_1,

$$
Q_{n+1} = \begin{bmatrix} Q_n & Q_n \\ Q_n & -Q_n \end{bmatrix} = Q_1 \times Q_n = Q_1^{[n+1]} \tag{16}
$$

where $A \times B$ denotes the Kronecker product of two matrices (Bellman, 1960). The following expansion theorem may be verified by direct multiplication:

$$
A^{[n]} = \prod_{k=0}^{n-1} (I^{[n-k-1]} \times A \times I^{[k]}) = \prod_{k=0}^{n-1} S_{[k]} \tag{17}
$$

where $\prod$ denotes regular matrix multiplication, and $I^{[k]}$ is the identity matrix of order 2^k. Now let $A = Q_1$ and $S_{[k]} = (I^{[n-k-1]} \times Q_1 \times I^{[k]})$. Then

$$
f^* = f \cdot S_{[1]} \cdot S_{[2]} \cdots S_{[n]} \tag{18}
$$

If the multiplications are done from left to right, this reduces to a simple case of the well-known fast Fourier transform algorithm (Good, 1958; Bergland and Hale, 1967). The matrix Q (the order 2^n will be understood) has been factored into the ordinary matrix product of n factors $S_{[k]}$ each of which is trivial and requires no storage of the q_{ij} coefficients. The kth factor involves 2^{n-k} submatrices of the form $\begin{bmatrix} 1 & 1 \\ 1 & -1 \end{bmatrix}$ along its main diagonal; each submatrix forms the sum and difference of a pair of 2^{k-1}-tuples. Thus, each of the n factors requires exactly 2^{n+1} integer additions or subtractions, and the entire transform for a 2^n-dimensional vector f requires $n2^{n+1}$ elementary operations.

Example 5. Figure 2 is a diagrammatic representation of the transform computation for a particular function of five arguments (from Table XI). A negative coefficient is indicated by an overbar. The last two rows of Fig. 2 permute the coefficients $f_N^*(w)$ into ascending order of their weight $|w|$ to show the lexicographic ordering of coefficients that makes this representative of prototype class 30B unique.

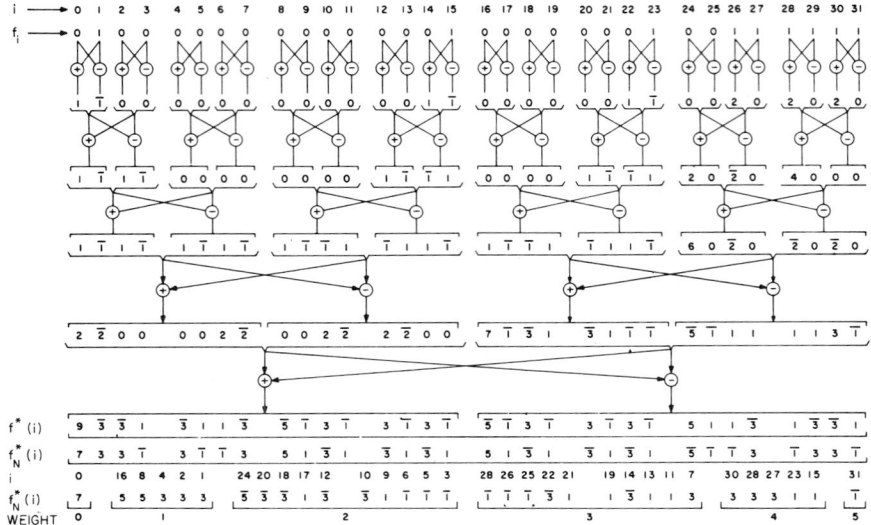

FIG. 2.   Computation of fast Fourier transforms.

5.   Range Translation to Improve Symmetry

The range of f up to now has been restricted to the integers 0 and 1. However, the Fourier expansion f^* represents any (real-valued) function f on $\mathbb{Z}^n$ as a (real) sum of (± 1)-valued kernel functions. The translation $f_N = \frac{1}{2} - f$ is a trivial modification of the range, but it produces a transform f_N^* with greatly improved symmetry properties. We append the subscript N because Ninomiya (1958) first derived this transform by reversing the sign of each component of f^* and adding 2^{n-1} to the first component $f^*(0)$. The following theorem shows that Ninomiya's transform is just the Fourier transform of $(\frac{1}{2} - f)$:

THEOREM 2.2.   Let $f_N(x) = \frac{1}{2} - f(x)$. Then $f_N^*(w) = 2^{n-1}\delta_{w0} - f^*(w)$ for all w in $\mathbb{Z}^n$.

Proof: Let $f_N(x) = -f(x) + \frac{1}{2}$. The linearity properties of the Fourier transform (as an expansion with respect to orthogonal basis functions in real Euclidean 2^n-space) imply that $(f + g)^* = f^* + g^*$. Therefore,

$$f_N^* = (-f)^* + (\tfrac{1}{2})^* = -(f)^* + (\tfrac{1}{2})(1)^*$$

The transform of a constant $(1)^*$ has jth coordinate $\sum_x (-1)^{w \cdot x^t}$; $w = w(j)$. In the proof of orthogonality of the Fourier basis functions, we observed that this expression reduces to $2^n \delta_{j0}$. Therefore, $f_N^* = -f_j^* + 2^{n-1} \delta_{j0}$.

In other words, the jth coefficient $f_N^*(w(j))$ of Ninomiya's coordinate representation for f is equal to f_j^* with its sign reversed for $0 < j < 2^n$, and $f_N^*(0) = 2^{n-1} - f_0^*$ for $j = 0$.

One way in which f_N^* is more symmetric than f^* is that mod 2 complementation of f produces complete sign reversal of all coefficients of f_N^*, but not of f^*. Using real addition, the binary complement of f becomes

$$\bar{f}(x) = 1 - f(x)$$

and its Fourier transform $(\bar{f})^* = (1)^* - f^* = 2^n \delta_{w0} - f^*(w)$. In other words, 2^n is added to the first component of f^* *after* sign reversal in each coordinate. On the other hand, $(\bar{f})_N = \frac{1}{2} - \bar{f} = \frac{1}{2} - (1 - f) = (f - \frac{1}{2}) = (-f_N)$. Therefore, $(\bar{f})_N^* = -f_N^*$ (i.e., the only effect of functional complementation is to multiply the entire transform f_N^* by (-1)). This symmetry property simplifies testing for equivalence of two functions under transformation groups which include functional complementation.

6. Relation to EXCLUSIVE OR Canonical Forms

Switching functions are often defined more compactly by algebraic expressions other than their disjunctive canonical forms. For example, $f(x)$ can be expressed as a sum of partial products of the variables. The variables may be complemented or uncomplemented independently from term to term. When the summation is Boolean, such forms are known as *normal forms*; when the summation is mod 2, they are known as (consistent or inconsistent) Δ-forms (Calingaert, 1961). By repeated use of the three identities below, such forms may be expressed as *real* sums of uncomplemented partial or *reduced product* functions $r_j(x)$ defined by

$$r_j(x) = \prod_{i=0}^{n-1} (x_i)^{b_i(j)} \tag{19}$$

where $b_i(j)$ is the jth component of the radix-two expansion of j, $x^0 = 1$, and $x^1 = x$. The required identities are (using ∨ for Boolean addition and Δ, the equivalent of ⊕, for mod 2 addition:

$$\bar{x} = 1 - x$$
$$x \, \Delta \, y = x + y - 2xy \tag{20}$$
$$x \lor y = x + y - xy$$

Example 6.   (Real expansion with respect to reduced product functions)

$$\begin{aligned}
f(x) &= x_0 \bar{x}_1 \Delta x_2 \\
&= x_0 \bar{x}_1 + x_2 - 2(x_0 \bar{x}_1 x_2) \\
&= x_0 - x_0 x_1 + x_2 - 2x_0 x_2 + 2x_0 x_1 x_2 \\
&= x_2 + x_0 - 2x_0 x_2 - x_0 x_1 + 2x_0 x_1 x_2 \\
&= r_1 + r_4 - 2r_5 - r_6 + 2r_7
\end{aligned}$$

which may be written as

$$f(x) = \sum_j g_j r_j(x) = rg^t \tag{21}$$

where

$$g = (0, 1, 0, 0, 1, -2, -1, 2)$$

and

$$r = (r_0(x), r_1(x), \ldots, r_7(x))$$

The Fourier expansion of $f(x)$ will now be obtained in terms of a general linear transformation relating the functions $r_j(x)$ to the Fourier basis set. The results will incidentally define the vector g and prove that the set of functions $\{r_j(x), 0 \le j < 2^n\}$ is linearly independent over the reals, hence is a basis (although not orthogonal); it follows then that the expansion of $f(x)$ with respect to the r_j basis over the *real* field is also unique. By definition,

$$Q_k(x) = (-1)^{xw^t} = \prod_1^n (-1)^{x_j w_j} = \prod_1^n (1 - 2x_t w_i(k)) \tag{22}$$

If the last expression is expanded as a real sum of partial products of (un-complemented) variables, only those r_j will appear which are factors of r_k; these will have as coefficients the integers (-2) to the power $\sum w_i(j) = |w(j)|$.

Now r_j is a factor of r_k if and only if $w(j) \le w(k)$ (i.e., $w_i(j) \le w_i(k)$ for $1 \le i \le n$). Calingaert (1961) has shown that $w(j) \le w(k)$ if and only if the binomial coefficient $\binom{k}{j}$ has odd parity. The expression for $Q_k(x)$ may therefore be written as[1]

$$Q_k(x) = \sum_{j=0}^{2^n-1} \left| \binom{k}{j} \right|_2 (-2)^{|w(j)|} r_j(x) \tag{23}$$

Calingaert (1961) defined A as the 2^n by 2^n matrix having $|\binom{k}{j}|_2$ as its (k,j)th element. The diagonal matrix having $(-2)^{|w(j)|}$, $0 \le j < 2^n$ as its jth diagonal element is denoted by W. In matrix form, $q(x) = r(x) WA^t$ where

---

[1] Here $|\binom{k}{j}|_2$ denotes the residue, mod 2, of the binomial coefficient.

the kth component of q or r is the function $Q_k(x)$ or $r_k(x)$, respectively, and the equation is an identity in the variables (i.e., true for each configuration of x).

The disjunctive canonical form of $r_j(x)$ is $r_j(x) = \sum b_{ji} p_i(x)$ where $b_{ji} = 1$ if and only if $w(i) \le w(j)$. Hence, $b_{ji} = |\binom{i}{j}|_2$ which is the (j, i)th element of A. Therefore

$$r_j(x) = \sum_i \left|\binom{i}{j}\right|_2 p_i(x) \tag{24}$$

or in matrix form,

$$r(x) = p(x)A$$

where the jth column of A is the vector representation of the disjunctive canonical form of $r_j(x)$. Now A is triangular with determinant 1 since $\binom{i}{i} = 1$ and $\binom{i}{j} = 0$ for $j > i$. Therefore, the rows of A are linearly independent and its inverse exists:

$$p(x) = r(x)A^{-1} \tag{25}$$

Furthermore, for every configuration of the variables x,

$$Q(x) = r(x)WA^t = p(x)AWA^t \tag{26}$$

$$f^*(w) = f(x)Q_w(x) \tag{27}$$

The above equation is an identity between the ith columns of AWA^t and Q, which proves the following theorem:

THEOREM 2.3.

$$Q = AWA^t \qquad \text{(over the reals)} \tag{28}$$

where

$$(A)_{ij} = \left|\binom{i}{j}\right|_2$$

and

$$(W)_{ij} = (-2)^{|w(i)|} \delta_{ij}$$

The linear transformations relating the bases $r(x)$, $p(x)$, and $q(x)$, or the coefficient vectors g, f, and f^* of the expansion of an arbitrary binary (or real) valued function with respect to these bases, are shown diagrammatically below in Fig. 3 (see Lechner, 1963a,b).

The transformation $f = gA^t$ (over the reals) is similar to the equation

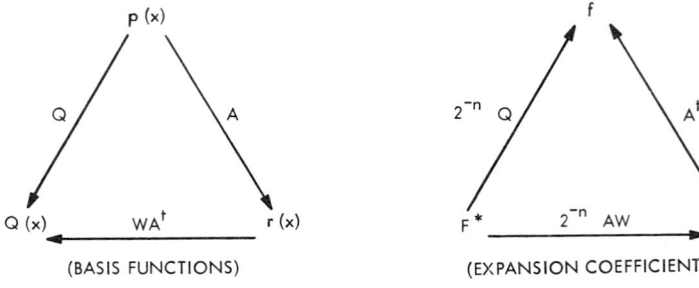

(BASIS FUNCTIONS)   (EXPANSION COEFFICIENTS)

FIG. 3.   Basis and coefficient transformations. $(u \overset{T}{\to} v$ implies $v = u\mathsf{T}$ and $u = v\mathsf{T}^{-1}$.)

$f = h\mathsf{A}^{t}$ (over $\mathbb{Z}_2$) which relates f to the expansion coefficients h of Calingeart's reduced product or Δ-sum canonical form (over $\mathbb{Z}_2$). Over $\mathbb{Z}_2$, A is self-inverse, so $h = f\mathsf{A}^{t}$ (mod 2); over the reals, however, $\mathsf{A} \neq \mathsf{A}^{-1}$

Example 7.   The Fourier transform f^* will be derived from the Δ-form expansion vector $g = (0, 1, 0, 0, 1, -2, -1, 2)$ of the preceding example of this section. From the diagram above, the appropriate transformation is

$$2^{-n} \cdot f^* = (g\mathsf{W}^{-1})\mathsf{A}^{-1} = y\mathsf{A}^{-1} = (\mathsf{A}^{t})^{-1}y^{t} \qquad (29)$$

The final product is expanded below (y having been precomputed by multiplying corresponding components of g and W^{-1}).

$$2^{-n} \cdot f^* =
\begin{bmatrix}
1 & -1 & -1 & 1 & -1 & 1 & 1 & -1 \\
0 & 1 & 0 & -1 & 0 & -1 & 0 & 1 \\
0 & 0 & 1 & -1 & 0 & 0 & -1 & 1 \\
0 & 0 & 0 & 1 & 0 & 0 & 0 & -1 \\
\hline
0 & 0 & 0 & 0 & 1 & -1 & -1 & 1 \\
0 & 0 & 0 & 0 & 0 & 1 & 0 & -1 \\
0 & 0 & 0 & 0 & 0 & 0 & 1 & -1 \\
0 & 0 & 0 & 0 & 0 & 0 & 0 & 1
\end{bmatrix}
\begin{bmatrix}
0 \\ -\tfrac{1}{2} \\ 0 \\ 0 \\ -\tfrac{1}{2} \\ -\tfrac{1}{4} \\ -\tfrac{1}{4}
\end{bmatrix}
=
\begin{bmatrix}
\tfrac{1}{2} \\ -\tfrac{1}{4} \\ 0 \\ \tfrac{1}{4} \\ 0 \\ -\tfrac{1}{4} \\ 0 \\ -\tfrac{1}{4}
\end{bmatrix}$$

$$(30)$$

The last column checks with the result $f^* = (4, -2, 0, 2, 0, -2, 0, -2)$ obtained by the direct transformation $f^* = f\mathsf{Q}$.

7.   Historical Note

Muller (1954) was apparently the first to use this transformation on switching functions. He derived the Fourier kernel functions and classified all

4-argument truth tables eight ways according to their empirically-determined expansion coefficients with respect to this basis. The Fourier basis vectors are closely related to the Reed–Muller class of error-correcting codes (Reed, 1954). Muller's pioneering work inspired Ninomiya to undertake a more analytic study of the properties of the Fourier expansion (without explicitly characterizing it as such). Ninomiya formalized Muller's concept of functional equivalence and determined many of its properties. Both Muller and Ninomiya were apparently unaware of the unifying principle which the restricted affine group over $\mathbb{Z}_2$ (see Section IV) provides for their notion of equivalence.

This author's interest in harmonic analysis was motivated chiefly by the extensive thesis research of Ninomiya (1958). Additional inspiration was drawn from earlier applications of Fourier analysis to coding theory, particularly by Zierler (1960) and Wells (1960). Berlekamp, in a private communication (1969) indicated that Ninomiya's prototype equivalence relation is now being applied to coding theory by Berlekamp *et al.* who have identified all 48 prototype classes for $n = 5$ (see Section IV,C). Menger (1969) considered a different type of transform for function spaces whose range was not a subset of the complex field, but the finite field $GF(p^n)$. The classical properties of Fourier transforms are not valid, and the utility of Menger's transform has yet to be determined.

Recent applications of the Rademacher–Walsh transform to communication theory have aroused much interest among engineers. This transform is a permutation of the Hadamard transform applied to a 2^n samples of a real-valued function (Whelchel *et al.*, 1968). If the sampled function is regarded as a function of the binary n-tuples which are radix-two expansions for the indices of their sample points, the theory developed in this chapter is applicable. However, our point of view is only relevant for functions with a discrete (e.g., 2-valued) range, because partitioning the function domain into level sets is essential to further theoretical development (see Section V,D).

B. SURVEY OF CLASSICAL PROPERTIES

This section proves three important properties of Fourier transforms for the domain $\mathbb{Z}^n$. Other miscellaneous properties are also mentioned herein. Discussion of the invariant properties of Fourier transforms under affine operators will be deferred until Section IV,C. The three major theorems presented herein are the Convolution theorem, the Quotient Group Character theorem, and the Poisson Summation theorem. Their proofs are simple consequences of the transform definition for the domain $\mathbb{Z}^n$ and could easily be extended to a direct sum of arbitrary cyclic groups (every finite Abelian group is isomorphic to one of this type). However, direct proofs for $\mathbb{Z}^n$ should expedite wider use of these new computational tools for combinational logic analysis. Most of this material was adapted from the text by Loomis (1953). Several recent books contain more extensive discussions of group theoretic applications (Rudin, 1962; Hewitt *et al.*, 1963).

1.   The Convolution Theorem

The convolution theorem is one of the most powerful tools of linear functional analysis. In 1963 the author completed an (unpublished) survey which interpreted abstract harmonic analysis in the context of switching theory. The material in this section is based on that survey. Applications of the convolution theorem were elusive until 1969 when the algorithm described in Section III was discovered. The convolution theorem for functions on Z^n can be generalized to arbitrary finite Abelian groups (direct sums of cyclic groups, or mixed radix number representations) in a straightforward way.

DEFINITION (*Convolution Sum*).   Let X be an n-dimensional vector space over the 2-element field. The *convolution sum* $f * g$ for any given pair of real- (or integer-)valued functions f and g on X is a real- (or integer-) valued function defined at every point c of X by the following sum over all elements of X:

$$(f * g)_c = \sum_{x \in X} f(x \oplus c)g(x) \tag{31}$$

Under certain summability conditions (always satisfied for finite domains), the convolution sum of a real- or complex-valued function defined on an Abelian group is identical to the inverse Fourier transform of the component-wise product of their Fourier transforms. Symbolically, $f * g = (f^* \cdot g^*)^*$. For any finite domain, this convolution theorem is a simple consequence of the transform definition. In ordinary Fourier analysis, this property eliminates a large fraction of the multiplications required for a convolution sum. In the direct-sum approach, this number grows as the square of the number of the points for which f and g both have significant nonzero values. Computational advantages of this theorem also occur for finite groups like Z_2^n, but for completely different reasons.

We will state and prove this theorem for the domain Z^n using transform notation of the previous section.

THEOREM 2.4 (*Convolution Theorem*).   In the notation of the preceding sections, let f and g be any two integer, real- or complex-valued functions on Z^n, let $(f * g)_c$ denote their convolution sum evaluated at $x = c$, and let

$(f*g*)$ denote the componentwise product $f*(w)g*(w)$ of their respective Fourier transforms $f*$ and $g*$. Then, for each c in $\mathbb{Z}^n$, $(f* * g)_c = 2^{-n}(f*g*)*$ evaluated at the point $x = c$.

Proof: Take the product of the two transform definitions

$$(f*g*)_w = \sum_x (-1)^{wx^t} f(x) \sum_y (-1)^{wy^t} g(y)$$

$$= \sum_{x,y} (-1)^{w(x+y)^t} f(x)g(y) \tag{32}$$

Now let $y = x \oplus c$, sum over x with c fixed, then sum over c:

$$(f*g*)_w = \sum_c (-1)^{wc^t} \sum_x f(x)g(x \oplus c)$$

$$= \sum_c (-1)^{wc^t} (f * g)_c = (f * g)_w^* \tag{33}$$

The final summation is merely the Fourier transform of $f * g$, evaluated at w. Transposing sides of this identity in w and taking the inverse transform on both sides produces the following result which is an identity for all c in $\mathbb{Z}^n$:

$$(f * g)_c = 2^{-n} \sum_w (-1)^{wc^t}(f*g*)_w = 2^{-n}(f* \cdot g*)_c^* \tag{34}$$

Example 8.   Let $f = (1, 1, 1, 0)$ and $g = (0, 1, 1, 1)$; $(f * g)$ is calculated directly below:

x	$f(x)$	$g(x \oplus c)$ versus c: $c = (0, 0)$	$(0, 1)$	$(1, 0)$	$(1, 1)$
$(0, 0)$	1	0	1	1	1
$(0, 1)$	1	1	0	1	1
$(1, 0)$	1	1	1	0	1
$(1, 1)$	0	1	1	1	0
$\sum f(x)g(x \oplus c) =$		2	2	2	3

The convolution theorem obtains the same result as follows:

$$f* = fQ = (1, 1, 1, 0)Q = (3, 1, 1, -1) \tag{35}$$

$$g* = gQ = (0, 1, 1, 1)Q = (3, -1, -1, -1) \tag{36}$$

$$(f*g*) = (9, -1, -1, 1)$$

Taking the inverse transform produces an identical result:

$$2^{-n}(f*g*)Q = \tfrac{1}{4}(9, -1, -1, 1)\begin{bmatrix} 1 & 1 & 1 & 1 \\ 1 & -1 & 1 & -1 \\ 1 & 1 & -1 & -1 \\ 1 & -1 & -1 & 1 \end{bmatrix} = (2, 2, 2, 3) \quad (37)$$

2. Quotient Group Character Theorem

Suppose V is a k-dimensional subspace of $\mathbb{Z}^n$ (an additive subgroup of order 2^k). Then $\mathbb{Z}^n$ has a disjoint decomposition into 2^{n-k} cosets $V + c$, one of which (for $c = 0$) is V itself. These cosets form a group under (mod 2) addition, denoted $\mathbb{Z}^n/V$, called the *quotient group* of $\mathbb{Z}^n$ (mod V). The theorem to be proved states that any function which is constant on each coset of V can be expanded in terms of a subset of 2^{n-k} Fourier basis functions called "quotient group characters," defined by $\mathrm{QG}(\mathbb{Z}^n/V) = \{Q_w(x) : Q_w(x) = 1$ for all $x \in V\}$. This theorem is a classical result of group character theory (Littlewood, 1940). Loomis (1953) gives an abstract proof. For a specific type of subspace, a direct proof is given herein, to bring out the significance of this theorem for our later applications.

This theorem will be useful in computing the Fourier transform of any function which is known *a priori* to be constant on each coset of a subspace of V. (In particular, the functions corresponding to k-cells or subcubes of $\mathbb{Z}^n$ are of this form.)

A "degenerate" function is one that depends on fewer than n linear combinations of its arguments. A corollary of the theorem states that degenerate functions (and their arguments) can be identified by computing the *rank* of the set of vectors w corresponding to the nonzero Fourier coefficients $f*(w)$.

THEOREM 2.5 (Quotient Group Characters). Let V be a subspace of $\mathbb{Z}^n$. Then the subset of Fourier basis functions $Q_V = \{Q_w(x) : xw^t = 0,$ for all $x \in V\}$ is a complete orthogonal basis for the subspace of $\mathscr{F}$ consisting of all functions that are constant on cosets of V. No other functions are generated by this basis. If V has dimension k, then Q_V consists of just those 2^{n-k} functions $Q_w(x)$ for which w is orthogonal to all $x \in V$. If V has a basis consisting of k unit vectors, then $Q_w \in Q_V$ iff w is in the space generated by that subset of $n - k$ unit vectors which are *not* in the basis for V. (In either case, the set of all vectors w_1 such that $w_1 x^t = 0$ for all x in V is called the complementary subspace or *nullspace* V' of V.)

Proof: For every k-dimensional subspace $V \subseteq Z^n$, there is a (nonunique) set S of 2^{n-k} elements (called coset leaders) $\{y_i \in Z^n : y_i + V \neq y_j + V$ unless $i = j\}$. Here $y + V = \{y \oplus z : z \in V\}$. Furthermore, every x in Z^n has a unique representation as $x = y \oplus z$, y in S and z in V. In terms of this decomposition, the Fourier transform of f is defined as follows:

$$f^*(w) = f^*(w_1) = \sum_{z \in V} \sum_{y \in S} f(y \oplus z)(-1)^{wy^t}(-1)^{wz^t} \qquad (38)$$

If f is constant on cosets of V, then $f(y \oplus z) = f(y)$ for all z in V and any y in S. Furthermore, if $w = w_1 \oplus w_2$ with $w_1 \in V'$ and $w_2 \in V$, then $w_1 z^t = 0$, and dimension (V') + dimension $(V) = n$. Therefore,

$$f^*(w) = \left[\sum_{z \in V} (-1)^{wz^t} \right] \left[\sum_{y \in S} f(y)(-1)^{wy^t} \right] = (2^k \delta_{w_2 0})(f \sqcap S)_w^* \qquad (39)$$

The first term arises because $z \to wz^t$ is a homomorphism from V into Z for every w (see the proof of orthogonality in Section II.A.3). Therefore, the first sum is zero unless $w \in V'$, 2^k for $w \in V'$. The second term $(f \sqcap S)^*$ denotes the Fourier transform of f as a function on the set S alone. The symbol $\sqcap$ denotes the *restriction* of f to the subset S of Z^n. Since S has 2^{n-k} elements, the function f restricted to S is a vector of dimension 2^{n-k}. Since $\delta_{w_2 0} = 0$ unless $w_2 = 0$, the only nonzero Fourier coefficients $f^*(w)$ in the transform of f are those for which $w = w_2 \oplus w_1 = w_1$ is in V'. In other words $f^* = 0$ except on V'.

So far, we have not used the condition that V has a basis consisting of unit vectors. In this case, S can be identified with V', because V' contains all the coset leaders of minimum weight.

To prove the converse (all functions generated by the orthogonal set $\{Q_w : w \in V'\}$ are constant on V-cosets), it is sufficient to show that every basis function $Q_w(x)$, $w \in V'$ (hence every linear combination of them) is constant on cosets of V. Each V-coset is of the form $y + V = \{y \oplus z : z \in V\}$, $y \in S$. On this coset, $Q_{w_1}(x) = (-1)^{w_1 x^t} = (-1)^{w_1(y+z)^t} = (-1)^{w_1 y^t}$ because $w_1 z^t = 0$ for every z in V if w_1 is in V'. Therefore, $Q_w(x)$ has the same value $(-1)^{w_1 y^t}$ on every point of $y + V$. Furthermore, there are 2^{n-k} linearly independent functions $Q_{w_1}(x)$, and these are a sufficient basis for the space of all functions on cosets of V.

Example 9: Let $n = 3$ and let V be the subspace of Z^3 consisting of all 3-tuples of even parity. That is, $V = \{(0, 0, 0), (0, 1, 1), (1, 0, 1), (1, 1, 0)\}$. The cosets of V are V itself and the set $V' = \{(0, 0, 1), (0, 1, 0), (1, 0, 0), (1, 1, 1)\}$. The subspace of functions that are constant on V and V' has a basis consisting of the function $f(x) = 1$ on V and the constant function $f = 1$.

Take the transform of f:

$(1, 0, 0, 1, 0, 1, 1, 0)\cdot$

$$
\begin{bmatrix}
1 & 1 & 1 & 1 & 1 & 1 & 1 & 1 \\
1 & -1 & 1 & -1 & 1 & -1 & 1 & -1 \\
1 & 1 & -1 & -1 & 1 & 1 & -1 & -1 \\
1 & -1 & -1 & 1 & 1 & -1 & -1 & 1 \\
\hline
1 & 1 & +1 & +1 & -1 & -1 & -1 & -1 \\
1 & -1 & +1 & -1 & -1 & 1 & -1 & 1 \\
1 & 1 & -1 & -1 & -1 & -1 & 1 & 1 \\
1 & -1 & -1 & 1 & -1 & 1 & 1 & -1
\end{bmatrix} = (4,0,0,0,0,0,0,4) \qquad (40)
$$

In other words, the only nonzero transform coefficients are those for which $w = (0, 0, 0)$ and $w = (1, 1, 1)$. Each of these vectors satisfies $wx^t = 0$ for all x in V, as stated in the theorem.

COROLLARY 2.1. Any function on $\mathbb{Z}^n$ whose nonzero transform coefficients occur for a subset $\{w : f^*(w) \neq 0\}$ of rank $n - k$ can be defined as a function of at most $(n - k)$ binary variables $(y_1, \ldots, y_{n-k})$ which are linear combinations of $x_1, \ldots, x_n$ (over $\mathbb{Z}$).

Proof: Let V' be the subspace generated by $\{w : f^*(w) \neq 0\}$. Since a subspace V is the nullspace of its nullspace V', the subspace V' must be the nullspace of some k-dimensional subspace V, and the preceding theorem implies that $f(x)$ is constant on cosets of V. That is, $f(x) = f(u)$ for any $x \in (u + V)$ and some unique $u \in S$ (the set of coset leaders for V). If B is an $(n - k)$ by n matrix whose rows are a basis for V', then every $y \in V'$ is a linear combination yB of B-rows for some $y \in \mathbb{Z}^{n-k}$. Now let $C = \begin{bmatrix} B \\ \cdots \\ A \end{bmatrix}$ where A is a basis for V. Then every $v \in V$ can be expressed as $v = z$A for some $z \in \mathbb{Z}^k$. Finally, $x = (y_1, \ldots, y_{n-k}; z_1, \ldots, z_k) \cdot C$ so that $(y; z) = xC^{-1}$. Let D be the first $(n - k)$ columns of C^{-1}. Then $y = xD$ defines 2^{n-k} vectors $y \in \mathbb{Z}^{n-k}$ in $1 : 1$ correspondence with elements of V'. The correspondence $g(y) = g(xD) = f(x)$ defines g (and f) uniquely for any f which is constant on cosets of V; therefore, $f(x)$ has been expressed as a (degenerate) function of $n - k$ essential arguments.

Example 10. The function $f(x) = 1$ on V, 0 elsewhere of the preceding example has a transform with only two nonzero coefficients: $f^*(w) = 4$ for $w = (0, 0, 0)$ and $(1, 1, 1)$. This set has rank $n - k = 1$; therefore, by the corollary, $f(x)$ is a function of one variable y. This is easily verified, because $x \in V$ iff $(xa^t) = 0$, where $a = (1, 1, 1)$. Therefore, $y = xa^t$ is a sufficient linear function, and $f(x) = \bar{y} = xa^t \oplus 1$.

COROLLARY 2.2. Let $V + c$ be any coset of any k-dimensional subspace V of Z^n, with $c \in V'$. Then the characteristic function $v_c(x)$ ($v_c(x) = 1$ on $V + c$ and 0 elsewhere) has the Fourier transform

$$v_c*(w) = 2^k \delta_{w_2 0}(-1)^{w_1 c^t} \qquad (41)$$

i.e., v_c* is 0 except on V', where it has magnitude 2^k and sign $(-1)^{w_1 c^t}$.

Proof: Following the theorem proof, let $w = w_1 \oplus w_2$, $x = y \oplus z$, with w_1 and $y \in V'$, w_2 and $z \in V$. Then $v_c(x) = \delta_{yc}$ and

$$v_c*(w) = \sum_z (-1)^{w_2 z^t} \sum_y (-1)^{w_1 y^t} \delta_{yc} = 2^k \delta_{w_2 0}(-1)^{w_1 c} \qquad (42)$$

COROLLARY 2.3. The characteristic function $v(x)$ of a subspace V of dimension k has the Fourier transform $v*(w) = 2^k$ on V' and 0 elsewhere.

Proof: Let $c = 0$ in the preceding corollary.

Example 11. The function $f(x)$ in the preceding example is unity on the subspace $V = \{(0, 0, 0), (0, 1, 1), (1, 0, 1), (1, 1, 0)\}$. Therefore its transform should be

$$f*(w) = v_c*(w) = 2^2 \delta_{w_2 0}(-1)^{w_1 c^t} \qquad (43)$$

Here $c = 0$ and $w_1 = (1, 1, 1)$ and $(0, 0, 0)$ are the only two vectors such that $w_1 x^t = 0$ for all x in V. Therefore, $f*(w) = 0$ for $w_2 \neq 0$ ($w \neq w_1$) and $f*(w) = 4$ for $w = w_1$. This is the same result calculated directly in the example following the main theorem.

The preceding theorem and its corollaries are used iteratively in the algorithms of Section III. Therefore, its computational implication will be mentioned here. By the quotient group character theorem, if a function is constant on cosets of V, then it can be sampled once on each coset, and its transforms can be generated on V'. In most of our applications, both V and V' will have a basis of unit vectors, and vectors y in V' are in $1 : 1$ correspondence with $(n - k)$-tuples. In this case, an algorithm to generate a list of

2^{n-k} pointers $i(j)$ such that $x(i(j)) \in V'$ can be constructed in approximately 2^{n-k} indexed add operations. A table lookup involving these 2^{n-k} components will permit $f_{i(j)}$ to be extracted from its 2^n-dimensional array and packed into a 2^{n-k}-dimensional array, after which its transform can be computed in $(n-k)2^{n-k+1}$ rather than $n2^{n+1}$ operations.

3. The Poisson Summation Theorem

Another important property of Fourier transforms, which plays only a minor role in classical applications but plays a major role in the theory of group characters, is the group-theoretic analog of the *Poisson summation theorem* (Loomis, 1953, p. 152). This theorem states that the average of a function's values over a subgroup H of the domain G may be computed by summing its Fourier transform over the quotient group G/H, with suitable normalization. It is far from obvious that the Poisson summation theorem for the real domain R is a specialization of this general group-theoretic statement.

There is a direct correspondence between subspaces of $\mathbb{Z}^n$ and additive subrings of R^n. This correspondence provides additional insight on the analogous roles of harmonic analysis for functions of binary n-tuples and functions on R^n. We will first state and interpret this theorem for the real domain R.

If $f(x)$ is a continuous bounded function whose squared norm ($\int |f|^2 \, dx$) is finite on R, then its Fourier transform $T_f(y)$ is defined for every y in R by

$$T_f(y) = \int_{-\infty}^{\infty} f(x)\overline{\phi_y(x)} \, dx \tag{44}$$

where $\phi_y(x) = \exp(2\pi i y x)$ is a character for every y in R. The Poisson summation formula states that, for any positive real number a and a suitably restricted function $f(x)$,

$$\sum_{n \in I} f(na) = (1/a)\sum_{k \in I} T_f(k/a) \tag{45}$$

where I is the ring of *all* integers.

One classical application of this theorem has been to convert a slowly convergent power series into a rapidly convergent one, by going to the transform domain. In our application, a sum over a rather large set of 2^k subspace elements $x \in V$ will be converted into a sum over the factor group $\mathbb{Z}^n/V$ whose size 2^{n-k} is inversely proportional to the size of V. To verify that this formula (for the real domain) actually makes a statement about subgroups and quotient groups, define $g(x) = \sum_{n \in I} f(x + na)$. Then $g(x)$ is periodic with period a, so $g(x)$ may be regarded as a function on the group

R/aR of real numbers under mod a addition. The set of numbers $\{ka, k \in I\}$ is a subgroup $H = aI$ of the group $G = R$, since it is closed under addition, and the domain of $g(x)$ is the quotient group $G/H = R/aI$ of residue classes of real numbers under addition modulo a. The character group of $G = R$ is

$$\chi_G = \{\phi_y : \phi_y(x) = \exp(2\pi i y x), \quad y \in R\} \tag{46}$$

The character group of $G/H = R/aI$ is defined (without proof) as:

$$\chi_{G/H} = \{\psi_k : \psi_k(x) = \exp(2\pi i (k/a) x), \quad k \in I\} \tag{47}$$

Since

$$\psi_k(x) = \phi_{k/a}(x) \tag{48}$$

$\chi_{G/H}$ may be identified with the following subset of χ_G :

$$\chi_{G/H} = \{\phi_y \in \chi_G : y = k/a \quad \text{for some} \quad k \text{ in } I\} \tag{49}$$

Note that $\chi_{G/H}$ consists of exactly those characters in χ_G which are constant on the subgroup $H = aI$ of all integral multiples of a in R. The next theorem is the equivalent statement for the finite Abelian group Z^n.

THEOREM 2.6 (Poisson Summation). If V is any subspace of Z^n, V' is the nullspace of V, and f is a real- or complex-valued function on Z^n, then

$$\sum_{x \in V} f(x) = 2^{k-n} \sum_{w \in V'} f^*(w) \tag{50}$$

We shall first prove a generalization of this theorem, which defines the sum of $f(x)$ over a *coset* $V + c$ of V. In Section III, this generalization will be derived from the convolution theorem. It provides a direct answer to the question "Which cosets of V are implicants of f or $\bar{f}$?".

Proof: Define $g(x) = \sum_{z \in V} f(x \oplus z)$. Note that g is constant on cosets of V since, for any u in V, $\sum f(x \oplus z) = \sum f(x \oplus u \oplus z)$ summed over V. As in the previous section, let $x = y \oplus z$, $w = w_1 \oplus w_2$ denote the unique decomposition of x and w into elements of S or V' and V, respectively. By the quotient group character theorem (Section II,B,2), the transform of g is defined by

$$g^*(w) = 2^k \delta_{w_2 0} \sum_{y \in S} (-1)^{w_1 y^t} g(y)$$

$$= 2^k \delta_{w_2 0} \sum_{y \in S} \sum_{z \in V} (-1)^{w_1 y^t} f(y \oplus z)$$

$$= 2^k \sum_{x \in Z^n} (-1)^{w x^t} \delta_{w_2 0} f(x) \tag{51}$$

using the identity which follows from the definition of y, z, w_1, and w_2: $\delta_{w_2 0}(-1)^{w_1 y^t} = \delta_{w_2 0}(-1)^{w x^t}$ for $x = y \oplus z$, $w = w_1 \oplus w_2$, and $z w_1^t = 0$.

Now multiply both sides by $(-1)^{wc^t}$ and sum over $\mathbb{Z}^n$ with respect to w (c is a constant vector in $\mathbb{Z}^n$):

$$\sum_w g^*(w)(-1)^{wc^t} = 2^k \sum_{w \in \mathbb{Z}^n} (-1)^{wc^t}\delta_{w20} \sum_{x \in \mathbb{Z}^n} f(x)(-1)^{wx^t} \tag{52}$$

The left side is $(g^*)^* = 2^n g(c)$, and the right side is

$$2^k \sum_{w \in \mathbb{Z}^n} \delta_{w20}(-1)^{wu^t}f^*(w) = 2^k \sum_{w_1 \in V'} (-1)^{w_1 u^t}f^*(w_1) \tag{53}$$

This proves the following *generalization* of the Poisson summation theorem:

$$2^n \sum_{z \in V} f(z \oplus c) = 2^{k-n} \sum_{w_1 \in V'} (-1)^{w_1 c^t}f^*(w_1) \tag{54}$$

The special case $c = 0$ is the statement of the theorem.

4. Miscellaneous Properties

Other properties of Fourier transforms on finite Abelian groups are either more specialized or will play only a minor role in the applications herein. Among the latter are the *Parseval identity* $(f, g) = (f^*, g^*)$ which relates inner products of functions on $\mathbb{Z}^n$ to inner products of their transforms. Another is *Bessel's inequality*, which states the monotone convergence of the mean squared error in approximating a function as a linear sum over a subset of the characters on G, as the subset is expanded to include more and more basis functions. A third is the *mean squared error minimization* property, which states that in any approximation of a function as a linear combination of some subset of the character functions, the Fourier transform coefficients are optimal in the sense of yielding minimum mean squared error.

All of these properties are important on infinite-dimensional function spaces, but are trivial consequences of the orthogonality property of the characters as a basis for a *finite*-dimensional vector space. In particular, note that the mean squared error minimization property is not significant for (0, 1)-valued functions, although it is for real-valued functions. When a binary function is approximated by a partial sum of its Fourier basis vectors, the error at any point is a multiple of 2^{-n}, and its sum of squares depends primarily on the *number* of points at which $f(x)$ disagrees with its "approximation."

C. FUNDAMENTAL THEOREM ON INVARIANCE

The theorems presented in this section were first applied to combinational logic by Ninomiya (1958). Their present form was introduced by Lechner (1968). The invariant properties of Fourier transforms of 2-valued functions under the restricted affine group (RAG) (to be defined in Section IV) clearly show that harmonic analysis provides the best context within which to analyze the equivalence relations induced by RAG and its subgroups (see Section IV,A,3).

The transform f_N^* of the translated function $f_N(x) = \frac{1}{2} - f(x)$ is the most convenient form from which to analyze prototype equivalence relations, because it is symmetric under functional complementation: $(\bar{f})_N = (\frac{1}{2} - \bar{f}) = -f_N$. The transform f^* of f and the invariants of Golomb (1959) possess closely related properties but do not exhibit them in such a convenient manner. Calculation of f^* or f_N^* by the fast Fourier transform algorithm of Section II,A,4 is straightforward.

Prototype transformations (elements of the group RAG defined in Section IV) apply affine transformations $y = xA \oplus b$ to the arguments of a function $g(y)$ and add linear polynomials $(xa^t \oplus c)$ to its output to produce another function $f(x) = g(xA \oplus b) \oplus xa^t \oplus c$ in the same prototype class. Both f_N^* and f^* have the same invariant properties under affine *domain* encodings $y = xA \oplus b$. However, the effect of adding $(xa^t \oplus c)$ to the function *output* is different for both. The first theorem on invariance will consider affine domain encodings alone. In this section, we will use $\oplus$ to represent mod 2 addition (the symmetric difference operation) on the *range* of a function, to avoid confusion with *real* addition of functions and their Fourier transform expansions. Of course, addition on the domain $X = \mathbb{Z}^n$ is *always* interpreted mod 2. However, the symbol $+$ is used in the exponent of (-1) because real and mod 2 addition have the *same* effect.

THEOREM 2.7 (Domain Encodings). Let $g(x)$ be a known function from X into $\mathbb{Z}$. If $f(x)$ is defined, for all x, by $f(x) = g(xA \oplus c)$, then the Fourier transforms of f and g are related as follows:

$$f^*(w) = (-1)^{wd^t}g^*(wA^{-t}); \qquad g^*(w) = (-1)^{wc^t}f^*(wA^t) \qquad (55)$$

$$f_N^*(w) = (-1)^{wd^t}g_N^*(wA^{-t}); \qquad g_N^*(w) = (-1)^{wc^t}f_N^*(wA^t) \qquad (56)$$

where A^{-t} is the transposed inverse of A and $d = cA^{-1}$.

Proof: In the definition of f^*, substitute $(y \oplus c)A^{-1}$ for x, $g(y)$ for $f(x)$, and sum over y instead of x.

$$f^*(w) = \sum (-1)^{wx^t}f(x)$$

$$f^*(w) = \sum (-1)^{wA^{-t}(y+c)^t}g(y)$$

$$= (-1)^{wA^{-t}c^t}\sum_y (-1)^{wA^{-t}y^t}g(y)$$

$$= (-1)^{wd^t}g^*(wA^{-t}) \qquad (57)$$

Replacing w by vA^t gives the converse relation after rewriting w for v and transposing:

$$f^*(vA^t) = (-1)^{vc^t}g^*(v) \tag{58}$$

$$g^*(w) = (-1)^{wc^t}f^*(wA^t) \tag{59}$$

The corresponding relation between f_N^* transforms is determined by substituting in the above expressions the definition

$$f_N^*(w) = (\tfrac{1}{2} - f)^* = 2^{n-1}\delta_{w0} - (-1)^{wd^t}g^*(wA^{-t})$$

where $\delta_{w0} = 1$ if $w = 0$ and 0 otherwise. Since

$$2^{n-1}\delta_{w0} = (-1)^{wd^t}2^{n-1}\delta_{w0}$$

for any d,

$$f_N^*(w) = (-1)^{wd^t}[2^{n-1}\delta_{w0} - g^*(wA^{-t})] = (-1)^{wd^t}g_N^*(wA^{-t}) \tag{60}$$

from which the converse is obtained as before.

Example 12. Let $g(x) = x_1x_2 \vee x_3$; that is, $g^{-1}(1) = \{1, 3, 5, 6, 7\}$. Suppose $f(x) = g(xA \oplus c)$. Then $f(x) = 1$ iff $g(y) = 1$ for $y = xA \oplus c$. The following table computes y from x for the specific transformation

$$A = \begin{bmatrix} 1 & 1 & 0 \\ 0 & 1 & 1 \\ 0 & 0 & 1 \end{bmatrix}, \qquad c = (1, 10) \tag{61}$$

i	$x(i)$	xA	$y(j) = xA \oplus c$	j	$g(y) = f(x)$
0	(0, 0, 0)	(0, 0, 0)	(1, 1, 0)	6	1
1	(0, 0, 1)	(0, 0, 1)	(1, 1, 1)	7	1
2	(0, 1, 0)	(0, 1, 1)	(1, 0, 1)	5	1
3	(0, 1, 1)	(0, 1, 0)	(1, 0, 0)	4	0
4	(1, 0, 0)	(1, 1, 0)	(0, 0, 0)	0	0
5	(1, 0, 1)	(1, 1, 1)	(0, 0, 1)	1	1
6	(1, 1, 0)	(1, 0, 1)	(0, 1, 1)	3	1
7	(1, 1, 1)	(1, 0, 0)	(0, 1, 0)	2	0

Thus, the function $f(x) = 1$ for $i = 0, 1, 2, 5, 6$.

Using the algorithm of Section II,A,4 we obtain g^* and f^*:

$$
\begin{aligned}
g &= (0, \quad 1, \quad 0, \quad 1, \quad 0, \quad 1, \quad 1, \quad 1) \\
h = g(Q_1 \times I^{[2]}) &= (1, -1, \quad 1, -1, \quad 1, -1, \quad 2, \quad 0) \\
r = h(I^{[1]} \times Q_1 \times I^{[2]}) &= (2, -2, \quad 0, \quad 0, \quad 3, -1, -1, -1) \\
g^* = r(I^{[2]} \times Q_1) &= (5, -3, -1, -1, -1, -1, \quad 1, \quad 1) \\
f &= (1, \quad 1, \quad 1, \quad 0, \quad 0, \quad 1, \quad 1, \quad 0) \\
h = f(Q_1 \times I^{[2]}) &= (2, \quad 0, \quad 1, \quad 1, \quad 1, -1, \quad 1, \quad 1) \\
r = h(I^{[1]} \times Q_1 \times I^{[1]}) &= (3, \quad 1, \quad 1, -1, \quad 2, \quad 0, \quad 0, -2) \\
f^* = r(I^{[2]} \times Q_1) &= (5, \quad 1, \quad 1, -3, \quad 1, \quad 1, \quad 1, \quad 1)
\end{aligned} \tag{62}
$$

The theorem states that $g^*(w) = f^*(w \cdot A^t)(-1)^{wc^t}$. The image wA^t and the sign change factor $(-1)^{wc^t}$ are calculated below for each w, with

$$A^t = \begin{bmatrix} 1 & 0 & 0 \\ 1 & 1 & 0 \\ 0 & 1 & 1 \end{bmatrix}, \qquad c = (1, 1, 0)$$

i	$w(i)$	wA^t	j	wc^t	$(-1)^{wc^t}$	$f^*(wA^t) = f_j^*$	$g^*(w) = g_i^*$
0	(0, 0, 0)	(0, 0, 0)	0	0	1	5	5
1	(0, 0, 1)	(0, 1, 1)	3	0	1	−3	−3
2	(0, 1, 0)	(1, 1, 0)	6	1	−1	+1	−1
3	(0, 1, 1)	(1, 0, 1)	5	1	−1	+1	−1
4	(1, 0, 0)	(1, 0, 0)	4	1	−1	+1	−1
5	(1, 0, 1)	(1, 1, 1)	7	1	−1	+1	−1
6	(1, 1, 0)	(0, 1, 0)	2	0	1	+1	1
7	(1, 1, 1)	(0, 0, 1)	1	0	1	+1	1

Thus the theorem produces the same relation between f^* and g^* that was verified by direct calculation.

The above conversion relations show that linear transformations $x \to xA$ and vector addition $x \to x \oplus c$ affect f^* (and f_N^*) in distinct and separate ways. When A permutes X, its transposed inverse A^{-t} permutes transform coordinates w. It is well-known that A and A^{-t} have the same cycle structure (Elspas, 1959). When c is added to x, points of X are permuted in pairs. However, the effect on w is merely to *change the signs* of those transform coefficients $f^*(w)$ for which the dot product wc^t has odd parity.

When A is restricted to be a permutation matrix P, the results are noteworthy. Consider the subset of $n!/(n-k)!k!$ points $(x_1, \ldots, x_n) = x$ in X that have weight k (i.e., exactly k of the x_i's have unit value and the remaining $(n-k)$ variables are zero). This set is closed under variable permutations $x \to xP$, but not under variable complementations $(x \to x \oplus c)$ or symmetry transformations $(x \to xP \oplus c)$. On the other hand, the corresponding set of points of weight k in W remains closed under general symmetry transformations. The effect of a nonzero c is merely to change the sign of f^* or f_N^* at certain points, not to permute the ordering of coefficients.

$$g^*(w) = (-1)^{wc^t} f^*(wP^t) \qquad \text{if} \quad f(x) = g(xP \oplus c) \qquad (63)$$

This property was used by Ninomiya and Golomb to detect the symmetry type to which a function belongs. It also provides necessary conditions for partial symmetry of a function with respect to a subset of variables or their complements and can even be applied to detect symmetry of f with respect to linear combinations of variables (mod 2).

The second fundamental property of Fourier transforms is their behavior under additive linear functions on their range. This property is more complicated to represent in terms of $f*$ so we consider only f_N*. Suppose we desire to find $(f \oplus g)_N*$ in terms of f_N*, where g is the *linear* function $g(x) = xa^t$, and addition is mod 2. We first convert $f \oplus g$ (mod 2) to the *real* expression

$$f \oplus g \text{ (mod 2)} = f(1 - g) + g(1 - f)$$
$$= f + g - 2fg \quad \text{(real arithmetic)} \tag{64}$$

Therefore, $1 - 2(f \oplus g) = (1 - 2f)(1 - 2g)$ is an identity for all x. We also note that the basis function $Q_w(x) = (-1)^{wx^t}$ can be represented as $Q_w(x) = (1 - 2wx^t)* = 2(wx^t)_N*$. (Here the dot product wx^t must be reduced mod 2 to 0 or 1.) The transform of $f \oplus g$ after translation now becomes

$$(f + g)_N* = [\tfrac{1}{2} - (f \oplus g)]*$$
$$= [\tfrac{1}{2}(1 - 2f)(1 - 2g)]*$$
$$= \sum \tfrac{1}{2}(-1)^{wx^t}(1 - 2f(x))(1 - 2g(x))$$
$$= \tfrac{1}{2}\sum_x (1 - 2f(x))(1 - 2xa^t)(1 - 2xw^t) \tag{65}$$

Now apply the relationship $1 - 2(f \oplus g) = (1 - 2f)(1 - 2g)$ to the last two factors:

$$(f \oplus g)_N*(w) = \tfrac{1}{2}\sum_x (1 - 2f(x))(1 - 2(xa^t \oplus xw^t))$$
$$= \tfrac{1}{2}\sum_x (1 - 2f(x))(-1)^{(w+a)x^t}$$
$$= f_N*(w \oplus a) \tag{66}$$

This, together with our previous observation that functional complementation merely reverses the sign of $f_N*(w)$ for all w, proves the second theorem on invariance:

THEOREM 2.8 (Range Encoding).   If $f(x)$ is defined for all x as $f(x) = g(x) \oplus xa^t \oplus d$ (mod 2), then $f_N*(w) = (-1)^d g_N*(w \oplus a)$. In other words, addition of linear functions $(f \to f \oplus xa^t)$, which has the effect of changing the signs of certain coordinates of f_N, permutes the transform coefficients in pairs (i.e., permutes the Fourier basis functions $Q_w(x)$. This effect of linear functions is dual to the effect of vector addition $(x \to x \oplus a)$ which permutes points of the function domain X (permutes the fundamental product or minterm basis for $f(x)$) but merely changes the signs of f_N* coefficients.

Example 13. The function $g(x) = x_1 \vee x_3$ with $g^{-1}(1) = \{1, 3, 5, 6, 7\}$ will be used again to illustrate range translation. Define $f(x) = g(x) \oplus xa^t \oplus d$ with $a = (1, 0, 1)$ and $d = 1$. The effect of this mapping on g is computed below.

i	$x(i)$	g_i	xa^t	$g_i \oplus xa^t \oplus 1 = f_i$
0	(0, 0, 0)	0	0	1
1	(0, 0, 1)	1	1	1
2	(0, 1, 0)	0	0	1
3	(0, 1, 1)	1	1	1
4	(1, 0, 0)	0	1	0
5	(1, 0, 1)	1	0	0
6	(1, 1, 0)	1	1	1
7	(1, 1, 1)	1	0	0

The transform of f is computed below:

$$
\begin{aligned}
f &= (1,\ 1,\quad 1,\quad 1,\ 0,\quad 0,\quad 1,\quad 0) \\
h = f(Q_1 \times I^{[2]}) &= (2,\ 0,\quad 2,\quad 0,\ 0,\quad 0,\quad 1,\quad 1) \\
r = h(I^{[1]} \times Q_1 \times I^{[1]}) &= (4,\ 0,\quad 0,\quad 0,\ 1,\quad 1,\ -1,\ -1) \\
f^* = r(I^{[2]} \times Q_1) &= (5,\ 1,\ -1,\ -1,\ 3,\ -1,\quad 1,\quad 1)
\end{aligned}
\tag{67}
$$

Using the range encoding theorem, $f_N^*(w) = (-1)^d g_N^*(w \oplus d)$; the same result is obtained from g_N^* as follows (taking g^* from the preceding example):

i	$w\ (i)$	g_i^*	$g_N^*(w)$	$w \oplus a$	$-g_N^*(w \oplus a) = f_N^*(w)$
0	(0, 0, 0)	5	−1	(1, 0, 1)	−1
1	(0, 0, 1)	−3	3	(1, 0, 0)	−1
2	(0, 1, 0)	−1	1	(1, 1, 1)	+1
3	(0, 1, 1)	−1	1	(1, 1, 0)	+1
4	(1, 0, 0)	−1	1	(0, 0, 1)	−3
5	(1, 0, 1)	−1	1	(0, 0, 0)	+1
6	(1, 1, 0)	1	−1	(0, 1, 1)	−1
7	(1, 1, 1)	1	−1	(0, 1, 0)	−1

From f_N^* we obtain $f^* = (5, +1, -1, -1, +3, -1, +1, +1)$ which agrees with the preceding direct computation.

We have discussed the separate effects of domain and range transformations which comprise the elements of RAG. From these two results, the following fundamental theorem is easily derived:

THEOREM 2.9 (*Fundamental Invariance Theorem*).   Let $g(x)$ be a known function from $X = \mathbb{Z}^n$ into $Z = \{0, 1\}$. Suppose $f(x)$ is defined for all x by

$$f(x) = g(x\mathsf{A} \oplus c) \oplus xa^t \oplus d \text{ (mod 2)}$$

where A is a nonsingular n by n matrix over GF(2), a and c are arbitrary binary n-tuples, and $d \in \mathbb{Z}$. Then the transforms of f_N and g_N are related as follows:

$$_N(x) = (-1)^{xa^t \oplus d} g_N(x\mathsf{A} \oplus c)$$

iff   (68)

$$g_N{}^*(w) = (-1)^{wc^t \oplus d} f_N{}^*(w\mathsf{A}^t \oplus a)$$

Proof: Let $h(x) = g(x\mathsf{A} \oplus c)$ and $f(x) = h(x) \oplus xa^t \oplus d$. Then by the second invariant property of $f_N{}^*$ (under range transformations)

$$f_N{}^*(w) = (-1)^d h_N{}^*(w \oplus a)$$ (69)

By the first invariant property of $f_N{}^*$ (under domain transformations):

$$h_N{}^*(w) = (-1)^{wb^t} g_N{}^*(w\mathsf{A}^{-t})$$ (70)

where $b = c\mathsf{A}^{-1}$. Combining the two expressions, we obtain

$$f_N{}^*(w) = (-1)^{d \oplus (w \oplus a)b^t} g_N{}^*((w \oplus a)\mathsf{A}^{-t})$$ (71)

Substituting $v = (w \oplus a)\mathsf{A}^{-t}$, we obtain

$$f_N{}^*(v\mathsf{A}^t \oplus a) = (-1)^{d \oplus v\mathsf{A}^t b^t} g_N{}^*(v)$$ (72)

which becomes the inverse relation upon replacing v by w and b by $c\mathsf{A}^{-1}$:

$$g_N{}^*(w) = (-1)^{wc^t \oplus d} f_N{}^*(w\mathsf{A}^t \oplus a)$$ (73)

To derive the symmetric companion statement relating f_N to g_N, apply the identity $1 - 2k = (-1)^k$ for $k = 0, 1$, to the function $k(x) = xa^t \oplus d \text{ (mod 2)}$. Then $f = h \oplus k = h + k - 2kh$ (real addition), so the theorem hypothesis becomes

$$f_N(x) = \tfrac{1}{2} - f(x) = (1 - 2k(x))(\tfrac{1}{2} - h(x))$$
$$= (-1)^{xa^t \oplus d}[\tfrac{1}{2} - g(x\mathsf{A} \oplus c)] = (-1)^{xa^t \oplus d} g_N(x\mathsf{A} \oplus c). \quad (74)$$

The symmetry of the theorem statement makes it easy to remember:

$$f_N(x) = (-1)^{xa^t \oplus d} g_N(x\mathsf{A} \oplus c) \qquad \text{iff} \quad g_N{}^*(w) = (-1)^{wc^t \oplus d} f_N{}^*(w\mathsf{A}^t \oplus a) \quad (75)$$

The second expression is deducible from the first one simply by interchanging f_N with g_N and a with c, then replacing f_N, g_N, and A by $f_N{}^*$, $g_N{}^*$, and

A^t, respectively. We have thus demonstrated a complete two-way duality between (a) adding a constant vector b to every point of X and (b) adding a linear function xa^t to every function in $\mathscr{F}$. This duality is obvious only when the range of $\mathscr{F}$ is translated from $\{0, 1\}$ to the symmetric set $\{\frac{1}{2}, -\frac{1}{2}\}$, in which case mod 2 addition of functions $f \oplus g$ becomes multiplication $(f \oplus g)_N = 2f_N g_N$. This is another justification for preferring the modified transform $f_N^* = (\frac{1}{2} - f)^*$ rather than the unmodified Fourier transform f^* to represent a 2-valued function f.

To recapitulate, the fundamental invariance theorem says that when an element $T_{A,a,c,d}$ of RAG is applied to a function f, it has the following effects on the transform f_N^*: Parameters A and a permute the components of the transform coefficient vector f_N^*, while parameters c and d merely change their signs. Furthermore, A does not move the 0-coefficient $f_N^*(0)$, while a interchanges $f_N^*(0)$ and $f_N^*(a)$. In Section V, these properties will be used to guide the selection of a sufficient encoding transformation to map any function into its prototype, or vice versa.

EXERCISES

1. Prove the recursion relation $Q_{n+1} = Q_1 \times Q_n$ by relating the radix-two expansions of the indices i and j of q_{ij} to the argument vectors x, w of $Q_w(x)$.

2. Generate Q_3 and take the Fourier transforms of the functions f, g, and h defined on $\mathbb{Z}^3$ by the modulo 2 sums $x_1 \oplus x_2$, $x_1 \oplus \bar{x}_3$, $\bar{x}_2 \oplus \bar{x}_3$. What can be said about the distribution of f^* coefficient magnitudes for these three functions?

3. Show that $\sum \{(-1)^{\Sigma v_i} : v \in \mathbb{Z}^n\} = 0$ when summed over all v in $\mathbb{Z}^n$.

4. Give a direct proof that $\sum \{(-1)^{bv^t} : v \in \mathbb{Z}^n\} = 2^n \delta_{b0}$. Hint: Express v as $v_1 \oplus v_2$ with $v_1 \leq b$ and $v_2 \leq \bar{b}$.

5. Define the matrix Q_4 and compute the Fourier transform $f^* = fQ_4$ of the 4-argument function $f(x)$ with $f^{-1}(1) = \{0, 2, 3, 5, 6, 8, 9, 11, 12\}$ by direct multiplication. Verify the matrix identity $F^* = Q_2 FQ_2$ where F and F^* are 4×4 matrices in which the elements of f and f^*, respectively, are inserted row by row (Hint: See Fig. 4).

6. Compute the Fourier transform of the function $g^{-1}(1) = \{7, 9, 12, 13, 15\}$ using the 4×4 matrix equation $G^* = Q_2 GQ_2$.

7. Find the convolution sum $f * g$ of the functions f and g of the two preceding problems by means of the formula $(f * g) = 2^{-4} Q_2 H^* Q_2$, where H^* is the componentwise product of the matrices F^* and G^*.

III. COMBINATORIAL APPLICATIONS

This section applies the Fourier transform properties described in Section II,B to three problems: prime implicant extraction, detection of disjunctive decompositions, and partial ordering of variables before factoring a logic function. These problems are called combinatorial because the classical approach to their solutions has generally been combinatorial in nature and because the cost of solving the problem (computation time, storage, or both) grows exponentially or factorially with the number of function arguments. Harmonic analysis has not yet been applied to the most significant problem of this type, the "minimal covering" problem (find the "least-cost" subset of prime implicants whose union includes all of $f^{-1}(1)$ or $f^{-1}(0)$). However, there is a remarkable consistency among the transform techniques required to detect prime implicants, detect disjunctive decompositions, or select variables to be factored out. This unifying tendency alone warrants further exploitation of harmonic analysis as a source of new algorithms.

The algorithms developed in this section do not use any of the invariant properties of Fourier transforms under affine operators developed in Section II,C. Section IV will examine these properties, and Section V will describe a synthesis technique based on them. Historically, that technique was developed before the techniques in this section were discovered. However, this section is recommended reading before Section V because it provides a rationale for the criteria used in that section to select an equivalent function of minimal complexity.

Sections III,A,1 and III,A,2 develop some elementary concepts and notations which may be skipped by the reader whose background includes some switching theory or modern algebra.

A. A TEST FOR IMPLICANTS OF f OR $\bar{f}$

This section and the next one will develop a new algorithm that can simultaneously detect all implicants of a fully defined function f and its complement $\bar{f}$. The algorithm is independent of the size of $f^{-1}(1)$. If f has "don't-care" conditions, most of the computation involved to find implicants of f must be duplicated to find the implicants of $\bar{f}$. In contrast to conventional methods, implicants are detected in decreasing rather than increasing order of their dimension. As each new implicant of dimension k is detected, it is compared to adjacent implicants (cosets of the same subspace), and redundant ones are discarded. Data is also kept from which the number of prime implicants or "PI's" which actually cover each mintern of f or $\bar{f}$ can be evaluated, without actually adding unity to each location in a truth table array covered by each PI.

A byproduct of the PI detection algorithm is a necessary condition for disjunctive decomposition of a function with respect to two complementary subsets of variables. This test will be described in Section III,A,3.

Section III,A,1 introduces standard notation. The reader who is unfamiliar with the terminology should refer to any standard text on switching theory such as Chapter 3 of Miller (1965) or Chapter 2 of Prather (1967). Section III,A,2 will be familiar to anyone with background in vector spaces over finite fields.

1. Introduction and Notation

The notation of Section II,A,1 will be used herein. A *subcube* of $\mathbb{Z}^n$ is a subset with the following special properties:

1. Certain "bound" argument variables are restricted to a fixed value which may be 0 or 1.
2. The remaining "free" variables can independently take on all combinations of 0 or 1 values.

(The term "subcube" arises from the geometric picture of $\mathbb{Z}^n$ as an n-dimensional unit cube (n-cube) having the 2^n binary n-tuples which represent elements of $\mathbb{Z}^n$ as vertex coordinates.) The *dimension* of a subcube is the number of *free* variables it contains. For convenience, we use the term k-cell (or k-cube) to denote a k-dimensional subcube of $\mathbb{Z}^n$. A subcube of dimension k includes 2^k points of $\mathbb{Z}^n$.

A unique k-cell is denoted by the ternary n-tuple $a = (a_1 \cdots a_n)$. The symbol "-" is assigned to the k components a_i for which x_i is a *free* variable in $\mathbb{Z}^n$; a_i is assigned the same 0 or 1 value as x_i for each of the $n - k$ bound variables x_i.[2] Since every coordinate of $\mathbb{Z}^n$ can be either free, fixed at 1, or fixed at 0, and every coordinate of a can independently assume one of three possible values (0, 1, or $-$), there are exactly 3^n ternary n-tuples corresponding to 3^n possible subcubes. The total number of k-cells in $\mathbb{Z}^n$ is $2^{n-k}C_{n,k}$, where $C_{n,k} = n!/k!(n-k)!$ is the number of ways to select k free variables among the n coordinates of $\mathbb{Z}^n$, and 2^{n-k} is the number of ways to assign values to the $n - k$ bound variables.

An *implicant* of a logical function f (or of its complement $\bar{f}$) defined on $\mathbb{Z}^n$ is a subcube of $\mathbb{Z}^n$ which is also a subset of the *level set* $f^{-1}(e) = \{x \in \mathbb{Z}^n : f(x) = e\}$ for $e = 0$ or 1. A *prime implicant* of a function f is an implicant which cannot be doubled in size by freeing any one of its bound variables. An

---

[2] The symbol "-" corresponds to "x" of Miller (1965) and "I" of Prather (1967).

important step in the simplification of logic equations is to generate all prime implicants of f while eliminating all redundant nonprime ones.

The list of all prime implicants for f contains the information necessary to derive a disjunctive normal formula (DNF) of minimal complexity, i.e., the simplest Boolean sum of products expression for f. Each prime implicant corresponds to an AND gate from which no literal can be omitted, because it will no longer correspond to a subset of $f^{-1}(1)$. Although no prime implicant contains redundant literals, some prime implicants may be redundant in the covering of f by a union of prime implicants.

The second step in simplifying logic formulas is to find a subset of the prime implicants which is minimal in some sense but which still covers all of f (or $\bar{f}$). In general, such a "minimal cover" is not unique. The problem of selecting a minimal cover from among the prime implicants of f will not be discussed here. However, the invariant properties of f^* under symmetries of the n-cube (Section II,C) suggests that harmonic analysis may have potential applications to the recognition of cyclic covers, which lead to difficulties in existing combinatorial approaches.

2. Cosets and Nullspaces

The algorithm to be described herein takes advantage of Fourier transform properties to separate the test "Is $V_b + c$ included in $f^{-1}(1)$?" into two parts. The first part operates on a subset of f^* and produces an array of integers corresponding to the cosets of V_b; the second part tests the integer corresponding to c, and on this basis decides whether $V_b + c$ implies f or $\bar{f}$.

The concept of a k-cell is inadequate to describe this algorithm, because it does not explicitly distinguish subspaces of $\mathbb{Z}^n$ from cosets of these subspaces. Subspaces (additive subgroups) of $\mathbb{Z}^n$ are merely implicants or k-cells with the added constraint that all bound variables are assigned zero value. This constraint insures that the subset so defined contains the zero vector (additive identity element) and is closed under addition. For example, $a = (\text{-}000\text{-})$ defines a subspace of $\mathbb{Z}^n$, while $a = (\text{-}101\text{-})$ does not.

DEFINITION. Let a denote a k-cell in which all bound variables are assigned zero value. The k-dimensional subspace defined by a will be represented by V_b where b is the binary n-tuple obtained from a by replacing the symbol - by 1. For example, if $a = (\text{-}000\text{-})$, then $b = (10001)$, and V_b is the following subset of $\mathbb{Z}^5$: (00000), (00001), (10000), (10001). Note that the unit vectors (10000), (00001) whose sum is b are a basis for V_b.

Each subspace of $\mathbb{Z}^n$ defines a disjoint partition of $\mathbb{Z}^n$ into disjoint *cosets*. The coset $V_b + c$ is obtained by adding some vector c in $\mathbb{Z}^n$ (but not in V_b) to every vector in V_b. This is called *translation* of V_b by c. (The notation $S + c$ denotes the set $\{x \oplus c : x \in S\}$, for any subset S of $\mathbb{Z}^n$.) For example, if $b = (10001)$ and $c = (01010)$, the coset $V_b + c = \{x \oplus c : x \in V_b\}$ is identical to the 2-cell $a = (\text{-}101\text{-})$. Furthermore, if V_b is a k-cell of $\mathbb{Z}^n$, then every one of the $2^{n-k} - 1$ nonzero assignments of values for the bound variables of V_b generates a *different* k-cell or coset of V_b, which together with V_b itself exhaust $\mathbb{Z}^n$. It is standard practice to call V_b the "identity" coset generated by the remaining (zero) assignment to all bound variables.

Since V_b is closed under addition, $V_b + (x \oplus c) = (V_b + x) \oplus c = V_b + c$ for any x in V_b. Therefore, 2^k different vectors $x \oplus c$ will translate V_b into the *same* coset $(V_b + c)$. However, only one of these vectors has zero values for each of the free variables in V_b; from now on we will use the symbol c to denote this (unique) translation vector.

The unique correspondence between a k-cell a and a coset $V_b + c$ defined above is specified by the following table of correspondences between components of a, b, and c:

a	b	c
0	0	0
1	0	1
-	1	0

Note that components of c corresponding to bound components of a must agree in value with a. Components of c which are free variables in a are assigned the value zero.

The *weight* of a vector x in $\mathbb{Z}^n$, denoted $|x|$, is defined as the number of coordinates with value 1, or the real sum $x_1 + \cdots + x_n$. Every vector except c itself in the coset $V_b + c$ (with c uniquely defined as above) is the translate of a nonzero element x of V_b; therefore, it has unit value for at least one of the free variables in V_b. In other words, $|x \oplus c| > |c|$ for every nonzero x in V_b. Because c is the unique vector of minimum weight in the coset $V_b + c$, it is called the *coset leader* (Peterson, 1961).

An alternate definition for V_b is the set $\{x : x \leq b\}$ of all elements of $\mathbb{Z}^n$ which satisfy the following partial ordering relation:

DEFINITION. For fixed $b \in \mathbb{Z}^n$, the notation $x \leq b$ means $x_i \leq b_i$ for $1 \leq i \leq n$.

Clearly $V_b = \{x : x \leq b\}$. If $|b| = k$, then V_b has k unit basis vectors and 2^k elements.

DEFINITION. The *characteristic function* of V_b, denoted $v_b(x)$, is defined by $v_b(x) = 1$ for x in V_b and 0 otherwise.

Now consider the set $V_{\bar{b}} = \{y : y \leq \bar{b}\}$. This set is also a subspace, the sum of whose unit basis vectors is $\bar{b}$, the logical complement of b. The unit bases for V_b and $V_{\bar{b}}$ generate all of $\mathbb{Z}^n$, and $xy^t = 0$ for any x in V_b and y in $V_{\bar{b}}$. Therefore, every x in $\mathbb{Z}^n$ has a unique decomposition $x = x_1 \oplus x_2$, $x_1 \in V_b$, $x_2 \in V_{\bar{b}}$, and $\mathbb{Z}^n$ is the direct sum of V_b and $V_{\bar{b}}$. In this notation, an alternate definition for the characteristic function of V_b is $v_b(x) = \delta_{x_2 0}$. The components x_1, x_2 are called the projections of x into V_b and $V_{\bar{b}}$, respectively (also denoted $x \lceil V_b$, $x \lceil V_{\bar{b}}$).

DEFINITION. For any subspace V of $\mathbb{Z}^n$, the set $\{y : yx^t = 0\}$ for all $x \in V$ is called the *nullspace* of V.

Clearly, $V_{\bar{b}}$ is the nullspace of V_b (and vice versa). Furthermore, $V_{\bar{b}}$ contains every coset leader for the partition of $\mathbb{Z}^n$ into cosets of V_b: $\mathbb{Z}^n = \{V_b + c : c \in V_{\bar{b}}\}$. Caution: unless V_b has a basis of unit vectors, $V_{\bar{b}}$ cannot always be identified with the set of coset leaders; and $\mathbb{Z}^n$ is *not* necessarily the direct sum of V_b and $V_{\bar{b}}$. For example, the subspace (00, 11) of $\mathbb{Z}^2$ is its own nullspace (mod 2). Its only coset is (01, 10).

The test for implicants will make use of another notational device: the *restriction* of a function f to a subset S of its domain.

DEFINITION. For any function f on $\mathbb{Z}^n$, and any subset $S \subseteq \mathbb{Z}^n$, *the restriction of f to S* (denoted $f \lceil S$) is defined as the set of ordered pairs

$$\{(y, f(y)) : y \in S\}.$$

If S is a *subspace* of the type V_b in $\mathbb{Z}^n$, it will be convenient to specialize this notation considerably, using the projection $x \lceil V_b$ defined previously. Suppose $|b| = k$. If we regard x as a function from the integer index set $(1, 2, \ldots, n)$ into $\mathbb{Z}$, then $x \lceil V_b$ picks out the free variables x_{r_j}, $1 \leq j \leq k$ which correspond to unit components of b ($b_{r_1}, b_{r_2}, \ldots, b_{r_k} = 1$). The projection $x \lceil V_b$ defines a new compact set of argument vectors in $\mathbb{Z}^k$ for the function $f \lceil V_b$; $V_b = (x_{r_1}, x_{r_2}, \ldots, x_{r_k}) = (y_1, \ldots, y_k)$. Then $f \lceil V_b$ can be defined as the set of ordered pairs $\{(y, g(y)) : y = x \lceil V_b, g(y) = f(x_1)\}$, where $x = x_1 \oplus x_2$, $x_1 \in V_b$, $x_2 \in V_{\bar{b}}$, and y merely compresses the vectors x_1 into k-tuples by deleting components of x which are identically zero.

3. Main Theorem on Implicant Extraction

The main theorem on implicant extraction will now be stated and proved using the theorems of Section II,B.

THEOREM 2.9 (Implicant Extraction). For any subspace $V_b = \{x : x \le b\}$ of $\mathbb{Z}^n$, and any coset leader c in the nullspace $V_{\bar{b}}$,

$$\begin{aligned} V_b + c &\subseteq f^{-1}(1) \qquad \text{iff} \quad (f^* \ulcorner V_{\bar{b}})_c^* = 2^n \\ V_b + c &\subseteq f^{-1}(0) \qquad \text{iff} \quad (f^* \ulcorner V_{\bar{b}})_c^* = 0 \end{aligned} \qquad (76)$$

Proof: Let $k = |b|$, the dimension of V_b. For $z = 0$ or 1, the condition $V_b + c \subseteq f^{-1}(z)$ is equivalent to

$$\sum \{f(x) : x \in V_b + c\} = z2^k \qquad (77)$$

which may be rewritten in terms of the characteristic function v_b:

$$\sum \{f(x) \oplus c); \quad x \in V_b\} = \sum \{f(x \oplus c)v_b(x) : x \in \mathbb{Z}^n\} = (f * v_b)_c \qquad (78)$$

by the definition of a convolution sum in Section II,B,1. By the convolution theorem (Section 11,B,1)

$$(f * v_b)_c = 2^{-n}(f^*(w)v_b^*(w))_c^* \qquad (79)$$

The transform of $v_b(x)$ is 2^k on $V_{\bar{b}}$ and 0 elsewhere by Corollary 2.3 of Section II,B,2. The remainder of the proof is clarified by adopting the following notation for the (unique) decompositions of x and w into direct sums of their projections into V_b and $V_{\bar{b}}$:

$$x = c \oplus u, \qquad c \in V_{\bar{b}}, \qquad u \in V_b \qquad (80)$$

$$w = w_1 \oplus w_2, \qquad w_1 \in V_{\bar{b}}, \qquad w_2 \in V_b \qquad (81)$$

Then

$$(f * v_b)_c = 2^{-n}((2^k \delta_{w_2 0})(f^*(w_1)))_c^* \qquad (82)$$

The inverse transform can also be decomposed into separate transforms on V_b and $V_{\bar{b}}$:

$$\begin{aligned} (f * v_b)_c &= 2^n \sum_{w_2} 2^k \delta_{w_2 0}(-1)^{w_2 u^t} \sum_{w_1} f^*(w_1)(-1)^{w_1 c^t} \\ &= 2^{-n} 2^k \sum_{w_1} f^*(w_1) Q_{w_1}(c) \\ &= 2^{k-n}(f^* \ulcorner V_{\bar{b}})_c^* \end{aligned} \qquad (83)$$

This result also follows from the Quotient Group Character theorem of Section II,B,2, but the derivation above avoids the notational problems involved in an unambiguous application of this theorem. It has been shown that for any $b \in Z^n$, $\sum \{f(x) : x \in V_b + c\} = 2^{k-n}(f^* \ulcorner V_c)_b^*$ for each c in $V_{\bar{b}}$. This sum will be 2^k or 0 iff $(f^* \ulcorner V_{\bar{b}})_c^* = 2^n$ or 0, respectively, which proves the theorem.

This theorem permits the question "Is the coset (subcube) $V_b + c$ an implicant of f or $\bar{f}$?" to be answered simultaneously for *every* coset of the subspace V_b merely by inspecting the results of a single inverse transform operation on the array $(f^* \ulcorner V_{\bar{b}})$, whose size is exactly equal to the number of cosets of V_b.

4. Example of Theorem Operation

The example shown in Fig. 4 should clarify this theorem and remove any mystery surrounding its operation. Figure 4 shows the truth table of a function $f(x)$ of five arguments arranged in a 4×8 array F, such that the first column of F corresponds to the subspace V_b, $b = (11000)$. This subspace has eight 4-element cosets. The top row of F corresponds to the subspace $V_{\bar{b}}$. Each column of F is the restriction $(f \ulcorner V_b + c)$ of f to a different coset of V_b.

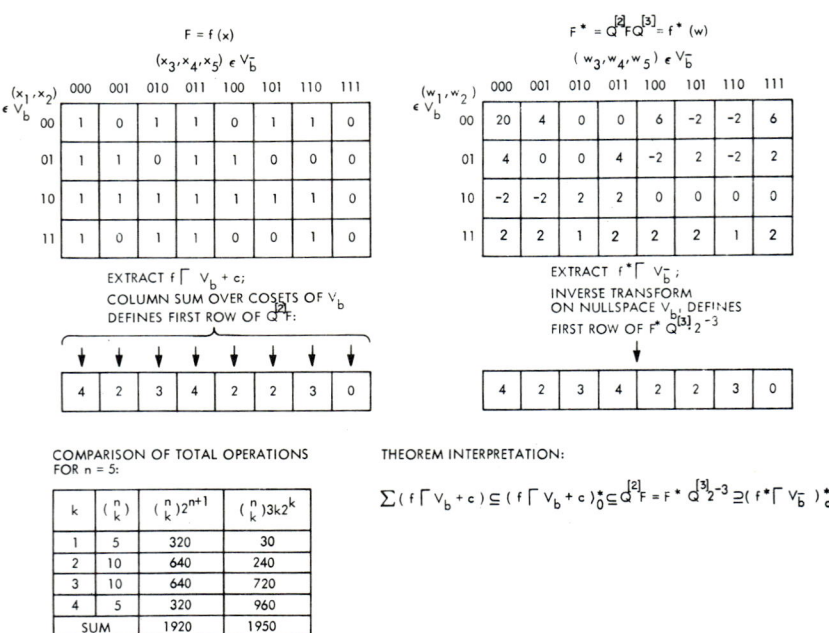

FIG. 4.   Comparison of direct and transform approaches to implicant extraction for the subspace V_{11000}.

On the left side of Fig. 4, the sum of f over each of these columns is computed directly. Notice that for $c = 0$ and $c = (00011)$, $V_b + c$ is an implicant of f, while for $c = (00111)$, $V_b + c$ is an implicant of $\bar{f}$ (since these are constant columns of F).

The array F* on the right side of Fig. 4 is the Fourier transform $f*$ of f. The Kronecker product definition of the fast Fourier transform algorithm in Section II,A,4 can also be written in the following matrix form, in which $Q^{[2]}$ transforms f on V_b and its cosets, $Q^{[3]}$ transforms f on $V_{\bar{b}}$ and its cosets, and the multiplication is associative:

$$F* = Q^{[2]}FQ^{[3]} \tag{84}$$

The restriction $(f* \ulcorner V_{\bar{b}})$ is merely the top row of F*, and its inverse transform is merely the top row of

$$2^{-3}F*Q^{[3]} = 2^{-3}Q^{[2]}F(Q^{[3]})^2 = Q^{[2]}F \tag{85}$$

since $Q^{[3]}Q^{[3]} = 2^3I$.

The direct approach (sum over cosets of V_b) is also equivalent to the top row of $Q^{[2]}F$, because the first row of $Q^{[2]}$ is merely the vector (1111).

In other words, identical results are obtained by both approaches. However, the transform algorithm required the definition of 2^k pointers followed by $k2^{k+1}$ additions to compute the inverse transform, while the direct approach requires 2^n pointers to be defined followed by 2^n additions to sum f over each coset.

A cumulative comparison of the cost $3k2^k$ for each subspace by the theorem versus the cost 2^{n+1} for each subspace by direct summation is also shown on Fig. 4. For $n = 5$, the two approaches are comparable. However, for $n \geq 10$, significant storage and computational advantages of the transform approach will be demonstrated in Sections III,B,5 and III,B,6.

B. ALGORITHM TO EXTRACT PRIME IMPLICANTS

This section describes an algorithm for extracting all *prime* implicants of f and $\bar{f}$ based on the preceding theorem and computes an accurate bound on its computation requirements. As a starting point for this algorithm, the Fourier transform $f*$ of f is computed and stored either in fast (core) memory or on a disc file. The array $f*$ has 2^n integer elements in the range $(-2^{n-1}, +2^n)$, and it is the starting point for all further calculations.

The basic iteration step is applied to subspaces of the type V_b of $\mathbb{Z}^n$. Subspaces can be processed one at a time in any order. However, it is preferable to process subspaces in descending order of their dimension which corresponds to ascending order of the number of bound variables in their cosets. Therefore, subspace indices b will be processed in decreasing order of their weight $|b| = n - k$. In other words, prime implicants containing $k = 1$ literals will be extracted first, those with $k = 2$ literals next, and so on. Prime implicants with n literals (minterms) will be extracted last.

This order of processing permits the algorithm to terminate after any value of k if it is found that f or $\bar{f}$ or both are covered completely by prime implicants containing k or fewer literals.

Another reason for this order of processing is that it permits reversion to a more direct approach if desired as k approaches n. The computational burden per subspace V_b increases as $(k + 1)2^k$; Section III,B,5 shows that a direct search approach involves less computation for the highest $(n/4)$ values of k. The labor of the direct approach can be further reduced by identifying all minterms previously covered as "don't care" argument states. The direct approach then proceeds by inspecting the n "nearest neighbors" of each remaining (still uncovered) minterm.

Functions with "don't care" conditions will be considered in the last paragraph of Section III,B.

1. Main Algorithm Definition

The basic iteration process for each V_b with $|b| = n - k$ consists of the following six steps:

(1) A list of pointers $\{i_j : x(i_j) \in V_{\bar{b}}$ and $x(i_j)\ulcorner V_b = y(j),\ 0 \le j < 2^k\}$ is constructed (e.g., using the algorithm of Section III,B,2 below) and saved for multiple uses.

(2) The subarray $f^* \ulcorner V_{\bar{b}} = \{f_j\ ; 0 \le j < 2^k\}$ is extracted from f^* by a table lookup process.

(3) The inverse transform $(f^* \ulcorner V_{\bar{b}})^*$ is computed by the fast Fourier transform algorithm of Section II,A,4.

(4) The array $(f^* \ulcorner V_{\bar{b}})^*$ is inspected at each argument y to detect implicants $V_b + c$, as proscribed in the preceding theorem.

(5) Elements of $(f^* \ulcorner V_{\bar{b}})^*$ corresponding to the k "nearest neighbor" cosets of $V_b + c$ are inspected to reject all redundant or nonprime implicants. All prime implicants are added to a list (which need not be retained in core memory).

(6) (Optional) The Fourier transforms of prime implicants corresponding to all cosets of V_b are accumulated in an auxiliary array of 2^n locations. This array defines the Fourier transform of the (real) sum of characteristic functions of prime implicants of f minus those of $\bar{f}$. By inverting this transform, the number of prime implicants (so far detected) which cover each minterm of f and $\bar{f}$ may be determined.

The next few paragraphs describe algorithms for steps (1), (5), and (6) in order to develop bounds on computation time for this procedure.

2. Generation of Subspace Pointers

In order to extract $f^* \vdash V_{\bar{b}}$ in step (1) and to update g^* from $\Delta g^* \vdash V_{\bar{b}}$ in step 6 of the PI extraction algorithm, a table of pointers $\{i_j, 0 \leq j < 2^k\}$ must be generated which indexes elements of the subspace V_b as a subset of $\mathbb{Z}^n$. The following algorithm requires approximately 2^k indexed additions to generate this table. The same principle can be used to generate $|x|$, $|x|$ mod 2, or similar quantities.

The first step is to identify the set of weights $\{2^{n-r} : b_r = 0\}$ whose sums (over subsets) define the integers i_j whose radix-two expansions are elements of $V_{\bar{b}}$. Let the jth power of two in *ascending* order from this set be $s_j = 2^{i_j}$, $1 \leq j \leq k$. Then the 2^k-dimensional vector P_k which contains the ordered list of pointers $\{i_j : x(i_j) \in V_{\bar{b}}\}$ is defined by the recursion relation

$$P_1 = (0, s_1)$$
$$P_{j+1} = (1, 1) \times P_j + (0, 1) \times [s_{j+1} \cdot E_j], \qquad \text{for} \quad 1 \leq j \leq k \tag{86}$$

Here $\times$ denotes the Kronecker product (see Section II,A,4), and E_j denotes a constant row vector containing 2^j unit entries.

The above formula may be verified directly. It states that the first 2^j entries of P_{j+1} and P_j are identical, while the last 2^j entries of P_{j+1} are obtained by adding s_{j+1} to corresponding entries of P_j :

$$P_{j+1}(2^j + i) = P_j(i) + s_{j+1} \qquad \text{for} \quad 0 \leq i < 2^j \tag{87}$$

3. Rejection of Nonprime Implicants

Step (4) of the above algorithm detects all implicants of f and $\bar{f}$ but does not identify *prime* implicants (those covered by larger ones). The following simple test to reject nonprime implicants is an adaptation of the "nearest neighbor" or "adjacent vertex" test used in recent algorithms for the direct buildup of larger implicants out of smaller ones (Morreale, 1967; Necula, 1967).

If a coset $V_b + c$ is included in a larger one, then the larger one must include all the free variables of $V_b + c$, plus at least one more free variable that was a bound variable in $V_b + c$. Suppose x_i was bound in $V_b + c$ but is free in this larger implicant; then the coset $V_b + (c \oplus e_i)$ is also an implicant of f or $\bar{f}$ (e_i is the unit vector corresponding to x_i, and the vector sum $c \oplus e_i$ is modulo two). Let $c = x(i_j)$ correspond to the argument $y(j)$ of $(f^* \vdash V_{\bar{b}})^*$. The above observation is formalized in the following theorem:

THEOREM 3.1. Let $(f^* \ulcorner V_{\bar{b}})_j{}^* = 2^n$ or 0 at the argument $y(j)$ corresponding to $c(j) = x(i_j)$ in $\mathbb{Z}^n$. Then the coset $V_b + c(j)$ is a prime implicant iff $(f^* \ulcorner V_{\bar{b}})^*$ does not have the same value (2^n or 0 respectively) at *any* of the k adjacent arguments $(y \oplus e_p)$, $1 \le p \le k$ (e_p is the pth unit vector in $\mathbb{Z}^k$).

If the test described above indicates that $V_b + c$ is redundant, then at least one other coset is also known to be redundant. If the second redundant coset is tagged, a duplicate test for redundancy can be avoided when the latter coset is scanned. The average number of tests (each requiring one operation to find the pointer $(i_j \pm 2^{k-p}) = q$ and another to test $(f^* \ulcorner V_{\bar{b}})^*$ at the argument $y(q)$) is reduced from k to $k/2$ per implicant.

4.   Identification of Essential Prime Implicants

An essential or "core" prime implicant (PI) of $f(x)$ is one which covers some point of $f(x)$ that is not covered by any other PI. Every minimal normal form which covers f must include all of the essential PI's. Although the PI extraction algorithm does not identify the essential PI's of f and $\bar{f}$, it does provide an efficient means of computing the integer-valued function $g(x)$

$$g(x) = \sum (\text{PI's of } f) - \sum (\text{PI's of } \bar{f}) \tag{88}$$

If this function is 1 or -1 at a particular point, it means that only one PI of f or $\bar{f}$, respectively, covers the point x.

The algorithm computes g^*, the transform of g, rather than g itself (2^n storage locations are required). Linearity permits the accumulation of g^* as a sum (over all b) of transforms $\Delta g_b{}^*(w)$, where $\Delta g_b(x)$ is the contribution to g of all PI's which are cosets of the *same* subspace V_b. The cost of adding $(\Delta g_b(x))^*$ to g^* is merely $(k + 3)2^k$, which (for small k at least) is much less than the $m2^{n-k}$ operations required to compute Δg_b directly. (The break-even point depends on the number m of PI's that are cosets of V_b.) Suppose that m PI's of f or $\bar{f}$ have been identified among the cosets of $V_{\bar{b}}$. Define $v_{b,c_i}(x)$ as the characteristic function of the ith prime implicant $V_b + c_i$ (i.e., $v_{b,c_i}^{-1}(1) = V_b + c_i$), and suppose also that $V_b + c_i \subseteq f^{-1}(z_i)$; for $1 \le i \le m$. Define $\Delta g_b(x)$ as the quantity contributed to the function $g(x)$ by cosets of $V_{\bar{b}}$:

$$\Delta g_b(x) = \sum_{i=1}^{m} (-1)^{1+z_i} v_{b,\,c_i}(x) \quad (\text{real sum}) \tag{89}$$

By linearity of the Fourier transform operation, it follows that

$$(g(x) + \Delta g_b(x))_w{}^* = g^*(w) + \Delta g_b{}^*(w) \quad (\text{real sum}) \tag{90}$$

The following theorem defines the transform of Δg_b:

THEOREM 3.2.   Let

$$\Delta g_b(x) = \sum_{i=1}^{m} (-1)^{1+z_i} v_{b,\,c_i}(x)$$

on $\mathbb{Z}^n$ and let

$$h(x) = \sum_{i=1}^{m} (-1)^{1+z_i} \delta_{x,\,c_i}, \qquad \text{for} \quad x \in V_{\bar{b}}$$

Then, if $w = w_1 \oplus w_2$, with $w_1 \in V_{\bar{b}}$ and $w_2 \in V_b$,

$$\Delta g_b{}^*(w) = 2^{n-k} \delta_{w_2 0}\, h^*(w_1) \tag{91}$$

where $n - k = |b|$ is the dimension of V_b.

Proof: By its definition, $\Delta g_b(x)$ is constant on cosets of V_b; therefore, the quotient group character theorem of Section II,B,2 applies:

$$\Delta g_b{}^*(w) = 2^{n-k} \delta_{w_2 0} (\Delta g_b(x) \ulcorner V_{\bar{b}})^* \tag{92}$$

The restriction of Δg_b to $V_{\bar{b}}$ is merely $h(x)$.

To apply this theorem, first define $h(y) = (-1)^{1+z_i}$ for $y(j)$ corresponding to the coset leader $c(i_j)$ of each PI ($h(y) = 0$ elsewhere on $\mathbb{Z}^k$). Next compute h^* on $\mathbb{Z}^k$, and finally use the table of subspace pointers $\{i_j, 0 \le j < 2^k\}$ to add $h^*(y(j))$ to $g^*(x(i_j))$. This accomplishes step (6) of the PI extraction algorithm. Note that $\delta_{w_2 0}$ implies $\Delta g_b{}^*(w) = 0$ except on $V_{\bar{b}}$.

Example 14.   The contribution $\Delta g(x)$ of the three PI's detected in the example of Section III,A,4 will be computed. The function $h(y)$ identifies two implicants of f and one of $\bar{f}$:

$$h(y) = (1, 0, 0, 1, 0, 0, 0, -1) \tag{93}$$

Using the matrix notation of Fig. 4, $\Delta g^* \ulcorner V_{\bar{b}} = 2^{n-k} h Q^{[3]} = (12, -4, -4, 12, 4, 4, 4, 4)$. The entire 4×8 array ΔG^* representing $\Delta g^*(x)$ is merely the product $(1, 0, 0, 0)^t (\Delta g^* \ulcorner V_{\bar{b}})$.

To take the inverse transform, merely premultiply by $Q^{[2]}$, postmultiply by $Q^{[3]}$, and multiply by 2^{-5}:

$$Q^{[2]}(1, 0, 0, 0)^t = (1, 1, 1, 1)^t \tag{94}$$

and

$$2^{-5}(\Delta g^* \ulcorner V_{\bar{b}}) Q^{[3]} = (1, 0, 0, 1, 0, 0, 0, -1) \tag{95}$$

The product of these two arrays is

$$G = 2^{-5}Q^{[2]}(\Delta G^*)Q^{[3]} = \begin{bmatrix} 1 & 0 & 0 & 1 & 0 & 0 & 0 & -1 \\ 1 & 0 & 0 & 1 & 0 & 0 & 0 & -1 \\ 1 & 0 & 0 & 1 & 0 & 0 & 0 & -1 \\ 1 & 0 & 0 & 1 & 0 & 0 & 0 & -1 \end{bmatrix} \tag{96}$$

This matrix accurately represents the number of PI's that cover each element x of $\mathbb{Z}^5$.

5. Computational Bounds

This section will derive an accurate bound on the number of elementary (indexed add or subtract) operations required to extract prime implicants and compare it with more direct approaches. Consider a subspace V_b of dimension $n - k$, whose 2^k cosets $V_b + c$ are to be tested to see which of them are k-literal implicants of f. $V_{\bar{b}}$ has dimension k, and the restriction of f^* to $V_{\bar{b}}$ has 2^k elements. Computation of the inverse transform f^{**} of $(f^* \sqcap V_{\bar{b}})$ requires $k2^{k+1}$ operations; 2^k additions are required to compute pointers to the subset $V_{\bar{b}} \subseteq \mathbb{Z}^n$; 2^k operations are required to extract the 2^k elements of f^* on $V_{\bar{b}}$, and 2^k more are required to test f^{**} for each coset leader. To verify "primeness," an average of k additional operations are required for each PI. A total of $(3k + 3)2^k$ operations are required to identify all PI's corresponding to cosets of V_b, if we assume the test for primeness is carried out for every one of the cosets $V_b + c$.

There are 2^n subspaces of $\mathbb{Z}^n$, $C_{n,k} = n!/(k!(n-k)!)$ of which have dimension k. Summing over all of these subspaces gives the following bound on total computation time to test *all* subcubes of $\mathbb{Z}^n$:

$$\sum_{k=1}^{n-1} C_{n,k} 3(k+1)2^k = 3^n(2n+3) \tag{97}$$

or an average of $2n + 3$ operations for every subcube of $\mathbb{Z}^n$. This bound grows at a rate $2n + 3$ times faster than 3^n, the total number of subcubes.

The exponential growth of this function, and its low cost for small values of k, suggests the possibility that another, direct approach might be more efficient after k passes some threshold. This possibility has been explored.

The two basic approaches to extraction of prime implicants in the literature involve comparison of each minterm x in $f^{-1}(1)$ or $f^{-1}(0)$ with either all of its adjacent vertices in $f^{-1}(1)$ or $f^{-1}(0)$ (Quine, 1955; McCluskey, 1956) or with all n adjacent vertices on the n-cube (Morreale, 1967; Necula, 1967). Knowledge about large subcubes of $f^{-1}(1)$ is built up gradually. The computational burden grows roughly as the square of the number of minterms in $f^{-1}(1)$ or $f^{-1}(0)$ and (generally speaking) prime implicants are only extracted for the smaller of these two sets. Since realistic bounds on computation time for these algorithms are difficult to obtain, exhaustive search was postulated as a reference procedure, similar to the direct summation over cosets that was illustrated in Fig. 4. Comparative results are shown in Table 1.

TABLE I
Computation Time (Millions of Operations)

N	K	DIRECT	TRANSFORM	OLD/NEW	SUMOLD	SUMNEW
5	1	0.00	0.00	5.333	0.0003	0.0001
5	2	0.00	0.00	1.778	0.0010	0.0004
5	3	0.00	0.00	0.667	0.0016	0.0014
5	4	0.00	0.00	0.267	0.0019	0.0026
10	1	0.02	0.00	170.667	0.0205	0.0001
10	2	0.09	0.00	56.889	0.1126	0.0017
10	3	0.25	0.01	21.333	0.3584	0.0133
10	4	0.43	0.05	8.533	0.7885	0.0637
10	5	0.52	0.15	3.556	1.3046	0.2088
10	6	0.43	0.28	1.524	1.7347	0.4911
10	7	0.25	0.37	0.667	1.9804	0.8597
10	8	0.09	0.31	0.296	2.0726	1.1707
10	9	0.02	0.15	0.133	2.0931	1.3243
15	1	0.98	0.00	5461.332	0.9830	0.0002
15	2	6.88	0.00	1820.444	7.8643	0.0040
15	3	29.82	0.04	682.667	37.6831	0.0476
15	4	89.46	0.33	273.067	127.1397	0.3752
15	5	196.80	1.73	113.778	323.0438	2.1050
15	6	328.01	5.73	48.762	651.0509	8.8317
15	7	421.72	19.77	21.333	1073.6750	28.5999
15	8	421.72	44.48	9.481	1495.3980	73.0786
15	9	328.01	76.98	4.267	1823.4060	149.9553
15	10	196.80	101.48	1.939	2020.2100	251.4325
15	11	89.46	100.64	0.889	2109.6660	352.0708
15	12	29.82	72.68	0.410	2139.4850	424.7542
15	13	6.88	36.13	0.190	2146.3660	460.8806
15	14	0.98	11.06	0.089	2147.3490	471.9395
20	1	41.94	0.00	**********	41.9430	0.0002
20	2	398.46	0.01	58254.220	440.4014	0.0071
20	3	2390.75	0.11	21845.330	2831.1520	0.1165
20	4	*********	1.16	8738.133	**********	1.2793
20	5	*********	8.93	3640.889	**********	10.2096
20	6	*********	52.09	1560.381	**********	62.3029
20	7	*********	238.14	682.667	**********	300.4436
20	8	*********	870.70	303.407	**********	1171.1460
20	9	*********	2579.86	136.533	**********	3751.0050
20	10	*********	6243.25	62.061	**********	9994.2540
20	11	*********	*********	28.444	**********	**********
20	12	*********	*********	13.128	**********	**********
20	13	*********	*********	6.095	**********	**********
20	14	*********	*********	2.844	**********	**********
20	15	*********	*********	1.333	**********	**********
20	16	*********	*********	0.627	**********	**********
20	17	2390.75	8068.78	0.296	**********	**********
20	18	398.46	2839.02	0.140	**********	**********
20	19	41.94	629.15	0.067	**********	**********

To compute $\sum_x \{f(x) : x \in V_b + c\}$ requires 2^{n-k} pointers to each of 2^k cosets of V_b and (2^{n-k}) table lookups to sum $f(x)$ over each of the 2^k cosets. The total cost for each k-dimensional subspace is 2^{n+1}. Multiplying by $C_{n,k}$ and summing over k gives a total of $2(2^n)^2 = 2(4)^n$ operations, or an average of $2(\frac{4}{3})^n > 2(3)^{n/4}$ operations for each of the 3^n subcubes of Z^n, rather than $2n + 3$ as in our algorithm.

The incremental cost for each k and the cumulative sum of each cost as k increases from 1 to $n - 1$ is shown for $n = 5, 10, 15, 20$ in Table I. The lowest value of k for which exhaustive search has a lower cost than the method herein is $k = 7$ for $n = 10$, $k = 11$, for $n = 15$, and $k = 16$ for $n = 20$. For $n = 15$ or 20, the cumulative saving by reverting to direct search for $k \geq 11$ or 16, respectively, is about 30% of the total time for implicant computation. At 500,000 indexed additions per second, the total time for $n = 15$ is 16 min for the transform algorithm, 11 min. for the hybrid (transform approach for $k \leq 10$, direct search for $k > 10$) and 75 min if direct search is used for all k.

6. Storage Bounds

The fact that all PI's need not be stored in core memory produces a very significant reduction in storage compared to previous methods. Meo (1968) observes that functions of ten or more variables may have many more than 2^n PI's. He estimated the size of memory required to compute the PI's of a function of n variables by either the Quine–McCluskey method (McCluskey, 1956) or the consensus rule (Quine, 1955). This size is half the maximum number of subcubes of any single dimension $(n - k)$, $C_M = \frac{1}{2}\max_k\{C_{n,k} 2^{n-k}, 1 \leq k < n\}$. In contrast, our method requires 2^n terms in the array f^* plus 2^{n-1} terms in $(f^* \sqcap V_b)$. Data from which to determine essential PI's doubles this to $3 \cdot 2^n$ locations. A comparison of C_M and $3 \cdot 2^n$ is given in Table II.

TABLE II
Storage Bounds C_M for Exhaustive Search and $3 \cdot 2^n$ for Transform Algorithm

n	C_M	$3 \cdot 2^n$	Ratio
5	40	96	1/2
10	7,680	3,072	2/1
15	1,540,000	98,704	15/1
20	317,500,000	1,048,566	300/1

The bound C_M grows by a factor of about 3 for each increment of n; the bound $3 \cdot 2^n$ doubles with each increment of n.

7. Functions with "Don't Care" Conditions

The preceding algorithm will now be extended to functions that are undefined for certain values of x in $\mathbb{Z}^n$. The extension is straightforward, but the symmetry that permitted simultaneous extraction of the implicants of both f and $\bar{f}$ is not possible. Suppose "-" denotes the set $(0, 1)$ of range elements into which each don't care argument state is mapped. Then the don't care set $f^{-1}(-)$ is defined along with $f^{-1}(1)$ and $f^{-1}(0)$. To obtain the prime implicants of f, we apply the algorithm just defined to the transform of the function $g(x)$ such that $g^{-1}(1) = f^{-1}(1) \cup f^{-1}(-) = K$; to determine implicants of $\bar{f}$, we must begin with the transform of $h(x)$ such that $h^{-1}(1) = f^{-1}(1) = L$. Since these are transforms of two different functions, the entire algorithm except for steps (1) and (6) must be repeated to get PI's of $\bar{f}$ as well as f. Instead of $3(k + 1)2^k$, the incremental cost for each subspace becomes $5(k + 1)2^k$, an increase of 60%. However, simultaneous computation of PI's for f and $\bar{f}$ will detect every coset that is an implicant of *both* sets $f^{-1}(1) \cup f^{-1}(-)$ and $f^{-1}(0) \cup f^{-1}(-)$, i.e., it is an implicant of $f^{-1}(-)$ *only*. Such implicants are *redundant* in any covering of *either* $f^{-1}(1)$ or $f^{-1}(0)$, and should be discarded.

C. TEST FOR DISJUNCTIVE DECOMPOSITION

A test for disjunctive decomposition is a byproduct of the transform approach to implicant extraction. This test relies on the property that

$$(f^* \sqcap V_b)_c^* = \sum \{f(x) : x \in V_b + c\} \tag{98}$$

and

$$(f^* \sqcap V_b)_c^* = \sum \{f(x) : x \in V_{\bar{b}} + c\}. \tag{99}$$

Ashenhurst (1959) has shown that a function which possesses a disjunctive decomposition $f(x) = g(z, \varphi(y))$, with $y = x \sqcap V_{\bar{b}}$ and $z = x \sqcap V_b$ has an easily recognized pattern to its truth table when the latter is shown on the cartesian product of the subspaces $V_{\bar{b}}$ and V_b (as in Fig. 4). This pattern arises because a decomposable function must either be constant at 0 or 1, equal to $\varphi(y)$, or equal to $\bar{\varphi}(y)$ on a particular coset of $V_{\bar{b}}$. In other words, every *column* sum over cosets of V_b must have at most two distinct values between 0 and 2^{n-k}. This test is easily implemented by keeping track of the number of distinct coefficient values possessed by the inverse transform $(f^* \sqcap V_{\bar{b}})^* \cdot 2^{k-n}$ during the implicant extraction algorithm.

Another necessary condition for functional decomposition can be derived from the row sums, which are obtained when cosets of V_b are processed by the PI extraction algorithm. A decomposable function of the form $f(x) = g(z, \varphi(y))$, $y = x \sqcap V_{\bar{b}}$, $z = x \sqcap V_b$ can have at most four distinct row sums over cosets of $V_{\bar{b}}$. Suppose that $f \restriction V_{\bar{b}} + z$ is a constant, independent of y; if $f = 0$ or 1, the row sum is 0 or 2^k, respectively; the only other possibilities are that $f \restriction V_{\bar{b}} + z = \varphi(y)$ or $f \restriction V_{\bar{b}} + z = \bar{\varphi}(y)$. The row sums in these two cases are $|\varphi|$ and $2^k - |\varphi|$, where $|\varphi|$ is the number of unit entries in the truth table for $\varphi(y)$, and $2^k - |\varphi|$ is the corresponding quantity for $\bar{\varphi}$.

The above two tests furnish two sufficient conditions that in practice may discriminate against a large fraction of the 2^{n-1} possible partitions of $\mathbb{Z}^n$ into two subspaces V_b and $V_{\bar{b}}$. Both tests should be applied to each subspace V_b as it is processed. A two-bit tag must be reserved for each of the 2^n possible vectors b to save the results of both tests for the subspace $V_{\bar{b}}$ so that later on when V_b is processed, both necessary conditions can be verified. The partition of variables for the decomposition, and in fact the truth table array on $V_b \times V_{\bar{b}}$ can then be printed out or otherwise used (e.g., to extract implicants of $g(z, \varphi)$ and $\varphi(y)$ rather than $f(x)$).

The advantage of this procedure depends directly on the fraction of the 2^n partitions of variables which are rejected by the test without requiring 2^n row and column sums or other pattern recognition techniques to verify sufficient conditions for a true decomposition.

A new test for disjunctive decomposition was described by Shen et al. (1969). This test is extremely efficient for randomly chosen functions but slow for functions which actually do possess a decomposition. The test herein is probably less efficient (for random functions) when performed alone; however, when performed during prime implicant extraction, it probably involves negligible labor compared to Shen's method.

D. CRITERIA FOR FACTORING VARIABLES

When a function has too many arguments to realize directly and does not possess a decomposition, then some (say, $n - k$) of the variables can be factored out and used to address a canonical decoding tree which selects one of 2^{n-k} functions of the remaining k arguments to be evaluated. Actually, mixed factoring can be used; for example, in the expression

$$f(x_1, x_2, x_3, x_4) = \bar{x}_1(x_2\, g_1(x_3, x_4) \vee \bar{x}_2\, g_0(x_3, x_4)$$
$$\vee\ x_1(x_3\, h_1(x_2, x_4) \vee \bar{x}_3\, h_0(x_2, x_4)) \tag{100}$$

In this expression, the choice of the second variable to be factored depends on whether $x_1 = 0$ or 1. In general, there are $n!$ ways to order the variables and many more ways to reorder them after some of them have been decoded.

To avoid exhaustive procedures, some complexity criterion is desirable by which to rank the variables. One such criterion is provided by the various sums computed by the PI extraction algorithm. For example, for $k = 1$, there are five subspaces V_b, and $(f^* \ulcorner V_b)^*$ identifies the sum $\sum \{f(x) : x \in V_b + c\}$ for each subspace. Now, generally speaking (but not always), the closer the sum of $f(x)$ over $V_b + c$ is to 0 or 2^{n-k}, the easier it is to construct the function. Therefore, the variable corresponding to the 1-dimensional subspace V_b, which produces the largest or smallest value of $(f^* \ulcorner V_b)_c^* = \sum \{f(x) : x \in V_b + c\}$ is the most promising candidate for initial factorization according to this *heuristic* criterion.

After x_i has been factored out, it is easy to rearrange f^* to separate $f^* \ulcorner V_b$ from $f^* \ulcorner V_b + c$, so that each can be processed separately. The preceding criterion can be applied on each subfunction separately to choose the next variable to be factored (independently for each subfunction).

EXERCISES

8. Find the characteristic functions of the subspaces of $\mathbb{Z}^5$ defined by the vectors $b_1 = (00111)$, $b_2 = (01010)$, $b_3 = (10000)$. That is, find

$$v_b^{-1}(1) = \{i : x(i) \in V_b\}$$

9. Find the nullspace for each subspace V_b in the preceding problem; define each nullspace by its ternary k-cube symbol.

10. Compute $(f^* \ulcorner V_b)^*$ for the 3-argument function $f(x) = x_1 x_2 \vee x_3$ and the two subspaces defined by $b_1 = (001)$, $b_2 = (110)$. Show that $x_1 x_2$ and x_3 are prime implicants of f by applying the PI extraction theorem to the result.

11. Show that the function $f^{-1}(1) = \{0, 1, 2, 3, 4, 5, 6, 7, 9, 10, 11\}$ has a disjunctive decomposition of the form $f = g(x_1, x_2, h(x_3, x_4))$ by applying the test of Section III,C.

IV. ANALYSIS OF THE PROTOTYPE EQUIVALENCE RELATION

This section defines the restricted affine group RAG, a subgroup of the symmetric group on the cartesian product or direct sum $X + Z$ of the vector space $X = \mathbb{Z}^n$ of arguments of a combinational logic function and the vector space $\mathbb{Z}$ of its outputs. Elements of this space are $(n + 1)$-tuples (x, z) with $z = f(x)$. The group RAG permutes this space under restrictions which preserve the single-valuedness of all functions from X into $\mathbb{Z}$. The equivalence relation induced on the space $\mathscr{F}$ of all such functions is called the *prototype equivalence relation*, after Ninomiya (1958), and the equivalence classes into which $\mathscr{F}$ is partitioned by RAG are called *prototype classes*. The group RAG is the largest subgroup of the affine group on $X + \mathbb{Z}$ that retains the loop-free or feedback-free constraint which prevents combinational logic from becoming asynchronous sequential logic circuitry.

Section IV,A gives a mathematical definition and physical interpretation of elements of RAG and locates each transformation group previously applied to combinational logic within a lattice of major subgroups of RAG. It also explicitly defines the weight-transforming property which distinguishes RAG from permutation groups acting on X alone.

Section IV,B gives results on the number of equivalence classes of functions under RAG for $n \leq 5$. For the case $n = 3$, an explicit breakdown is given of prototype classes into generic classes (under argument permutation, argument, and/or function complementation). More detailed results are also provided on the number of classes of functions with a specific value of their weight or size of $f^{-1}(1)$, under linear and affine groups acting on X.

Section IV,C contains explicit data on 46 out of the 48 possible 5-variable prototypes. These 46 classes were obtained by a nonexhaustive method, and the remaining ambiguity has not been resolved.

A. DEFINITION, INTERPRETATION, AND CONTEXT

In this section, the prototype equivalence relation is interpreted physically, defined algebraically, and placed within the historical context of other equivalence relations of interest in switching theory.

1. Notation

The notation at the beginning of Section II,A will be used. The full linear group and the full affine group on $\mathbb{Z}^n$ will be denoted LG(n) and AG(n), respectively. Elements of LG(n) and AG(n) will be represented as follows (Birkhoff and Maclane, 1953):

$$T_A(x) = x\mathsf{A} \text{ is in LG}(n) \text{ iff } \mathsf{A} \text{ nonsingular over } \mathbb{Z} \tag{101}$$

$$T_{A,c}(x) = x\mathsf{A} \oplus c \text{ is in AG}(n) \text{ iff } T_A \text{ is in LG}(n) \text{ and } c \in \mathbb{Z} \tag{102}$$

The direct product of AG(n) and AG(m) is denoted AG(n) × AG(m) and similarly with LG(n) × LG(m).

2. Engineering Motivation and Definition

In order to motivate our definition of prototype equivalence, some engineering considerations will be introduced. Consider first the group $AG(n + 1)$ of all affine transformations $T_{\bar{A}\bar{c}}$ which operate on vectors z in $\mathbb{Z}^{n+1}$ by post-multiplication: $zT_{\bar{A}\bar{c}} = z\bar{A} \oplus \bar{c}$. We represent these transformations as matrices which operate on ordered pairs (x, z) where x is the input to, and $z = f(x)$ is the output of, an element of $\mathscr{F}$. To show the effects of $T_{\bar{A}\bar{c}}$ on x and z explicitly, we partition the matrix $\bar{A}$ and translation vector $\bar{c}$:

$$T(x, z) = (x; z)\begin{bmatrix} A & a^t \\ b & e \end{bmatrix} \oplus (c; d)$$
$$= (xA \oplus zb \oplus c; ze \oplus xa^t \oplus d) \tag{103}$$

Note that a, b, and c are binary n-tuples, while d and e are binary scalars; furthermore, A can be singular and e can be zero as long as the complete matrix is nonsingular.

If a and b are restricted to be identically zero (and $e = 1$) to retain non-singularity in Eq. (103), we obtain a representation for the direct product group $AG(n) \times AG(1)$ which includes all the subgroups previously treated in the literature of switching theory. The restricted affine group RAG also requires b to be zero and $e = 1$, but a may range over all of $\mathbb{Z}^n$. In other words, the following proper inclusion relation applies:

$$AG(n) \times AG(1) \subset RAG \subset AG(n + 1) \tag{104}$$

Our definition of RAG will be motivated by first studying the effects of $AG(n + 1)$ and $AG(n) \times AG(1)$.

The direct product of two affine groups acting on X and Z, respectively, produces mutually independent encodings on the domain X and the range Z of a switching function, as illustrated in Fig. 5a. Since $AG(1)$ has only two elements, range transformations at most double the size of the transformation group and reduce the number of equivalence classes by a factor of at most two. However, for multiple-output functions, Z becomes $\mathbb{Z}^m$, $AG(n)$ becomes $AG(m)$, and such a direct product extension is nontrivial.

Figure 5a gives a physical interpretation of the condition for equivalence of two functions f and g under the group $AG(n) \times AG(1)$. The domain encoding is a transformation of the form $y = xA \oplus c$, A nonsingular over Z and c arbitrary in $\mathbb{Z}^n$. With the exception of Ninomiya (1958) and Stone and Jackson (1966), the literature on functional equivalence under transformation groups is devoted exclusively to direct product groups with a subgroup of $AG(n)$ acting on X and a subgroup of $AG(m)$ acting independently on the range.

Figure 5b illustrates why the full affine group $AG(n + 1)$ is not a proper object for study. The value of $y = xA \oplus zb \oplus c$ in Eq. (103) depends on the

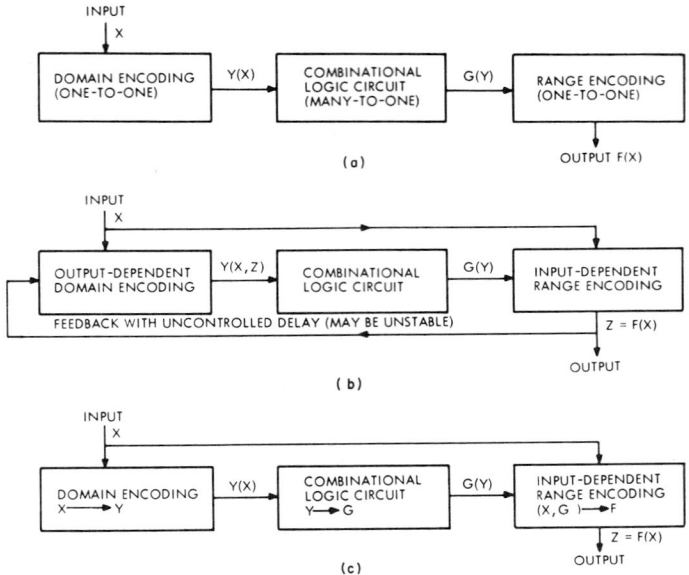

FIG. 5.   Models for single-valued functions imbedded in encoding transformations on $X + \mathbb{Z}$. (a) Fixed domain and range encodings. (b) Mutually-interacting domain and range encodings. (c) Input-dependent range encoding; no feedback.

function output z. Unfortunately, adding zb to the function input implies ambiguities and possible instabilities usually associated with uncontrolled feedback loops within combinational logic circuits. In sequential circuits, feedback is controlled by inserting delays in the feedback path. In our (combinational) logic model, feedback must be avoided by forcing b to be zero in Eq. (103).

On the other hand, feed-forward of the sum xa^t can still be used to modify the output of an arbitrary switching function. This does not violate our restriction to feedback-free combinational logic circuits, yet it generalizes the direct product group $AG(n) \times AG(1)$ by introducing n more free variables (a_1 through a_n). This intermediate case is illustrated in Fig. 5c. In this model, an affine transformation operates on the domain or space of inputs x to produce the (encoded) input $y = x\mathrm{A} \oplus c$ to the logic circuit. Its output $g(y)$ together with the function arguments $x_1, x_2, \ldots, x_n$ are linearly combined by a range transformation which defines $f(x)$ in the following way:

$$f(x) = g(y) \oplus x_1 a_1 \oplus \cdots \oplus x_n a_n \oplus d = g(x\mathrm{A} \oplus c) \oplus xa^t \oplus d$$

where d and each a_i are binary constants. In effect, we have produced $f(x)$ by adding a linear polynomial $d \oplus xa^t$ to the output of a logic circuit g whose input y was defined by an affine transformation of x. The complete expression for $f(x)$ is

$$f(x) = g(x\mathsf{A} \oplus c) \oplus xa^t \oplus d \qquad (105)$$

In matrix form,

$$(y, g) = (x, f)\begin{bmatrix} \mathsf{A} & a^t \\ 0 & 1 \end{bmatrix} \oplus (c, d) \qquad (106)$$

(Eq. (103) with $b = 0$).

DEFINITION. The *Restricted Affine Group* on $X + \mathbb{Z}$ is that subgroup of affine transformations on $X + \mathbb{Z}$ whose partitioned matrix forms obey the constraint $b = 0$.

That RAG is actually a subgroup can be verified by direct multiplication. A more rigorous characterization of RAG will be given in a later section (in terms of invariant subspace restrictions). The name RAG was given to this group by Lechner (1963a,b) (see Section A.4). The group RAG was first studied by Ninomiya (1958), who gave the name "prototype equivalence" to the relation it induces on the function space $\mathscr{F}$.

The term "restricted affine equivalence" is also reasonable for the equivalence relation induced by RAG as a subgroup of $AG(n + 1)$ (although "generalized" affine equivalence is just as valid if we regard RAG as a group which includes the direct product $AG(n) \times AG(1)$). However, we will retain Ninomiya's term "prototype equivalence", not only because there is little likelihood of applying larger transformation groups to $\mathscr{F}$, but also to avoid possible confusion with the recent use of the term "affine equivalence" to refer to the subgroup of $AG(n) \times AG(1)$ which induces Ninomiya's "family" equivalence relation (Stone and Jackson, 1969). In this subgroup, A is restricted to be a permutation matrix, and "affine equivalence" without further qualification has a broader connotation than the family relation.

The reader who desires additional engineering content can now skip to Section V, which describes encoded-input logic, a new logic synthesis technique based on the group RAG. In this approach, the additions implied by Eq. (106) are actually performed by a physical array of cells containing 2-input, single-output EXCLUSIVE OR operators, one for each unit component of A, a, c, and d. In this way, any function f which is equivalent to g under the group RAG can be realized by embedding g within the proper encoding transformation (see Fig. 5c).

The group RAG is actually a special case of a broad class of affine subgroups which will be defined in a later section. The zero submatrix restriction ($b = 0$) will be generalized and related to invariant subspace restrictions when such a group acts on $\mathbb{Z}^{n+1}$.

3.   Lattice of Classical Subgroups

The group RAG includes as subgroups most of the transformation groups whose induced equivalence relations have been studied in the literature of switching theory. Figure 6 illustrates the partial ordering among the subgroups

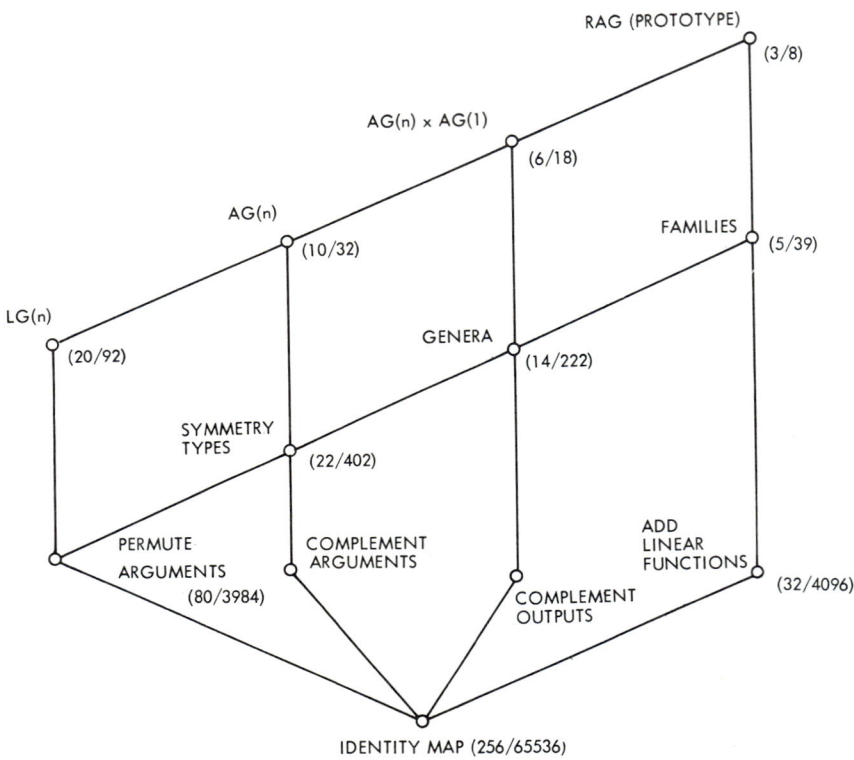

FIG. 6.   Lattice of classical subgroups of RAG. The number of equivalence classes for $(n = 3)/(n = 4)$ are shown in parenthesis.

of RAG. Table III defines each subgroup by specific restrictions on A, a, c, and d. Bibliographic references to enumeration of equivalence classes under these groups are given in the text.

The two left hand columns of this lattice involve transformation groups on the domain X alone. The function output is left untouched. The first column includes the group of input permutations and its generalization to the full linear group (Hellerman, 1961; Harrison, 1964; Slepian, 1960). Slepian's results on group code equivalence under $LG(n)$ have also been translated into switching theory terminology (Lechner, 1967).

The lowest point in the second column is the group that complements

TABLE III
Restrictions on RAG to Obtain Subgroups

Subgroup or Equivalence Relation	A	c	a	d
Prototype	A	c	a	d
Family	P	c	a	d
Add Linear Functions	I	$\emptyset$	a	$\emptyset$
AG(n) x AG(1)	A	c	$\emptyset$	d
Genus	P	c	$\emptyset$	d
Complement Function	I	$\emptyset$	$\emptyset$	d
AG(n)	A	c	$\emptyset$	$\emptyset$
Symmetry Types	P	c	$\emptyset$	$\emptyset$
Complement Arguments	I	c	$\emptyset$	$\emptyset$
LG(n)	A	$\emptyset$	$\emptyset$	$\emptyset$
Permute Arguments	P	$\emptyset$	$\emptyset$	$\emptyset$

Notation:

$\quad$ A = Nonsingular Matrix from LG(n)

$\quad$ P = Permutation Matrix

$\quad$ I = Identity Matrix

a, c = Arbitrary Vector in Z^n

$\quad$ d = Arbitrary Constant (0 or 1) in Z

$\quad$ $\emptyset$ = Zero Vector

input variables by adding a constant vector to them. The product of this group with the group of input variable permutations produces the group whose equivalence classes induced on $\mathscr{F}$ are called "symmetry types" by Polya (1940), Slepian (1967), Singer (1952), and Ashenhurst (1952). When the linear group on X is combined with argument complementations or vector addition on X, we obtain the full affine group AG(n) on the domain of switching functions (Nechiporuk, 1958). Harrison (1963) counted equivalence classes under this group, which has also been studied in the context of finite automata (Gill, 1966; Daykin, 1963).

$\quad$ The third column of Fig. 6 includes those transformation groups which are direct products of domain transformations with range transformations. As previously stated, only two nonsingular range transformations are possible when the function has a single output; either the identity or the functional complementation. The group generated by functional complementation on Z plus variable permutations and complementations on X defines a set of equivalence classes called genera (Ninomiya, 1961). The group AG(n) × AG(1) is the direct product of the affine groups on X and on Z, and Harrison (1963) enumerated the number of equivalence classes under this group in 1962. Harrison used a generalization of Polya's counting theorem (DeBruijn, 1959). This theorem is applicable to direct products of transformation groups on the domain and range of arbitrary mappings (see Chapter IV).

The groups in the right-hand column of Fig. 6 are no longer direct products of domain and range transformation groups, because they involve the addition of linear functions to the output of an arbitrary logic function. Consequently, DeBruijn's theorem does not apply. Ninomiya (1958) was the first to study such transformation groups, the smallest of which translates the function range in an argument-dependent way by adding linear functions (modulo two sums of arguments). When function complementation, argument complementation, and argument permutations are combined with range translation by linear functions, the result is a subgroup of RAG which partitions functions into equivalence classes that Ninomiya called "families". Stone and Jackson (1969) discussed properties of certain graphs which interrelate these families; they applied the rather broad term "affine equivalence" to this family relation which is a subgroup of RAG.

The composition of the direct product group AG(n) × AG(1) with addition of linear functions produces the restricted affine group shown on the upper right corner of Fig. 6. This group was first defined by Ninomiya (1958).

From the previous section, $f(x)$ is represented in terms of $g(y)$ and affine mappings on X and Z as follows:

$$f(x) = g(x\mathsf{A} \oplus c) \oplus xa^t \oplus d \quad (\text{mod } 2) \tag{107}$$

4.   Groups with Invariant Subspace Restrictions

The preceding paragraph justified the definition of RAG from physical considerations (the need to avoid feedback of a function output back to its input). This section defines a general class of linear and affine subgroups which includes RAG as a special case.

Let $\mathbb{Z}^n = \mathbb{Z}^{n_1} + \mathbb{Z}^{n_2} + \cdots + \mathbb{Z}^{n_r}$ be the direct sum of r subspaces $\mathbb{Z}^{n_i}$ and consider the subset of all similarly partitioned matrices in LG(n) with the special quasi-triangular form

$$\mathsf{A} = \begin{bmatrix} \mathsf{A}_{11} & \mathsf{A}_{12} & \cdots & \mathsf{A}_{1r} \\ 0 & \mathsf{A}_{22} & \cdots & \mathsf{A}_{2r} \\ \vdots & \vdots & \ddots & \vdots \\ 0 & 0 & \cdots & \mathsf{A}_{rr} \end{bmatrix} \tag{108}$$

with r nonsingular matrices A_{ii} of dimension n_i along the main diagonal $(n = n_1 + \cdots + n_r)$.

DEFINITION. The Restricted Linear Group RLG $(n_1/n_2/\cdots/n_r)$ is the subgroup of LG(n) which consists of all quasitriangular matrices of the above form.

The motivation for the term *restricted* linear group is the following fundamental property of RLG: Let $s_k = n_1 + n_2 + \cdots + n_k$ for $1 \le k \le r$. Then, for each k, the set of all n-tuples whose first s_k components are zero is a subspace of $\mathbb{Z}^n$ and is closed under the group RLG $(n_1/n_2/\cdots/n_r)$; i.e., it is an invariant subspace under this RLG. Thus each rectangular zero submatrix restriction that can be defined for RLG matrices produces a corresponding invariant subspace restriction.

DEFINITION. Let $n_1 + n_2 + \cdots + n_r$ be any partition of the integer n. The *Restricted Affine Group* RAG$(n_1/n_2/\cdots/n_r)$ is that subgroup of AG(n) formed by combining RLG$(n_1/n_2/\cdots/n_r)$ with the additive group of all translations $x \to x + c$ on $\mathbb{Z}^n$.

In our applications, $r = 2$, $n_2 = 1$, and $n_1 = n$ will be supplied by the context. Therefore, we will drop the indices from RAG. In the general case of a switching function with m binary outputs (Lechner, 1963), the notation RAG(n/m) is used to identify m and n explicitly. The quantities a^t and d expand to m columns, and the unit element in the lower right corner of T becomes a nonsingular $m \times m$ matrix.

This zero restriction on submatrices of elements of RAG leads to considerable theoretical difficulty; in fact, no algorithm known to us can reduce an arbitrary member of the restricted affine group to rational canonical form if the number m of output variables is greater than one. The problem here is that inner automorphisms of RAG (to be admissable as similarity transformations which can produce canonical forms) must also obey this zero submatrix restriction; otherwise, they are not physically realizable as feedback-free encoding transformations. Fortunately, rational canonical forms of restricted affine transformations have been completely characterized in the special case of a single output function in which $m = 1$ and a is a row vector, not a matrix of $m > 1$ rows. This characterization is constructive in the sense that it defines an algorithm by which a single prototype encoding transformation can be reduced to rational canonical form (as far as it is possible to do so) by similarity transformations that are inner automorphisms of the group RAG.

This construction was applied to count equivalence classes of prototype transformations, but heuristic techniques based on harmonic analysis may actually be more effective in actual synthesis of the associated encoding transformations.

5. An Isomorphism between $RLG(1/n/1)$ and $RAG(n/1)$

A natural isomorphism exists between $RAG(n/1)$ operating on $\mathbb{Z}^{n+1}$ and $RLG(1/n/1)$ acting on the set of all vectors $(1; x; z)$ with $x \in \mathbb{Z}^n$ and $z \in \mathbb{Z}$. This set is a coset of the invariant subspace $(0; X; \mathbb{Z})$ of $\mathbb{Z}^{n+2}$. The correspondence preserves the identity of each component A, a, c, and d of the matrix form of RAG (See Eq. (103)), and it permutes the x and z components of the coset $(1; x; z)$ in the same manner as the corresponding element of RAG.

$$T = \begin{bmatrix} 1 & c & d \\ 0^t & A & a^t \\ 0 & 0 & 1 \end{bmatrix} = T_{A,\,a,\,c,\,d} \tag{109}$$

To verify that this matrix has the desired effect on $(1; X; \mathbb{Z})$, premultiply it by $(1; x; z)$ and observe the desired result.

$$(1; x; z)T = (1; xA \oplus c; z \oplus xa^t \oplus d) \tag{110}$$

THEOREM 4.1. The subgroup $RLG(1/n/1)$ of $LG(n + 2)$ is isomorphic to the subgroup $RAG(n/1)$ of $AG(n + 1)$.

Proof: The correspondence defined by inserting components A, a, c, d of Eq. (106) into the matrix $T_{A,\,a,\,c,\,d}$ is obviously one to one. Preservation of the group operation can be verified directly by matrix multiplication which is left as an exercise.

This isomorphism is an important step in the reduction of RAG elements to rational canonical form (a necessary preliminary to enumeration of equivalence classes). Generalization to groups with more general partitions of X and (for $m > 1$) of $\mathbb{Z}^m$ is straightforward.

6. The Weight-Transforming Property of RAG

Transformation groups acting on X alone merely permute points. In contrast, a general element of RAG does not preserve the weight of f (size or cardinality of $f^{-1}(1)$) unless a and d are zero. The following theorem (really a corollary of the fundamental invariance theorem of Section II,C) defines a specific set of different possible weights which characterizes the members of a given prototype class. This set of weights is defined explicitly by the difference between 2^{n-1} and each member of the spectrum or set of integer magnitudes which is included in the range of the Fourier transform f_N^*.

THEOREM 4.2.   The prototype class which contains f contains at least one function g whose weight (size of $g^{-1}(1)$) is m iff m is a member of the set $\{2^{n-1} \pm f_N^*(a) : a \in \mathbb{Z}^n\}$.

Proof: By the transform construction of Section II,A,4, $|f^{-1}(1)| = f^*(0) = 2^{n-1} - f_N^*(0)$. If $g = \bar{f}$, then $g = f \oplus 1$ and $g_N^*(0) = 2^n - f^*(0) = 2^{n-1} + f_N^*(0)$. Now let $g = f \oplus xa^t \oplus d$. By the preceding theorem, $g_N^*(0) = (-1)^d f_N^*(a) = 2^{n-1} - g^*(0) = 2^{n-1} - |g^{-1}(1)|$.

Therefore, $g^{-1}(1) = 2^{n-1} - g_N^*(0) = 2^{n-1} - (-1)^d f_N^*(a) = m$ iff $m = 2^{n-1} \pm f_N^*(a)$ for some $a \in \mathbb{Z}^n$. The value of d (which is arbitrary) determines the $\pm$ sign in this equation.

Using this theorem, it is a simple matter to compute the set of possible weights of functions which are derivable from a given function f by adding a linear function xa^t. The subset of vectors a for which $f \oplus xa^t$ has this weight is also directly identifiable from f_N^*.

B. ENUMERATION OF PROTOTYPE EQUIVALENCE CLASSES

The problem of simply counting equivalence classes of n-input, single-output functions under RAG is complicated by the fact that RAG is not merely a domain transformation group like the symmetry group studied by Slepian and many others. DeBruijn's generalization of Polya's theorem was used successfully by Harrison to count classes under direct products of domain and range transformation groups (see Chapter IV). However, RAG is not a direct product group, so DeBruijn's theorem does not apply. Yet not only the counting problem (how many classes are there?), but also the recognition problem (to which class does a given function belong?) have been solved completely for RAG for $n \leq 4$ (Ninomiya, 1958). Unfortunately, Ninomiya's semiexhaustive technique is difficult to extend to five variables.

The case $n = 5$ was enumerated by a new technique for counting functional equivalence classes under RAG (Lechner, 1963a,b). The main result was that 48 prototype classes are now known to exist for $n = 5$. Unfortunately, the counting procedure does not explicitly define representatives for equivalence classes. Therefore, another approach to identification of class representatives is described in Section IV,C, followed by a tabulation of 46 out of the 48 prototype equivalence classes for $n = 5$.

1.   Application of Polya's Theorem

The counting technique employed by Lechner (1963) appears to involve less restrictive conditions than DeBruijn's theorem. A function f from X to $\mathbb{Z}$ is represented uniquely by its truth table, which is the set of all ordered pairs (x, z) such that $z = f(x)$. Now f can also be regarded as an abstract relation (i.e., as a subset of $X + \mathbb{Z}$ which is called the "graph" of f). An alternate representation of $f(x)$ is by the characteristic function of its graph, i.e., a binary function $C_f(x, z)$ such that $C_f(x, z) = 1$ if $z = f(x)$, and zero otherwise. These two representations for f are illustrated for a 3-input, 2-output example in Table IV.

Since RAG is a subgroup of the general affine group on $X + \mathbb{Z}$, it permutes points in the *domain* of C_f. It is easy to show that the set of all single-valued

TABLE IV
Decomposition into Equivalence Classes

A. Truth Table				B. Graph $C_f(x, z)$							
$x \in X$			$z \in \mathbb{Z}^2$	X:			$\mathbb{Z}^2$: $f_1(x)$	0	0	1	1
$(x_1$ x_2 $x_3)$			$(f_1(x), f_2(x))$	x_1 x_2 x_3			$f_2(x)$	0	1	0	1
0	0	0	0 0	0	0	0		1	0	0	0
0	0	1	1 0	0	0	1		0	0	1	0
0	1	0	1 0	0	1	0		0	0	1	0
0	1	1	0 1	0	1	1		0	1	0	0
1	0	0	1 0	1	0	0		0	0	1	0
1	0	1	0 1	1	0	1		0	1	0	0
1	1	0	0 1	1	1	0		0	1	0	0
1	1	1	1 1	1	1	1		0	0	0	1

relations (i.e., functions) is closed under RAG. Therefore, the following form of Polya's theorem can be applied to enumerate equivalence classes, provided that the required parameters can be evaluated:

THEOREM 4.3 (Polya, 1937). The number $N_{F/G}$ of equivalence classes into which the set F of all subsets of a finite set S is partitioned under a finite group G of permutations on S is defined by:

$$N_{F/G} = \sum_{C_i \in G} [N_{F/C_i}/N_{C_i}] \qquad (111)$$

where the sum is over all equivalence classes C_i of elements in G under its group of inner automorphisms, $N_{F/C_i} = 2^{N_{S/C_i}}$ is the number of subsets of S left invariant by any one element T of class C_i, N_{S/C_i} is the number of transitive subsets or cycles into which S is partitioned by repeated application of T, and N_{C_i} is the number of inner automorphisms of G which leave T invariant.

This theorem is derivable from Harrison's statement of a theorem of Frobenius (Theorem 2.1) in Chapter IV of this book. The symbols α, S, and $I(\alpha)$ of Harrison correspond directly to T, F, and N_{F/C_i} of our Theorem 4.3. Harrison's sum over all group elements α can be replaced by a sum over conjugate classes C_i of group elements, because all group elements α in the same class C_i leave invariant the same number $I(\alpha) = N_{F/C_i}$ of subsets of S. The N_{C_i} T-invariant inner automorphisms of G themselves form a subgroup (see Harrison's Fact 2.3 in Chapter IV) with n_i cosets in G. Therefore, $g = n_i N_{C_i}$ and Harrison's formula becomes:

$$(1/g) \sum_{\alpha} I(\alpha) = (1/g) \sum_{i} N_{F/C_i} n_i = \sum_{i} N_{F/C_i}/N_{C_i} \qquad (112)$$

which is Theorem 4.3.

In our application, $S = X + \mathbb{Z}$, $G = $ RAG, and F is the set of characteristic functions corresponding to truth tables. This form of Polya's theorem depends on the fact that every element of an equivalence class C_i of RAG under its group of inner automorphisms leaves invariant the same number of *single-valued* functions (Theorem 6 of Lechner, 1963). To apply Polya's theorem, it is necessary to determine the cycle structure of a typical element of every class C_i and the number N_{C_i} of T-invariant inner automorphisms of G. Methods for constructively defining these parameters for the group RAG were evolved by Lechner (1963). However, for the sake of brevity, details will be omitted herein.

Fortunately, the number of classes is generally far smaller than the number of elements of G. The parameter N_{F/C_i} is zero for all but a small fraction of the C_i. This helps to reduce the computations.

2. Summary of Results

Table V gives a lower bound for the order of RAG, the total number of classes C_i in RAG, and finally the number of classes whose members leave single-valued functions invariant for $2 \leq n \leq 5$.

Table VI summarizes the results obtained for the number of classes into which RAG partitions F, together with previously published data on other groups for comparison.

TABLE V
Structure of Classes in RAG

n =	2	3	4	5
Order (RAG)	192	21,504	10,321,924	$20 \cdot 10^9$
Number of Classes C_i	13	28	62	124
Classes with $N_{F/C_i} \neq 0$	7	13	32	62

TABLE VI
Comparison of Equivalence Class Counts

Group n =	1	2	3	4	5	6	7
RAG	1	2	3	8	48	>69,000	$>10^{19}$
AG(n) x AG(1)	2	3	6	18	206	7,888,299	$>10^{21}$
AG(n)	3	5	10	32	382	15,768,919	$>10^{21}$
LG(n)	4	8	20	92	2,744	950,998,216	$>10^{24}$
(n-cube symmetries)	3	6	22	402	1,228,158	$>10^{14}$	

Table VII lists representatives for the eight prototype classes when $n = 4$, as chosen by Ninomiya (1958). Ninomiya's table of symmetry types and their class memberships has been reprinted by Harrison (1965). All but one prototype is linearly separable. The numbers of families, domain classes, genera, and functions in each prototype class are also tabulated, along with the Fourier coefficient magnitudes (spectrum) which identifies prototype class membership for $n \leq 4$ (see Fig. 2).

The Fourier spectrum of a function $f(x)$ with range (0, 1) is defined as the unordered set of positive integers which are the absolute values of the coefficients of the Fourier transform f_N^* (i.e., the transform of the (real-valued function $(\frac{1}{2} - f(x))$ defined in Section III. The notation k^n in the last column

TABLE VII
Decomposition into Equivalence Classes

Prototype Index	Representative Function	Threshold Gate Realization	Number of:					Fourier Coefficient Sets
			Prototypes	Families	Domain Classes[a]	Genera	Functions	
8	0		1	1	2	5	32	$8^1 0^{15}$
7	$y_1 y_2 y_3 y_4$	$y_1+y_2+y_3+y_4 \geq 4$	1	1	2	5	512	$7^1 1^{15}$
6	$y_1 y_2 y_3$	$y_1+y_2+y_3 \geq 3$	1	4	3	30	3840	$6^1 2^7 0^8$
5	$y_1 y_2 (y_3 \vee y_4)$	$2y_1+2y_2+y_3+y_4 \geq 5$	1	6	3	52	17,920	$5^1 3^3 1^{12}$
4: 4A	$y_1 y_2$	$y_1+y_2 \geq 2$	1	6	2	19	1120	$4^4 0^{12}$
4B	$y_1 \mathrm{Maj}(y_2 y_3 y_4)$	$2y_1+y_2+y_3+y_4 \geq 4$	1	13	3	74	26,880	$4^2 2^8 0^6$
3	$y_1 \mathrm{Maj}(y_2 y_3 y_4)$ $\vee y_2 y_3 y_4$	$y_1+y_2+y_3+y_4 \geq 3$	1	4	2	33	14,336	$3^6 1^{10}$
2	$y_1 \mathrm{Maj}(y_2 y_3 y_4)$ $\vee y_2 y_3 y_4 \vee \bar{y}_1 \bar{y}_2 \bar{y}_3 \bar{y}_4$	Sum of Index 3 and Index 7 Prototypes[b]	1	4	1	4	896	2^{16}

[a] Under the affine group on the domain plus functional complementation.
[b] Index 7 prototype with complemented variables as inputs.

of Table VII means there are n coefficients equal to $\pm k$ among the transform coefficients. For example, $8^1 0^{15}$ means 15 of the coefficients are 0-valued, and one of them is equal to ± 8.

3. The Three-Argument Case

Ninomiya (1961) studied the correspondence between symmetry types and prototypes in detail for functions of $n \leq 4$ arguments. For $n = 3$, the transformations can be determined by inspection. Particular solutions for the encoding transformation that converts symmetry types into prototypes are shown in Table VIII. The "$+$" sign represents Boolean INCLUSIVE OR addition in this table.

For the 3-input, 1-output situation, only 14 different mappings are required. Each of these converts a generic class into a prototype class; each genus is a union of symmetry types under the equivalence relation induced by functional complementation. Each generic class is a different row of Table VIII. For example, generic class 3 is in prototype class 2 and contains 12 different functions. The 12 different functions in this class may be obtained

PROTOTYPE CLASS K	GENERIC CLASS N	FUNCTIONS PER CLASS	STANDARD SUM (MINTERMS IN $f_N(x)$)	NORMAL FORM	NUMBER OF INVERTERS PLUS AND/OR GATES	LINEARLY ENCODED FORM	NUMBER OF AND GATES PLUS EXCLUSIVE OR
4	1	1	–	0	0	0	0
4	9	6	4,5,6,7	x_1	0	x_1	0
4	10	6	2,3,4,5	$x_1\bar{x}_2 \oplus \bar{x}_1 x_2$	5	$x_1 \oplus x_2$	1
4	11	2	1,2,4,7	$x_1\bar{x}_2\bar{x}_3 + \bar{x}_1 x_2\bar{x}_3 + \bar{x}_1\bar{x}_2 x_3 + x_1 x_2 x_3$	7	$x_1 \oplus x_2 \oplus x_3$	2
3	2	8	7	$x_1 x_2 x_3$	1	$x_1 x_2 x_3$	1
3	6	24	5,6,7	$x_1 x_2 + x_1 x_3$	3	$x_1\bar{x}_2\bar{x}_3 \oplus x_1$	2
3	7	24	1,6,7	$x_1 x_2 + \bar{x}_1\bar{x}_2 x_3$	5	$\bar{x}_1\bar{x}_2\bar{x}_3 \oplus x_1 \oplus x_2 \oplus 1$	4
3	8	8	3,5,6	$x_1\bar{x}_2 x_3 + \bar{x}_1 x_2 x_3 + x_1 x_2\bar{x}_3$	7	$\bar{x}_1\bar{x}_2\bar{x}_3 \oplus x_1 \oplus x_2 \oplus x_3 \oplus 1$	8
2	3	12	6,7	$x_1 x_2$	1	$x_1 x_2$	1
2	14	24	0,5,6,7	$x_1 x_2 + x_1 x_3 + \bar{x}_1\bar{x}_2\bar{x}_3$	7	$\bar{x}_2\bar{x}_3 \oplus x_1$	4
2	4	12	4,7	$x_1\bar{x}_2\bar{x}_3 + x_1 x_2 x_3$	5	$x_1(x_2 \odot x_3 \oplus 1) \oplus x_1$	3
2	13	24	3,4,6,7	$x_2\bar{x}_3 + x_1\bar{x}_3 + x_1 x_2 x_3$	4	$(x_1 \oplus x_2)(x_3) \oplus 1)$	3
2	5	4	0,7	$x_1 x_2 x_3 + \bar{x}_1\bar{x}_2\bar{x}_3$	6	$(x_1 \oplus x_2 \oplus 1)(x_1 \oplus x_3 \oplus 1)$	5
2	12	8	3,5,6,7	$x_1 x_2 + x_2 x_3 + x_1 x_3$	4	$(x_1 \oplus x_2 \oplus 1)(x_1 \oplus x_3 \oplus 1) + x_1 \oplus x_2 \oplus x_3 \oplus 1$	9
		TOTAL: 256			AVERAGE COST: 2.7		2.1

TABLE VIII
Generic Classes of 3-Input Functions

by complementing the function and/or permuting and/or complementing input variables in all possible ways.

Some of the linearly encoded "prototype" functions in Table VIII require one 2- or 3-input AND gate in addition to the EXCLUSIVE OR's, but no function requires more than one. Based on the duality theorem ($\bar{x} \vee \bar{y} = xy \oplus 1$), logical *sum* rather than logical *product* prototype forms appear to be more economical (e.g., generic class 7 becomes $(x_1 \vee x_2 \vee x_3) \oplus x_1 \oplus x_2)$. However, this depends on the generic class representative, since variable and functional complementation are used to map functions *within* each generic class.

The average number of gates and/or inverters is also shown on Table V-8 for each generic class. The products of these numbers with the number of functions per class, summed over all 14 classes and divided by 256 gives an average cost (for randomly chosen functions) of 2.7 gates or inverters for the normal form realization and 2.1 gates or inverters for the linearly encoded form. (However, inversions *within* each generic class were not considered in the average.) Inverters were equated with gates for the normal form realization, because inversion is accomplished by EXCLUSIVE OR gates in the linearly coded version. EXCLUSIVE OR gates were equated with AND/OR gates because circuits of equivalent complexity are expected to result for batch fabricated (LSI) arrays of EXCLUSIVE OR circuits (fanout restrictions compensate for the inherently greater complexity of EXCLUSIVE OR gates).

4. Number of Classes versus Function Weight

Since affine domain transformations merely permute points of X, it is of interest to identify the number of equivalence classes of functions under $AG(n)$ on X, separately for each *weight* $m = |f^{-1}(1)|$ (size or cardinality of the subset $f^{-1}(1)$). This has been done by Nechiporuk (1958) for $n \le 5$ and all m. A similar computation was carried out by Slepian (1960) for the linear group $LG(n)$. Slepian's data (for group codes) was modified by Lechner (1967) for the context of switching functions. Table IX compares the results of Nechiporuk and Slepian for $n \le 5$.

Slepian's data actually extended to $n \le 9$ and $m \le 19$. Table X from Lechner (1967) presents this data in its entirety.

Since $RLG(1/n/1)$ or $RAG(n/1)$ is a subgroup of both $LG(n + 2)$ and $AG(n + 1)$, the data in Tables IX and X might help to establish bounds on the number of equivalence classes under $RAG(n/1)$. By the theorem on weight-transformation in Section IV,A,6, equivalence classes under $RAG(n/1)$ are actually unions of a small number of equivalence classes under $AG(n)$ of different weights. The latter are directly determined by the number of distinct coefficient magnitudes in $f_N{}^*$.

TABLE IX

Number of Equivalence Classes of Weight m under the Groups $AG(n)$ and $LG(n)$

n / m	1		2		3		4		5[b]	
0	1	1	1	1	1	1	1	1	1	1
1	1	2	1	2	1	2	1	2	1	2
2	1	1	1	2	1	2	1	2	1	2
3			1	2	1	3	1	3	1	3
4			1	1	2	4	2	5	2	5
5					1	3	2	7	2	8
6					1	2	3	9	4	14
7					1	2	3	11	5	23
8					1	1	4	12	8	35
9							3	11	9	55
10							3	9	15	84
11							2	7	16	117
12							2	5	23	158
13							1	3	24	204
14							1	2	30	242
15							1	2	30	274
16							1	1	38	290
Totals:	3	4	5	8	10	20	32	92	382	2744

[a] Data on $N_{Am, n}$ came from Nechiporuk (1958); data on $N_{m, n}$ came from Table X.
[b] For $n = 5$, the identity $N_{Am, n} = N_{A2^n - m, n}$ is used; similarly for $N_{m,n}$.

TABLE X

The Number of Equivalence Classes of Switching Functions of n Arguments and Weight m or $2^m - m$.[a]

m \ n	1	2	3	4	5	6	7	8	9
0	1	1	1	1	1	1	1	1	1
1	2	2	2	2	2	2	2	2	2
2	1	2	2	2	2	2	2	2	2
3		2	3	3	3	3	3	3	3
4		1	4	5	5	5	5	5	5
5			3	7	8	8	8	8	8
6			2	9	14	15	15	15	15
7			2	11	23	29	30	30	30
8			1	12	35	54	61	62	62
9				11	55	107	135	143	144
10				9	84	227	329	368	377
11				7	117	495	887	1,070	1,122
12				5	158	1,131	2,728	3,700	4,010
13				3	204	2,698	9,615	16,337	17,748
14				2	242	6,615	39,039	85,028	101,210
15				2	274	16,488	180,773	538,627	770,334
16				1	290	41,016	923,922	4.43067	7.92075
17					274	99,965	1.19623	42.9759	105.524
18					242	235,547	27.3633	459.395	1817.13
19					204	530,998	147.945	5103.84	35432.4

[a] Integer larger than 10^6 are rounded off to six significant figures; entries containing decimal points should be multiples by 10^6.

C. EXPLICIT DEFINITION OF 5-VARIABLE PROTOTYPES

In order to use the prototype equivalence relation for synthesis, it is convenient to have an explicit tabulation of prototype class representatives and a simple means of identifying the class to which an arbitrary function belongs. For $n = 5$, there are only 48 classes (Lechner, 1963a,b); therefore, an exhaustive tabulation was attempted, and the following results are reproduced from Lechner (1968). For $n = 6$, exhaustive tabulation of the (more than 69,000) classes might also be feasible, although its utility is doubtful. For $n \geq 7$, there are too many classes to tabulate. However, other synthesis techniques which require neither tabulation nor recognition will be proposed in Section V. The catalog of prototypes in Table XI of this section verifies the conjecture of Ninomiya (1958) that the spectrum of a function of five or more inputs does not uniquely determine its prototype class. Using this table, we can determine the prototype class of an arbitrary five-input function to within a small ambiguity.

A Monte Carlo technique was used to identify 40 different spectra; an argument based on invariance of level set rank under $RLG(1/n/1)$ showed that six of these spectra are associated with at least two distinct prototype classes ("split" classes). Two of the 48 prototype classes are still not identified. These two classes may have spectra that are distinct from the 40 varieties already found, or they may result from additional class-splitting of one of the six identified split classes, or breakup of one of the 34 other classes.

1.   Monte Carlo Approach

The Monte Carlo analysis used a pseudorandom binary sequence generator to produce a total of 132,000 distinct 5-input functions. The truth tables for most of these functions were defined by adjacent but nonoverlapping 32-bit segments of the sequence generator's output. The Fourier transforms of all these functions were computed and stored on tape. This tape was then re-sorted and scanned to produce a list of the distinct spectra and a count of the number of times each spectrum appeared among the entire sample of functions.

The first computer run generated 130,000 different functions starting from a randomly chosen initial state of the sequence generator and detected 36 distinct spectra. Next, we made a shorter run (2000 functions) starting from an initial condition that was specially chosen for its rarity: a zero value was assigned to 31 of its 32 bit positions. Four new spectra were discovered, but the previously discovered spectra reappeared with consistently greater frequency. The apparent degeneracy of this special initialization process into pseudorandom behavior discouraged us from further attempts at exhaustive sampling.

The initial sample (130,000 functions) is an extremely small fraction of the total number of 5-input functions (approximately four billion). The percentage with which functions having the same spectrum reappeared in this sample are also tabulated. (An asterisk means the spectrum appeared only in the smaller sample of 2000 functions.) These frequencies vary widely and give some idea of the relative sizes of different prototype equivalence classes (or pairs of classes if they have the same spectrum).

We do not know the extent to which a choice for the sequence generator's initial conditions affects the delay before the first appearance of a function with a given spectrum. Neither do we know the sequence length necessary to identify all nontrivial spectra. From an engineering standpoint this is not too important since the functions found in practical applications also depart from randomness to an unknown, but probably more significant, extent. For example, the number of 1's and 0's in a randomly chosen truth table has a binomial distribution, but real-life examples from logic design seem to be biased toward a much lower or higher number of 1's.

2.   Detection of Split Classes

The first split class was detected accidentally while attempting to map a class representative for each spectrum into a unique prototype. To produce a unique prototype, we followed Ninomiya's lead and tried to derive an encoding transformation which permuted the transform coefficients into a maximal lexicographic ordering within the transform vector f_N^*. Standard sums $f^{-1}(1)$ for the particular prototype class representatives that resulted from this attempt are listed in Table XI at the end of this section.

In several cases, we found it impossible to map one function into another with the same spectrum. A constructive proof was then derived to show that such a transformation could not exist within the restricted affine group. Thus, the existence of six split classes was demonstrated. These split classes are identified in the table as pairs A, B within the same type number designation. The Fourier transform coefficient vectors are listed for each split class pair in Table XII below. Proof that functions with these spectra actually split into two equivalence classes under RAG is based on the invariance of level set ranks under the group $RLG(1/n/1)$. It is not known whether the spectrum (size of each level set) plus the ranks of each level set (as a subset of $\mathbb{Z}^n$) is always sufficient to resolve the ambiguity.

3.   Explanation of Results

Each row of Table XI contains (from left to right) a prototype class number, the spectrum of all members of that class, the approximate size of that class, the standard sum of the prototype function chosen as the class representative, and a remarks column.

Class numbers have been assigned in descending lexicographic order of the spectrum of the class. The six cases in which two prototype classes have the same spectrum have been labeled A and B and contain two standard sums. Table XII also identifies the complete Fourier transform coefficient vector $f_N^* = (\frac{1}{2} - f)^*$ for both prototype representatives in each split class.

TABLE XI
5-Variable Prototype Class Representatives

Prototype Class No.	Spectrum	Size of Class	Standard Sum $\left\{i:f_i = 1\right\}$	Remarks
1	$16^1 0^{31}$	•	(no terms)	Degenerate class (n'=0)
2	$15^1 1^{31}$	•	31	
3	$14^1 2^{15} 0^{16}$	•	30, 31	Degenerate class (n'=4)
4	$13^1 3^7 1^{24}$	0.01	29, 30, 31	
5	$12^1 4^7 0^{24}$	<.01	28, 29, 30, 31	Degenerate class (n'=3)
6	$12^1 4^3 2^{16} 0^{12}$	0.05	27, 29, 30, 31	
7	$11^1 5^3 3^4 1^{24}$	0.04	27, 28, 29, 30, 31 ˙	
8	$11^1 5^1 3^{10} 1^{20}$	0.24	23, 27, 29, 30, 31 ˙	
9	$10^1 6^3 2^{12} 0^{16}$	0.02	26, 27, 28, 29, 30, 31	Degenerate class (n' 4)
10	$10^1 6^1 4^4 2^{14} 0^{12}$	0.60	23, 27, 28, 29, 30, 31	
11	$10^1 6^1 2^{30}$	0.05	16, 23, 27, 29, 30, 31	
12	$10^1 4^6 2^{15} 0^{10}$	0.64	15, 23, 27, 29, 30, 31	
13	$9^1 7^3 1^{28}$	< 0.01	25 thru 31	
14	$9^1 7^1 5^2 3^6 1^{22}$	0.57	23, 26 thru 31	
15	$9^1 7^1 3^{12} 1^{18}$	0.40	19, 23, 27, 28, 29, 30, 31	
16	$9^1 5^3 3^9 1^{19}$	3.3	15, 23, 27, 28, 29, 30, 31	
17	$9^1 5^1 3^{15} 1^{15}$	0.66	15, 16, 23, 27, 29, 30, 31	
18	$8^4 0^{28}$	•	24 thru 31	Degenerate class (n'=2)
19	$8^2 6^2 2^{14} 0^{14}$	0.07	23, 25 thru 31	
20	$8^2 4^8 0^{22}$	0.03	22, 23, 26, 27, 28, 29, 30, 31	Degenerate class (n'=4)
21	$8^2 4^4 2^{16} 0^{10}$	0.45	21, 23, 26, 27, 28, 29, 30, 31	
22	$8^1 6^2 4^4 2^{14} 0^{11}$	4.9	15, 23, 26, 27, 28, 29, 30, 31	
23	$8^1 6^1 4^6 2^{15} 0^9$	6.7	15, 19, 23, 27, 28, 29, 30, 31	
24	$8^1 4^{12} 0^{19}$	0.41	3, 15, 23, 27, 28, 29, 30, 31	
25	$8^1 4^8 2^{16} 0^7$	2.5	14, 15, 16, 23, 27, 29, 30, 31	
26	$7^3 5^1 3^7 1^{21}$	0.50	15, 23, 25 thru 31	
27	$7^2 5^4 3^4 1^{22}$	0.62	15, 22, 23, 26 thru 31	
28	$7^2 5^2 3^{10} 1^{18}$	7.4	15, 21, 23, 26 thru 31	
29	$7^1 5^5 3^7 1^{19}$	9.8	7, 15, 23, 26 thru 31	
30	$7^1 5^3 3^{13} 1^{15}$	13.3		Split class
30A			14, 15, 20, 23, 25, 27, 29, 30, 31	
30B			1, 15, 23, 26 thru 31	

The spectrum of a function is a list of the number of occurrences of each integer magnitude as a coefficient of the transform $f_N{}^* = (\frac{1}{2} - f)^*$. The spectrum is invariant for all members of a given class. The notation $16^1 0^{31}$ for the spectrum of class number 1 means that the coefficient ± 16 appears once and 0 appears 31 times in the Ninomiya transform of the function. The positions and signs of these coefficient magnitudes vary over the functions within each class. This is illustrated by the split classes (e.g., 30A and B) for

TABLE XI (Continued)

Prototype Class No.	Spectrum	Size of Class	Standard Sum $\{i:f_i = 1\}$	Remarks
31	$6^6 2^{10} 0^{16}$	0.02	14, 15, 22, 23, 26 thru 31	Degenerate class (n'=4)
32	$6^4 4^4 2^{12} 0^{12}$	2.7		Split class
32A			13, 15, 22, 23, 26 thru 31	
32B			7, 15, 23, 25 thru 31	
33	$6^4 2^{28}$	0.77		Split class
33A			14, 15, 21, 23, 24, 27 thru 31	
33B			1, 15, 23, "4, 26 thru 31	
34	$6^3 4^6 2^{13} 0^{10}$	10.0	11, 15, 22, 23, 25, 27 thru 31	
35	$6^2 4^8 2^{14} 0^8$	15.7		Split class
35A			13, 15, 19, 22, 23, 24, 27, 29, 30, 31	
35B			1, 15, 22, 23, 26 thru 31	
36	$6^1 4^{10} 2^{15} 0^6$	3.9	13, 14, 15, 18, 22, 23, 25, 27, 28, 31	
37	$5^6 3^{10} 1^{16}$	10.5		Split class
37A			7, 11, 15, 22, 23, 25, 27 thru 31	
37B			1, 14, 15, 22, 23, 26, 27 thru 31	
38	$5^4 3^{16} 1^{12}$	2.5	3, 12, 15, 21, 23, 26 thru 31	
39	$4^{16} 0^{16}$	0.28		Split class
39A			6, 7, 11, 15, 22, 23, 24, 25, 27, 28, 29, 31	
39B			0, 1, 14, 15, 22, 23, 26 thru 31	Class 39B is also Degenerate (n'=4)
40	$4^{12} 2^{16} 0^4$	0.32	1, 3, 13, 14, 22, 23, 26 thru 31	

TABLE XII
Fourier Transform Coefficient Vectors for Representatives of Split Classes $(f_N^*(w(j))$ versus $j)$

Prototype Class No.	Value of j→0	16	8	4	2	1	24	20	18	17	12	10	9	6	5	3	28	26	25	22	21	19	14	13	11	7	30	29	27	23	15	31
30A	7	5	5	5	3	3	$\bar{1}$	$\bar{1}$	1	$\bar{3}$	$\bar{1}$	$\bar{3}$	$\bar{3}$	3	$\bar{1}$	$\bar{1}$	$\bar{3}$	$\bar{1}$	3	$\bar{1}$	$\bar{1}$	1	3	$\bar{1}$	$\bar{3}$	1	1	1	3	$\bar{1}$	3	$\bar{3}$
30B	7	5	5	3	3	3	$\bar{5}$	$\bar{3}$	$\bar{3}$	1	$\bar{3}$	$\bar{3}$	1	$\bar{1}$	$\bar{1}$	$\bar{1}$	$\bar{1}$	$\bar{1}$	$\bar{1}$	$\bar{3}$	1	1	$\bar{3}$	1	1	3	3	3	3	1	1	$\bar{1}$
32A	6	6	6	6	4	2	$\bar{2}$	$\bar{2}$	$\bar{4}$	2	$\bar{2}$	0	$\bar{2}$	0	$\bar{2}$	0	$\bar{2}$	0	$\bar{2}$	0	$\bar{2}$	0	$\bar{4}$	2	0	0	4	2	0	0	0	0
32B	6	6	6	4	4	4	$\bar{6}$	0	0	0	0	0	0	$\bar{2}$	$\bar{2}$	$\bar{2}$	0	0	0	$\bar{2}$	$\bar{2}$	$\bar{2}$	$\bar{2}$	$\bar{2}$	$\bar{2}$	4	2	2	2	0	0	0
33A	6	6	6	6	2	2	$\bar{2}$	$\bar{2}$	2	2	$\bar{2}$	$\bar{2}$	2	2	$\bar{2}$	$\bar{2}$	$\bar{2}$	$\bar{2}$	$\bar{2}$	$\bar{2}$	2	2	2	$\bar{2}$	$\bar{2}$	2	2	2	$\bar{2}$	2	2	$\bar{2}$
33B	6	6	6	2	2	2	$\bar{6}$	$\bar{2}$	$\bar{2}$	$\bar{2}$	2	$\bar{2}$	2	2	$\bar{2}$	$\bar{2}$	$\bar{2}$	$\bar{2}$	$\bar{2}$	$\bar{2}$	2	2	$\bar{2}$	2	2	2	2	2	2	2	2	$\bar{2}$
35A	6	6	4	4	4	4	0	0	$\bar{4}$	0	$\bar{2}$	2	$\bar{2}$	$\bar{2}$	$\bar{2}$	$\bar{2}$	$\bar{2}$	$\bar{2}$	$\bar{2}$	$\bar{2}$	2	2	0	4	0	$\bar{4}$	0	0	0	4	2	$\bar{2}$
35B	6	6	4	4	4	2	$\bar{4}$	$\bar{4}$	$\bar{4}$	2	$\bar{2}$	$\bar{2}$	0	$\bar{2}$	0	0	$\bar{2}$	$\bar{2}$	0	$\bar{2}$	0	0	4	2	2	2	4	2	2	2	0	0
37A	5	5	5	5	5	5	$\bar{3}$	$\bar{3}$	1	1	1	1	$\bar{3}$	$\bar{3}$	1	$\bar{3}$	1	$\bar{3}$	1	1	$\bar{3}$	$\bar{3}$	$\bar{3}$	$\bar{3}$	1	1	1	1	1	1	1	1
37B	5	5	5	5	5	1	$\bar{3}$	$\bar{3}$	$\bar{3}$	1	$\bar{3}$	$\bar{3}$	1	$\bar{3}$	1	1	$\bar{3}$	$\bar{3}$	1	$\bar{3}$	1	1	$\bar{3}$	1	1	1	5	1	1	1	1	1
39A	4	4	4	4	4	4	$\bar{4}$	0	4	0	4	$\bar{4}$	$\bar{4}$	4	0	4	0	$\bar{4}$	0	0	0	0	$\bar{4}$	0	4	0	0	0	0	0	0	0
39B	4	4	4	4	4	0	$\bar{4}$	$\bar{4}$	$\bar{4}$	0	$\bar{4}$	$\bar{4}$	0	$\bar{4}$	0	0	$\bar{4}$	$\bar{4}$	0	$\bar{4}$	0	0	$\bar{4}$	0	0	0	4	0	0	0	0	0

which Table XII lists two different complete transform vectors having the same spectrum.

The size of each class is the percentage of occurrences of functions with the given spectrum out of the first pseudorandom sample of 130,000 functions. Four spectra that appeared only in the second sample of 2000 functions are identified by an asterisk in this column.

The last column of Table XII identifies both split classes and degenerate classes, i.e., classes whose prototype representatives are functions of $n' \leq 4$ variables. There are exactly eight degenerate classes; these correspond to the eight prototype classes of four variables listed in Table VII.

EXERCISES

12. In each block of Fig. 5a–c, supply the matrix equation which defines the appropriate component of an element of $AG(n) \times AG(1)$, $AG(n + 1)$, or $RAG(n/1)$, respectively.

13. Using the cartesian product representation of $X + \mathbb{Z}^m$ as on part B of Table IV (with x indexing rows and $y \in \mathbb{Z}^m$ indexing columns), demonstrate the proper subgroup inclusion relationships $RAG(n/m) < S_n \times S_m{}^n < S_{n+m}$ where S_n is the symmetric group of permutations P of the 2^n rows of $X + \mathbb{Z}^m$, and $S_m{}^n$ is the direct product of 2^n permutation groups Q_x, $x \in X$, where Q_x permutes the elements within row x of $X + \mathbb{Z}^m$.

14. By postmultiplying (x, g) first by $T_{A,a,c,d}$ and then by $T_{B,b,e,f}$, derive the multiplication rule for elements of RAG.

15. Define a partial ordering among the symbols I, P, A, $\emptyset$, a, c, d on Table III which will produce the subgroup ordering on the lattice diagram of Fig. 6.

V. SYNTHESIS OF ENCODED INPUT LOGIC

In this section, the prototype equivalence relation is proposed as a basis for actual synthesis. A transformation from the restricted affine group (called an "encoding" transformation) is used to map each desired function f into a simpler prototype function g in the same equivalence class. If "don't care" states appear, they are selected so as to yield the simplest prototype or (more generally) to minimize total cost for prototype plus encoding transformation.

The prototype is designed according to conventional methods and imbedded within the encoding transformation to realize the desired function f. The encoding transformation can be used to simplify either a minimal covering of the function by prime implicants, or the number of threshold elements, if a threshold logic realization is desired. Of course, functions with iterative decompositions, such as adders or counters, and others which are already in simplest prototype form, are identified. In other words, such functions are imbedded within the (trivial) identity transformation.

Section V,A gives some background on alternate approaches to large-scale-integrated (LSI) circuit design. (Some of these approaches are described in more detail in other chapters of this text.) A knowledge of Section IV,A is assumed. The synthesis problem for encoded-input logic is defined in Section V,B. Its solution depends directly on the invariant properties of Fourier transforms from Section II,C. Examples are presented in Section V,C. These examples compare encoded-input logic to alternate approaches within the context of LSI circuit technology.

At its present stage of development, synthesis of encoded-input logic is partly analytic, partly heuristic. The last section (V,D) identifies key mathematical problems whose solutions would remove some of the heuristic aspects of the solution. A knowledge of Section IV is assumed.

A. REVIEW OF LSI LOGIC DESIGN APPROACHES

As background for the encoded input approach to combinational logic design, this section will briefly discuss the impact of LSI technology on logic design criteria and approaches. The LSI circuit technology produced a revolution in methods of manufacturing logic circuitry. Using LSI, large amounts of digital logic (e.g., many hundreds of gates) can be produced as a single LSI *array*, on a semiconductor substrate of microscopic dimensions (e.g., 0.1 to 0.01 in.^{2}).

There are significant differences between the two major semiconductor technologies, bipolar transistors and metal–oxide–silicon (MOS) in current use. In the interest of brevity, these differences will be ignored in the following comparison, which is relative and not absolute. Generally speaking, the advantages and disadvantages cited below apply to both technologies, although not to the same degree.

1. Classification of LSI Design Approaches

Table XIII classifies LSI design approaches by the nature of their application. The economics of LSI manufacturing is quite sensitive to overall complexity of an array and to production volume. Reliability decreases more than inversely as complexity increases, while design and development costs

TABLE XIII
Classification of LSI Approaches

| TYPE | UNIVERSAL (MULTIPLE USES PER SUBSYSTEM) | | SPECIALIZED (ONE-OF-A-KIND PER SUBSYSTEM) | | |
| | VARIABLE MEMORY ARRAYS | UNIVERSAL LOGIC PACKAGES | HOMOGENEOUS ARRAYS | | SPECIALIZED DEVICES AND-OR LAYOUTS |
			LOGIC-ORIENTED	MEMORY-ORIENTED	
APPLICATIONS	PROGRAMS DATA BASE REGISTERS FF ARRAYS ASSOCIATIVE MEMORIES	DIODE MATRICES AND GATE ARRAYS FOR SWITCHING, ENCODERS/DECODERS, AND EXTERNALLY-WIRED LOGIC FUNCTIONS	SPECIAL-PURPOSE LOGIC DESIGN (WITH INTERNAL WIRING)	READ-ONLY PROGRAMS AND TRUTH TABLE LOOKUP	HYBRID A/D APPLICATION AND INTERFACE CIRCUITRY (EXTERNAL OR BORDER LSI ARRAY)
DEVICE LAYOUT:	STANDARD	STANDARD	STANDARD	STANDARD	VARIES WITH APPLICATION
INTERCONNECTIONS 1ST LEVEL (HIGH RESOLUTION)	STANDARD	STANDARD	STANDARD	CUSTOM (IF USED)	
2ND AND 3RD LEVEL (LOW RESOLUTION)	STANDARD	STANDARD	CUSTOM	CUSTOM (IF USED)	

are three orders of magnitude greater than per unit production costs. Consequently, the first LSI packages to be introduced were "universal" types, which could be used repetitively within a system and sold to more than one customer.

Universal arrays are of two basic types: variable (erasable) memories and universal logic arrays. Memory arrays can be produced in high volume without changing the production tooling, either at the substrate level which defines the layout of solid-state devices (transistor amplifiers or diodes) or at the several levels of deposited wiring for device interconnection. Flip-flop arrays used as registers also include logic for transfer and selection gating. This logic is simple and highly repetitive. Its design is straightforward, and its LSI implementation tends to be pin-limited by the word size used on the parallel bus structure for input/output data transfer.

Generally speaking, it is impossible to provide universal arrays of purely combinational logic (not transfer gating mixed with register memory as described above) of a size and complexity comparable to that of variable memories, without introducing an excessive number of external connections. This is self-defeating from the standpoints of reliability and efficiency of semiconductor utilization. Therefore, varying degrees of specialization are introduced into LSI arrays for combinational logic applications.

The specialized approach to LSI design is shown on the right half of Table XIII. It is subdivided according to the degree to which the device layout and/or interconnections must be tailored to the particular application environment and function. Because specialization reduces production quantities, design, development, and testing costs have a significant effect on unit selling price.

The most highly specialized requirements are imposed by interface circuitry which must satisfy electrical (signal) as well as logical compatibility requirements (e.g., high power output drivers which control external devices and low-level signal detectors (sense amplifiers) in core memories). Packages of this type may be required on the borders of LSI logic arrays. The design of these packages varies so greatly with the application that they cannot be discussed in general terms herein.

The other kind of specialized LSI approach is called a *homogeneous array* in Table XIII. Homogeneous arrays may be either logic-oriented or memory-oriented and are further subdivided in the next section.

2. Comparison of Homogeneous Arrays

Table XIV is a more detailed comparison of the memory-oriented approach with three different logic-oriented approaches to homogeneous LSI arrays. These four categories are not mutually exclusive or even partially ordered, as shown later.

Information-theoretic arguments show that for a given n, solution complexity is quite sensitive to the size of the smallest subset $f^{-1}(1)$ or $f^{-1}(0)$ to be recognized. All practical techniques must exploit this sensitivity for economical solutions. The number 2^n can easily exceed the capacity of the table-lookup (read-only memory) approach. When the size of $f^{-1}(1)$ or $f^{-1}(0)$ is so small that it can be recognized (i.e., its argument vector x can be decoded) at much less cost than storing 2^n bits of data (i.e., when the entropy of the problem corresponds to much less than 2^n bits of information), then another approach should be used. In principle, both macrocellular and microcellular arrays are "universal" (given enough components). However, a practical approach must find some way to avoid the disjunctive canonical form's exponential growth factor of 2^n while at the same time providing a computationally efficient synthesis algorithm.

The read-only memory approach has a growing number of applications in which the required number of words (2^n) does not exceed physical limits on array size and/or speed. (Read-only storage access to one of 2048 bits in 1 μsec. or one of 256 bits in 200 nsec, was available by 1969 using MOS and bipolar LSI technology, respectively, at a cost below \$100 in quantity.)

When the topology of the system requires repetitive stages of logic with only a few arguments (e.g., adding and counting circuits, inter-register transfer gates, and address decoders), then the most popular approach is the conventional macrocellular array or its microcellular equivalent. Usually logic for these applications is closely coupled with register memory functions and cannot be simplified further by prototype encoding transformations.

TABLE XIV
Comparison of Homogeneous Arrays

Major Area of Application	Nonrepetitive Logic (Nondecomposable - many inputs, few outputs)	Iterative Arrays (Multistage parallel or sequential operation, with few inputs and outputs per stage)
Conventional or Direct Approach	Read-only Memories	Macro-Cellular Arrays
	(Storage of the truth tables of m output or next-state functions of n arguments)	Custom interconnected flip-flops and Nand gates for inter-register transfers and/or next-stage mappings)
State-of-the-Art Limitation	(Device count limited)	(I/O or pin-limited)
Cost/Size Ratio	$(m+n) \cdot 2^n$	$\underset{(PI)}{Min} \left\{ (mnw) \right\}$
Synthesis Algorithms	Not Needed	Combinatorial Topology
Unconventional or Indirect Approach	Encoded-Input Logic	Micro-Cellular Arrays
	(Nand or threshold gates imbedded within prototype encoding transformations)	(Standardized interconnections among cells with limited fanin/fanout)
State-of-the-Art Limitation	(I/O or pin-limited)	(Device count limited)
Cost/Size Ratio	$\underset{(RAG)}{Min} \left\{ d(n+m)^2 + \underset{(PI,\ TF)}{Min} (mnw) \right\}$	$\underset{(CA)}{Min} \left\{ (mnw) \right\}$
Synthesis Algorithms	Harmonic Analysis	Canonical

Legend:
  m = number of function outputs   PI = Cover by Prime Implicants
  n = number of arguments   CA = Cellular Array Groundrules
  w = Min $(f^{-1}(1))$, $(f^{-1}(0))$ RAG = Restricted Affine Group
  d = average density of encoding matrix TF = Cover by Threshold Functions

By suitable definition of cell complexity and array layout, the macrocell and microcellular array approaches can be made to merge into one another. They actually represent points on a spectrum with coarse (macro) structure at one end and fine (micro) structure at the other. There is an inverse relationship between cell complexity and required number of cells, and a similar inverse relation between interconnection pattern complexity and number of interconnections. The microcellular approach reduces cell size and interconnection pattern to simplest form; it pays for this advantage by a larger number of cells and interconnection paths. Unfortunately, the quantitative dependency of the required number of cells and interconnections on the number and variety of types and intercell connection paths is unknown. Only gross and overly conservative bounds (such as $n2^n$) are known for $\min_{(CA)} (mnw)$ (the minimum cost of cellular arrays subject to a particular set (CA) of ground rules). Synthesis techniques which might improve on this bound are not yet available. For an extensive discussion, see Elspas *et al.*, 1967.

3. Encoded Input Logic

This approach to logic design is illustrated in Fig. 7 and 8. The desired function f is replaced by a simpler function g imbedded in a prototype encoding transformation. This approach combines both macro- and microcellular approaches in a mathematically tractable way. The prototype encoding transformation uses a regularly connected array of simple cells to reduce a given function with n inputs and m outputs to its simplest equivalent form. This imbedded prototype function could be trivial (for example, if the desired function generates the parity check matrix for an error-detecting code). Then, encoded input logic reduces to a *micro*cellular array of 2-input, 1-output cells of a *single type* (EXCLUSIVE OR gates or mod 2 adders). On the other hand, the encoding transformation itself may be trivial (e.g., the identity matrix); for example, if the desired function already is the simplest member

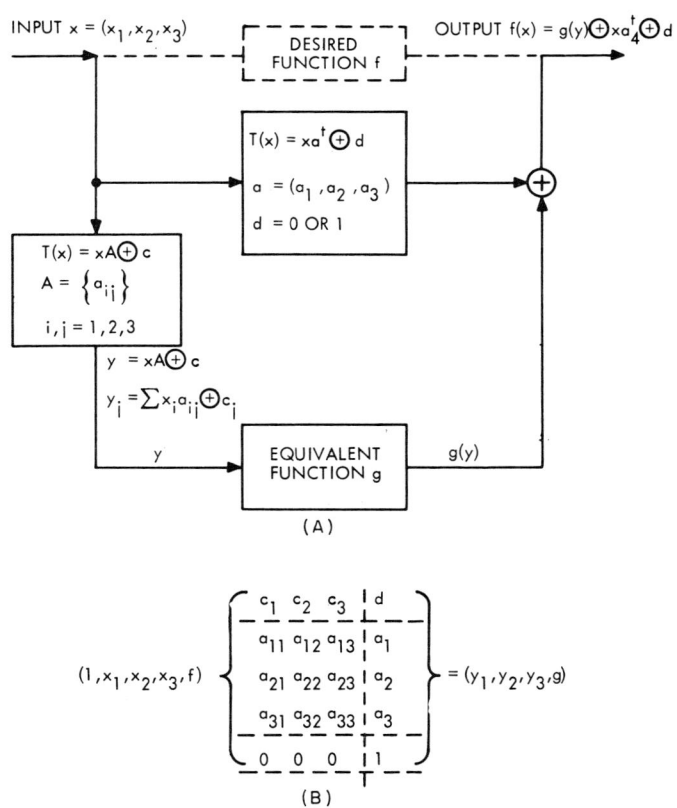

FIG. 7. Successive steps in the process of encoding a single output function. **(a)** Block diagram. **(b)** Matrix form.

SCHEMATIC DIAGRAM :

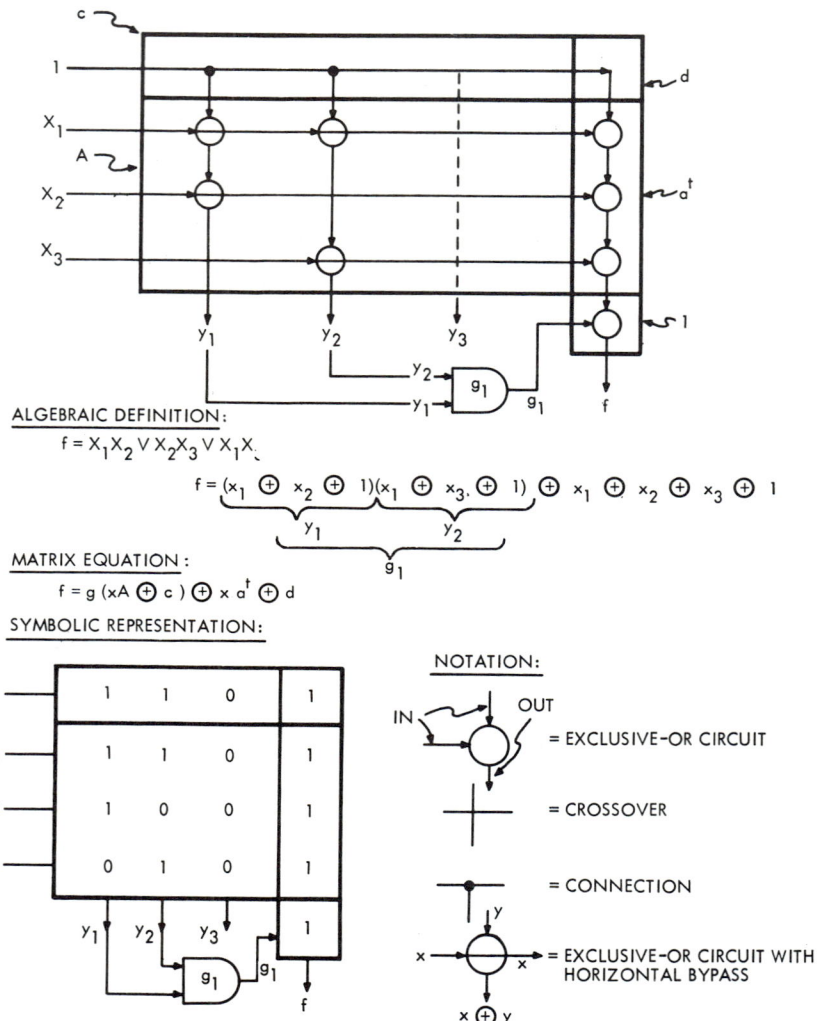

ALGEBRAIC DEFINITION:

$$f = X_1 X_2 \vee X_2 X_3 \vee X_1 X_.$$

$$f = \underbrace{(x_1 \oplus x_2 \oplus 1)}_{y_1}\underbrace{(x_1 \oplus x_3 \oplus 1)}_{y_2} \oplus x_1 \oplus x_2 \oplus x_3 \oplus 1}_{g_1}$$

MATRIX EQUATION :

$$f = g \, (xA \oplus c) \oplus x \, a^t \oplus d$$

SYMBOLIC REPRESENTATION:

NOTATION:

FIG. 8. Equivalence of abstract and schematic representations.

of its prototype equivalence class. In this case, $g = f$, and the logic reduces to a conventional *macro*cellular array. Of course, the prototype can be realized as a microcellular array to complete one full circle of inclusion relationships (which shows that these approaches cannot be partially ordered). In particular, logic which is highly repetitive and decomposable (for example, interregister transfer gating or iterative counting and adding circuits) tends to have macrocellular logic realizations that cannot be improved upon by prototype encoding transformations. The particular example shown in Fig. 8 is for illustrative purposes only. It would not be practical since the cost of encoding outweighs the difference between the normal form for the majority function and its prototype (a 2-input AND gate).

The microcellular approach to LSI can be reduced to linear-input logic but only if *additional* structure is imposed on one section of the array (the encoding transformation), while at the same time intracell complexity and intercell wiring constraints are relaxed in another direction (to permit macrocellular realization of the imbedded function). These qualifications have a very important mathematical consequence. Well-defined synthesis techniques are available both for the encoding transformation (as described herein) and for the imbedded prototype function (either of the approaches in the top row of Table XIV can be used). A significant *physical* implication goes hand in hand with this mathematical tractability: The array size can be strictly bounded by $(n + 1)^2$ $\min_{RAG} \{d\}$, and the imbedded macrocellular array is strictly bounded by $\min_{RAG} \min_{PI, TF} \{mnw\}$. (To some extent, these two cost elements vary inversely and can be traded against one another.) The notation used here and in Table XIV has this significance: $\min_{RAG} \{d\}$ indicates that the array density (fraction of EXCLUSIVE OR cells actually utilized as contrasted to mere wiring crossovers) depends on the encoding transformation from within RAG; $\min_{PI, TF} \{\min_{RAG} \{mnw\}\}$ signifies that prototype encoding transformations can be used to reduce the complexity of the imbedded function. This complexity is characterized exactly by its cubical complex, for which the parameters n, w, and m are merely a convenient shorthand. As a practical matter, however, it is fortunate that w and m are (empirically) related to the cost of the minimal covering by prime implicants. (In principle, we would like to find the minimum PI or TF cost for every function in the prototype equivalence class, and then select the minimum. However, the number of equivalent functions grows as 2^{n^2}, and exhaustive search is impossible. Therefore, we are forced to use simpler criteria, such as maximize the correlation between f or $\bar{f}$ and the $(n + 1)$ Fourier basis functions of zero or one argument variable. Selection of the proper encoding transformation can also be based on reducing the number of levels of threshold logic needed to realize the function. Again, heuristic criteria are used for selection, based on the Fourier coefficients.

Ninomiya (1958) used an empirical criterion for prototype selection in which the Fourier coefficient magnitudes were arranged lexicographically in descending order of their size, after first arranging their argument vectors $w \in \mathbb{Z}^n$ in *increasing* order of their weight. Unfortunately, the prototype thus selected does not always have a least-cost minimal cover, compared to other possible rearrangements of the transform coefficients, as a counterexample herein will demonstrate. Perhaps more complicated measures of complexity will produce more accurate relative predictions of the cost of prototype realization. Some possibilities for further research are discussed in Section V,D.

B. SYNTHESIS TECHNIQUES FOR ENCODED INPUT LOGIC

The problem to be considered herein is the following: given a function $f(x)$ from $X = \mathbb{Z}^n$ to $\mathbb{Z}$, how can we find a simplest realization of the form

$$f(x) = g(x\mathsf{A} \oplus c) \oplus xa^t \oplus d \qquad (113)$$

This expression is an identity for all x if and only if f and g are in the same prototype equivalence class. In other words, the synthesis problem may be restated as follows: Given f, identity the prototype class to which f belongs, then find the simplest combination of a prototype class representative g and encoding transformation components A, a, c, and d which satisfy Eq. 113.

A solution to this problem is feasible using the invariant properties of the Fourier transform under prototype transformations. These properties are restated here from the fundamental invariance theorem of Section II,C:

$$f(x) = g(x\mathsf{A} \oplus c) \oplus xa^t \oplus d$$

iff

$$f_\mathrm{N}(x) = (-1)^{xa^t \oplus d} g(x\mathsf{A} \oplus c) \qquad (114)$$

iff

$$g_\mathrm{N}^*(w) = (-1)^{wc^t \oplus d} f_\mathrm{N}^*(w\mathsf{A}^t \oplus a)$$

This identity relates the migration of the subset $f^{-1}(1)$ through the domain X (and its combination with corresponding subsets of the linear function $(xa^t \oplus c)$) to the corresponding permutation (or sign changes) of the transform coefficients $f_\mathrm{N}^*(w)$. However, this is useful only to the extent that the coefficient pattern f_N^* can be related to function complexity. Ninomiya (1958) found empirically that the simplest prototype representative normally had an arrangement of Fourier coefficients $f_\mathrm{N}^*(w)$ which (as far as possible) concentrated the largest coefficient magnitudes at points $w = e_i$ with unit weight $|w| = \sum w_i = 1$. The next two subsections will show that the weight-defining property of $f^*(0)$ and the implicant-determining property of $(f^*(0) \pm f^*(e_i))$ provide not only a rationale for this empirical approach but also an explicit iterative approach for reducing a function to simplest form.

The cubical complex defined by $f^{-1}(1)$ is invariant to symmetry transformations (argument permutations and complementations). However, prototype transformations have a significant effect on the complexity of the subcube structure of $f^{-1}(1)$. This is obvious from the fact that prototype transformations compute parity functions, all of whose prime implicants are essential 0-cells or minterms, reducing them to trivial functions of a single variable. Before discussing this effect, the non-weight-preserving property of range translations ($f \oplus xa^t$) will be discussed.

1.   Altering the Function Weight by Addition of Linear Functions

Prototype transformations have the ability to change the *weight* of f (size of $f^{-1}(1)$) since they include the addition of linear functions as a subgroup. In this way, a function pair $(f, \bar{f})$, both with many minterms (size of $f^{-1}(1)$ approaching 2^{n-1}) can often be reduced to another pair $(g, \bar{g})$ with $|g|$ less (or more) and $|\bar{g}|$ more (or less) than 2^{n-1}. It has been observed (empirically) that functions with small or large values of $|f|$ are often easier to implement than functions whose weight approaches 2^{n-1}.

An algorithm to find the linear function which maximizes or minimizes the weight of f or $\bar{f}$ is easily defined:

Let $|f|$ denote the weight of f or cardinality of $f^{-1}(1)$. Then $|f| = f^*(0)$ because all entries are $(+1)$ in the first column of the matrix Q such that $fQ = f^*$. From Section IV,A,6, the first coefficients $f^*(0)$ and $f_N^*(0)$ are related as follows:

$$f_N^*(0) = 2^{n-1} - f^*(0) = 2^{n-1} - |f| = -(2^{n-1} - |\bar{f}|) \qquad (115)$$

$$-2^{n-1} \le f_N^*(0) \le 2^{n-1} \qquad (116)$$

In other words, $|f|$ will be reduced to a minimum by maximizing $f_N^*(0)$, and $|f|$ will be maximized by minimizing $f_N^*(0)$.

The next question is how to select a linear function $xa^t \oplus d$ which, when added to f, produces a function g of minimum or maximum weight. Section IV,A,6 provides the answer. Substituting $w = 0$, we have

$$g_N^*(0) = 2^{n-1} - |g| = (-1)^d f_N^*(a) \qquad (117)$$

That is, $|g| = 2^{n-1} \pm f_N^*(a)$, depending on the value of d. In other words, either the maximum or the minimum magnitude of $|g|$ or $|\bar{g}|$ is produced by the same value of a, which may be any w such that $f_N^*(w)$ has the largest *absolute value*.

$$|\bar{g}|_{\max} = |g|_{\max} = 2^{n-1} + \max_a |f_N^*(a)| \qquad (118)$$

Furthermore, *which* of g or $\bar{g}$ has maximum weight is determined solely by the value of the (arbitrary) binary constant d. (This is obvious because $g(x) \oplus d = \bar{g}$ iff $d = 1$.)

2. A 4-Argument Example

The following example shows that adding a linear function can provide a valid alternate to a normal form realization, at least if speed is not a limiting factor.

Figure 9a is the truth table of a function $f(x)$ with $n = 4$, shown as a

f	0	1	2	3 = j
i = 0	1	0	1	0
4	0	1	0	0
8	1	1	0	1
12	0	0	0	1

(a)

f*	0	1	2	3
0	7	-1	1	1
4	3	3	1	1
8	-1	3	1	-3
12	-1	3	-3	1

(b)

g	0	1	2	3
0	1	1	1	1
4	1	1	1	0
8	0	1	1	1
12	0	1	0	0

(c)

g*	0	1	2	3
0	11	-1	1	-3
4	3	-1	-3	1
8	3	3	1	1
12	-1	-1	1	1

(d)

FIG. 9. A 4-argument example. (a) Truth table of f. (b) Fourier transform f^*. (c) Truth table of g. (d) Fourier transform g^*. Alternate forms for f: $f(x) = \bar{x}_1\bar{x}_2\bar{x}_4 \vee x_1x_2\bar{x}_3 \vee x_1x_3x_4 \vee \bar{x}_1x_2\bar{x}_3x_4$; $f(x) = g(x) \oplus xa^t = (\bar{x}_1\bar{x}_4 \vee \bar{x}_3x_4 \vee \bar{x}_2x_3) \oplus x_1 \oplus x_2 \oplus x_4$.

matrix F_{ij}. (The row and column indices i and j have $(x_1x_2 00)$ and $(00x_3 x_4)$ for their binary codes, and the (i, j)th entry is $f(x(i + j))$.) Figure 9b is the Fourier transform of $f(x)$, computed as $Q_2 FQ_2$. (Q_2 is defined in Section II,A,4.) A minimal normal form for f contains 4 prime implicants (three have 3 and one has 4 literals). The total cost of f is five gates with 17 gate inputs.

From the preceding discussion, it is apparent that we can increase the weight $|f|$ from $7 = (2^3 - 1)$ to $(2^3 + 3) = 11$ by adding the linear function xa^t to f, where a is any of the four vectors w such that $|f^*(w)| = + 3$. Increasing $|f|$ *may* increase the chance of finding larger implicants with four gate inputs. For example, suppose $a = (1101)$, and define $g(x)$ as the sum of f and the linear function xa^t: $g(x) = f(x) \oplus x_1 \oplus x_2 \oplus x_4$. The functions g and g^* are defined in Figs. 9c and 9d. Notice that $|g| = g^*(0) = 11$. Note also that the dot products of the first row and first column of g^* with $(1, -1, 1, -1)$ and $(1, 1, 1, 1)$, respectively, add up to 16, indicating that at least two 4-element cosets are implicants of g by the test of Section III,A,3.

The minimal cover of g contains three prime implicants $\{(0\text{-}\text{-}0), (\text{-}\text{-}01),$ $(\text{-}01\text{-})\}$. In other words, $g(x) = \bar{x}_1\,\bar{x}_4 \vee \bar{x}_3\,\bar{x}_4 \vee \bar{x}_2\,x_3$, which requires only four gates and nine gate inputs. Another three (EXCLUSIVE OR) gates of two inputs each are required to add xa^t to g; thus, the total cost becomes seven gates with 15 inputs, rather than five gates with 17 inputs. If gate inputs, rather than gate count, is the more important criterion, then the second approach is an improvement over the first one. If a single 4-input EXCLUSIVE OR gate is available, the total cost becomes five gates with 13 gate inputs. For LSI logic, the number of gate inputs is an important factor since it affects the complexity of deposited interconnections and the number of wiring crossovers.

The selection of this particular value of a was fortuitous. Not all choices reduce f to such a simple form; some must be followed by a linear transformation $x \rightarrow xA \oplus c$ on the domain of f, and the total number of gate inputs is greater. As of now, the choice of a must proceed by trial and error exhaustion of the possibilities.

3. Selection of Linear Domain Transformations

In this section, we give a heuristic technique for selecting a linear transformation which tends to maximize the size and minimize the number of prime implicants in a minimal cover of g or $\bar{g}$. A related problem is to select a transformation which minimizes the number and complexity of threshold functions required to realize g. This problem will not be considered herein, but it appears to be a fruitful direction for future research.

The implicant detection theorem of Section III,A,3 provides the basis for heuristic selection. That theorem says

$$V_b + c \subseteq f^{-1}(z) \quad \text{iff} \quad (f^* \ulcorner V_{\bar{b}})_c{}^* = 2^n z \tag{119}$$

If b is any subspace of dimension $n - 1$, then $\bar{b}$ is a unit vector e_i, and $(f^* \ulcorner V_{\bar{b}})$ is just the 2-tuple $\{f^*(0), f^*(e_i)\}$. Its inverse transform has $f^*(0) \pm f^*(e_i)$ for its two components. Now suppose $\bar{b}$ is a vector of weight 2 ($b = e_i \oplus e_j, 1 \leq i \leq j \leq n$). Then $(f^* \ulcorner V_{\bar{b}})$ is $(f^*(0), f^*(e_i), f^*(e_j), f^*(e_i \oplus e_j))$, and the four components of its inverse transform are the dot products of this vector with $(1, 1, 1, 1), (1, -1, 1, -1), (1, 1, -1, -1)$, and $(1, -1, -1, 1)$, respectively. This clarifies the role of $f^*(0)$ and $f^*(e_i)$ in determining prime implicants. An implicant of f (or $\bar{f}$) with exactly k literals will exist if and only if one of the 2^k components of $(f^* \ulcorner V_{\bar{b}})^*$ is 2^n (or 0). Since $f^*(0)$ enters every such sum, adding xa^t to $f(x)$ (thus moving the largest coefficient $f^*(a)$ to the zero position) tends to maximize the probability that g or $\bar{g}$ will possess large implicants.

Further progress toward large implicants can be made by mapping the n next largest spectral coefficients of $f_N^*(w \oplus a)$ into the positions $w = e_i$, $1 \le i \le n$. Let the original positions of these large coefficients be at $w = a_1, a_2, \ldots, a_n$. To produce $g_N^*(e_i) = f_N^*(e_i A^t \oplus a)$, it is only necessary to solve the equation

$$e_i A^t \oplus a = a_i, \qquad 1 \le i \le n \tag{120}$$

Since $e_i A^t$ is merely the ith row of A^t, the matrix A is defined by

$$i\text{th row of } A^t = a_i \oplus a, \qquad 1 \le i \le n \tag{121}$$

In other words, the ith *column* of A is merely the vector $(a_i \oplus a)^t$, where xa^t is the linear function which maximized g_0^*, and a_i is the location of the coefficient $f^*(w)$ which ranks ith in magnitude after $f^*(a)$ itself.

4. A Pathological Example

Before going to a realistic problem, another, more or less pathological example, will be considered. This indicates how a function which would be very costly to implement as a minimal normal form becomes trivial with linear domain encodings.

The classical procedure for simplifying a Boolean expression is to find a minimal normal form, which is a union of prime implicants whose logical sum has the same truth value as the desired function. The effectiveness of this simplification process depends on the nature of the original Boolean expression and the minimization criteria. One example for which this approach runs into serious difficulty is the function represented by the 16×8 binary array in Table XV. Each column of this array corresponds to one of eight input variables to a function $f(x)$, and each row of the array represents an assignment of 0 or 1 values to all variables x_i in the vector x of input arguments such that $f(x) = 1$. For all vectors not listed on the table, $f(x) = 0$. This function has 16 essential prime implicants, each one a product of all eight argument variables or their negations. The minimal normal form Boolean expression for $f(x)$ is the logical sum of these 16 products.

As a matter of practical convenience, the total number of occurrences of the argument variables is often used as a measure of the cost of mechanizing such an expression. For 2-level logic, the number of input signals to individual logic gates is 144.

An alternate representation for the domain of $f(x)$ in this example is a vector space $X = \mathbb{Z}^8$. The 16 vectors in $f^{-1}(1)$ form a closed subset of this vector space under modulo two addition. In other words, $f^{-1}(1)$ is a 4-dimensional subspace of X, and our problem becomes one of testing an arbitrary input configuration for membership in this subspace.

TABLE XV
8-Variable Example

	x_1	x_2	x_3	x_4	x_5	x_6	x_7	x_8
0	0	0	0	0	0	0	0	0
1	1	1	0	1	0	0	1	0
2	1	1	1	0	1	0	0	0
3	0	1	1	1	0	1	0	0
4	0	0	1	1	1	0	1	0
5	1	0	0	1	1	1	0	0
6	0	1	0	0	1	1	1	0
7	1	0	1	0	0	1	1	0
8	0	0	1	0	1	1	0	1
9	0	0	0	1	0	1	1	1
10	1	0	0	0	1	0	1	1
11	1	1	0	0	0	1	0	1
12	0	1	1	0	0	0	1	1
13	1	0	1	1	0	0	0	1
14	0	1	0	1	1	0	0	1
15	1	1	1	1	1	1	1	1

To simplify $f(x)$, we note that there must exist a 4-dimensional subspace V of X which is the nullspace of $f^{-1}(1)$. For this example, $V = f^{-1}(1)$ (i.e., $f^{-1}(1)$ is its own nullspace), and each row vector of the table is orthogonal (mod 2) to every other one. Therefore, any linearly independent set of four 8-tuples from the table can be selected as the columns of an 8×4 binary matrix $A = \{a_{ij}\}$. The matrix equation $y = x A$ represents a many-to-one transformation from X into a space Z^4 of binary vectors $(y_1 y_2 y_3 y_4)$. This transformation has the unique property that $y = x A = (0000)$ if and only if $x \in f^{-1}(1)$. Therefore, for $y = x A$ define $g(y) = (\bar{y}_1 \bar{y}_2 \bar{y}_3 \bar{y}_4)$. Then, for any x in X, $g(y) = 1$ if and only if $f(x) = 1$. In other words, $f(x)$ can be mechanized in two steps:

$$f(x) = g(y(x)) \tag{122}$$

where $y(x) = x A$ and $g(y) = (\bar{y}_1 \bar{y}_2 \bar{y}_3 \bar{y}_4)$.

The cost of $g(y)$ is only four gate inputs. Suppose that the circuit complexity of two 2-input mod 2 sum operators or EXCLUSIVE OR gates is comparable to that of one 4-input AND/OR gate. Since each column of A contains four unit entries, each linear equation $y_j = \sum x_i a_{ij}$ (mod 2) adds six gate inputs (two for each nonzero a_{ij} except the first one in each column). There are many different choices for the matrix A, but all of them require 32 gate inputs because row 15 of Table XV must be included in A. The total of 28, compared to the original cost of 144, gate inputs represents a very significant reduction in complexity.

C. A 6-VARIABLE EXAMPLE

The following example is a function of six arguments which is the kernel of the 8-variable truth table exhibited on page 79 of Elspas *et al.* (1967). This truth table defines a universal logic module (ULM) (see Chapter VI), which is capable of realizing any 4-argument function by properly permuting and selecting signal inputs. It has the advantage of being complex and also realistically motivated. The truth table, Fourier transform, and spectrum of this 6-argument function, which has 22 minterms in the subset $f^{-1}(1)$, is shown in Table XVI. A list of prime implicants for f and $\bar{f}$ is shown in Table XVII. Both tables were generated by a preliminary version of the prime implicant extraction algorithm described in Section III,B. This program was written in the BASIC language for an interactive time-sharing system and is relatively inefficient (e.g., it did not implement the test of Section III,B,4 but searched the entire list of prime implicants to reject redundant ones); 5- or 6-argument problems require 5 or 11 sec, respectively, of central processor time.

A minimal cover was manually generated from the list of 12 PI's of f (those whose ternary symbol A is followed by $F = 1$ rather than $F = 0$ on Table XVII), plus eight minterms or 0-cells of f that are not covered by any of the PI's. This cover is defined by Table XVIII; all of the 0-cells and all but the last two 1-cells are essential PI's. The total number of inputs to the ten AND gates which realizes these terms is $48 + 40 = 88$. Adding the OR gate and its 16 inputs gives 11 gates and 104 gate inputs as a measure of complexity for the 2-level normal form of this function.

The first step in synthesis of encoded input logic is to add a linear function which will not only minimize the weight of $f \oplus xa^t \oplus d$ but also move large coefficient magnitudes into the position $f^*(e_i)$ where e_i is a unit vector. The possible values for $f^*(0) = |f|$ or $2^n - |f|$ are $32 \pm v, v = 2, 4, 6, 8, 10$. (From Section IV,A,6, if $h(x) = f(x) \oplus xa^t$, then $2^{n-1} - h^*(0) = h_N^*(0) = f_N^*(a) = -f^*(a)$ for $a \neq 0$.) Thus, $|f| = 22$ or 42 is the best attainable value,

TABLE XVI
Truth Table, Transform, and Spectrum for 6-Input Example

		TRUTH TABLE							
ROW I	■COL J INDICES:								
I:	J= 0	1	2	3	4	5	6	7	W(I)
0	1	0	0	1	0	0	0	0	0
8	0	1	1	0	0	0	0	1	1
16	0	0	1	0	0	0	0	1	1
24	0	0	0	0	0	1	1	0	2
32	1	0	1	1	0	1	1	0	1
40	0	0	1	0	1	1	0	0	2
48	0	0	0	0	0	1	1	0	2
56	0	1	0	0	1	0	0	1	3

		FOURIER TRANSFORM:							
ROW I	■COL J INDICES:								
I:	J= 0	1	2	3	4	5	6	7	W(I)
0	22	0	-2	-4	-2	4	-2	0	0
8	0	2	-4	2	4	2	0	10	1
16	4	2	0	2	8	2	-4	2	1
24	2	0	2	4	2	-4	2	8	2
32	-4	-2	-4	2	4	2	4	-2	1
40	-2	0	2	8	2	0	-2	-8	2
48	-2	-4	2	4	2	-4	6	-4	2
56	-4	2	4	-2	-4	-2	-4	10	3

SPECTRUM HAS   7 MAGNITUDES

SPECTRUM:

INDEX	ABSVAL	#POS	#NEG	TOTAL#
1	0	8	0	8
2	22	1	0	1
3	2	18	10	28
4	4	8	12	20
5	10	2	0	2
6	8	3	1	4
7	6	1	0	1

and the vector a may be zero or either of the two vectors w where $f^*(w) = 10$ in Table XVI. Choosing $a = (001111)$ will not only retain $|f| = 22$, but it will also interchange the coefficient pairs (f_{16}^*, f_{31}^*), (f_{32}^*, f_{47}^*), and (f_{48}^*, f_{63}^*), which puts 8, 8, and 10 into the unit vector locations $(i + j) = 16, 32$, and their sum $(i + j) = 48$.

Table XIX is the truth table, transform, and spectrum of the sum $f \oplus xa^t$. Note that the spectrum is identical, but the transform coefficient locations have been permuted pairwise by adding a linear function to f. Table XX shows that the new function pair $(g, \bar{g})$ has fewer prime implicants (45 instead of 53); furthermore, f now has some 2-cells, and $\bar{f}$ has some 3-cells.

TABLE XVII
Prime Implicants for 6-Input Example

1-CELLS (PI'S WITH 5 LITERALS)			2-CELLS (PI'S WITH 4 LITERALS)		
PI# 28	A=10001-	F= 1	PI# 1	A=0001--	F= 0
PI# 29	A=10100-	F= 0	PI# 2	A=0110--	F= 0
PI# 30	A=10110-	F= 1	PI# 3	A=1100--	F= 0
PI# 31	A=10111-	F= 0	PI# 4	A=010-0-	F= 0
PI# 32	A=1000-0	F= 1	PI# 5	A=00-10-	F= 0
PI# 33	A=1010-1	F= 0	PI# 6	A=01-00-	F= 0
PI# 34	A=000-10	F= 0	PI# 7	A=11-01-	F= 0
PI# 35	A=011-11	F= 0	PI# 8	A=00-1-0	F= 0
PI# 36	A=100-10	F= 1	PI# 9	A=01-0-1	F= 0
PI# 37	A=101-11	F= 0	PI# 10	A=11-0-0	F= 0
PI# 38	A=110-11	F= 0	PI# 11	A=01--00	F= 0
PI# 39	A=111-10	F= 0	PI# 12	A=0-010-	F= 0
PI# 40	A=10-001	F= 0	PI# 13	A=0-01-0	F= 0
PI# 41	A=10-010	F= 1	PI# 14	A=0-0-01	F= 0
PI# 42	A=10-101	F= 1	PI# 15	A=0-1-00	F= 0
PI# 43	A=10-111	F= 0	PI# 16	A=0--100	F= 0
PI# 44	A=1-0101	F= 1	PI# 17	A=-1000-	F= 0
PI# 45	A=1-0110	F= 1	PI# 18	A=-1101-	F= 0
PI# 46	A=1-0111	F= 0	PI# 19	A=-100-1	F= 0
PI# 47	A=1-1100	F= 1	PI# 20	A=-110-0	F= 0
PI# 48	A=1-1110	F= 0	PI# 21	A=-10-00	F= 0
PI# 49	A=-00000	F= 1	PI# 22	A=-1-000	F= 0
PI# 50	A=-00011	F= 1	PI# 23	A=-1-011	F= 0
PI# 51	A=-00111	F= 0	PI# 24	A=--0001	F= 0
PI# 52	A=-01010	F= 1	PI# 25	A=--0100	F= 0
PI# 53	A=-01110	F= 0	PI# 26	A=--1000	F= 0
			PI# 27	A=--1011	F= 0

TABLE XVIII
Minimal Cover by Prime Implicants

PI Number	One-Cells	O-Cells
49	-00000	001001
50	-00011	001111
52	-01010	010010
44	1-0101	010111
45	1-0110	011101
47	1-1100	011110
42	10-101	111001
28	10001-	111111

TABLE XIX
Truth Table, Transform, and Spectrum after Adding Linear Function

TRUTH TABLE

ROW I	COL J INDICES: J= 0	1	2	3	4	5	6	7	W(I)
0	0	0	0	0	0	1	1	0	0
8	0	0	0	0	1	0	0	0	1
16	1	0	1	1	0	1	1	1	1
24	0	1	1	0	1	1	1	1	2
32	0	0	1	0	0	0	0	1	1
40	0	1	0	0	0	1	0	1	2
48	1	0	0	1	0	0	0	0	2
56	0	0	1	0	0	0	0	0	3

FOURIER TRANSFORM:

ROW I	COL J INDICES: J= 0	1	2	3	4	5	6	7	W(I)
0	22	0	-2	-4	-2	4	-2	0	0
8	0	2	-4	2	4	2	0	10	1
16	-8	-2	4	-2	-4	-2	0	-2	1
24	-2	4	-2	-8	-2	0	-2	-4	2
32	8	2	0	-2	-8	-2	0	2	1
40	2	-4	-2	-4	-2	4	2	4	2
48	-10	4	2	4	2	-4	-2	4	2
56	4	-6	4	-2	-4	-2	4	2	3

SPECTRUM HAS 7 MAGNITUDES

SPECTRUM:

INDEX	ABSVAL	#POS	#NEG	TOTAL#
1	0	8	0	8
2	22	1	0	1
3	2	10	18	28
4	4	12	8	20
5	10	1	1	2
6	8	1	3	4
7	6	0	1	1

The positions of the larger coefficients can be further improved by the linear transformation $y = xA$ with A defined below.

$$A = \begin{bmatrix} 1 & 0 & 0 & 1 & 0 & 1 \\ 0 & 1 & 1 & 0 & 0 & 0 \\ 0 & 0 & 1 & 0 & 0 & 1 \\ 0 & 0 & 0 & 1 & 0 & 0 \\ 0 & 0 & 0 & 0 & 1 & 0 \\ 0 & 0 & 0 & 0 & 0 & 1 \end{bmatrix} \qquad (123)$$

This matrix A was chosen by requiring A^t to permute certain coordinates of f^*. Namely, A^t leaves invariant the unit vectors e_1, e_2, and e_5 ($w(i)$ for $i = 32, 16$, and 2) and interchanges 4 with 36. In order to make A^t nonsingular

TABLE XX
Prime Implicant Table after Adding Linear Function

```
5-CELLS (PI'S WITH 1 LITERALS)


4-CELLS (PI'S WITH 2 LITERALS)


3-CELLS (PI'S WITH 3 LITERALS)

PI# 1              A=00-0--     F= 0
PI# 2              A=11-1--     F= 0
PI# 3              A=1-01--     F= 0
PI# 4              A=1--1-0     F= 0

2-CELLS (PI'S WITH 4 LITERALS)

PI# 5              A=0111--     F= 1
PI# 6              A=001-1-     F= 0
PI# 7              A=010-1-     F= 1
PI# 8              A=100-0-     F= 0
PI# 9              A=111-0-     F= 0
PI# 10             A=001--1     F= 0
PI# 11             A=100--1     F= 0
PI# 12             A=101--0     F= 0
PI# 13             A=111--1     F= 0
PI# 14             A=01-11-     F= 1
PI# 15             A=01-1-1     F= 1
PI# 16             A=00--11     F= 0
PI# 17             A=01--10     F= 1
PI# 18             A=10--00     F= 0
PI# 19             A=11--01     F= 0
PI# 20             A=1-0-01     F= 0
PI# 21             A=1-1-00     F= 0
PI# 22             A=--0000-    F= 0
PI# 23             A=--0101-    F= 0
PI# 24             A=--000-1    F= 0
PI# 25             A=--010-0    F= 0
PI# 26             A=--00-00    F= 0
PI# 27             A=--00-11    F= 0
PI# 28             A=--01-10    F= 0
PI# 29             A=--0-000    F= 0
PI# 30             A=--0-011    F= 0
PI# 31             A=---0001    F= 0
PI# 32             A=---0100    F= 0
PI# 33             A=---1000    F= 0
PI# 34             A=---1011    F= 0

1-CELLS (PI'S WITH 5 LITERALS)

PI# 35             A=0100-0     F= 1
PI# 36             A=1011-1     F= 1
PI# 37             A=011-01     F= 1
PI# 38             A=101-01     F= 1
PI# 39             A=110-10     F= 0
PI# 40             A=0-0101     F= 1
PI# 41             A=0-0110     F= 1
PI# 42             A=0-1100     F= 1
PI# 43             A=--10000    F= 1
PI# 44             A=--10011    F= 1
PI# 45             A=--11010    F= 1
```

and triangular, rows 3 and 6 were also chosen to interchange 8 with 24 and 1 with 41 (coefficients of f^*). The most obvious effect of A^t is to replace $f^*(e_3) = 2$ by $g^*(e_3) = h^*(e_3 A^t) = h^*(w(36)) = 8$. In this way we obtain still a third truth table for $g(y)$ with $y = xA$, $g(y) = h(x) = f(x) \oplus xa^t$ for all x, and a transform $g^*(w) = h^*(wA^t)(g^*(e_i) = h^*$ evaluated at (ith row of A^t)) (see Table XXI). Table XXII shows that a further simplification has been achieved (g has eleven 3-cells, h had only four). Only 5 PI's are essential now, and only 1 of them is a 0-cell. After eliminating points covered by the essential PI's, other PI's become essential. The minimal cover in Table XXIII requires only 8 AND gates with 37 inputs.

TABLE XXI
Truth Table, Transform, and Spectrum for Encoded Input Logic

TRUTH TABLE OF F(X):

ROW I ∎COL J INDICES:

I: J= 0	1	2	3	4	5	6	7	W(I)
0	0	0	0	0	0	1	1	0	0
8	0	0	0	0	0	1	0	0	1
16	1	0	0	1	1	1.	1	1	1
24	1	0	1	1	0	1	1	1	2
32	0	0	0	0	0	0	0	1	1
40	0	1	0	1	0	1	0	0	2
48	0	0	0	0	0	0	1	0	2
56	0	0	0	0	0	1	1	0	3

TRANSFORM OF F(X):

ROW I ∎COL J INDICES:

I: J= 0	1	2	3	4	5	6	7	W(I)
0	22	-4	-2	-4	-8	2	0	10	0
8	-2	4	-2	4	-4	-2	4	-2	1
16	-8	-6	4	-2	2	0	-2	-4	1
24	0	2	-4	-2	-2	4	2	0	2
32	8	2	0	2	-2	4	-2	4	1
40	4	-2	4	-2	-2	-4	-2	4	2
48	-10	4	2	-8	-4	-2	0	2	2
56	2	0	-2	-4	4	-2	0	2	3

SPECTRUM HAS   7 MAGNITUDES:

INDEX	ABSVAL	#POS	#NEG	TOTAL#
1	0	8	0	8
2	22	1	0	1
3	4	12	8	20
4	2	10	18	28
5	8	1	3	4
6	10	1	1	2
7	6	0	1	1

END OF FXFM & SPCTRM

TABLE XXII
Prime Implicant Table for Encoded Input Logic

```
5-CELLS (PI'S WITH 1 LITERALS)

4-CELLS (PI'S WITH 2 LITERALS)

3-CELLS (PI'S WITH 3 LITERALS)

PI#  1              A=00-0--    F= 0
PI#  2              A=11-0--    F= 0
PI#  3              A=10---0    F= 0
PI#  4              A=1-00--    F= 0
PI#  5              A=1-0-0-    F= 0
PI#  6              A=1--0-0    F= 0
PI#  7              A=1---00    F= 0
PI#  8              A=-000--    F= 0
PI#  9              A=-01--0    F= 0
PI# 10              A=-0-0-0    F= 0
PI# 11              A=-0--00    F= 0

2-CELLS (PI'S WITH 4 LITERALS)

PI# 12              A=0101--    F= 1
PI# 13              A=001-1-    F= 0
PI# 14              A=011-1-    F= 1
PI# 15              A=110--1    F= 0
PI# 16              A=01-11-    F= 1
PI# 17              A=01-1-1    F= 1
PI# 18              A=00--11    F= 0
PI# 19              A=01--11    F= 1
PI# 20              A=11--11    F= 0
PI# 21              A=0--001    F= 0
PI# 22              A=0--101    F= 1
PI# 23              A=-0111-    F= 0
PI# 24              A=-1-001    F= 0
PI# 25              A=-1-110    F= 1
PI# 26              A=--0001    F= 0
PI# 27              A=--0010    F= 0
PI# 28              A=--1100    F= 0
PI# 29              A=--1101    F= 1

1-CELLS (PI'S WITH 5 LITERALS)

PI# 30              A=0110-0    F= 1
PI# 31              A=1010-1    F= 1
PI# 32              A=010-00    F= 1
PI# 33              A=101-01    F= 1
PI# 34              A=01-000    F= 1
PI# 35              A=0-0110    F= 1
PI# 36              A=1-1111    F= 0
```

TABLE XXIII
Minimal Cover of $g(y)$

0 1 - - 1 1	1 0 1 0 - 1
0 - - 1 0 1	0 1 0 - 0 0
- 1 - 1 1 0	0 - 0 1 1 0
- - 1 1 0 1	1 0 0 1 1 1

Four EXCLUSIVE OR gates (8 inputs) for xa^t and four more (8 inputs) for A gives a total of 16 gates and 53 inputs rather than 11 gates with 104 gate inputs for the original function. Clearly, which realization is preferred depends on the *relative* weighting assigned to number of gates versus gate inputs. However, gate inputs tend to be more important in LSI because they correspond to signal paths (deposited wiring), and a decrease in their number is likely to correspond to a decrease in the number of wiring crossovers as well.

The conjecture that minimum-cost prototypes are produced by mapping larger coefficient magnitudes $g^*(w)$ into locations with smallest weight $|w| = \sum w_i$ is disproved by the following counterexample. The distribution of spectrum coefficients of the three largest magnitudes $|g^*(w)|$ on Table XXI is shown below versus the weight of the argument w.

$|g^*(w)|$	$\{w\}$	$|w|$
22	0	0
10	7, 48	2, 3
8	4, 16, 32, 51	1, 1, 1, 4
6	17	2

We now construct a further mapping of g into a new prototype. It is obvious that the two largest coefficients, $|g^*| = 10$, can be relocated to locations of weight 1 (say, to $w = 16, 32$) by selecting the first two rows of a new matrix A^t. After this is done, it is possible to select only *two* more linearly independent A-rows from among the 4 vectors w where $|g^*(w)| = 8$. These vectors (indices 32, 4) become the third and fourth rows of A^t. Finally, the vector $w = 17$ for which $|g^*(w)| = 6$ becomes row 5, and $w = 40$ for which $|g^*(w)| = 4$, becomes its last row. The resulting matrix A_2 will encode y as $z = yA_2$ and supply inputs to a new prototype function $p(z)$ such that $p(yA_2) = g(y) = f(x)$ for $z = yA_2 = xAA_2$, $x \in \mathbb{Z}^n$. The minterms of p are as follows:

$p^{-1}(1) = \{4, 6, 7, 12, 14\text{--}23, 26, 30, 51\text{--}54, 58, 63\}$. The matrix A_2 and the locations of large transform coefficients $p^*(w)$ are as follows:

$$A_2 = \begin{bmatrix} 0 & 1 & 1 & 0 & 0 & 1 \\ 0 & 1 & 0 & 0 & 1 & 0 \\ 0 & 0 & 0 & 0 & 0 & 1 \\ 1 & 0 & 0 & 1 & 0 & 0 \\ 1 & 0 & 0 & 0 & 0 & 0 \\ 1 & 0 & 0 & 0 & 1 & 0 \end{bmatrix}$$

$p^*(w)$	$\{w\}$	$|w|$
10	16, 32	1, 1
8	4, 8, 24, 52	1, 1, 2, 4
6	2	1
4	1, 10, ...	1, 2, ...

A comparison of $g^*(w)$ to $p_*(w)$ for $|w| = 1$ is given below:

w	32	16	8	4	2	1
$|g^*(w)|$	8	8	2	8	2	4
$|p^*(w)|$	10	10	8	8	6	4

Clearly, p^* achieves a better score by this empirical measure of complexity (largest coefficient values at vectors w of lowest weight). Unfortunately, however, the minimal covering of $p(z)$ turns out to be more costly than that for $g(y)$ in Table XXII above, in spite of the fact that p and $\bar{p}$ together have only 24 PI's, two of which are dimension 4, while g and $\bar{g}$ have 36 PI's, all of dimension 3 or less (see Table XXI). Specifically, a minimal two-level cover for $p(z)$ or $\bar{p}(z)$ requires 48 or 43 gate inputs, respectively, while $g(y)$ can be covered by a two-level form with only 37 gate inputs.

D. FUNDAMENTAL PROBLEMS AND EXTENSIONS

One cannot fail to be impressed with the conceptual insight and unifying principles provided by harmonic analysis. It provides a common basis for hitherto unrelated techniques for implicant extraction, functional decomposition, and the analysis of linear encodings. Unfortunately, the theory developed in this Chapter can be directly applied only to the space $\mathscr{F}$ of all *single-output* logic functions. Consequently, extensions to *multiple-output* functions, and then to *sequential* machines or finite state automata have great potential value. This section identifies several key problem areas which are likely candidates for further research, namely, encoded input threshold logic synthesis, the probability distribution of spectral coefficient magnitudes for randomly chosen functions, canonical forms for elements of RAG(n/m) with $m > 1$, multiple-output function synthesis, and applications of harmonic analysis to sequential circuits.

1. Synthesis Techniques for Encoded Input Threshold Logic

Recent literature on threshold logic makes an increasing number of references to the Hadamard (abstract Fourier) transform basis for Z^n (Hwa and Sheng, 1969). However, no one has yet attempted to use prototype transformations to reduce a function (as much as possible) toward a threshold-realizable form. Consequently, the utility of Fourier transforms as a tool to aid in this reduction has also gone unrecognized. Perhaps the invariant properties of the Fourier transform and its implicant detection capability will lead to an algorithm for extraction of linearly separable subfunctions and for an encoding transformation which identifies the simplest threshold-realizable prototype class representatives.

2. Probability Distribution of Spectral Coefficient Values

Section IV,A,6 showed that adding xa^t to f (mod 2) produces a function g whose weight is $f^*(a)$ rather than $f^*(0)$. A smaller or larger weight (number of points in $f^{-1}(1)$) often leads to a simpler realization of f; therefore, the largest magnitude $f^*(w)$, or more generally, the distribution of spectral coefficient magnitudes, is of interest for a randomly chosen function. It may be that large weight reduction factors cannot be expected for randomly chosen functions. An intuitive argument might be based on an analysis which shows that there are neither enough functions of low weight to go around, nor enough linear functions to add, as n becomes large.

If $f(x) = 0$ or 1 with probability $\frac{1}{2}$, then an integer-valued spectral coefficient $f_N^*(w)$ is the sum of 2^n variables with mean 0 and variance ($\frac{1}{4}$). Therefore, the standard deviation of $f_N^*(w)$ is $2^{(n/2)-1}$. In other words, $f_N^*(w)$ coefficients tend to cluster within a fraction $2^{-n/2}$ of the midpoint of their range $\pm 2^{n-1}$. Another approach to this problem would be to use order statistics (probability that the *largest* of 2^n Fourier coefficients has a certain value). More generally, the distribution of the $n + 1$ largest coefficient values is of interest to evaluate the general effectiveness of the techniques of Section V,B.

3. Canonical Forms for Elements of RAG(n/m)

The constructive definition of primary rational canonical forms for proto-type transformations appears to be a fundamental unsolved problem of affine (and linear) group theory when the number of output variables is generalized from 1 to m. The problem can be stated mathematically as follows: Given a subgroup RAG(n/m) of the general affine group, find a canonical representative for each equivalence class of matrices in RAG(n/m) under similarity transformations which are inner automorphisms of RAG(n/m). The single constraint on this subgroup is that it leaves invariant a given m-dimensional subspace of the space $\mathbb{Z}^{n+m}$ on which it acts. This constraint generalizes the $b = 0$ restriction in the derivation of RAG($n/1$) in Section IV,A,2. For $m = 1$, unique canonical forms were derived for proto-type transformations by Lechner (1963). However, this was a specialized construction which assumed an output space of one dimension.

4. Multiple Output and Many-Valued Functions

In Section III, prime implicants, essential prime implicants, and decom-posability conditions were identified for a single-output function using har-monic analysis. One powerful generalization of harmonic analysis techniques would be their extension to m-output combinational logic. Of course, this means generalizing the group RAG to RAG(n/m) as described in Section IV,A,4.

Of course, the covering problem for multiple-output functions can apply the technique of Section III one function at a time. More generally, since RAG(n/m) includes AG(m) on the space of output m-tuples $f = (f_1, \ldots, f_m)$, 2^m different functions can be produced from any linearly independent subset of m functions $(g_1, \ldots, g_m)$. For m sufficiently small, it is possible to evaluate the cost of each of the 2^m linear combinations of f_1 through f_m, then select a basis $g = g_1, \ldots, g_m$ for this subspace of $\mathscr{F}$, and a mapping B such that $g\mathsf{B} = f$ is an identity for every argument x. Of course, this technique is also applicable to a 2^m-valued function under a suitable coding of its outputs into binary m-tuples.

One question of interest here, particularly if n is larger than m, is whether some of the functions to be implemented are actually modulo two linear combinations of other functions that can be realized in a more efficient man-ner. Thus, we might realize a set of functions $g_1, \ldots, g_m$, then apply an $m \times m$ affine transformation to these m functions and produce $f_1, \ldots, f_m$. In matrix form, $f = g\mathsf{B} \oplus d$, with $\mathsf{B} \in LG(m)$ and $d \in \mathbb{Z}^m$. The prototype trans-formation group RAG(n/m) includes such transformations on the outputs of imbedded prototype functions g_1 through g_m. A straightforward generalization will handle functions $f_1, \ldots, f_m$ of rank $k < m$ by k imbedded functions $g_1, \ldots, g_k$.

An exhaustive evaluation of the 2^m candidate basis functions is possible for reasonable values of m (say, $m \leq 10$). The evaluation would proceed by analyzing the Fourier spectrum (or even solving the minimal covering problem if m is small enough) for each of the 2^m possible linear combinations (mod 2) of the functions $f_1, \ldots, f_m$ (regarded as 2^n-tuples over $\mathbb{Z}$). From among these 2^m functions, a basis $(g_1, \ldots, g_m)$ of lowest cost would be selected. The only restriction on this basis is that (as 2^n-tuples over $\mathbb{Z}$) they have the same rank $k \leq m$ as the original set of functions $f_1, \ldots, f_m$. In other words, this multiple-output synthesis approach works backward from the required m functions and attempts to find $k \leq m$ other functions, which may belong to other prototype classes. Then a least-cost representative of each prototype class is mechanized, and finally the domain encodings which supply input variables to these class representatives are combined and redundant columns of their respective matrix representation are discarded. This technique appears worthwhile even though it does not represent a general solution to the problem in the group theoretic sense.

5. The State Assignment Problem for Sequential Circuits

A synchronous sequential circuit or finite-state automata is a 6-tuple $M = (S, I, \emptyset, T, f, g)$ in which I, S, and $\emptyset$ represent the space of inputs $X = \mathbb{Z}^k$, states $S = \mathbb{Z}^m$, and outputs $u = \mathbb{Z}^p$, respectively, (Hartmanis and Stearns, 1966). For convenience, we also identify T as the space of "next state" values and assume that T is mapped back into S by the identity transformation after a finite time delay. The functions $f(X \oplus S \to T)$ and $g(X \oplus S \to U)$ are identical to δ and λ of Hartmanis and Stearns (1966) and define the next-state and output mappings, respectively. As usual, when the sequential structure of f is being analyzed, we concentrate on f and ignore g. The next-state mapping f for a sequential circuit with k inputs, p outputs, and m state variables is defined by an m-output combinational logic function of $n = m + k$ arguments. Therefore, harmonic analysis may be a useful synthesis tool provided that the theory presented in this chapter can be generalized to multi-output or vector-valued functions.

In their text, Hartmanis and Stearns (1966) pointed out that isomorphic machines are identical except for renaming the states, inputs, and outputs. The group RAG($k/m/m$) defined in Section IV,A,4 is a subgroup of the permutation subgroup on $(X \oplus S \oplus T)$ which induces the "isomorphic machine" equivalence relation. Therefore, each class of isomorphic machines is a *union* of prototype equivalence classes.

Suppose that we start with a completely arbitrary state assignment and attempt to decompose the permutation of $(X \oplus S \oplus T)$ which produces the "best" equivalent machine (best reassignment of state names) into two parts: (1) a recoding of the initial assignment into a coset of the prototype class which contains the "best" assignment and (2) a member of RAG which produces the "best" assignment. The second map has a well-defined algebraic structure, and harmonic analysis can be used as a synthesis tool. The first map still requires the state partitioning techniques of Hartmanis and Stearns (1966).

One important question is whether the decomposability properties of the next state mapping are more sensitive (less invariant) to the between-coset mappings or to the intracoset prototype transformation. If decomposability is more affected by mappings *between* cosets, then harmonic analysis complements state partitioning techniques and should follow them in the synthesis procedure. On the other hand, if decomposability is more sensitive to elements of RAG, then arbitrary state coding followed directly by harmonic analysis may replace current state assignment techniques as a direct synthesis approach. The example of Section V,C produced a triangular matrix A which corresponds to a cascade-type realization. This suggests that RAG does affect decomposability of next state mappings.

6.   Orbits and Stability Groups

Generally speaking, a function f which possesses symmetries under some reasonable transformation group on X is simpler to realize than a function which does not. The same is true for functions which are invariant to certain elements of $AG(n)$ on X. These elements must permute the subsets $f^{-1}(1)$ and $f^{-1}(0)$ within themselves (without mixing them up). Actually, RAG permutes the corresponding subsets of the characteristic function of f as a subset of $X \oplus Z$—see Section IV,A,2. However, for simplicity, we will consider only $AG(n)$ rather than $RAG(n/1)$ herein.

The set of all elements of $AG(n)$ which leave a given function f invariant form a group, called the *stability group* of f under $AG(n)$, denoted G_f. For notational simplicity, the terminology and properties of the partition lattice theory as described by Hartmanis (1960) and two other concepts "level set" and "orbit" borrowed from topological dynamics (via Hellerman, 1961) will be used in this paragraph.

Each element t in G_f generates a cyclic subgroup which permutes the set $X \oplus Z$; the cycle sets of this permutation define a *partition* of X denoted π_t. The operation of combining overlapping cycles from all partitions π_t, $t \in G_f$, produces the lattice-theoretic join, or upper bound of the partitions π_t; this join is denoted π_f^G, and the disjoint subsets of X which are its components are called *orbits* of G_f. Note that $\pi_t \leq \pi_f^G$ for all $t \in G_f$ ($\leq$ denotes lattice inclusion here). Two elements x and y of X are in the same orbit of G_f iff some element of G_f maps x into y.

The two subsets $f^{-1}(1)$ and $f^{-1}(0)$ are called *level sets* of f. They also define a partition of X, called the *level set partition* and denoted π_f^L. These two partitions induced by f are fundamentally related as follows: $\pi_f^G \leq \pi_f^L$.

It is known that similarity transformations preserve the lattice of invariant subspaces of a subgroup of $LG(n)$ (Thrall, 1952). Examples indicate that a similarity transformation on $AG(n)$ which maps one or more generators of G_f into rational canonical form is equivalent to an encoding transformation which maps f into a prototype class representative g of "simplest" form (Lechner, 1963a,b). The heuristic explanation for this is that a canonical matrix in $LG(1/n)$ (isomorphic to $AG(n)$) has a quasidiagonal form, and therefore the invariant subspaces of this form have bases consisting of unit vectors. Since $\pi_g^G \leq \pi_g^L$, the prototype g corresponding to f must be constant on the orbits of π_f^G whose unions are these subspaces. We conjecture that more prime implicants are likely to exist for g than for any other member of the prototype class and that this holds true in general.

In a short paper on coding theory, Wells (1960) showed how harmonic analysis could be used to successively refine π_f^L so that the sequence of resulting partitions of X had π_f^G for their lower bound. (The invariance theorems of Section II,C show that π_f^{L*}, the level set partition of f^*, and π_f^G, the stability group partition of G_f, are intimately related.) Lechner (1963a,b) showed that the orbits of G_f provide useful information which helps to identify the group G_g. Although Well's algorithm never became useful in the coding theory context, we conjecture that it can be extended to affine groups and will then have far reaching theoretical and practical consequences in switching theory as an aid to prototype class identification and synthesis of encoded input logic.

7. Simplification of Encoding Transformations

One important tradeoff in synthesis of encoded-input logic can be posed as follows: How does d (the density of 1's in the encoding transformation $t \in RAG$) vary with the complexity of the prototype class representative g chosen to implement the function? One extreme is to imbed a *unique* prototype $g = ft$, in which case the encoding transformation t can still be varied within the coset tG_g of the stability group of g defined in Section V,D,6.

The opposite extreme is to permit a prototype representative from any of the symmetry types in the prototype equivalence class. For one of these symmetry types (the one to which f belongs), the encoding transformation of the input arguments reduces to a (trivial) permutation matrix. The examples of Sections V,B,4 and V,C show that a relatively simple encoding transformation (a sparse matrix with only 15 out of a possible 51 unit elements) can achieve impressive results in some cases.

A similar problem can be posed for multiple-output functions: How many distinct columns are needed in the matrix A in order to transform x into the most appropriate vectors $y_1, \ldots, y_m$ such that m different functions f_1 to f_m can be realized from $g_1(y_1)$ through $g_m(y_m)$? One extreme is to restrict A to n columns, in which case only one of the m functions may be completely reducible to prototype form; the other extreme is to transform each function independently to prototypes $g_1, \ldots, g_m$, by elements t_1 through t_m in RAG, then search through the cosets $t_1 G_{g_1}, \ldots, t_m G_{g_m}$ of the stability groups of g_1 through g_m to find the combination of A-matrices with the smallest total number of unit entries or the smallest number of distinct columns.

EXERCISES

16. Derive an equivalent algebraic formula for the majority function illustrated on Fig. 8 in terms of the input arguments $\bar{x}_1$, x_2, and x_3 and the imbedded function $g(y) = y_1 y_2$. Define A, a, c, and d for the encoding transformation relating f to g. Does the option of inputting complemented or uncomplemented arguments affect the cost of the encoding transformation?

17. Derive another equivalent formula for the majority logic function of Fig. 8 using x_1, x_2, and x_3 as inputs with a different prototype function $g(y) = y_1 \lor y_2$. Define the components A, a, c, and d of the encoding transformation. Does the prototype selection influence the cost of the encoding transformation?

18. Starting with the function $g(y)$ in Fig. 15, find a matrix A such that $z = yA$, $p(z) = g(y)$, and $p^*(w) = 3$ at $w = 1, 2, 4, 8$, and 15.

19. Starting with the function $f(x)$ in Fig. 9, find a matrix A^t such that $h^*(w) = f^*(wA^t) = \pm 3$ for $w = w(i)$, $i = 1, 2, 4, 8, 6, 9$, $f(x) = h(xA)$, $y = xA$, and $\bar{h}(y) = y_1 y_2 \lor y_1 \bar{y}_3 \lor \bar{y}_2 \bar{y}_4 \lor \bar{y}_3 \bar{y}_4$.

REFERENCES

AIKEN, H., *et al.* (1951). Synthesis of electronic computing and control circuits. *Ann. Comput. Lab. Harvard Univ.* **27**, Harvard Univ. Press, Cambridge, Massachusetts.

ASHENHURST, R. (1952). The application of counting techniques. *Proc. Ass. Comput. Machinery*, 293–305.

ASHENHURST, R. (1953). A method for determining functional invariance. *In* Theory of Switching. Rept. No. BL-2, Section II, Harvard Computation Laboratory, pp. 1–12.

ASHENHURST, R. (1959). The decomposition of switching functions. *Ann. Comput. Lab. Harvard Univ.* **29**, No. 30. Harvard Univ. Press, Cambridge, Massachusetts.

BACHMAN, G. (1964). "Elements of Abstract Harmonic Analysis." Academic Press, New York.

BARTEE, T. C. (1961). Computer design of multiple-output logical networks. *IEEE Trans. Electron. Comput.* **EC-10(1)**, 21–31.

BELLMAN, R. (1960). "Introduction to Matrix Analysis." McGraw-Hill, New York.

BERGLAND, G. D., and HALE, H. W. (1967). Digital real-time spectral analysis. *IEEE Trans. Electron. Comput.* **EC-16(2)**, 180–185.

BIRKHOFF, G., and MACLANE, S. (1953). "A Survey of Modern Algebra," Macmillan, New York.

CALINGAIRT, P. (1961). Switching function canonical forms based on commutative and associative binary operations. *Trans. Amer. Inst. Elec. Eng., Part 1*, **52**, 808–814.

COHN, M. (1960). Switching Function Canonical Forms over Integer Fields. Ph.D. Thesis, Harvard Univ. Cambridge, Massachusetts.

CURTIS, H. (1959). A functional canonical form. *ACM J.* **6(2)**, 245–258.

CURTIS, H. (1962). "A New Approach to the Design of Switching Circuits." Van Nostrand, Princeton, New Jersey.

DAYKIN, D. E. (1963). On Linear sequences over a finite field. *Amer. Math. Mon.* **70(6)**, 637–642.

DEBRUIJN, N. G. (1959). Generalization of Polya's theorem in Enumerative combinatorial analysis. *Kon. Ned. Akad. Wetersch. Ser. A* **62(2)**, 59–69.

DIETMEYER, D. L., and Duley, J. R. (1968). Generating prime implicants via decimal encoding and decimal arithmetic *ACM Comm.* **11(7)**, 520–523.

ELPAS, B. (1959). The theory of autonomous linear sequential networks. *IRE Trans. Circuit Theory* **CT-6(1)**, 45–60.

ELPAS, B., GOLDBERG, J., JACKSON, C. L., KAUTZ, W. H., and STONE, W. S. (1967). Properties of Cellular Arrays for Logic and Storage. Rep. No. AFCRL-67-0463, Stanford Res. Inst., Menlo Park, California.

FRIEDMAN, A. D. (1967). Comments on Luccio (1966) paper. *IEEE Trans. Electron. Comput.* **EC-16(2)**, 221–223.

GILL, A. (1966). Graphs of affine transformations with applications to sequential circuits, *In* Conf. Rec., 7th Annual Symp. on Switching and Automata Theory, 127–135.

GOLDBERG, R. R. (1961). "Fourier Transforms." Cambridge Tract in Math. and Mathematical Physics, No. 52 (Appendix). Oxford Univ. Press, London and New York.

GOLOMB, S. W. (1959). On the classification of Boolean functions. *IRE Trans. Circuit Theory* **CT-6**, Special Supplement, 176–86.

GOLOMB, S. W. (1964). (Ed.). "Digital Communications with Space Applications." Prentice-Hall, Englewood Cliffs, New Jersey.

GOOD, I. J. (1958). The interaction algorithm and practical fourier analysis. *J. Roy. Statist. Soc. Ser. B.* **20(2),** 361–372.

HARRISON, M. (1963). On the number of classes of (n, k) switching networks. *J. Franklin Inst,* **276(4),** 313–327.

HARRISON, M. A. (1964). On the classification of Boolean functions by the general linear and affine groups. *J. Soc. Appl. Math.* **12,** 285–299.

HARRISON, M. A. (1965). "Introduction to Switching and Automata Theory." McGraw-Hill, New York.

HARTMANIS, J. (1960). Symbolic analysis of a decomposition of information processing machines. *Infor. Contr.* **3,** 154–178.

HARTMANIS, J., and STEARNS, R. E. (1966). "Algebraic Structure Theory of Sequential Machines." Prentice-Hall, Englewood Cliffs, New Jersey.

HELLERMAN, L. (1961). Equivalence classes of logical functions. *IBM Tech. Publ.* TR 00.819.

HEWITT, E., and ROSS, K. A. (1963). "Abstract Harmonic Analysis," Vol. I, Academic Press, New York.

HWA, H. R., and Sheng, C. L. (1969). An approach for the realization of threshold functions of order r. *IEEE Trans. Comput.* **C-18(10),** 923–939.

JACOBSON, N. (1951). "Lectures in Abstract Algebra," Vol. 1, Chapter VI. Van Nostrand, Princeton, New Jersey.

KELLERMAN, E. (1968). A formula for logical network cost. *IEEE Trans. Comput.* **C-17(9),** 881–884 (reviewed in *IEEE Trans. Comput.* **C-18(2),** 204).

LAWLER, E. L. (1956). Minimization of switching circuits subject to reliability conditions. *IRE Trans. Electron. Comput.* **EC-10(4),** 781–782.

LECHNER, R. J. (1963a). Affine Equivalence of Switching Functions. Ph.D. Thesis, Harvard Univ., Cambridge, Massachusetts (Submitted to Bell Telephone Labs. as Theory of Switching Rep. BL-33 by Harvard Computation Laboratory).

LECHNER, R. J. (1963b). On transformations among switching function canonical forms. *IRE Trans. Electron. Comput.* **EC-12(2),** 129–130.

LECHNER, R. J. (1967). A correspondence between equivalence classes of switching functions and group codes. *IRE Trans. Electron. Comput.* **EC-16(5),** 621–624.

LECHNER, R. J. (1968). A transform approach to logic design. *Proc. 1968 9th Symp. on Switching & Automata Theory,* pp. 213–234 (also in *IEEE Comput. Trans.* **C-19(7),** 672-640).

LITTLEWOOD, D. (1940). "The Theory of Group Characters and Matrix Representations of Groups." Oxford Univ. Press, London and New York.

LOOMIS, L. M. (1953). "An Introduction to Abstract Harmonic Analysis." Van Nostrand, Princeton, New Jersey.

LUCCIO, F. (1966). A method for the selection of prime implicants. *IEEE Trans. Electron. Comput.* **EC-15,** 205–212; see also Friedman, A. D. (1967). *IEEE Trans. Electron. Comput.* **EC-16(2),** 221–222.

McCLUSKEY, E. J. (1956). Minimization of Boolean functions. *Bell. Syst. Tech. J.* **35,** 1417.

McNAUGHTON, R., and PATTERSON, G. W. (1963). Combinational elements and fiducial-state assignments. *In* Proc. of the Symp. on Mathematical Theory of Automata, Vol. XII, pp. 459–481. Polytechnic Press, Brooklyn, New York.

MENGER, K. L., Jr. (1969). A transform for logic networks. *IEEE Trans. Comput.* **C-18,** 241–250.

MEO, A. R. (1968). Synthesis of many-variable functions. *In* "Networks and Switching Theory" (G. Biorci, ed.), Chapter VI, pp. 470–482. Academic Press, New York.

MILETO, F., and PUTZOLO, G. (1964). Average values of quantities appearing in Boolean function minimization. *IEEE Trans. Electron. Comput.* **EC-13(2),** 87–92.

MILLER, R. E. (1965). Switching Theory, Vol. I, Combinational Circuits, Wiley, New York.

MINNICK, R. C. (1961). Linear-input logic. *IEEE Trans. Electron. Comput.* **EC-10(1)**, 6–17.

MINNICH, R. C. (1964). Cutpoint cellular logic. *IRE Trans. Electron. Comput.* **EC-13(6)**, 685–697.

MORREALE, E. (1967). Partitioned list algorithms for prime implicant determination from canonical forms. *IEEE Trans. Electron. Comput.* **EC-16**, 611–620.

MULLER, D. E. (1953). Metric Properties of Boolean Algebra and their Application to Switching Circuits. Rep. No. 46, Digital Computer Lab., Illinois, Univ. Urbana, Illinois.

MULLER, D. E. (1954). Boolean algebras in electric circuits design. *Amer. Math. Mon.* **61(7)**, Part II, 27–28.

NECHIPORUK, E. (1958). On the synthesis of networks using linear transformations of argument variables. *Dokl. Akad. Nauk. SSSR* **123(4)**, 610–612 (English transl. *Automat. Express*, April 1959, 12–3).

NECULA, N. N. (1967). A numerical procedure for determination of the prime implicants of a Boolean function. *IEEE Trans. Electron. Comput.* **EC-16(5)**, 687–689.

NINOMIYA, I. (1958). A Theory of the coordinate representation of switching functions. *Mem. Fac. Eng., Nagoya University* **10(2)**, 175–190. (Rev. *IEEE Trans. Electron. Comput.* **EC-12(2)**, 152, (1963).

NINOMIYA, I. (1961). A study of the structures of Boolean functions and its application to the synthesis of switching circuits. *Mem. Fac. Eng., Nagoya Univ.* **13(2)**, Ph.D. Thesis, Univ. Tokyo (Tables reprinted in Harrison, 1965).

PETERSON, W. W. (1961).

POLYA, G. (1940). Sur les types des propositions composees. *J. Symbolic Logic* **5**, 98–103.

PRATHER, R. C. (1966). Introduction to Switching Theory: A Mathematical Approach, Allyn & Bacon, Boston.

QUINE, W. V. (1955). A way to simplify truth functions. *Amer. Math. Mon.* **59(9)**, 627–631.

REED, I. S. (1954). A class of multiple error-correcting codes and the decoding scheme. *Trans. IRE* **PGIT-4**, 38–49.

ROTH, J. P. (1959). Algebraic, topological methods in synthesis. *In* Proceedings of 1957 International Symposium on the Theory of Switching, Annals of the Computation Lab, Vol. 29, pp. 57–73. Harvard Univ. Press (Review in *IRE Trans. Electron. Comput.* **EC-9(4)**, 516, 1960).

ROTH, J. P. and KARP, R. M. (1962). Minimization over Boolean graphs. *IBM J. Res. Develop.* **6**, pp. 227–238. (Review in *IEEE Trans. Electron. Comput.* **AC-12(2)**, 150–151, 1963.)

RUDIN, W. (1962). "Fourier Analysis on Groups." Wiley (Interscience) New York.

SHEN, V. Y., McKELLER, A. C., and WEINER, P. (1969). A fast algorithm for testing switching functions for disjunctive decompositions. *In Proc. 3rd Ann. Princeton Conf. Information Sci. Systems*, pp. 480–483. Princeton Univ., Princeton, New Jersey.

SINGER, T. (1952). The theory of counting techniques. *Proc. Ass. Comput. Machinery*, 287–291.

SLEPIAN, D. (1954). On the number of symmetry types of Boolean functions of n variables. *Can. J. Math.*, **5(2)**, 185–193.

SLEPIAN, D. (1956). A class of binary signalling alphabets. *Bell Syst. Tech. J.* **35(1)**, 203–234.

SLEPIAN, D. (1960). Some further theory of group codes. *Bell. Syst. Tech. J.* **39(5)**, 1219–1952.

SLEPIAN, D. (1967).

SMITH, D. R. (1968). A partitioning method for combinational synthesis. *IEEE Trans. Comput.* **C-17**, 72–75.

STONE, H. S., and JACKSON, C. (1969). Structures of affine classes. *IEEE Trans. Comput.* **C-18**, 252–257.

THRALL, R. (1962). On a Galois connection between algebras of linear transformations and lattices of subspaces of a vector space. *Can. J. Math.* 227–239.

WEISS, C. D. (1969). The characterization and properties of cascade-realizable switching functions. *IEEE Trans. Comput.* **C-18(7)**, 624–633.

WELLS, W. W. (1960). Classification of Binary Sequences. Rep. 22G-0036 (AD 236796), MIT Lincoln Laboratory, Cambridge, Massachusetts.

WHELCHEL, J. E., Jr., and GUINN, D. F. (1968). "The fast Fourier–Hadamard transform and its use in signal representation and classification. *In* "EASTCON 1968 Record."

ZIERLER, N. (1960). On decoding linear error-correcting codes. *IRE Trans. Informa. Theory* **IT-6**, 450–458.

Chapter VI

UNIVERSAL LOGIC MODULES

I.	STATEMENT OF THE PROBLEM	230
II.	BOUNDS FOR $M(n)$	231
	A. Lower Bound	231
	B. Upper Bound	235
III.	THE CONSTRUCTION OF ULM'S FOR SMALL n	241
	A. ULM's for $n \geq 3$	241
	B. The Construction of a $(7, 4)$ ULM	244
IV.	OTHER APPROACHES TO THE UNIVERSAL MODULE PROBLEM	249
V.	HISTORICAL REFERENCES	252
	REFERENCES	253

ABSTRACT. A Universal Logic Module (ULM) is a combinational logic module with m input terminals that is capable of realizing every n-variable switching function, $n < m$, where function specialization is achieved by distributing the n variables, their complements, and the constants 0 and 1 freely among the m input terminals. Several methods for constructing ULM's are described in this chapter as well as formulas that show the dependency of m on n. In particular a lower bound for m is derived that shows that it grows as $2^n/\log_2 n$. An upper bound on m is obtained from the Preparata–Muller construction technique that yields modules for which m grows as $2^n/\log_2 n$, although the actual number of terminals is slightly greater than the number predicted by the lower bound. For small n, ULM's in three and four variables are given. The 3-variable ULM is known to have the minimum number of terminals while the 4-variable ULM is not known to be optimum.

Related studies by Lechner, Yau, and Tang are also reported. Lechner's approach is to use a universal set of modules rather than a single module. Yau and Tang modify the basic ULM assumptions somewhat and allow terminal to terminal jumpers on their universal modules. Such modules have a number of terminals that approach $2^n/n$ asymptotically which is somewhat better than the Preparata–Muller modules. However, for n in the range of practical interest the Preparata–Muller modules have fewer terminals.

The development of integrated circuit technology has focused attention on the use of complex modules as the building blocks of digital systems in place of the NAND and NOR circuits that in recent years have been the building blocks. The study of complex modules is not a new one, however, in the annals of switching theory literature. Von Neumann in the early 1950's studied

229

iterative networks that are constructed from complex sequential cells and proved that these iterative networks can simulate Turing machines (Von Neumann, 1966). Moore (1962) discusses self-reproduction in a similar context. The theory of cellular logic (Chapters VII–IX) is very much concerned with the study of complex cells and their relation to the functional behavior and complexity of iterative networks.

In this chapter we study the design of complex logic modules that are "universal" with respect to n variables in the sense that a universal module can realize any switching function of n or fewer variables. These devices are generally known as *universal logic modules* (ULM's). The view that we adopt here is that such modules can serve as building blocks in any logic network, not necessarily just iterative logic networks. The primary results of this chapter are related to the design of such modules, although there is some discussion in the literature of the design of networks of ULM's.

The primary difference between ULM's and modules for cellular networks is that the function selection mechanism for ULM's is entirely external to the module, whereas for cellular building blocks the mechanism is entirely internal to the cell. The selection mechanism for ULM's, in fact, is the way in which n variables, and the constants 0 and 1 are distributed among m input terminals where $m > n$. Since the number of input/output terminals on a module strongly influences the cost of that module in newer technologies, one of the objectives of the work reported in this chapter is to minimize the number of terminals of ULM's. To this end several constructive techniques for synthesizing ULM's are reported, as well as theoretical bounds on the minimum number of terminals that such modules can have. One method of construction yields a number of terminals that asymptotically approaches the lower bound as the number of variables increases.

Section I of this chapter gives a statement of the problem, and Section II presents the bounds and general methods of construction. In Section III we treat the problem of designing optimum ULM's for three and four variables. Section IV discusses other approaches to the problem of designing complex modules. One approach described in Section IV involves a function selection mechanism that is slightly different from the ULM selection mechanism. The second approach uses a collection of modules rather than a single module to achieve universality.

I. STATEMENT OF THE PROBLEM

Consider a combinational logic net with m input terminals and two complementary output terminals. The m input terminals may be connected freely to any of $2(n + 1)$ source wires, carrying the variables $x_1, \bar{x}_1, \ldots, x_n, \bar{x}_n$ and the constant signals 0 and 1, respectively. If under arbitrary connections of this sort the output terminals, together, produce all n-variable Boolean functions (and their complements) $f, \bar{f}$, then we refer to the net as an (m, n) ULM. The principal problem associated with such nets are:

(1) Find ULM's with minimum m for every n.
(2) Determine the dependence on n of the minimum terminal count $M(n)$.
(3) Alternatively, find good estimates (upper and lower bounds) on $M(n)$ as a function of n.

Before attempting to answer the questions at hand, it is necessary to relate the problem statement to the notions of equivalence that have been introduced in Chapter IV. Among the permissible connections to a ULM are those that include the group G_n of arbitrary permutations and complementations of the n free variables. Hence, if a function $f(x_1, x_2, \ldots, x_n)$ is realizable by any logic module under the allowed interconnections, then all other functions of the same symmetry type are also realizable. The availability of complementary outputs allows us to state that if a function f is realizable then all functions of the same symmetry genus are also realizable because the genus equivalence relation includes the effect of negation of the range of the function. We make frequent use of the symmetry type and genus relations in this chapter, particularly when it is necessary to show the universality of a module under test. In fact universality is established by testing the module for only one function in each genus.

II. BOUNDS FOR $M(n)$

In this section we calculate lower and upper bounds for the function $M(n)$, the minimum number of input terminals of a ULM that is universal in n variables. The lower bounds are calculated by combinatorial arguments, and for $n \geq 4$ it is not known at present if they are "sharp." The upper bounds are derived by constructive techniques, so that there exist ULM's that meet the upper bound. Although the upper and lower bounds differ for $n \geq 4$, they have the same asymptotic behavior as n becomes large.

A. LOWER BOUND

A lower bound on $M(n)$ can be found by counting the number of distinct connections that can be made from m terminals to the $2(n + 1)$ source wires. This number is compared to the total number of complementary pairs of n-variable functions, and the lower bound on m is the smallest m such that the number of connections is at least as large as the number of function pairs.

Since there are $(2n + 2)^m$ ways of distributing the source wires among the m terminals and there are 2^{2^n-1} n-variable function pairs, we obtain

$$(2n + 2)^m \geq 2^{2^n - 1} \tag{1}$$

or by taking logarithms,

$$m \geq (2^n - 1)/[1 + \log_2(n + 1)] \tag{2}$$

Table I gives the values of m, the lower bound for $M(n)$, for small values of n.

TABLE I
A Lower Bound for $M(n)$

n	2	3	4	5	6	7	8	9	10
lower bound for $M(n)$	2	3	5	9	17	32	62	119	230

A slightly more accurate bound can be obtained by taking into account in the enumeration only those connection types that use all n variables. We compute the number, $C(m, n)$, of these connection types (under permutation and complementation of the signal variables) and compare $C(m, n)$ to the number G_n of (nondegenerate) genera. Before attempting this count, we note here that we can obtain these numbers asymptotically by dividing the left side of Eq. (1) by $n!\,2^n$, to obtain the number of nonequivalent connections, and the right side of Eq. (1) by $n!\,2^{n+1}$, the number of group operations that determine the equivalence partition. Carrying this through yields an equation that displays essentially the same asymptotic behavior as Eq. (4) below. Nevertheless, we shall carry through the exact count here because the specific numbers that result will be useful in later sections of this chapter.

Consider a pattern of connections from the m-input terminals of a module to the $2(n + 1)$ source wires for which k module terminals are connected to the constant sources (0 and 1) and $m - k$ terminals are connected to variable sources $x_j, \bar{x}_j$, where $0 \leq k \leq (m - n)$. For the moment, let us regard each pair of complementary variable wires $x_j, \bar{x}_j$ (and also the pair 0, 1) as a single object. Our freedom to choose true or complemented variables can be brought in later by multiplying by 2^m. Now, the number of ways in which we may choose $m - k$ terminals from the set of m and partition these among the n boxes $(x_j, \bar{x}_j)$, so that no box is left empty, is given by the expression:

$$n! \left\{ \begin{matrix} n \\ m - k \end{matrix} \right\} \left(\begin{matrix} m \\ m - k \end{matrix} \right)$$

where $\{^n_r\}$ is a Stirling number of the second kind and represents the ways in which r objects may be assigned to n nonempty subsets (Abramowitz and Stegun, 1964). The factor $n!$ appears simply because we wish to assign terminals to distinguished "boxes," whereas the Stirling numbers refer only to indistinguishable subsets; hence, we must introduce all $n!$ permutations of the subsets.

The total number of different patterns of connections between the m terminals and the $2(n + 1)$ wires is then given by summing the above expression over k for $k = 0, 1, \ldots, (m - n)$, and by multiplying the result by 2^m, since in any selection of subsets we may independently assign each of the m terminals to a "true" or complemented source wire. Hence, we have the result below for the number of different connections to the terminals of the module, wherein each variable is represented.

$$N(m, n) = n!\, 2^m \sum_{k=0}^{m-n} \binom{m}{k} \left\{ {n \atop m-k} \right\} = n!\, 2^m \left\{ {n+1 \atop m+1} \right\} \tag{3}$$

(See Abramowitz and Stegun, 1964, p. 825). Now, each connection counted in Eq. (3) differs only in the permutation and complementation of source variables from $n!\,2^n - 1$ other connections. That is to say, any permutation and/or complementation of the source wires would not alter the genus of resulting output functions $(f, \bar{f})$. Hence, we may divide the above expression by $n!\,2^n$, obtaining

$$C(m, n) = \frac{N(m, n)}{n!\, 2^n} = 2^{m-n} \left\{ {n+1 \atop m+1} \right\} \geq G_n \tag{4}$$

A simpler, and much more elegant, derivation of this same result can be obtained by treating the $n + 1$ pairs of source wires alike (regardless of whether they represent constant or variable sources) and by introducing a dummy terminal $\#0$ in addition to the m actual input terminals of the module. We now have $n + 1$ variables, $x_0, \ldots, x_n$ (and their complements) and $m + 1$ terminals, $\#0, 1, \ldots, m$. We ask, "In how many ways can we assign to each variable a nonempty subset of the $m + 1$ terminals?" The answer (by our previous reasoning) is $(n + 1)!\, \{^{n+1}_{m+1}\}$. Of course, the dummy terminal $\#0$ must be ignored now, and thus our count drops by a factor of $n + 1$, since given any real connection to the m terminals we may reinsert the dummy terminal in $n + 1$ different ways by assigning it to different subsets. When the only connection to the source pair $(x_0, \bar{x}_0)$ is to the dummy terminal, then deletion of the dummy terminal means, of course, that the constant sources are unused in this particular connection scheme. As before, we divide by $n!\,2^n$ to allow for source variable transformations, and we multiply by 2^m to account for complementations to the module input terminal signals. The result is as already stated.

Table II shows some values of the function $2^{m-n} \{^{n+1}_{m+1}\}$ obtained from Abramowitz and Stegun (1964).

At the bottom of Table II are shown the values for G_n, the number of nondegenerate genera of Boolean functions of n variables. The first entry in each column for which $C(m, n) \geq G_n$ determines the lower bound for $M(n)$ and is marked ⊓.

This yields the lower-bound values:

n	2	3	4	5	6	7
lower bound for $M(n)$	3	4	6	10	17	32

Table II improves on Table I only for $n \leq 5$. Using the asymptotic formula (Abramowitz and Stegun, 1964)

$$\{^{n+1}_{m+1}\} \sim (n + 1)! (n + 1)^{m+1}$$

we obtain the asymptotic expression

$$m > [2^n - 2 \log_2(n + 1)!]/[1 + \log_2(n + 1)]$$

which substantiates the remarks made earlier.

TABLE II
Values of $C(m, n)$

m =No. Module Terminals	n =Number of Source Variables						
	1	2	3	4	5	6	7
1	⌐1⌐						
2	6	1					
3	28	⌐12⌐	1				
4	120	100	⌐20⌐	1			
5	496	720	260	30	1		
6	2016	4816	2800	⌐560⌐	42	1	
7	8128	30912	27216	8400	1064	56	
⋮					⋮	⋮	
10					⌐5.7×10^6⌐	⋮	
⋮						⋮	
17						⌐4×10^{14}⌐	
⋮						⋮	
32							⌐4.7×10^{32}⌐
$G_n =$	1	2	10	208	615,904	2×10^{14}	1.31×10^{32}

B. UPPER BOUND

Here we describe a constructive method for designing (m, n) ULM's such that the asymptotic behavior of m as a function of n is the same as the behavior of the lower bound $M(n)$. The method is due to Preparata and Muller (1970).

The method of construction depends on a partition of the Boolean n-cube with the following properties. Let $P = \{P_1, P_2, \ldots, P_t\}$ be a partition of the minterms of a Boolean n-cube such that for each nonempty proper subset of minterms of each block P_i, there is a variable x_j such that $\dot{x}_j P_i$ is a function that is true only on the specified subset of minterms. (The notation $\dot{x}_j$ denotes either x_j or $\bar{x}_j$.) For example, the three minterms

x_1	x_2	x_3	x_4
0	0	0	0
0	1	1	0
1	0	1	0

satisfy the property since the six ways of selecting a complemented or uncomplemented variable from among the first three variables account for the six possible ways of selecting a proper nonempty subset of the minterms.

The ULM construction method yields modules of the form shown in Fig. 1. The universal function shown there is the function $U = z_1 P_1 + z_2 P_2 + \cdots + z_t P_t$. Since the sets P_i are disjoint, every function of n-variables has a unique decomposition as a sum of subsets of the P_i. To realize the function $f(x_1, x_2, \ldots, x_n)$ we calculate fP_i for each i. If $fP_i = 0$, we set $z_i = 0$; if $fP_i = P_i$, we set $z_i = 1$. Otherwise we find the variable x_j such that $\dot{x}_j P_i = fP_i$ and we set $z_i = \dot{x}_j$.

We now show how to find the partition P for all n and for each n we shall calculate t, the number of blocks in the partition P. The total number of inputs to the ULM is $n + t$.

The total number of ways of selecting subsets of a given block of the partition P is $2n + 2$ since each of the n-variables can be complemented or uncomplemented, and the constants 0 and 1 can be used to select the empty subset of the entire block. Hence, no block of the partition may have more than s members where $s \leq \log_2(2n + 2)$, and the total number of blocks in P is given by

$$t(n) \geq 2^n/\log_2(2n + 2) = 2^n/(1 + \log_2(n + 1)) \tag{5}$$

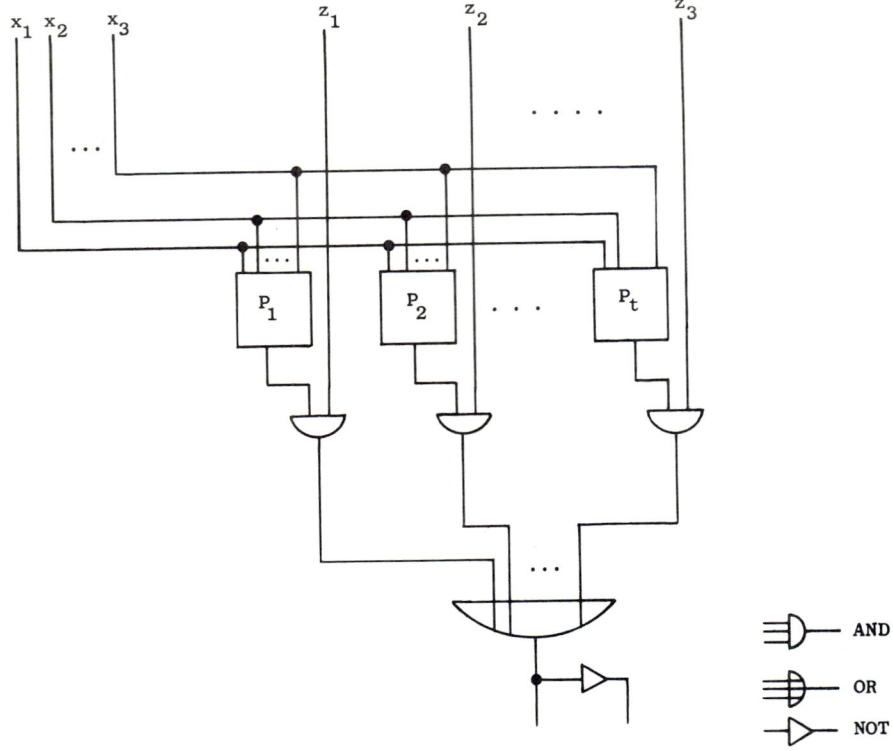

FIG. 1. A ULM of the Preparata–Muller type.

Our construction method produces a partition of the minterms that comes very close to meeting the bound in Eq. (5). Note that if t behaves as given in Eq. (5) then $m = n + t$, which asymptotically approaches the lower bound given in Eq. (2).

In the sequel, a matrix A is said to be a *block of s minterms* if it is an $s \times n$ matrix and if there is a subset of $2^{s-1} - 1$ columns of A such that these columns and their complements encode the binary integers in the range $1 \le i \le 2^s - 2$. If A is a block then any nonempty proper subset of rows of A can be selected by choosing a variable or the complement of a variable that encodes one of the integers between 1 and $2^s - 2$.

Example. The following is a block of three minterms. Columns 1, 3, and 4 and their complements encode the integers 1 through 6.

$$A = \begin{matrix} 1 & 0 & 0 & 1 \\ 0 & 0 & 1 & 1 \\ 0 & 0 & 0 & 0 \end{matrix}$$

Let v and w be n-tuples of 0's and 1's. We define $v \oplus w$, the *vector sum* of v and w, to be the 0, 1 vector whose components are the mod 2 sum of the corresponding components of v and w. If $v = (0, 0, 1, 1)$ and $w = (1, 0, 0, 1)$ then $v \oplus w = (1, 0, 1, 0)$.

If v is an arbitrary n-tuple and A is a block of s-minterms whose ith row is a_i, then it follows from the definition of block that the matrix

$$A \oplus v = \begin{matrix} a_1 \oplus v \\ a_2 \oplus v \\ \vdots \\ a_s \oplus v \end{matrix}$$

is also a block. The effect of addition by v on A is to complement a subset of columns of A, but this does not change its block property. Since blocks are mapped into blocks under arbitrary vector sums, we shall seek a partition by finding a set of vectors V and a block A such that the vectors in V generate the blocks of the partition when added to A.

In particular, consider V to be a subspace of V_n, the vector space of n-tuples with elements in GF(2). If we pick the block A such that the s-elements of A lie in distinct cosets of V with respect to V_n, then the blocks generated by V and A are disjoint. To see this is true let v_1 and v_2 be vectors in the subspace V, and suppose it happens that there exist n-tuples a_1 and a_2 in A such that $v_1 \oplus a_1 = v_2 \oplus a_2$, that is, two blocks generated by v_1, v_2 and A intersect. Then $v_1 \oplus v_2 = a_1 \oplus a_2$ which shows that the sum of a_1 and a_2 is in the subspace V. But then a_1 and a_2 are in the same coset which leads to a contradiction.

We use the property above in the construction procedure. In particular, we choose s so that $2^{s-1} \leq n < 2^s - 1$. The details of the construction are quite straightforward when s is a power of two. When s is not a power of two the construction is slightly more complex.

We shall use the following matrix to define the vector space V. The matrix H_s is defined to be an $r \times s$ matrix whose columns are the binary encodings of the integers $1 \leq i \leq s$. We choose $r = \lceil \log_2 s \rceil$.

Example.

$$H_5 = \begin{matrix} 1 & 0 & 1 & 0 & 1 \\ 0 & 1 & 1 & 0 & 0 \\ 0 & 0 & 0 & 1 & 1 \end{matrix}$$

The vector space $V^{(1)}$ is defined to be the null space of the $r \times n$ matrix $H^{(1)} = [H_{s-1} | 0]$ where the 0 submatrix in $H^{(1)}$ is an $r \times n - s + 1$ matrix.

The matrix $A^{(1)}$ is the block that has the following structure:

$$A^{(1)} = \left[\begin{array}{c|c|c} I_{s-1} & \begin{array}{c}\text{columns are all binary } s\text{-tuples}\\ \text{with 2 or more 1's}\end{array} & 0 \\ \hline 0 & 0 & 0 \end{array}\right]$$

$$= \left[\begin{array}{c|c|c} I_{s-1} & U_{s-1} & 0 \\ \hline 0 & 0 & 0 \end{array}\right]$$

where I_s is an $s \times s$ identity matrix, and U_s is any $s \times 2^s - (s+1)$ matrix whose columns are the binary s-tuples with two or more 1's. The 0 submatrices in $A^{(1)}$ are such that the bottom row and rightmost $n - 2^{s-1} + 1$ columns of $A^{(1)}$ are all 0. Clearly, $A^{(1)}$ is a block. The subspace $V^{(1)}$ and $A^{(1)}$ come close to forming the partition that we seek.

THEOREM 2.1. $V^{(1)}$ and $A^{(1)}$ generate a set of 2^{n-r} disjoint blocks of length s. Moreover, this set of blocks is a partition if and only if s is a power of two.

Proof: We have shown that if v is any vector in $V^{(1)}$, then $v \oplus A^{(1)}$ is a block. To show that the 2^{n-r} blocks generated by $V^{(1)}$ and $A^{(1)}$ are disjoint it is sufficient to show that the rows of $A^{(1)}$ are in distinct cosets of V_n with respect to $V^{(1)}$. To show the latter we must show the r-tuples of the form $a_i^{(1)}H^{(1)t}$ are distinct because the coset in which $a_i^{(1)}$ lies is determined by $a_i^{(1)}H^{(1)t}$. From the definition of $H^{(1)}$ and $A^{(1)}$ we see that $a_i^{(1)}$ is the ith column of $H^{(1)}$ for $1 \le i \le s - 1$ and is the 0 r-tuple if $i = s$. Each of these r-tuples is distinct so that blocks generated by $A^{(1)}$ and $V^{(1)}$ are disjoint.

There are $2^{n-r} = 2^{n-\lceil \log_2 s \rceil}$ vectors in $V^{(1)}$ and $s = 2^{\log_2 s}$ vectors in $A^{(1)}$. Hence there are $2^{n-\lceil \log_2 s \rceil + \log_2 s} = 2^{n-(\lceil \log_2 s \rceil - \log_2 s)}$ distinct n-tuples in the collection of blocks generated by $V^{(1)}$ and $A^{(1)}$. This number is 2^n if and only if $\lceil \log_2 s \rceil = \log_2 s$ which occurs if and only if s is a power of two. This completes the proof of the theorem.

Example. If $n = 7$, then $s = 4$ since this is the unique value that satisfies $2^{s-1} - 1 \le n < 2^s - 1$. For this case $r = \lceil \log_2 s \rceil = 2$. The construction yields the matrices below.

$$A^{(1)} = \begin{array}{ccccccc} 1 & 0 & 0 & 1 & 1 & 0 & 1 \\ 0 & 1 & 0 & 1 & 0 & 1 & 1 \\ 0 & 0 & 1 & 0 & 1 & 1 & 1 \\ 0 & 0 & 0 & 0 & 0 & 0 & 0 \end{array}$$

$$H^{(1)} = \begin{array}{ccccccc} 1 & 0 & 1 & 0 & 0 & 0 & 0 \\ 0 & 1 & 1 & 0 & 0 & 0 & 0 \end{array}$$

$V^{(1)}$, the null space of $H^{(1)}$, has dimension 5. The 2^5 vectors of $V^{(1)}$, when added to the 2^2 vectors of $A^{(1)}$ in all possible ways form 2^7 distinct vectors that span a 7-dimensional space.

A typical vector in $V^{(1)}$ has its first three coordinates equal and the other coordinates are arbitrary. Hence if we pick the vector $v = (1, 1, 1, 0, 0, 0, 0)$ in $V^{(1)}$ the block generated by v is the block below.

$$
v \oplus A^{(1)} =
\begin{matrix}
0 & 1 & 1 & 1 & 1 & 0 & 1 \\
1 & 0 & 1 & 1 & 0 & 1 & 1 \\
1 & 1 & 0 & 0 & 1 & 1 & 1 \\
1 & 1 & 1 & 0 & 0 & 0 & 0
\end{matrix}
$$

Theorem 2.1 shows that we can generate the required partition whenever s is power of two, and that we cover all but $2^{\lceil \log_2 s \rceil} - s$ cosets when s is not a power of two. We can obtain a cover of the cosets left over by using the construction technique iteratively in the following way. Suppose s is not a power of two and let $r = \lceil \log_2 s \rceil$. Now consider the $2^r - s$ cosets that are not covered in the cosets generated by $A^{(1)}$ and $V^{(1)}$. We shall attempt to cover the uncovered cosets with blocks of $s' < s$ elements each.

From an intuitive point of view, the cosets of V_n relative to $V^{(1)}$ form too coarse a partition to cover all of V_n in the manner that we desire. Hence we cover as much of V_n as possible with cosets relative to $V^{(1)}$, and we attempt to cover the rest of V_n with a collection of smaller cosets. The smaller cosets are constructed by finding $V^{(2)}$, a proper subspace of $V^{(1)}$. We select the dimension of $V^{(2)}$ to be just sufficiently less than that of $V^{(1)}$ so that we can find as few cosets of $V^{(2)}$ as possible to cover the rest of V_n. If the resulting partition is still too coarse then we shall find $V^{(3)}$, $V^{(4)}$, etc., until at last all of V_n is covered.

Let s' be the unique integer satisfying the inequalities

$$
2^{s'} \le 2(n + 1 - (s - 1)) < 2^{s'+1}
$$

Let $A^{(2)}$ be the $(s' - 1) \times n$ matrix

$$
A^{(2)} = \left[\begin{array}{c|c|c|c} 0 \\ \overrightarrow{s-1} & I_{s'} & U_{s'} & 0 \end{array} \right]
$$

where the first 0 submatrix is $(s' - 1) \times (s - 1)$, and let $V^{(2)}$ be the null space of $H^{(2)}$ where

$$
H^{(2)} = \left[\begin{array}{c|c|c} H_{s-1} & 0 & 0 \\ 0 & H_{s'-1} & 0 \end{array} \right]
$$

If we let $r' = \lceil \log_2 s' \rceil$, then $H^{(2)}$ is an $(r + r') \times n$ matrix.

THEOREM 2.2. The blocks generated by $A^{(2)}$ and $V^{(2)}$ are disjoint blocks and disjoint from the blocks generated by $A^{(1)}$ and $V^{(1)}$.

Proof: To prove that the blocks generated by $A^{(2)}$ and $V^{(2)}$ are disjoint, it is sufficient to show that the rows of $A^{(2)}$ lie in distinct cosets of V_n relative to $V^{(2)}$. This follows because the products of the form $a_i^{(2)}H^{(2)t}$ are distinct for each of the rows $a_i^{(2)}$ of $A^{(2)}$.

To prove that no block generated by $A^{(2)}$ and $V^{(2)}$ has any n-tuple in common with a block generated by $A^{(1)}$ and $V^{(1)}$ it is sufficient to show that all of the vectors in $A^{(1)}$ and $A^{(2)}$ lie in distinct cosets of V_n relative to $V^{(2)}$ But this follows directly since products of the form $a_i^{(1)}H^{(2)t}$ and $a_j^{(2)}H^{(2)t}$ are distinct by the construction of $A^{(1)}$, $A^{(2)}$, and $H^{(2)}$. This proves the theorem.

If s' is a power of two the construction is complete. If not, the iterative process continues by selecting s'' to be the unique integer that satisfies the inequalities $2^{s''} \leq 2(n + 1 - (s - 1) - (s' - 1)) < 2^{s''} + 1$. If s'' is not a power of two, another iteration is required. The iterative process must terminate because the sequence $s, s', s'', \ldots$ is derived from the logarithm of a decreasing sequence of integers.

Example. For $n = 18$, we obtain the matrices shown below. For this case $s = 5$, $r = 3$, $s' = 4$, $r' = 2$.

$$
A^{(1)} = \begin{matrix}
1 & 0 & 0 & 0 & 1 & 1 & 1 & 0 & 0 & 0 & 1 & 1 & 1 & 0 & 1 & 0 & 0 & 0 \\
0 & 1 & 0 & 0 & 1 & 0 & 0 & 1 & 1 & 0 & 1 & 1 & 0 & 1 & 1 & 0 & 0 & 0 \\
0 & 0 & 1 & 0 & 0 & 1 & 0 & 1 & 0 & 1 & 1 & 0 & 1 & 1 & 1 & 0 & 0 & 0 \\
0 & 0 & 0 & 1 & 0 & 0 & 1 & 0 & 1 & 1 & 0 & 1 & 1 & 1 & 1 & 0 & 0 & 0 \\
0 & 0 & 0 & 0 & 0 & 0 & 0 & 0 & 0 & 0 & 0 & 0 & 0 & 0 & 0 & 0 & 0 & 0
\end{matrix}
$$

$$
H^{(1)} = \begin{matrix}
1 & 0 & 1 & 0 & 0 & 0 & 0 & 0 & 0 & 0 & 0 & 0 & 0 & 0 & 0 & 0 & 0 & 0 \\
0 & 1 & 1 & 0 & 0 & 0 & 0 & 0 & 0 & 0 & 0 & 0 & 0 & 0 & 0 & 0 & 0 & 0 \\
0 & 0 & 0 & 1 & 0 & 0 & 0 & 0 & 0 & 0 & 0 & 0 & 0 & 0 & 0 & 0 & 0 & 0
\end{matrix}
$$

$$
A^{(2)} = \begin{matrix}
0 & 0 & 0 & 0 & 1 & 0 & 0 & 0 & 1 & 1 & 1 & 0 & 0 & 0 & 1 & 1 & 1 & 0 \\
0 & 0 & 0 & 0 & 0 & 1 & 0 & 0 & 1 & 0 & 0 & 1 & 1 & 0 & 1 & 1 & 0 & 1 \\
0 & 0 & 0 & 0 & 0 & 0 & 1 & 0 & 0 & 1 & 0 & 1 & 0 & 1 & 1 & 0 & 1 & 1 \\
0 & 0 & 0 & 0 & 0 & 0 & 0 & 1 & 0 & 0 & 1 & 0 & 1 & 1 & 0 & 1 & 1 & 1
\end{matrix}
$$

$$
H^{(2)} = \begin{matrix}
1 & 0 & 1 & 0 & 0 & 0 & 0 & 0 & 0 & 0 & 0 & 0 & 0 & 0 & 0 & 0 & 0 & 0 \\
0 & 1 & 1 & 0 & 0 & 0 & 0 & 0 & 0 & 0 & 0 & 0 & 0 & 0 & 0 & 0 & 0 & 0 \\
0 & 0 & 0 & 1 & 0 & 0 & 0 & 0 & 0 & 0 & 0 & 0 & 0 & 0 & 0 & 0 & 0 & 0 \\
0 & 0 & 0 & 0 & 1 & 0 & 1 & 0 & 0 & 0 & 0 & 0 & 0 & 0 & 0 & 0 & 0 & 0 \\
0 & 0 & 0 & 0 & 0 & 1 & 1 & 0 & 0 & 0 & 0 & 0 & 0 & 0 & 0 & 0 & 0 & 0
\end{matrix}
$$

Table III shows the value of $t(n) + n$ that is obtained from the iterative construction. The entries in Table III can be obtained easily by using the recursion formula

$$t(n) = 2^{n-r} + (2^r - s)2^{s-1-r}t[n - (s - 1)] \tag{6}$$

TABLE III
An Upper Bound for $M(n)$

n	2	3	4	5	6	7	8	9	10
upper bound for $M(n)$	4	6	10	16	28	39	72	137	266

If s is a power of two, then the second term in Eq. (6) vanishes and the first term is seen to be the correct value of $t(n)$ by construction. If s is not a power of two, then the first term accounts for the minterms covered by $A^{(1)}$ and $V^{(1)}$, and the second term accounts for the minterms not covered by $A^{(1)}$ and $V^{(1)}$. The factor $t[n - (s - 1)]$ appears because the second step of the iterative construction is essentially the solution of a problem of dimension $n - (s - 1)$. The multiplicative factor $(2^r - s)2^{s-1-r}$ accounts for the number of occurrences of the blocks for the $n - (s - 1)$ variable problem in the n-variable problem.

In the next section, we shall sharpen the upper bound for small n.

III. THE CONSTRUCTION OF ULM'S FOR SMALL n

The upper and lower bounds for $M(n)$ that have been derived are asymptotically equal for large values of n, but differ for all values of n that are of practical interest. We are most interested in obtaining accurate values of $M(n)$ for $n \leq 5$, because for $n \leq 5$ the number of pins per module is sufficiently small to make feasible the construction of ULM's for $n \leq 5$ within present integrated circuit technology. In this section we find exact values for $M(2)$ and $M(3)$, and we show that $6 \leq M(4) \leq 7$ by construction. The problem of finding tight bounds for $M(n)$ for $n \geq 5$ is still open.

A. ULM'S FOR $n \geq 3$

For $n = 1$, the ULM construction problem is trivial, since the four source wires carry the four 1-variable functions, 0, 1, x, and x'. Hence $M(1) = 1$, and the ULM function is the function $f(x) = x$.

For $n = 2$, $M(2) = 3$ since the function $F_3(a, b, c) = ab \oplus c$ for example, yields the two nontrivial genera represented by $x \oplus y$ and xy, as well as the trivial genera represented by $(0, 1)$ and x. It is clear that no 2-variable functions $F_2(a, b)$ would suffice to yield all 2-variable genera, since the parity functions would be unobtainable if F_2 were chosen as the product ab or the sum $a + b$ while the sum and product functions are unobtainable if $F_2 = a \oplus b$.

For $n = 3$ the problem becomes a bit more difficult. There are 14 genera, of which 10 are nondegenerate. Table II shows that there are 20 nonequivalent ways of distributing three variables, their complements, and the constants 0 and 1 among four terminals of a ULM. Hence, conceivably there could be a (4, 3) ULM; however, the following theorem rules this out.

THEOREM 3.1.   Let $f(x_1, x_2, x_3, x_4)$ be an arbitrary 4-variable function. Let $S(f)$ be the set of 3-variable function genera producible by f under the set of permissible input arrangements. Then $S(f)$ does not contain all three of the genera whose representatives are the 3-variable parity function, majority function, and AND function, respectively.

Proof: The proof is based on a weight argument. The 20 nonequivalent terminal connections correspond to 20 projections of the 4-cube onto the 3-cube. Any two of these projections either lie on disjoint halves of the 4-cube or intersect on four points. (This can be checked easily by examining the maps of the projections.) Now, consider how two projections may realize the 3-variable parity function and the 3-variable AND function, respectively. Suppose the projections are nondisjoint; i.e., they intersect in four points. It is easily verified (again by checking the maps) that any such intersection with the parity function must be of exactly weight two. Therefore, any 2-variable function lying on a projection that intersects the projection containing the parity function must have weight of at least two and no more than six. But the genus containing the AND function has members of weight 1 and 7 only; hence, the AND and the parity functions must lie on disjoint projections.

If this is the case, all of the other 18 remaining projections intersect both the AND function and the parity function in four points, and the eight intersection points form each projection. These intersections are of weight 0, 1, 3, or 4 with the AND (or 3-variable OR) function, and two with the parity function. Then, no function that intersects both the parity and the AND (or OR) functions can have weight 4. But the majority function has weight 4. Hence, there can be no 4-variable function that contains representatives of the 3-variable parity, AND, and majority function genera in its set $S(f)$. This proves the theorem.

From Theorem 3.1 we know that $M(3) \geq 5$. We now show the construction of a $(5, 3)$ ULM. A natural starting point for the construction is to identify the 4-variable genera which are most nearly $(4, 3)$ ULM's. This is easily done by an exhaustive search that involves classifying 20 projections of each of the 222 4-variable genera onto 3-variable functions. This search reveals that three genera, H_{87}, H_{76}, and H_{77} (N_{92}, N_{90}, and N_{91}, respectively)[1] have the property of containing nine of the ten nondegenerate 3-variable functions among their projections. For example H_{87} does not realize the genus containing the 3-variable AND function, and the other two functions fail to realize the genus containing the 3-variable parity function. Maps of these functions are given in Fig. 2a.

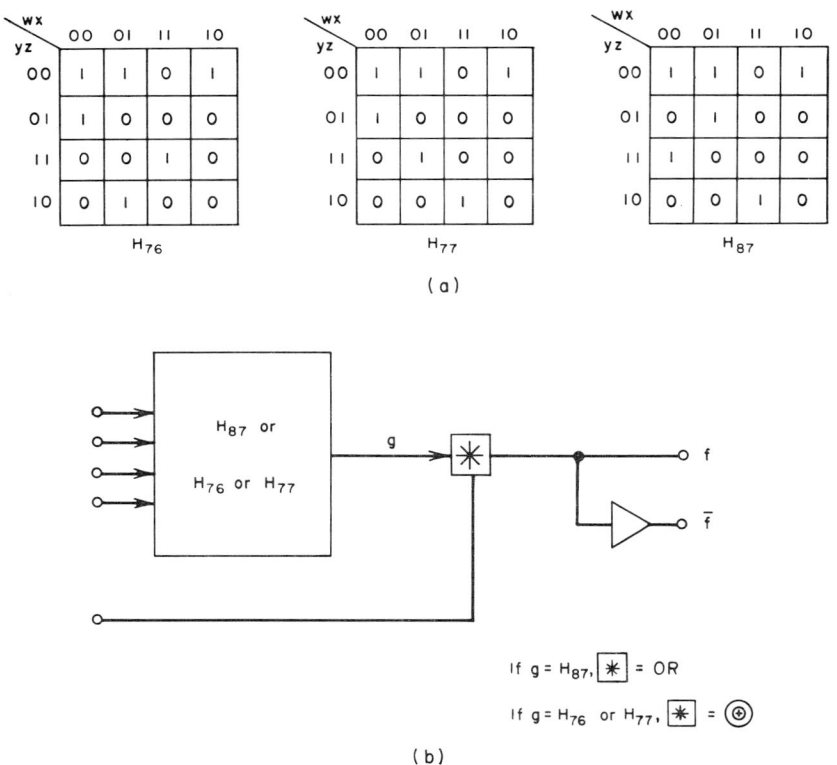

FIG. 2. (a) Karnaugh maps of the Harvard functions H_{76}, H_{77}, and H_{87}. (b) A $(5, 3)$ ULM.

---

[1] The notation H_{87} refers to the 87th function in the Harvard classification Aiken, *et al.*, (1951). Ninomiya's classification of the same function (Harrison, 1965) is denoted by N_{92}.

To construct a (5, 3) ULM from each of the three "almost" universal functions, a fifth input is introduced, as shown in Fig. 2b. The fifth input and the output of H_{87}, H_{76}, or H_{77} are used to drive a 2-variable function cell, an OR for H_{87}, and an EXCLUSIVE OR for the other functions. With a logical zero applied to the fifth input, nine of the ten genera can be produced on the network output by using the properties of the 4-variable functions. The tenth genus is produced by tying the fifth input to a variable, say x, and connecting the 4-variable functions to realize the 2-variable function $y + z$, if the module is H_{87}, and $y \oplus z$ otherwise. Note that complementing an input to an EXCLUSIVE OR gate is equivalent to complementing the output of the gate. Therefore, using the fifth input to control complementation, the (5, 3) ULM's constructed from H_{76} and H_{77} have the additional property that all 3-variable functions can be produced on a preselected output line of the module.

B. THE CONSTRUCTION OF A (7, 4) ULM

The construction of a 4-variable ULM has proved to be a difficult problem for several reasons. In order to establish the universality of a function, one may either exhibit the projections of the function that yield the 222 4-variable genera, or one may assume something about the structure of the function and show from the structure that all 4-variable genera are realizable without necessarily enumerating the 222 projections. An example of the latter technique is the iterative construction that we used earlier in this chapter to establish that $M(4) \leq 10$. Iterative constructions and other assumptions on the structure of a universal function tend to be inefficient in the use of input terminals. Hence, experience has shown that it is unlikely that one can find good 4-variable ULM's merely by postulating their structure.

The technique that has been used most successfully is essentially a search technique, usually used in conjunction with weak assumptions on the structure of the ULM function. The searches can be programmed for digital computer, but they are rather cumbersome. We see from Table II that the number of projections to be examined for each test function is 540 if the test function is a 6-variable function, and the number is 8400 if the test function has seven variables. Moreover, if the genus of each projecti on has to be identified, this computation is highly nontrivial. A new algorith m has been developed for the latter computation that greatly increases the effi ciency of the testing procedure (Ankerlin and Jackson, 1967). Forslund and Waxman (1966) used a hill-climbing technique in their quest for a 4-variable U LM. They came close to finding a (7, 4) ULM but failed in their search. Elsp as *et al.* (1967) used a search technique based on symmetry families which also came close to yielding a (7, 4) ULM. Using an "almost" universal 7-input module they constructed an (8, 4) ULM.

As an outgrowth of the work reported by Elspas *et al.* (1967), Ankerlin refined the search technique and found the (7, 4) ULM shown in Table IVa.

TABLE IVa
A (7, 4) ULM

$x_1\ x_2\ x_3$	x_4 x_5 x_6 x_7	0000 0000 0011 0110	0000 1111 1100 0110	1111 1111 0011 0110	1111 0000 1100 0110
0 0 0		1110	0110	0111	1011
0 0 1		1000	0101	1110	0111
0 1 1		1001	1101	1001	0001
0 1 0		0000	1111	0010	0̸111
1 1 0		1001	1011	0100	0001
1 1 1		0001	1010	0001	0010
1 0 1		1001	1101	0001	0000
1 0 0		0100	0110	1010	0101

This ULM is the best 4-variable ULM known, and establishes that $6 \le M(4) \le 7$.

Ankerlin used a hill-climbing technique of an unusual type to find the (7, 4) ULM. His iterative procedure required that the following information be gathered for a 7-variable test function. For each genus, the number of projections of the 7-variable function that realize this genus is tabulated, and if the genus is realizable, one of the many possible projections that realize this genus is also tabulated.

The iteration proceeds by identifying some genus that is not realized among the projections. An auxiliary table is searched to find a list of all genera that contain functions that differ by a single bit in their truth tables from a function in the unrealizable genus. This list contains at most 16 entries. Using the list, the list of realizable genera is searched to find if one of the realizable functions differs from the unrealizable function by a single bit. The object of the search is to discover if there is a single bit of the test function that can be changed such that the unrealizable genus becomes realizable. In case a realizable genus is found that differs by a single bit from the unrealizable genus, and if the realizable genus is realized by more than one projection, it is selected as a candidate for a single bit change. Using the projection information stored with the realizable genus, the algorithm next determines which bit of the test function should be changed to realize the unrealizable genus. The new test function is then analyzed and the iteration repeated until a function is found that realizes all genera. The algorithm is clearly dependent on the initial test function. Ankerlin chose a carefully constructed test function that realized all totally symmetric functions among its projections. The choice was a fortunate one since it took less than one minute to complete the search on a CDC-6400.

Table IVb gives the ULM connections for each of the 222 genera of 4-variable switching functions. The index of each genus is represented by its Ninomiya number (Harrison, 1965). The "Octal" column contains the representation of a 16 bit vector that gives the true and false minterms of the equivalence class representative. In this representation the right-most bit gives the value of minterm 0, that is, the minterm with all variables complemented. The next three columns indicate how to connect the ULM terminals so that no more than four distinct variables and the constants 0 and 1 are distributed among its seven inputs. The designations $x1, x2, \ldots, x7$ represent the seven input terminals and the variable $x0$ represents the constant 0. A ULM terminal tied to the constant 0 is shown by equating it to $x0$. Thus the entry $x0 = x1$ under $N46$ indicates that terminal 1 is connected to a constant 0. Similarly, a terminal tied to a constant 1 is indicated by showing it to be unequal to $x0$ as is done, for example, for the sixth terminal under $N2$. Two terminals are connected to the same variable if their corresponding designations are equated. See, for example, terminals 2 and 4 under $N1$. A pair of terminals connected to complementary signals is indicated by showing them to be unequal as is the case for terminals 5 and 6 under $N1$.

The table specifies explicitly how to reduce the number of free terminals from 7 to 4, but the connections of the four free variables to the ULM is indicated implicitly. To find this connection, select the four independent terminals with smallest indices in their dependency sets, arrange these in order and associate these in order with four variables of the 4-variable function to be realized. For genus $N2$ the dependency sets are $\{x1, x3\}$, $\{x2, x7\}$, $\{x4\}$, and $\{x5\}$. (The terminal $x6$ is connected to a constant.) Hence the four terminals to receive the switching function arguments are terminals 1, 2, 4, and 5. If we label the switching function arguments as y_1, y_2, y_3, and y_4 then we must connect y_1 to terminal 1, y_2 to terminal 2, y_3 to terminal 4, and y_4 to terminal 5.

The upper and lower bounds for $M(n)$ together with the best known ULM's for $n \leq 4$ are summarized in Table V. It is interesting to note, in particular, that the current state of knowledge leaves the case for $n = 5$ still open. There is a great discrepancy between the lower bound of 10 inputs and the upper bound of 16 inputs for a 5-variable ULM and the difference could be of some practical importance. Since the construction of the upper bound is based on a method that leaves n terminals permanently wired to the variables

TABLE IVb
The (7, 4) ULM Connections

N					N				
N 1	000000	X2=X4,	X3=X7,	X5'X6	N 2	002000	X0'X6,	X1'X3,	X2=X7
N 3	140000	X0'X1,	X0'X7,	X2=X6	N 4	177557	X0=X1,	X0=X6,	X3'X7
N 5	177576	X0=X1,	X5=X6,	X5=X7	N 6	100001	X1'X2,	X1'X4,	X3=X7
N 7	124000	X0'X1,	X0'X2,	X6=X7	N 8	120002	X0'X1,	X0'X4,	X2=X6
N 9	060020	X0'X1,	X0'X4,	X5'X6	N 10	005020	X0'X1,	X0=X2,	X3=X7
N 11	010050	X0'X1,	X0'X3,	X0'X7	N 12	001204	X0'X1,	X0'X4,	X2'X3
N 13	120240	X0'X1,	X2'X3,	X2=X6	N 14	006300	X0'X1,	X0=X7,	X3'X6
N 15	060220	X2'X4,	X2'X5,	X2=X7	N 16	127765	X0'X3,	X0=X4,	X0=X7
N 17	173157	X0=X3,	X0=X6,	X5=X7	N 18	014201	X0'X1,	X0=X5,	X3'X6
N 19	010121	X0'X1,	X0'X3,	X2'X7	N 20	005042	X0'X1,	X0=X3,	X2=X7
N 21	021050	X0'X1,	X0'X2,	X0'X7	N 22	157635	X0=X1,	X0=X2,	X5=X6
N 23	167636	X0=X3,	X2'X4,	X2=X5	N 24	175437	X0=X1,	X0=X2,	X0'X7
N 25	057375	X0=X1,	X0=X3,	X4'X5	N 26	120044	X0'X1,	X0'X3,	X0=X5
N 27	156657	X0=X1,	X0=X2,	X0=X6	N 28	163735	X0=X1,	X0=X2,	X0'X4
N 29	112010	X0'X1,	X0'X7,	X3=X5	N 30	120024	X0'X1,	X0'X2,	X3'X6
N 31	102102	X0'X1,	X0'X2,	X0'X4	N 32	120242	X0'X1,	X0'X2,	X0'X6
N 33	037517	X0=X1,	X0=X7,	X3'X6	N 34	150140	X0'X1,	X0'X3,	X5'X7
N 35	000652	X0=X1,	X0'X2,	X4=X5	N 36	021121	X0'X1,	X0'X2,	X3'X7
N 37	114402	X0'X1,	X0=X2,	X0'X4	N 38	166732	X0=X1,	X0'X4,	X3=X5
N 39	064220	X0'X1,	X0=X6,	X4=X5	N 40	012640	X0'X1,	X0=X2,	X3'X6
N 41	055736	X0=X1,	X0'X5,	X2=X6	N 42	023102	X0'X1,	X0'X2,	X0'X3
N 43	111204	X0'X1,	X0'X4,	X2'X5	N 44	157153	X0'X5,	X1'X7,	X2=X6
N 45	152100	X0'X4,	X0'X6,	X3'X5	N 46	137613	X0=X1,	X0'X4,	X2'X7
N 47	176670	X0=X1,	X0'X2,	X4'X5	N 48	042604	X0'X1,	X0=X3,	X2=X4
N 49	130022	X0'X1,	X0'X2,	X4=X6	N 50	064210	X0'X1,	X0=X5,	X2=X6
N 51	135333	X0=X1,	X0'X4,	X3'X7	N 52	152747	X0=X1,	X0=X2,	X0=X3
N 53	054773	X0=X1,	X0=X4,	X0'X5	N 54	157361	X0=X1,	X0=X3,	X2=X5
N 55	157235	X0=X1,	X0=X2,	X3=X5	N 56	102046	X0'X1,	X0'X2,	X3'X4
N 57	044640	X0'X1,	X0=X2,	X0=X5	N 58	060430	X0=X1,	X0=X5,	X0'X6
N 59	127652	X0=X1,	X0'X3,	X0=X6	N 60	125656	X0'X1,	X0=X4,	X2'X3
N 61	170770	X0'X1,	X0=X7,	X2'X3	N 62	002760	X0'X1,	X0=X2,	X4=X7
N 63	055240	X0'X1,	X0=X2,	X0=X6	N 64	153652	X0=X1,	X2=X3,	X4=X5
N 65	075532	X0=X1,	X3'X6,	X5=X7	N 66	015241	X0'X1,	X0=X4,	X3'X6
N 67	112141	X0=X3,	X0'X4,	X5'X6	N 68	053672	X0=X1,	X0'X5,	X2=X3
N 69	155645	X0=X1,	X0=X2,	X4=X6	N 70	152653	X0=X1,	X0=X4,	X2=X3
N 71	164627	X1'X4,	X3=X5,	X6'X7	N 72	125402	X0'X1,	X0=X2,	X0'X3
N 73	105657	X0=X1,	X0=X2,	X4=X7	N 74	056437	X0=X1,	X0=X2,	X4'X6
N 75	017420	X0=X1,	X0'X3,	X0'X7	N 76	034373	X0=X1,	X0'X2,	X0=X7
N 77	153515	X0=X1,	X0'X6,	X2'X7	N 78	132761	X0'X1,	X0=X2,	X3'X7
N 79	170672	X0=X1,	X0'X2,	X0'X5	N 80	121222	X0'X1,	X0'X4,	X3=X6
N 81	157215	X0=X1,	X0=X2,	X0=X5	N 82	163417	X0=X1,	X0=X3,	X2'X4
N 83	156465	X0=X1,	X0=X3,	X2=X6	N 84	053635	X0=X1,	X0=X2,	X0'X5
N 85	064212	X0'X1,	X0'X5,	X2=X6	N 86	176432	X0=X1,	X0=X2,	X4'X7
N 87	130144	X0'X1,	X0'X3,	X4'X5	N 88	006650	X0=X1,	X0'X2,	X0=X5
N 89	015260	X0'X1,	X0=X2,	X4'X6	N 90	032242	X0=X1,	X0'X2,	X5'X6
N 91	045602	X0'X1,	X2=X4,	X3=X5	N 92	057235	X0=X1,	X0=X2,	X4'X5
N 93	064244	X0'X1,	X0=X6,	X3=X5	N 94	042446	X0'X1,	X0=X4,	X2=X3
N 95	054412	X0=X1,	X0=X4,	X2'X5	N 96	036672	X0=X1,	X0'X2,	X5=X6
N 97	054640	X0'X1,	X0=X2,	X3'X5	N 98	044640	X0'X1,	X0=X3,	X0=X5
N 99	051224	X0'X1,	X0'X4,	X2'X6	N100	116672	X0=X1,	X0'X3,	X0'X5
N101	154475	X0=X1,	X0=X3,	X6=X7	N102	044642	X0'X1,	X0=X2,	X4=X5
N103	044644	X0'X1,	X0=X3,	X4=X5	N104	055746	X0'X1,	X0=X4,	X0'X5
N105	131044	X0'X1,	X0'X3,	X2'X5	N106	133615	X0=X1,	X0=X2,	X4=X5
N107	110305	X0'X1,	X2=X5,	X2'X7	N108	123014	X0=X3,	X0=X7,	X1=X6
N109	170362	X0'X1,	X2'X3,	X4=X7	N110	052642	X0'X1,	X0=X2,	X4=X6
N111	114446	X0'X1,	X2=X3,	X2'X4	N112	064622	X0'X4,	X2'X6,	X5'X7
N113	131242	X0'X1,	X0'X3,	X0'X6	N114	007653	X0=X1,	X0'X2,	X0=X4
N115	153401	X0=X1,	X0=X3,	X4=X5	N116	156204	X0'X1,	X3'X7,	X4'X5
N117	170130	X0=X1,	X0'X3,	X4=X7	N118	104417	X0=X1,	X0=X4,	X2'X3
N119	106437	X0=X1,	X0=X2,	X3=X7	N120	125632	X0'X1,	X3'X4,	X3=X6
N121	156306	X0'X2,	X1'X5,	X4=X7	N122	161460	X0=X1,	X0=X5,	X0'X7
N123	104735	X0=X1,	X0=X2,	X0'X3	N124	152250	X0'X1,	X0'X5,	X0'X7
N125	055465	X0=X1,	X0=X3,	X0'X6	N126	064324	X0'X1,	X0'X5,	X0=X6
N127	005361	X0'X1,	X0=X2,	X0=X7	N128	155072	X0=X1,	X0=X4,	X0=X6
N129	036272	X0=X1,	X0'X2,	X4'X6	N130	162142	X0'X1,	X0'X2,	X0'X5
N131	124644	X0'X1,	X0'X2,	X4'X6	N132	042631	X0'X1,	X0=X2,	X0=X3

TABLE IVb (*Continued*)
The (7, 4) ULM Connections

N133	161456	X0=X1, X0=X5, X2'X7	N134	163611	X0=X1, X0=X2, X3'X4
N135	015251	X0'X1, X0=X3, X4'X6	N136	016306	X0'X1, X0=X5, X4=X7
N137	060736	X0=X1, X0'X6, X2=X5	N138	150612	X0=X1, X0'X2, X0'X4
N139	054432	X0=X1, X0'X5, X2=X4	N140	054650	X0'X1, X0=X3, X2'X5
N141	054642	X0=X1, X2=X3, X5'X6	N142	055271	X0'X1, X0=X3, X0=X6
N143	117303	X0=X1, X0'X4, X5'X7	N144	044662	X0'X1, X0=X2, X3=X5
N145	111145	X0=X1, X0'X6, X5'X7	N146	135060	X0=X1, X0=X5, X2=X7
N147	127204	X0'X1, X0'X2, X0=X3	N148	157015	X0=X1, X0=X3, X0=X5
N149	106650	X0=X1, X0'X3, X0=X5	N150	125604	X0'X1, X2'X3, X2=X4
N151	127102	X0'X1, X0'X2, X3=X4	N152	125502	X0'X1, X0'X3, X2=X4
N153	015317	X0'X1, X0'X5, X0=X7	N154	053615	X0'X1, X0=X2, X3'X5
N155	071171	X0'X1, X0'X2, X6'X7	N156	034033	X0=X1, X0'X2, X4'X7
N157	120155	X0=X1, X0=X7, X3=X6	N158	134424	X0'X1, X0'X2, X0=X6
N159	055662	X0=X1, X0'X6, X2=X3	N160	034752	X0'X1, X0=X3, X5=X7
N161	044744	X0'X1, X0=X3, X0=X6	N162	106470	X0=X1, X0=X3, X0=X7
N163	015624	X0'X1, X2=X4, X2'X6	N164	075311	X0'X3, X2'X5, X4'X6
N165	170360	X0'X3, X1'X2, X4=X7	N166	055132	X0=X1, X2=X4, X3'X6
N167		DEGENERATE	N168	113151	X1'X4, X1'X6, X3=X5
N169	121272	X0'X1, X2'X4, X3=X6	N170	125612	X0=X1, X0'X2, X0=X3
N171	157600	X0=X1, X0'X4, X3=X7	N172	066314	X0'X2, X1=X6, X4=X7
N173	132522	X0=X1, X0'X4, X0'X6	N174	042656	X0'X1, X0=X3, X0=X4
N175	155241	X0=X2, X0'X4, X0=X7	N176	117111	X0'X4, X0=X5, X2'X6
N177	007721	X0=X1, X0'X2, X0=X3	N178	055261	X0'X1, X0=X4, X0=X6
N179	053650	X0=X1, X2=X3, X2'X5	N180	144474	X0'X1, X0=X3, X5'X7
N181	113145	X0=X1, X4=X6, X5'X7	N182	127412	X0=X1, X0=X2, X6'X7
N183	154330	X0=X1, X0=X5, X2=X4	N184	102355	X0'X1, X2=X5, X6'X7
N185	076501	X0'X2, X3'X7, X4=X6	N186	036063	X0=X1, X0'X2, X5'X7
N187	036252	X0=X1, X0'X2, X0=X6	N188	131216	X0=X3, X1=X5, X6'X7
N189	065232	X0'X1, X3=X6, X4=X5	N190	115246	X0'X3, X0'X5, X4=X6
N191	051712	X0=X1, X3'X5, X4'X6	N192	053152	X0'X6, X2'X5, X4=X7
N193	036132	X0=X3, X0=X6, X1'X2	N194	113245	X0=X2, X4=X6, X5'X7
N195	170521	X0'X1, X0'X3, X0=X7	N196	053415	X0'X1, X0=X3, X2'X5
N197	172105	X0'X1, X0'X6, X4'X7	N198	055425	X0'X1, X0=X2, X0'X6
N199	127046	X0'X1, X0'X2, X0=X4	N200	174121	X0'X1, X0'X2, X0=X7
N201	175201	X0=X1, X0'X4, X6'X7	N202	156244	X0'X1, X3'X5, X3'X7
N203	064364	X0'X1, X0=X6, X5'X7	N204	171130	X0'X1, X0'X2, X4=X7
N205	032662	X0=X1, X0'X2, X0'X6	N206	155441	X0'X2, X0'X4, X3'X7
N207	057212	X0'X1, X0'X5, X0'X6	N208	152702	X0=X1, X0'X4, X3=X6
N209	065432	X0=X1, X0=X5, X3'X6	N210	152611	X0=X1, X0=X2, X0=X4
N211	036262	X0=X1, X0'X2, X3=X6	N212	045346	X0'X1, X0=X4, X2'X5
N213	054447	X0'X1, X0=X2, X5'X6	N214	154413	X0=X1, X0=X4, X3=X5
N215	055342	X0'X1, X0=X4, X3'X5	N216	054750	X0'X1, X0=X3, X4'X5
N217	103545	X0'X1, X5'X6, X5'X7	N218	054662	X0'X1, X0=X2, X0'X5
N219	066251	X0'X1, X2'X5, X4'X6	N220	045646	X0'X1, X0=X4, X3=X5
N221	034730	X0=X1, X0'X2, X5=X7	N222	055306	X0'X1, X0'X5, X4=X6

TABLE V
Best known upper and lower bounds for $M(n)$

n	2	3	4	5	6	7	8	9	10
lower bound	3	5	6	10	17	32	62	119	230
upper bound	3	5	7	16	28	39	72	137	266

$x_1, x_2, \ldots, x_n$, it is conjectured that the upper bound can be lowered by properly taking into account all of the connections possible. Consequently, the true value of $M(5)$ is probably closer to 10 than to 16.

IV. OTHER APPROACHES TO THE UNIVERSAL MODULE PROBLEM

The ULM investigation is based upon the assumption that a particular class of module interconnections can be used to determine the module function. At least two other approaches have been treated by switching theorists that merit description in this chapter. The first approach uses a different interconnection scheme to specialize universal modules. The second approach assumes that the class of connections available are symmetry transformations on the input or outputs and that, by necessity the, "core" module may have to be selected from a small set of modules. In this approach no single module has the universal property, but a set of modules is universal in the sense that any function can be realized by selecting the appropriate module in the set and then selecting an input and output arrangement from the class of permissible symmetry operations.

The first approach is due to Yau and Tang (1970). Their universal modules use the property that any function can be expanded about any r variables, $r \leq n$ into the form

$$f(x_1, x_2, \ldots, x_n) = \sum_{i_1, i_2, \ldots, i_r} x_1^{i_1} x_2^{i_2} \cdots x_r^{i_r} f(i_1, \ldots, i_r, x_{r+1}, \ldots, x_n) \quad (7)$$

where the notation $x_1^{i_1}$ means x_1 if $i_1 = 1$ and $\bar{x}_1$ if $i_1 = 0$. The universal module is constructed as shown in Fig. 3. Input variables are partitioned into two sets, one of size r and the other of size $k = n - r$. The k variables enter a logic network that produces almost all of the 2^{2^k} functions of k variables. Actually it produces $2^{2^k} - 2(k + 1)$ functions where the $2(k + 1)$ that are not produced are the $2k$ functions of one variable and the constant functions. The second section of the universal module is a tree in r variables that forms the expansion given in Eq. (7). The universal module is specialized to produce any function of n variables by connecting combinations of intermediate output functions to the 2^r collector tree inputs.

The total number of input and function specialization terminals on the universal module, $T(n)$, is

$$T(n) = n + 2^{2^k} - 2(k + 1) + 2^{n-k} \quad (8)$$

Since $T(n)$ depends on k, we wish to select k as a function of n so as to minimize $T(n)$. The minimum is found by finding the derivative of Eq. (8) with

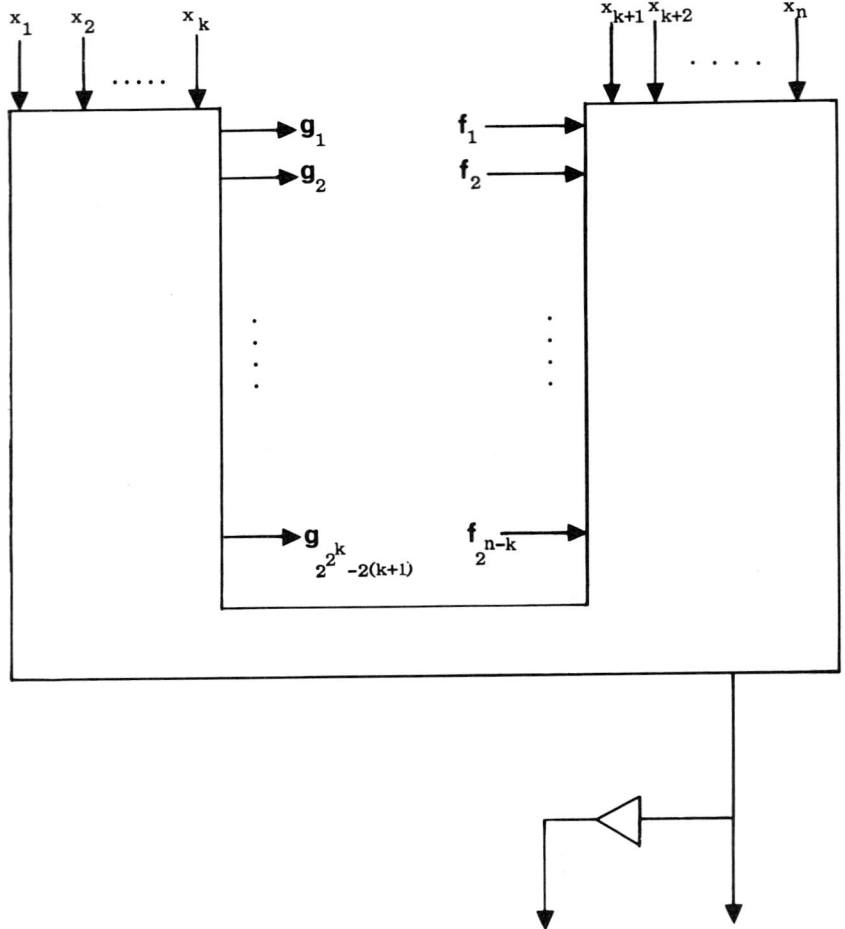

FIG. 3.   A universal module of the Yau–Tang type.

respect to k and equating it to 0. $T(n)$ is minimized by choosing k approximately equal to $\log_2(n - 2 \log_2 n)$. For this choice Eq. (8) becomes

$$T(n) \cong n + 2^n/n^2 - 2(1 + \log_2 n) + 2^n/n$$

For sufficiently large n, $T(n)$ is proportional to $2^n/n$, which is somewhat smaller than $M(n) \cong 2^n/(1 + \log_2(n + 1))$. Table VI gives values of $T(n)$ for small n and gives the value of k for which minimum $T(n)$ is obtained.

The enumeration of $T(n)$ in Table VI differs slightly from Eq. (8) for $k = 1$. If $k = 1$, the universal module can be constructed from a collector tree alone, and therefore the front part of the module is unnecessary. This

TABLE VI
$T(n)$

n	2	3	4	5	6	7	8	9	10
k	1	1	1	1	2	2	2	2	2
$T(n)$	3	6	11	20	32	49	82	147	276

reduces $T(n)$ by 1. Comparison of Tables V and VI show that the Yau–Tang modules have more terminals than the Muller–Preparata ULM's for $n \leq 10$.

The second approach to be considered here is due to Lechner (1970) who applied a theory developed by Ninomiya (1961) and by Lechner (1963). The problem is to find a universal set of modules with respect to a group of symmetry operations. A set of modules is universal in n variables if there is a module in the set and a symmetry operation on the connections to the module for each n-variable function such that the function is realizable by the module when the connections are made as specified by the symmetry operation. For example, it is well known that the group G_4 of arbitrary permulations and complementations acting on the arguments of 4-variable switching functions partitions the set of 4-variable functions into 402 disjoint classes (Harrison, 1965). Hence, there exist universal sets of 402 modules that are universal with respect to G_4. If we enlarge the symmetry operations to include negation on the output, then the number of classes is reduced to 222 with an equivalent reduction in the size of the universal set of modules.

Lechner (1970) reports that the Restricted Affine Group (RAG) is a group of symmetry operations for which it is practical to find a universal set of modules for the 5-variable switching functions. The equivalence classes induced by this group of symmetries have been called *prototypes* by Ninomiya (1961). We can describe the RAG operations succinctly in the following way. Let A be a nonsingular $n \times n$ matrix with entries in GF(2), and let a and b be $1 \times n$ matrices over GF(2), and c be a constant in GF(2). Then two n-variable functions $f(x)$ and $g(x)$ are equivalent with respect to RAG_n if and only if there exist A, a, b, and c such that

$$f(x) = g(x\text{A} \oplus a) \oplus xb^t \oplus c$$

Here we treat x as an n-dimensional row vector so that the term $x\text{A}$ is a matrix product, xb^t is the inner product of x and b, and all arithmetic is in GF(2).

The number of classes induced by RAG_n for small n is given in Table VII. We see there that there are only eight classes for $n = 4$ and only 48 for $n = 5$. Thus, a universal set for the 5-variable functions need contain only 48 modules. Clearly, the RAG symmetry operations are not sufficiently numerous to follow this approach for $n > 5$.

TABLE VII
The number of classes induced by RAG_n

n	2	3	4	5	6	7
Number of classes	2	3	8	48	$>69,000$	$>10^{19}$

For $n = 4$ the problem has been solved in the sense that the prototypes of all 4-variable symmetry types have been tabulated. (Ninomiya, 1961.) A universal set of modules consists of a set of eight functions, one function from each of the eight prototypes. Function synthesis requires that a universal module be embedded in a logic network such that the input variables can be transformed by an affine transformation of the form $xA \oplus a$ and then passed to the universal module, while the module output is added to the affine function $xbi \oplus c$. Affine transformations require only EXCLUSIVE OR gates to implement. These gates do the addition in GF(2). Multiplication in GF(2) can be realized without logic gates by selectively making or breaking interconnections. For example, $(x_1, x_2, x_3) \cdot (1, 0, 1)^t = 1 \cdot x_1 \oplus 0 \cdot x_2 \oplus 1 \cdot x_3 = x_1 \oplus x_3$. Hence, in this example, multiplication by $(1, 0, 1)^t$ is realized by connecting x_1 and x_3 to an EXCLUSIVE OR gate while x_2 is disconnected from that gate.

Lechner (1970) reports that the situation for $n = 5$ is almost solved but is still open. The 5-variable functions are far too numerous to enumerate explicitly. A Monte Carlo analysis was used to produce a random sample of 5-variable functions which were categorized by prototype. This led to the identification of members of 46 of the 48 prototypes. There remains only to identify members of the two remaining prototypes to solve this part of the problem completely. Since the random sample included less than 0.01 % of the 5-variable functions, a potentially large search may have to be conducted to find the last two prototypes.

V. HISTORICAL REFERENCES

Universal modules similar to ULM's were first studied by Dunham (1957) and Dunham et al. (1959). Those studies assumed that complements of variables are not available for module inputs. The problem studied in this chapter was formulated by W. H. Kautz, and many of the results are drawn from his work and the work of his colleagues, R. Ankerlin, B. Elspas, C. L. Jackson, and H. S. Stone. (Elspas et al., 1967). Forslund and Waxman (1966) reported some of the results that appeared in Elspas et al. (1967). The statement of Theorem 3.1 appears in Forslund and Waxman (1966) but the proof of the theorem is incomplete there. King (1966) discusses the construction of modules that are universal with a restricted subset of switching functions, which is somewhat different than our notion of universality with respect to all n-variable switching functions. Other contributors to the theory of universal logic modules include Patt (1967), Yau and Tang (1968), and Meo (1968). The contributions of Yau and Tang (1970) and Preparata and Muller (1970) are particularly important in the sense that they establish the tightest known upper bounds on the number of terminals for universal modules.

EXERCISES

1. Construct a 3-variable Karnaugh map for each of the 20 projections of the 4-variable map into a 3-variable map. Label the cells of the 3-variable maps by their corresponding labels in the 4-variable map and indicate which variables are connected to constants or connected together.

2. Show that H_{96} yields nine of the ten 3-variable genera among its projections.

3. Show the matrices $A^{(i)}$, $V^{(i)}$, and $H^{(i)}$ that appear in the Preparata–Muller construction method for $n = 20$.

4. Assume that complements of input variables are not available and that a universal module produces only a single output. Derive asymptotic formulas for the minimum number of terminals on universal modules under these assumptions. [Hint: As n grows large, nearly all equivalence classes of functions are maximal size when the equivalence relation is induced by the group of permutations of inputs.]

REFERENCES

ABRAMOWITZ, M., and STEGUN, I. A. (1964). "Handbook of Mathematical Functions." U.S. Dept. of Commerce, NBS Appl. Math. Ser., Vol. 55.

AIKEN, H., *et al.* (1951). Synthesis of electronic computing and control circuits. *Ann. Comput. Lab. Harvard Univ.*, Harvard Univ. Press, Cambridge, Massachusetts.

ANKERLIN, R. A., and JACKSON, C. L. (1967). A rapid method for the identification of the type of a four-variable Boolean function. *IEEE Trans. Electron. Comput.* **EC-16**, 870–871.

DUNHAM, B. (1957). The multipurpose bias device, part I: the commutator transistor. *IBM J. Res. Develop.* **1**, 119–129.

DUNHAM *et al.* (1959). The multipurpose bias device, part II: the efficiency of logical elements. *IBM J. Res. Develop.* **3**, 46–53.

ELSPAS, B., GOLDBERG, J., JACKSON, C. L., KAUTZ, W. H., and STONE, H. S. (1967). Properties of Cellular Arrays for Logic and Storage, Sci. Rep. No. 3, Contr. AF-19-628-5828 (DDC AD-658832), pp. 59–83. Stanford Res. Inst., Menlo Park, California.

FORSLUND, D. C., and Waxman, R. (1966). The universal logic block (ULB) and its application to logic design. *Proc. 7th IEEE Ann. Symp. Switching and Automata Theory*, 236–250.

HARRISON, (1965). " Introduction to Switching and Automata Theory." McGraw-Hill, New York.

KING, W. F., III (1966). The synthesis of multipurpose logic devices. *Proc. 7th IEEE Ann. Symp. Switching and Automata Theory*, 236–250.

LECHNER, R. J. (1963). Affine Equivalence of Switching Functions. Ph.D. Thesis, Harvard Univ., Cambridge, Massachusetts.

LECHNER, R. J. (1970). A transform approach to logic design. *IEEE Trans. Comp.*, **C-19**, 627–640.

MEO, A. R. (1968). Modular tree structures. *IEEE Trans. Electron. Comput.* **EC-17**, 432–442.

MOORE, E. F. (1962). Machine models of self-reproduction. *Proc. Symp. Appl. Math.* **14**, 17–33.

NINOMIYA, I. (1961). A study of the structures of boolean functions and its application to the synthesis of switching circuits. Ph.D. thesis, *Mem. Fac. Nagoya Univ.* **13**, No. 2.

PATT, Y. N. (1967). A complex module for the synthesis of combinational switching circuits. *Proc. 1967 AFIPS, Spring Joint Comput. Conf.* **20**, 699–705.

PREPARATA, F. P., and Muller, D. E. (1970). Generation of near-optimal universal boolean functions. *J. Comput. Syst. Sci.*, **4**, 93–102

VON NEUMANN, J. (1966). " The Theory of Self-Reproducing Automata " (A. W. Burks, ed.). Univ. of Illinois Press, Urbana, Illinois.

YAU, S. S., and TANG, C. K. (1968). Universal logic circuits and their modular realizations. *Proc*, 1968 *AFIPS, Spring Joint Comput. Conf.* **32**, 297–305.

YAU, S. S., and TANG, C. K. 1970. Universal logic modules and their applications. *IEEE Trans. Comp.*, **C-19**, 141–149.

Chapter VII

CELLULAR LOGIC

AMAR MUKHOPADHYAY
and
HAROLD S. STONE

I. INTRODUCTION   256

II. SINGLE-RAIL CASCADES   258
    A. The Canonical Form of a Redundant Cascade   259
    B. Derivation of Algorithm for Testing Cascade Realizability 263
    C. Statement of the Algorithm   270
    D. The Length of Redundant Cascades in Canonical Form   273
    E. The Number of Realizable Functions   278
    F. Historical References   280

III. TWO-RAIL CASCADES   281
    A. The Basic Two-Rail Cascade   281
    B. Efficient Two-Rail Cascades   283

IV. TWO-DIMENSIONAL ARRAYS   285
    A. Cutpoint Arrays   286
    B. q-Function or Adder Array   289
    C. 2^n-Cell Arrays   293
    D. NOR Arrays   296
    E. NOR–NAND Arrays for Combinational Functions   297
    F. Majority-Gate Arrays   300

V. MINIMIZATION OF CELLULAR ARRAYS   300
    A. Unate Cellular Logic   301
    B. A More General Algorithm   305
    C. Other Minimization Problems   306

VI. REVIEW OF OTHER WORKS IN CELLULAR AREA   307
    A. Introduction   307
    B. Work of Hennie   307
    C. Work of Unger   308
    D. Work of Holland   309
    E. Work of von Neumann   310

REFERENCES   311

ABSTRACT. The theory of logical design with cellular logic arrays is presented in this chapter. Mathematical models of some of the simple cullular structures are studied and synthesis algorithms for arbitrary switching functions are developed. The theory of the fundamental 1-dimensional cascade, the Maitra cascade, is discussed in detail. The elementary theory of two-rail cascades is also presented. Logical design techniques for arbitrary combinational functions using several 2-dimensional arrays are discussed and the efficiency of different structures with respect to design parameters like cell complexity, interconnection structure, size, etc., are evaluated. A discussion on the minimization problems that arise in cellular arrays and their partial solutions is included. This chapter concludes with a review of research in the macrocellular area.

I. INTRODUCTION

Cellular logic is a branch of switching theory which deals with mathematical models and the techniques for the analysis and synthesis of digital networks in the form of cellular arrays. A *cellular array* consists of a 1-, 2-, or 3-dimensional iterative arrangement of similar or identical cells with a uniform interconnection pattern on the cells. Digital circuits have been designed in the past which exhibited cellularity. Important among them are the cross bar switch used in telephone systems, the diode arrays used as decoders and function matrices in early computer networks, and the magnetic core storage (Minnick, 1967); but cellularity in these circuits is more a spatial convenience rather than a reflection of a trend in the technology. Study of digital circuits in cellular forms took on a new importance at the beginning of the sixties with the advent of integrated circuits and batch fabricated semiconductor technologies. In this technology, an individual logical element may contain a large number of circuit components (like transistors, resistors, or capacitors, etc.) interconnected on a monolithic substrate called a "chip." Furthermore, in large-scale integration (LSI) technology, it is possible to have a number of such chips produced and interconnected on a single wafer. From manufacturing points of view, it is advantageous to have the wafer produced in the form of a cellular array of chips. Besides manufacturing yield, fabrication of logic circuits in such array forms offers several advantages. Packing density is higher, not only because of the reduction of size of the individual components but also because of elimination of much of the interconnection wiring between the chips which are usually needed to produce a functional module using integrated circuit packages. The reliability of devices produced by LSI is higher because of the fact that the physical parameters of the circuit can be controlled within specified tolerances by carefully designed semiconductor processes. Also the faults due to the failure of interconnection wiring, which is rather frequent phenomenon in lumped digital circuits, are very much reduced due to the reduction of the wiring itself. Probably the most distinctive feature of array circuits produced by LSI is the flexibility in performance. The flexibility can be incorporated by either varying the function of each cell of the array or by varing the interconnection pattern. Specialization can be achieved in two ways: first, by built-in logical and interconnection parameters which can be preset at the final stages of production for specialized custom-designed arrays; second, by associating with each cell a set of parameters which can be electronically programmed by control signals applied to the terminals of the array. Cellular arrays of the latter types are called programmable arrays (see Chapter IX).

There are certain disadvantages of the cellular arrays, too. The major ones come from maintenance and standardization. A difficult problem arises when one has to locate and correct a faulty cell in an array. Since the number of test points or the input/output pins available is very small compared to what can be expected of a circuit of similar complexity built with lumped components, the difficulties in diagnosis problems have just been compounded. It seems unlikely that good and practically manageable algorithmic solutions to this probem will be developed in the near future because of what seems to be an inherent contradiction in the objective: programmability and flexibility which can be achieved by increasing the cell complexity. This implies an exponential growth of fault types and correspondingly astronomical growth of the number of tests to be applied to a very limited number of test points. The logical designer of cellular arrays will have to accept a certain amount of failure in the circuit and will have to invent synthesis procedures for fault-tolerant circuits. Another major problem in cellular logic is standardization. We now have at hand a technology which can produce arrays of cells of very large complexity, but we do not yet know how to use these devices efficiently in practical designs. This is because of the rapid growth of the number of logic functions of a cell with the number of input/outputs, so that when a cell has more than four or five inputs, one simply does not know what to put into the cell in order to obtain a cell that may be widely used in digital circuits. One proposed solution is to make each cell a universal logic module or ULM's. But the research on ULM's (see Chapter VI) does not promise us a bright picture because of the rapid growth in the number of input pins to a ULM with the increase of the number of variables.

Because of the above major difficulties, cellular arrays have yet not received universal acceptance by digital circuit designers. While experimenting with cellular logic, manufacturers still tend to adhere to IC circuit packages of standard "cell types" with customized metallization interconnection or a hybrid combination of cellular and noncellular arrays. But, LSI is still evolving, and some of its major technical problems may be satisfactorily solved within this decade while its full impact on the logical and system design methods for the next generation computers are being evaluated (Walter *et al.*, 1968). Some of the distinguishing trends of this impact are: decentralization of memory and control, more emphasis on software implementation by hardware using logic-in-memories, microprogrammed read-only store, and massive amounts of parallel logic and programmable reconfigurability. It seems rather certain that cellular logic arrays will play a significant role in the implementation of these ideas.

This chapter is intended to form a foundation to a theory of logical design with cellular arrays. In particular, mathematical models of some of the simple cellular structures will be studied and synthesis algorithms for arbitrary switching functions will be developed. The efficiency of different structures with respect to design parameters like cell complexity, interconnection structure, size, etc., will be evaluated. Application of this body of knowledge to the actual logical design of digital systems is then a matter of routine procedures and will not be discussed here. Section II presents the theory of a fundamental one-dimensional cellular structure, called the Maitra cascade. Section III presents the elementary theory of two-rail cascades. The general framework of the theory of multirail cascades is presented in Chapter VIII. Section IV presents some of the well known logical design techniques for the synthesis of arbitrary combinational switching function using two-dimensional arrays. Section V briefly describes some of the minimization problems that arise in cellular arrays and their partial solutions. Section VI gives a brief review of other research done in the cellular area.

II. SINGLE-RAIL CASCADES[1]

This section covers the theory of the simplest of all cellular structures, cascades of 2-input 1-output cells. The salient feature of such networks is that they are functionally incomplete in that there exist switching functions that cannot be realized by any single rail cellular cascade.

The principal result of this section is the derivation of a canonical form for single-rail cascades under the assumption that any input variable may drive an arbitrary number of cells. The canonical form provides a basis for an algorithm that can be used to test an arbitrary function for cascade realizability and to synthesize cascades when possible. A bound of $(n^2 + n - 4)/2$ is found to be the length of the longest cascade required to realize a function of n-variables, and it is shown that some functions attain this bound. The functional incompleteness of single rail cascades is a strong incompleteness, for the fraction of n-variable functions realizable by single cascades becomes vanishingly small as n grows large.

---

[1] This section is largely drawn from Stone and Korenjak (1965) and reproduced by permission of the IEEE.

A. THE CANONICAL FORM OF A REDUNDANT CASCADE

A cell C is defined to be a logic element with two inputs x and y and an output $z = f(x, y)$ where f is one of the 16 Boolean functions of two variables. A typical cell is shown in Fig. 1. A cellular *cascade* is a structure composed of m cells C_i, $1 \leq i \leq m$, such that the ith cell output z_i is connected to the

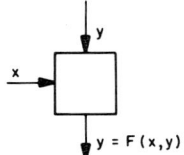

FIG. 1.   A Maitra cell.

$(i + 1)$th cell input y_{i+1}. Figure 2 shows a typical cascade. Since there are m cells, there are $m + 1$ external inputs to the cascade (counting the m horizontal inputs and the single vertical input). If these inputs are connected to $m + 1$ distinct Boolean variables, the cascade is said to be an *irredundant cascade,* wheras if one or more inputs are driven by the same Boolean variable the cascade is said to be a *redundant cascade.* Both redundant and irredundant cascades are Maitra cascades, and the cells are Maitra cells (Maitra, 1962). Figure 2 gives an example of a redundant cascade. Since the x_1 variable in Fig. 2 drives two cells, it is said to be a redundant input, while the remaining

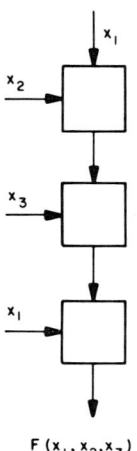

FIG. 2.   A redundant Maitra cascade.

inputs to the cascade are irredundant inputs since each of them drives a single cell.

For the purpose of this discussion, we shall assume that each cell can produce one of the three cell functions $\{x \cdot y, x + y, x \oplus y\}$ where the operators are the AND, OR, and EXCLUSIVE OR operators, respectively. It is well known (Ashenhurst, 1959) that these three functions are necessary and sufficient to characterize all decompositions of functions into combinations of 2-argument functions. In terms of Maitra cascades, any function that can be realized by a cascade of Maitra cells can be realized by a cascade of three function cells of the same length provided external cell inputs of the latter may be complemented arbitrarily (Sklansky, 1963).

The principal result of this section is a proof that the network shown in Fig. 3 is the most general configuration of a redundant cascade in the following sense. For $n \geq 3$, every realizable n-variable function can be realized by a cascade of the form shown, i.e., the cascade contains a subcascade that produces an $(n - 1)$-variable function and that is followed by a cell driven by the nth variable x_n. If the cell driven by x_n is " $\cdot$ " or " $+$ ", it may be followed by no more than $(n - 1)$ " $\oplus$ " cells driven by variables other than x_n, and such

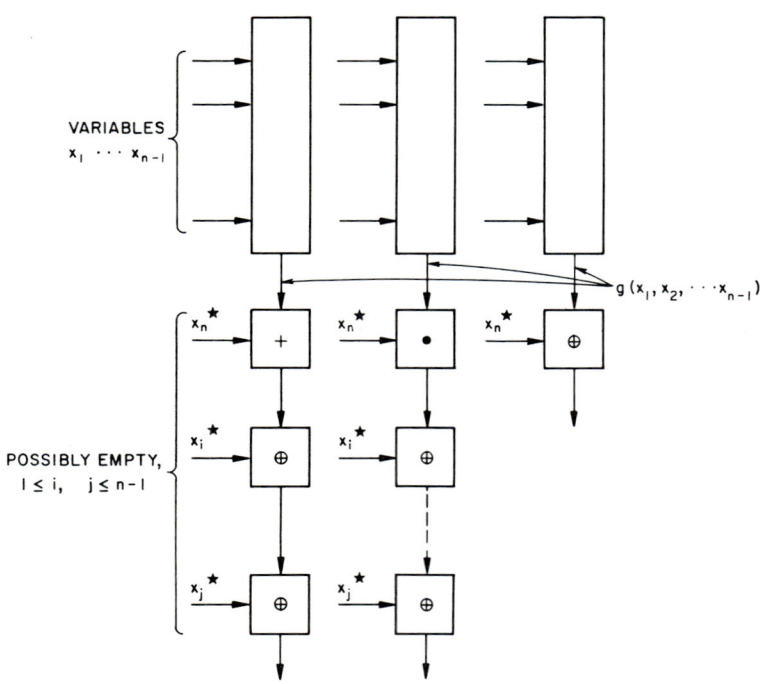

FIG. 3.   Canonical form of a redundant cascade.

that no two of the "⊕" cells are driven by the same variable. This description is recursive in that the $(n-1)$-variable cascade in Fig. 3 is itself of the same form as that of the entire cascade. That the network shown in Fig. 3 is indeed the most general cascade depends on the following three lemmas.

LEMMA 2.1. Let $f(x_1, x_2, \ldots, x_n)$ be an arbitrary n-variable Boolean function. Then $\bar{f}(x_1, x_2, \ldots, x_n)$, the complement of f, is cascade realizable if and only if f is realizable. Furthermore, if both are realizable, then the minimal length realization of f is the same length as the minimal length realization of $\bar{f}$.

Proof: The lemma follows immediately from the property that a function and its complement have the same decomposition structure in the sense described by Ashenhurst (1959).

In the material that follows we use the notation $\dot{x}_i$ to denote either x_i, or $\bar{x}_i$.

LEMMA 2.2. In a cascade realization of a function f, if x_i is the input to the jth cell, then $\dot{x}_i$ need never be the input of the $(j+1)$th cell.

Proof: Consider two adjacent cells for which the condition is violated, as shown in Fig. 4. Then $y_{j+1} = f(x_i, y_{j-1})$. But every function of two variables is producible by a single logical cell if we permit complementation of the cell inputs. Hence, for all possible values of the operation R_j and R_{j+1}, the two cells shown in Fig. 4 could be replaced by a single cell, although this may require that $\bar{y}_{j-1}$ be produced by the preceding cells in the network. However, by Lemma 2.1, if y_{j-1} is producible by a cascade of length $(j-1)$, then $\bar{y}_{j-1}$ is also producible by a cascade of length $(j-1)$, Hence, the cascade in Fig. 4 can be shortened by one cell and one of the two adjacent x_i inputs may be eliminated.

LEMMA 2.3. In any cascade realization of a function f, if the jth cell is a " · " or " + " cell with input x_i, then there need never be any cells driven by $\dot{x}_i$ among the first $(j-1)$ cells in the cascade.

Proof: Assume that the output of the $(j-1)$th cell is y_{j-1}. Then expanding about x_i we obtain

$$y_{j-1} = \bar{x}_i\, y_{j-1}\,(x_i = 0) + x_i\, y_{j-1}\,(x_i = 1)$$

where $y_{j-1}\,(x_i = 0)$ is the function obtained by evaluating y_{j-1} for $x_i = 0$.

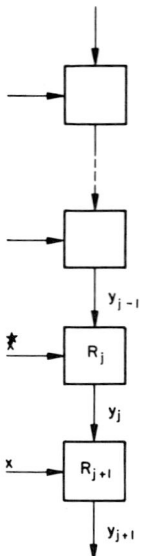

FIG. 4.   A cascade described by Lemma 2.2.

If the jth cell is a " $\cdot$ " cell then

$$y_j = x_i y_{j-1} = x_i y_{j-1} \, (x_i = 1)$$

Hence, among the first $(j - 1)$ cells we can set $x_i = 1$ for all those cells driven by $\dot{x}_i$ without altering the output of the jth cell. In fact, all cells among the first $(j - 1)$ that are driven by constants introduced by this manipulation may be eliminated, and the network shortened. Eliminating $\dot{x}_i$ cells requires, at most, the complementation of intermediate cascade functions (where x_i drives a "$\oplus$" cell) or replacement of an intermediate cascade output with a constant 0 or 1 (where $\bar{x}_i$ drives a " $\cdot$ " cell or x_i drives a "$+$" cell, respectively). In all other cases the $\dot{x}_i$ cell may simply be deleted from the cascade.

If the jth cell is a "$+$" cell, the result follows immediately by duality and the lemma is proved.

It is now possible to demonstrate the generality of the network shown in Fig. 3 by collecting together the results of Lemmas 2.2 and 2.3.

THEOREM 2.1 (Canonical Form Theorem).   For $n \geq 3$, every nondegenerate Maitra function of n-variables may be realized by a cascade that consists of

1.   a Maitra cascade that realizes a nondegenerate function of $n - 1$ variables and that drives;
2.   (a) either a single "$\oplus$" cell driven by the nth variable x_n, or (b) a " $\cdot$ "

or "+" cell driven by x_n followed by a possibly empty string of "⊕" cells driven by variables other than x_n and such that no two "⊕" cells are driven by the same variable.

This statement is recursive to the extent that the cascade in 1 is also in canonical form.

Proof: We proceed by induction on the number of nondegenerate variables of the function.

Note that for $n = 2$, a cascade consists of a single cell because all 2-variable functions are producible by a single cell. The remainder of the proof for the case $n = 3$ follows as a consequence of the induction step below.

For $n \geq 3$, it is always possible to partition a cascade into two cascades, one of which produces a nondegenerate function of $(n - 1)$ variables and drives a second cascade whose first cell is driven by the nth variable. We choose to label our inputs so that this variable is labeled x_n. The cell driven by x_n may either be "+", "·", or "⊕". Assume the cell is "·" or "+". To this cell we may append only redundant "⊕" cells driven by any of the n variables without violating Lemma 2.3. Since "⊕" is commutative and $x \oplus x = 0$, we can always shorten the string of "⊕" cells by two cells if two cells are driven by the same variable. Hence, no variable may appear more than once in the string. Since "⊕" is commutative, if x_n drives a "⊕" cell anywhere in the string we may always interchange inputs so that x_n drives the "⊕" cell immediately following the "·" or "+" cell driven by x_n. Lemma 2.2 applies in this case and one x_n cell may be eliminated. Hence, x_n need not drive a "⊕" cell in this string. This proves 2(b) of the theorem.

Suppose that the top cell in the second cascade is a "⊕" cell. Lemma 2.3 and the commutativity of "⊕" permit us to append to this cell only "⊕" cells with inputs chosen from the set of $(n - 1)$ variables. Using commutativity again, we may interchange inputs and place x_n at the bottom of the redundant "⊕" cells. In this form, a single "⊕" cell driven by x_n is driven by a cascade that produces a nondegenerate function of $(n - 1)$ variables. This completes the proof of the theorem.

Thus we have demonstrated that the network of Fig. 3 is the most general configuration of a cascade realization of a function of n variables. We will now proceed to the main results of the study using this general configuration as our basic tool.

B. DERIVATION OF ALGORITHM FOR TESTING CASCADE REALIZABILITY

The derivation of the canonical form of a Maitra cascade permits one to formulate a procedure to test a function for cascade realizability and to generate a Maitra cascade for the function if one exists. The method we choose here is one that determines the bottom cell or cells of a cascade realization of the function, and determines the intermediate cascade function that must drive the bottom cells. The procedure is applied iteratively to the new intermediate function determined at each step of the process, until either a trivial intermediate function is found or no cells at the bottom of any cascade may be found that possibly can produce the intermediate function. In the latter case, the function is not realizable, while in the former the method has terminated with the synthesis of a complete cascade.

We may apply the procedure in any one of three forms that have been described previously in the literature. Maitra's (1962) method is essentially a map technique, Sklansky's (1963) is tabular, and Levy's et al. (1964) is algebraic. Each of these may be generalized to be valid synthesis techniques for redundant cascade synthesis. Because the tabular method is the most easily programmed it will be described in detail. An example of synthesis using maps will also be shown to give some insight into the character of the testing algorithm.

The basis of the tests for the irredundant cells has been described in the literature. For the sake of completeness, we shall rederive, in Lemma 2.4, the necessary and sufficient conditions for the existence of an irredundant cell of specified form at the bottom of a cascade, although the techniques closely parallel Sklansky (1963).

DEFINITION.   The weight of a Boolean function is the number of combinations of its arguments for which the function takes on the value 1, i.e., the number of 1's in the function's truth table. A function of n variables is said to have neutral weight if precisely half of the 2^n entries in the truth table or the characteristic vector are 1's.

LEMMA 2.4.   Let $f(x_1, x_2, \ldots, x_n)$ and $g(x_1, x_2, \ldots, x_{n-1})$ be nontrivial Boolean functions. Then

(1) $f = x_n g$ iff $\bar{x}_n f = 0$ and $g = f(x_1, x_2, \ldots, x_{n-1}, 1)$

(2) $f = \bar{x}_n g$ iff $x_n f = 0$ and $g = f(x_1, x_2, \ldots, x_{n-1}, 0)$

(3) $f = x_n + g$ iff $x_n \bar{f} = 0$ and $g = f(x_1, x_2, \ldots, x_{n-1}, 0)$

(4) $f = \bar{x}_n + g$ iff $\bar{x}_n \bar{f} = 0$ and $g = f(x_1, x_2, \ldots, x_{n-1}, 1)$

(5) $f = x_n \oplus g$ iff $g = f(x_1, x_2, \ldots, x_{n-1}, 0)$

$\qquad\qquad\qquad\qquad = \bar{f}(x_1, x_2, \ldots, x_{n-1}, 1)$

In cases (1) and (2), the weight of f is less than neutral, in cases (3) and (4) the weight of f is greater than neutral, and in (5) the weight of f is neutral.

Proof: To prove part (1), expand f about the variable x_n.

$$f = \bar{x}_n f(x_1, x_2, \ldots, x_{n-1}, 0) + x_n f(x_1, x_2, \ldots, x_{n-1}, 1)$$
$$= \bar{x}_n g_0 + x_n g_1$$

Then

$$\bar{x}_n f = x_n g_0$$

which is identically 0 if and only if $g_0 = 0$. But if $g_0 = 0$, then $f = x_n g_1$, which is the form required. The converse follows immediately from

$$\bar{x}_n f = \bar{x}_n(x_n g) = 0.$$

The weight of f is immediately seen to be less than or equal to neutral, with equality if and only if $f(x_1, x_2, \ldots, x_{n-1}, 1) = 1$. Since we require a nontrivial g, the weight of f is less than neutral. Parts (2), (3), (4), and (5) follow similarly.

A complete test of redundant cascade realizability requires a test for redundant "$\oplus$" cells that may follow irredundant "$+$" or "$\cdot$" cells. This test is described in the following lemma.

LEMMA 2.5. Let $f(x_1, x_2, \ldots, x_n)$ be an n-variable function such that either $f(x_1, x_2, \ldots, x_{n-1}, 0)$ or $f(x_1, x_2, \ldots, x_{n-1}, 1)$ is a parity function of one or more variables. Then if p is this residual parity function, f has a decomposition of the form

$$f = p \oplus (\dot{x}_n h) = \bar{p} \oplus (\bar{x}_n + \bar{h})$$

where h is a uniquely determined function of $(n-1)$ variables. The converse is also true.

Proof: By expanding f about x_n, we obtain

$$f = \bar{x}_n g_0 + x_n g_1$$

and observe that, by hypothesis, either g_0 or g_1 is a parity function. Let us assume that g_0 is the parity function. Then we have

$$f = \bar{x}_n p + x_n g_1$$
$$= \bar{x}_n p \oplus x_n g_1$$
$$= p \oplus x_n(p \oplus g_1)$$
$$= p \oplus x_n(h)$$
$$= \bar{p} \oplus (\bar{x}_n + \bar{h})$$

Since g_0 and g_1 are both uniquely determined functions, h is unique. If g_1 rather than g_0 is the parity function, we obtain the same generic form. The converse follows immediately by noting that either $f(x_1, x_2, \ldots, 0)$ or $f(x_1, x_2, \ldots, 1)$ is the parity function p when f has the form stated in the hypothesis. This completes the proof of the lemma.

Lemmas 2.4 and 2.5 provide the tools necessary to describe a complete realizability algorithm. If a function satisfies the conditions stated in either of the lemmas, then a partial decomposition of the function has been found. The partial decomposition determines one or more cells at the bottom of a cascade realization of the function, and a residual function that drives this portion of the cascade. The residual function depends on one fewer variable than the original function. Hence, the partial decomposition of a function may be viewed as the " extraction " of a single variable for the function. Complete cascade decompositions may be obtained by iterative application of the lemmas to successive residual functions until a trivial residual function is found.

At various points during the test procedure, several different partial decomposition conditions may be satisfied simultaneously. For example, one may have a choice between extracting x_1 giving a residual function g, or extracting x_2 yielding a residual function h. In these cases, we may wonder if it is necessary to test all residual functions in order to determine if the function under consideration is cascade realizable. The following lemmas show that we never have to test more than one function at any given step in the process, i.e., in the previous example, g is cascade realizable if and only if h is cascade realizable.

First we consider the case in which two variables satisfy the irredundant cell tests stated in Lemma 2.4.

LEMMA 2.6.   Let $f(x_1, x_2, \ldots, x_n)$ have partial decompositions

$$f = \dot{x}_i R_i g = \dot{x}_j R_j h$$

where g and h are not functions of x_i and x_j, respectively, and R_i and R_j are operators in the set $\{\cdot, +, \oplus\}$. Then g is cascade realizable if and only if h is cascade realizable.

Proof: The weight conditions stated in Lemma 2.4 indicate that R_i and R_j must be the same operator. Let us assume that f has weight greater than neutral so that R_i and R_j are "$+$". Then

$$f = \dot{x}_i + g = \dot{x}_j + h$$

and both $\dot{x}_i$ and $\dot{x}_j$ are implicants of f. Therefore f must have the decomposition

$$f = \dot{x}_i + \dot{x}_j + k$$

where k is not a function of x_i and x_j and

$$g = \dot{x}_j + k$$
$$h = \dot{x}_i + k$$

Evidently, g is realizable if and only if k is realizable, and similarly for h. Therefore, g is realizable if and only if h is realizable. Similar arguments apply if the operator is " $\cdot$ " or " $\oplus$." In case $\bar{x}_i = x_j$, we find that R_i must be "$\oplus$" if g and h are to be nontrivial. Then

$$f = x_i \oplus g = \bar{x}_i \oplus \bar{g}$$

and g is realizable if and only if $\bar{g}$ is realizable by Lemma 2.1. This proves the lemma.

The following lemma applies to the case in which two or more redundant decompositions (as described in Lemma 2.5), are possible.

LEMMA 2.7.   Let $f(x_1, x_2, \ldots, x_n)$ have the decompositions

$$f = p_1 \oplus (\dot{x}_i g) = p_2 \oplus (\dot{x}_j h)$$

where p_1 and p_2 are parity functions of one or more variables and g and h are not functions of x_i and x_j, respectively. Then g is cascade realizable if and only if h is cascade realizable.

Proof: The proof divides naturally into two cases, depending on whether $x_i = \bar{x}_j$ or $x_i \neq \bar{x}_j$.
   Case I. $x_i = \bar{x}_j$. We expand f about x_i and obtain

$$f = \bar{x}_i f_0 + x_i f_1$$

where f_0 is f evaluated for $x_i = 0$, and f_1 is f evaluated for $x_i = 1$. By hypothesis, and Lemma 2.5, $f_0 = p_1, f_1 = p_2$, and p_1 and p_2 are parity functions of variables not including x_i. Then

$$f = p_1 \bar{x}_i + p_2 x_i$$
$$= p_1 \bar{x}_i \oplus p_2 x_i$$
$$= p_1 \oplus [x_i(p_1 \oplus p_2)] = p_2 \oplus [\bar{x}_i(p_1 \oplus p_2)]$$

But these are both complete cascade decompositions of f. Hence f is realizable, and the residual function is $p_1 \oplus p_2$ regardless of whether x_i or $\bar{x}_i$ is extracted from f.

Case II. $x_i \neq \bar{x}_j$. By hypothesis, f has the partial decompositions

$$f = p_1 \oplus (\dot{x}_i g) = p_2 \oplus (\dot{x}_j h)$$

Then, by Lemma 2.5, p_1 and p_2 are residual functions obtained by evaluating f on x_i and x_j, respectively. Without loss of generality, let

$$p_1 = f(x_i = 0)$$
$$p_2 = f(x_j = 0)$$

where $f(x_i = 0)$ is f evaluated for $x_i = 0$. Expanding f about both x_i and x_j we obtain

$$f = \bar{x}_i \bar{x}_j f_{00} + \bar{x}_i x_j f_{01} + x_i \bar{x}_j f_{10} + x_i x_j f_{11}$$

where $f_{00} = f(x_i = 0, x_j = 0)$, etc. Then

$$p_1 = \bar{x}_j f_{00} + x_j f_{01}$$

and

$$p_2 = \bar{x}_i f_{00} + x_i f_{10}$$

Since p_1 and p_2 are both parity functions, we must have

$$p_1 = \begin{cases} f_{00} = f_{01} & \text{or} \\ f_{00} \oplus x_j = f_{01} \oplus \bar{x}_j \end{cases}$$

and

$$p_2 = \begin{cases} f_{00} = f_{10} & \text{or} \\ f_{00} \oplus x_i = f_{10} \oplus \bar{x}_i \end{cases}$$

where f_{00}, f_{01}, and f_{10} must be parity functions in $n - 2$ or fewer variables. Two possible forms arise for each of the functions p_1 and p_2 because they may or may not depend on x_i and x_j, respectively. To prove the lemma we must treat three cases, according to the dependence of p_1 and p_2 on x_i and x_j. Either (a) p_1 and p_2 are independent of x_i and x_j, (b) one of the two functions depends on x_i or x_j, or (c) p_1 and p_2 depend on x_i and x_j, respectively.

Case II(a). $p_1 = f_{00}, p_2 = f_{00}$. Since this occurs if and only if $f_{00} = f_{01} = f_{10}$, we may substitute for f_{01} and f_{10} and obtain

$$f = \bar{x}_i \bar{x}_j f_{00} + \bar{x}_i x_j f_{00} + x_i \bar{x}_j f_{00} + x_i x_j f_{11}$$

Since the terms are disjoint, we may change the "+" to "$\oplus$" and bring the equation into the following form after some additional manipulation:

$$f = f_{00} \oplus [x_i x_j (f_{00} \oplus f_{11})]$$

Then f has the decompositions

$$f = f_{00} \oplus (x_i g) = f_{00} \oplus (x_j h)$$

where $g = x_j (f_{00} \oplus f_{11})$ and $h = x_i (f_{00} \oplus f_{11})$. Clearly, g is cascade realizable if and only if h is cascade realizable, for both depend on the realizability of the function $f_{00} \oplus f_{11}$.

Case II(b). One of the two residual functions does not depend on x_i or x_j. Without loss of generality, assume it is p_1 so that

$$p_1 = f_{00}$$
$$p_2 = f_{00} \oplus x_i$$

Then $f_{00} = f_{01} = \bar{f}_{10}$. Substituting and replacing "+" by "$\oplus$" as before, we obtain

$$f = f_{00} \oplus x_i \oplus x_i x_j \oplus x_i x_j (f_{00} \oplus f_{11})$$

Substituting for p_1 and p_2 gives

$$f = p_1 \oplus x_i [\bar{x}_j + (f_{00} \oplus f_{11})] = p_1 \oplus (x_i g)$$
$$= p_2 \oplus x_j [x_i (f_{00} \oplus f_{11})] = p_2 \oplus (x_j h)$$

where $g = \bar{x}_j + (f_{00} \oplus f_{11})$ and $h = x_i (f_{00} \oplus \bar{f}_{11})$. By Lemma 2.1, we know that $f_{00} \oplus \bar{f}_{11}$ is realizable if and only if $f_{00} \oplus f_{11}$ is realizable since they are complementary functions, so that g is cascade realizable if and only if h is realizable.

Case II(c). $p_1 = f_{00} \oplus x_j$, $p_2 = f_{00} \oplus x_i$. In this case, $f_{00} = \bar{f}_{01} = \bar{f}_{10}$. Proceeding as before, we may manipulate f into form

$$f = f_{00} \oplus x_i \oplus x_j \oplus [x_i x_j (f_{00} + f_{11})]$$

Then f has the decompositions

$$f = p_1 \oplus x_i [\bar{x}_j + (f_{00} \oplus \bar{f}_{11})] = p_1 \oplus (x_i g)$$
$$= p_2 \oplus x_j [\bar{x}_i + (f_{00} \oplus \bar{f}_{11})] = p_2 \oplus (x_j h)$$

where $g = \bar{x}_j + (f_{00} \oplus \bar{f}_{11})$ and $h = x_j + (f_{00} \oplus \bar{f}_{11})$. But in this case, as in the prior cases, g and h are related such that g is cascade realizable if and only if h is realizable. This proves the lemma.

One more lemma is required to handle the case in which a function may have an irredundant and a redundant decomposition.

LEMMA 2.8. Let $f(x_1, x_2, \ldots, x_n)$ have the decompositions $f = \dot{x}_i R g = p_1 \oplus (\dot{x}_j h)$ where R is one of the operators "$\cdot$", "$+$", or "$\oplus$" and g and h are not functions of x_i and x_j, respectively. Then g is cascade realizable if and only if h is cascade realizable.

Proof: If the operator R is " $\cdot$ " or " $+$ ", then either $f(x_i = 0)$ or $f(x_i = 1)$ is a constant. We apply the proof of Lemma 2.7 to this case, replacing p_2 by a constant, and we find that the proof still holds. If the operator R is " $\oplus$ ", then Lemma 2.4 implies that $f(x_i = 0) = \bar{f}(x_i = 1)$. The redundant decomposition involving x_j requires that $f(\dot{x}_j = 0)$ be a parity function by Lemma 2.5. These two conditions can be satisfied if and only if f has the decomposition

$$f = x_i \oplus [q \oplus (\dot{x}_j h)]$$

where q is a parity function and q and h are not functions of x_i and x_j. (The proof of this statement follows in a manner similar to the proof of Lemma 2.7.) Therefore,

$$f = x_i \oplus g = p \oplus (\dot{x}_j h)$$

where $g = q \oplus (\bar{x}_j h)$ and $p = x_i \oplus q$. Hence, g is realizable if and only if h is realizable and the proof is complete.

Lemmas 2.6, 2.7, and 2.8 establish the fact that cascade realizability is not sensitive to the order in which variables are extracted from a function. With this result we are prepared to state a detailed testing and synthesis algorithm.

C. STATEMENT OF THE ALGORITHM

The complete algorithm for determining cascade realizability consists of the application of Lemmas 2.4 and 2.5 in a systematic way. We shall present an example in which the algorithm is applied using both map and tabular methods to illustrate the details of the algorithm.

It is convenient to treat Lemma 2.4 as a set of "subroutines," i.e., independent algorithms that can be applied at various steps in the main algorithm. We specify these tests first. Assume f is a function to be tested, x_i is an argument of f, and $f(x_i = 1)$ is f evaluated for $x_i = 1$.

Test for " $\cdot$ " *Cell:* If $\bar{x}_i f = 0$, then extract a " $\cdot$ " cell with an x_i input, and the residual function is $f(x_i = 1)$.

If $x_i f = 0$, then extract a " $\cdot$ " cell with an $\bar{x}_i$ input, and the residual function is $f(x_i = 0)$.

Test for " $+$ " *Cell:* If $x_i \bar{f} = 0$, then extract a " $+$ " cell with an x_i input, and the residual function is $f(x_i = 0)$.

If $\bar{x}_i \bar{f} = 0$, then extract a " $+$ " cell with an $\bar{x}_i$ input, and the residual function is $f(x_i = 1)$.

Test for " $\oplus$ " *Cell:* If $f(x_i = 0) = \bar{f}(x_i = 1)$, then extract a " $\oplus$ " cell with an x_i input, and the residual function is $f(x_i = 0)$.

Main Algorithm:

Step 1. Enter the function f and a list of its arguments.

Step 2. Compute the weight of f. If the weight is less than neutral, i.e., less than 2^{n-1}, where n is the number of arguments, then test for a " $\cdot$ " cell for each argument x_i. If f has weight greater than neutral, apply the "$+$" cell test for each argument. If f has neutral weight, apply the test for "$\oplus$" cell for each argument.

Step 3. If the respective tests attempted in Step 2 failed for all x_i, then go to Step 5; if the " $\cdot$ " or "$+$" test succeeded, delete the input variable x_i from the argument list and return to Step 2 to treat the residual function.

Step 4. If the residual function from the extraction of a "$\oplus$" cell is a constant 0 or 1, then the decomposition is complete. If the constant is 1, then complement the last input found. In either case, the last "$\oplus$" cell discovered is deleted and replaced by an $\dot{x}_i$ input that drives the top of the cascade directly.

If the residual function is not constant, then delete the input variable from the argument list and return to Step 2 to apply the algorithm to the residual function.

Step 5. For each argument of the function, calculate the residual functions $f(x_i = 0)$ and $f(x_i = 1)$. If a residual function has neutral weight then apply the "$\oplus$" cell test iteratively until (a) the residual function is completely decomposed into a parity function of one or more variables, or (b) the residual function fails to decompose completely into a parity function after applying the "$\oplus$" cell test for all possible variables.

If (a) is the case go to Step 6. If all residual functions fail to satisfy (a), then the function is not cascade realizable, and the test may be terminated at this point.

Step 6. Extract x_i and the redundant "$\oplus$" cells from the function according to one of the relations

$$f = p \oplus \dot{x}_i g$$
$$= \bar{p} \oplus [\bar{x}_i + \bar{g}]$$

where p is the residual parity function, $g = f(x_i = 0) \oplus f(x_i = 1)$, and $\dot{x}_i = x_i$ if $f(x_i = 0)$ is the parity function, $\dot{x}_i = \bar{x}_i$ if $f(x_i = 1)$ is the parity function. Apply the process from Step 3 to the function g or $\bar{g}$ with x_i deleted from the argument list.

Example. To illustrate the use of the algorithm we choose the function $f(x_1, x_2, x_3, x_4) = \sum(0, 1, 6, 7, 10)$, which is the leader for Harvard class #50 (Aiken *et al.*, 1951). The map and truth table of the function are shown in Fig. 5a. Since f has less than neutral weight, we attempt to extract a " $\cdot$ " cell and note that no variable satisfies the test. The application of Step 5 yields

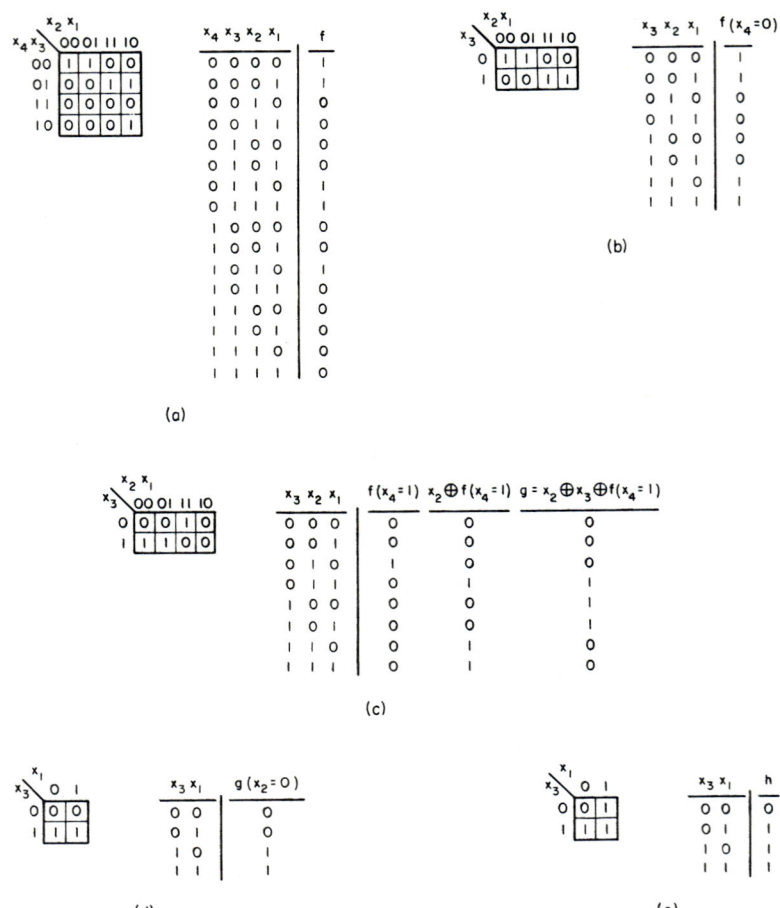

FIG. 5.   The application of the algorithm to a test function. (a) Map and truth table of $f(x_1, x_2, x_3, x_4)$. (b) Evaluation of $f(x_4 = 0)$. (c) Calculation of $g(x_1, x_2, x_3)$. (d) Evaluation of $g(x_2 = 0)$. (e) Residual function $h(x_1, x_2)$.

$f(x_4 = 0)$ as a residual function with neutral weight. This is tabulated and mapped in Fig. 5b. Application of the "$\oplus$" cell test or inspection of the map shows that $f(x_4 = 0)$ is the parity function $\bar{x}_2 \oplus x_3$. At this point, we choose to extract x_4 using the relation

$$f = x_2 \oplus x_3 \oplus [\bar{x}_4 + (x_2 \oplus x_3 \oplus f(x_4 = 1))]$$

Let $g(x_1, x_2, x_3)$ be the residual function $x_2 \oplus x_3 \oplus f(x_4 = 1)$. Figure 5c shows the calculation of g from the tabular representation. Applying Step 2 to g, we attempt to extract a "$\cdot$" cell because g has weight less than neutral. This test fails for all variables. Proceeding to Step 5, we find that $g(x_2 = 0)$ has neutral weight. The map in Fig. 5d shows that this is a parity function of 1 variable, $g(x_2 = 0) = x_3$. We may extract x_2 according to the relation in Step 6

$$g = x_3 \oplus [x_2(x_3 \oplus g(x_2 = 1))]$$

Let $h(x_1, x_3)$ be the new residual function $x_3 \oplus g(x_2 = 1)$. Since h is a 2-variable function, we know immediately that it is realizable by a single cell that can be determined immediately from the map of h given in Fig. 5e. However, we choose to apply Step 2 of the algorithm to show how the algorithm terminates. Since h has weight greater than neutral, we apply the "$+$" cell test and successfully extract a "$+$" cell driven by x_1. The residual function $h(x_1 = 0)$ has neutral weight and, therefore, must be tested for "$\oplus$" cell extraction. After extracting x_3 as the input to a "$\oplus$" cell, we find a trivial residual function, indicating that the function has been completely decomposed. Replacing $x_3 \oplus 0$ by x_3 we collect the previous results to give

$$f(x_1, x_2, x_3, x_4) = x_3 \oplus x_2 \oplus \{\bar{x}_4 + x_3 \oplus [x_2(x_1 + x_3)]\} \qquad (*)$$

The algorithm described before has been programmed and applied to the 238 Harvard functions (1951) of weight 8 or less. The results are listed in Table I. The variables in the table are coded so that x_i is represented by i and $\bar{x}_i$ by $i + 4$. Cascades are shown rotated a quarter turn, with the top cell and inputs appearing on the left and the bottom cell and output on the right. Comparison of $(*)$ with the 50th entry of the table should clear up any ambiguity of interpretation of the table. Cascade realizations of functions with weight greater than 8 may be generated from the cascade realizations of the complementary functions in a straightforward manner. Since all functions in a Harvard class are cascade realizable if a single member is realizable, Table I gives enough information to solve the realizability problem completely for functions of four or fewer variables.

D. THE LENGTH OF REDUNDANT CASCADES IN CANONICAL FORM

In this section we shall show that a cascade that produces a function of n variables need contain no more than $(n^2 + n - 4)/2$ cells, and that the bound is attained by some functions for all $n \geq 2$. We shall also show that a cascade in canonical form is also a minimal length realization except for one general case. Finally, it is noted that the synthesis algorithm can extract variables in such a way as to violate Lemma 2.2, thereby producing noncanonical

TABLE I
Redundant Decompositions of 4-Variable Functions[a]

	0	1	2	3	4	5	6	7	8	9
0	TRIVIAL	8765 ·:·	876 · ·	6187 ●··	*	*	6587 +··	31638 ·+●·	*	765238 +·●●·
10	*	*	87 ·	*	72538 ●+●·	6538 +●·	*	*	*	728 ●·
20	*	*	*	*	*	7218 ●●·	*	*	*	*
30	*	6578 ·+·	21474 +·+●	761238 ·+●●·	*	*	*	*	76128 ·+●·	*
40	*	*	*	*	*	*	*	*	*	*
50	3123823 +·●+●●	872345234 ·+●●+●●●	*	*	*	321238123 +·●●+●●●	8534634 ·+●+●●	*	*	6178 ●+·
60	*	*	*	65474 ·●+●	416174 ●+●+●	*	4274 ·+●	21474 ●·+●	768 +·	*
70	*	*	*	*	*	*	*	*	*	*
80	*	*	*	*	7181634 ●+●+●●	*	*	*	436534 ●++●●	*
90	*	*	8721234 +●+●●●	*	*	*	6147124 ●·+●●●	*	7658 ++·	472124 +●+●●
100	*	87634 +·●●	321823 ●·+●●	*	*	865274 +·●+●	*	*	*	*
110	857464 +·●+●	41764 ·++●	*	4252474 ·+●●+●●	*	*	*	*	*	*

274

275

120	*	*	476124 + . + ●●	*	*	764 + ●	82174 ●● + ●	*	*	*	654712 . ● . ●
130	*	*	875634 + . + ●●	7654 + + ●	81364 + ● + ●	85274 . ● + ●	*	*	*	76124 . + ●
140	42174 . . + ●	*	*	8763124 + . ● + ●●	83654 ● + + ●	318364 ● + ● + ●	8353234 ● + ● + ●●	*	*	8321 ●●●
150	46574 + . + ●	*	*	4363534 . + ● + ●●	*	*	*	*	*	832 ●●
160	825274 . ● + ●	*	*	*	*	761234 . + ●●●	*	*	*	6543 + ●●
170	8252474 . + ●● + ●	83631234 . + ● + ●●●	*	*	8364 ● + ●	*	764123 . ● + ●●	*	83 ●●	72534 ● + ●●
180	456474 + . ● + ●	83634123 . + ●● + ●●	*	*	4325234 . . . + ●●●	*	6574 . + ●	*	*	*
190	*	*	82524712 . + ●● . ●●	875634 + + . ●	*	*	82574 ● . + ●	43532134 ● + ● + ●●●	*	*
200	*	*	82527124 . + ● . ●●●	476534 . + ● + ●	*	*	*	*	*	*
210	*	*	*	8761234 + . + ●●●	*	6174 ● + ●	*	836123 ● . + ●●	*	4171214 ● + ● + ●●

a (*) No decomposition.

cascades. To generate a minimal length cascade from one obtained by application of the synthesis algorithm, it is necessary to bring the cascade into canonical form by eliminating the cells that violate Lemma 2.2. Then the cascade
can be checked for the exceptional nonminimal case. Both of these procedures
are rather simple to apply.

We are first interested in determining $C(n)$, the maximum number of
cells necessary to produce an arbitrary cascade realizable function of n-
variables. This bound is easily obtained as the solution to the difference
equation suggested by the Canonical Form theorem. Since for every $n \geq 3$,
no more than n cells need be appended to an $(n - 1)$-variable cascade to produce an n-variable function we have the relation

$$C(n) \leq C(n - 1) + n$$

with boundary conditions $C(2) = 1$. The solution to the above equation with
equality is

$$C(n) = 1 + \sum_{i=3}^{n} i = (n^2 + n - 4)/2$$

Since we used equality rather than the inequality, this is an upper bound. The
bound is firm, however, because we can construct an n-variable function that
requires $C(n)$ cells in its realization for all n. Consider the function realized
by a cascade whose top cell is "$\cdot$" producing $x_1 x_2$. To this is appended a "$+$"
cell driven by x_3 followed by "$\oplus$" cells driven by x_1 and x_2. Each successive
variable x_i is introduced as the input to a "$+$" cell followed by $i - 1$ "$\oplus$"
cells driven by the other preceding variables, as shown in Fig. 6. This cascade
is clearly a maximum length n-variable canonical cascade. Straightforward
application of the testing algorithm demonstrates that this realization of the
function is unique to within the ordering of the variables in a sequence of
adjacent "$\oplus$" cells, complementation of intermediate functions, and the permutation of x_1, x_2, and $\bar{x}_3$.[2]

Given a cascade in canonical form, is it guaranteed to be a minimal
length realization of the function? In almost all cases, the answer is yes. We
have only to examine Lemmas 2.6, 2.7, and 2.8 to see that this is true. Because
the lemmas consider all possible cases in which there are alternate ways of
extracting a pair of variables from a function, we can discover those cases that
lead to nonminimal canonical forms.

The cases indicated in Lemma 2.6 show that the length of a cascade is
independent of the order of extraction of a pair of irredundant variables.

---

[2] The overbar on x_3 is intentional since the function is symmetric in x_1, x_2 and $\bar{x}_3$ but
not in x_1, x_2, and x_3.

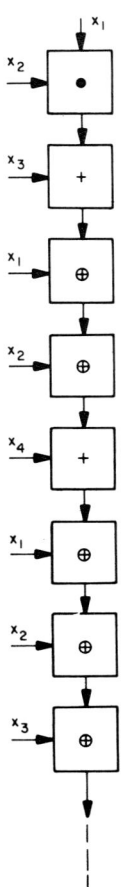

FIG. 6.   A cascade in canonical form that achieves the maximum length bound.

Lemma 2.7 is considered in more detail later. Two cases of Lemma 2.8 are special cases of Lemma 2.7, while the third is another example of the length of a cascade being insensitive to the order of extraction of a pair of variables.

Lemma 2.7 treats four cases, two of which are symmetric in the variables that may be extracted. As one might expect, these Cases, II(a), and II(c), lead to realizations that are independent of the order of extraction. Case I illustrates a pair of alternate decompositions which may lead to cascades of different lengths, while both forms of the decomposition are canonical. The difference in the lengths is equal to the difference in the number of redundant "$\oplus$" cells in the cascade. Finally, Case II(b) illustrates alternate partial decompositions that lead to realizations whose lengths differ by one cell. The longer of these decompositions is not canonical, however, for it violates Lemma 2.2. To see this, note that

$$f = (f_{00} \oplus x_i)\{x_j x_i(f_{00} \oplus f_{11})\}$$
$$= f_{00} \oplus x_i \oplus \{x_i x_j(f_{00} \oplus f_{11})\}$$

In this form, a pair of x_i variables are adjacent which violates the condition that no two adjacent cells in a canonical cascade are driven by the same variable.

Since the forms of Cases I and II(b) are the only forms that lead to non-minimal cascades, these are the only cases necessary to check to minimize cascades synthesized by application of the algorithm. Case II(b) arises if a variable in a sequence of redundant "$\oplus$" cells also drives a cell in a sequence of identical cells that immediately precedes the "$\oplus$" cells. For example, a string of redundant "$\oplus$" cells may be immediately preceded by several "$+$" cells, and x_i may drive a cell in both sets. Since inputs to a sequence of identical cells may be permuted arbitrarily without altering the cascade output, a permutation can be applied to each set of cells so that x_i drives the top-most "$\oplus$" cell and the bottom-most "$+$" cell. This is the general form of Case II(b). Then one x_i cell can be eliminated with appropriate modification of the cascade according to the relations given in Case II(b).

To check for Case I requires only an examination of the top several cells of the cascade to see if they produce a function of the form

$$f = p \oplus x_i q$$

where p and q are parity functions. Then p can be replaced by $p \oplus q$ and x_i replaced by $\bar{x}_i$ if this leads to a shorter cascade.

E. THE NUMBER OF REALIZABLE FUNCTIONS

It is clear from Table I that most 4-variable functions are not cascade realizable, although the realizable functions are relatively numerous. In this section we show that the fraction of functions that are cascade realizable becomes vanishingly small as n grows large.

Let $W(n)$ be the number of realizable symmetry types of n-variables, where symmetry type is defined in Chapter IV. We select a representative function from each class, $W(n)$ in all, and determine a canonical cascade for each. To these cascades we append cells driven by $x_1, x_2, \ldots, x_{n+1}$ according to the constraints of the canonical form and consider the set of $(n + 1)$-variable functions which are generated. Evidently all $W(n + 1)$ realizable $(n + 1)$-variable symmetry types are represented in this set.

Of the functions in this set there are $W(n)$ of the form

$$f_{n+1} = x_{n+1} \oplus f_n$$

where f_n and f_{n+1} are n- and $(n+1)$-variable functions, respectively. The remaining functions are equivalent to functions of the form

$$f_{n+1} = \dot{p}_n \oplus (x_{n+1} f_n) = \bar{\dot{p}}_n \oplus (\bar{x}_{n+1} + \bar{f}_n)$$

where p_n is a parity function in n or fewer variables corresponding to a set of redundant "$\oplus$" cells. The Canonical Form theorem guarantees that these two forms exhaust the functions that have been generated. Since there are 2^{n+1} ways of picking $\dot{p}_n$ and $W(n)$ ways of picking f_n, we obtain the bound

$$W(n+1) \le W(n)(2^{n+1} + 1)$$

The actual number of realizable classes and functions for 2, 3, and 4 variables is tabulated in Table II. Here we see that the bound is quite poor.

TABLE II
Cascade Realizable Functions

	$n=2$	3	4
Realizable classes			
a. Irredundant	5	14	35
b. Redundant	5	19	135
Total classes	5	22	402
Realizable functions			
a. Irredundant	16	152	2368
b. Redundant	16	242	21,216
Total functions	16	256	65,536

Nevertheless, we replace it by the poorer bound

$$W(n) \le W(n)2^{n+2}$$

which has a solution

$$W(n) \le (1) \prod_{i=2}^{n} 2^{i+1} = W(1)2^{(n+4)(n-1)/2}$$

But $W(1) = 3$ (the functions 0, 1, and $\dot{x}$) so that

$$W(n) \le 3 \cdot 2^{(n+4)(n-1)/2}$$

Recalling that no symmetry type may contain more than $n!2^n$ members, the number of realizable n-variable functions $T(n)$ is bounded by

$$T(n) < 3n!2^n \cdot 2^{(n+4)(n-1)/2}$$

The number of n-variable functions is 2^{2^n}, so that the fraction of functions that are cascade realizable is bounded by $T(n)/2^{2^n}$. As n becomes large, this fraction becomes vanishingly small. Yet many more functions are realizable by redundant cascades than by irredundant cascades, as indicated in Table II.

F. HISTORICAL REFERENCES

Many of the notions presented in this section are strongly influenced by the work of Ashenhurst (1959). Irredundant cascade decompositions, for example, are iterated simple disjunctive decompositions. Maitra (1962) first formalized the idea of irredundant cascades, and his contribution was followed by the contributions of Sklansky (1963), Levy *et al.* (1964), Minnick (1964), and Weiss (1969b). Each of these works treat the problem of irredundant cascade synthesis and the problem of determining realizability. Their results are subsumed for the most part by Lemma 2.4. Stone (1965) and Sklansky *et al.* (1964) enumerated the functions realizable by irredundant cascades thereby correcting an erroneous result that appeared in Sklansky (1963).

The theory of single rail redundant cascades appeared in a paper by Stone and Korenjak (1965) who each discovered the theory independently and simultaneously. This section is drawn largely from their work. Fantauzzi (1968a,b) presented the material on redundant cascades in a somewhat different framework, but essentially all of his results are in the Stone and Korenjak paper.

EXERCISES

1. (Stone and Korenjak) show that for all n, precisely 12 totally symmetric switching functions are realizable by redundant cascades.

2. Show that "realizability of a redundant cascade" is a property of symmetry families in the following sense. If one function in a family is redundantly realizable then the entire family is redundantly realizable. (Use the definition of family that appears in Chapter V.)

3. (Sklansky, Korenjak, and Stone) For each n we define the following 3 functions:
   A_n = the number of irredundantly realizable functions of n-variables with weight greater than neutral weight.
   B_n = the number of irredundantly realizable functions of n-variables with neutral weight.

$C_n =$ the number of irredundantly realizable functions of n-variables with weight less than neutral weight.

(a) Show that the functions obey the following recursion formulas.

$$A_n = C_n = A_{n-1} + B_{n-1} + C_{n-1}$$
$$B_n = A_{n-1} + B_{n-1}$$

(b) Solve the recursion formulas above by finding the values of A_1, B_1, and C_1.

4. Show that the number of families of n-variable functions that are realizable by redundant cascades is equal to the number of types of $(n-1)$-variable functions that are realizable by redundant cascades.

III. TWO-RAIL CASCADES

A. THE BASIC TWO-RAIL CASCADE

There are several ways we can attempt to increase the logical capabilities of a cascade:

(1) Increase the number of external inputs to each cell.
(2) Increase the number of arterial connections; that is, make the cascade *multirail*.
(3) A combination of (1) and (2).

All these approaches require increased complexity of the cell function and therefore it is clear intuitively that they augment further the number of producible functions.

The first of these approaches has been investigated by Lendaris and Stanley (1963). It is assumed that there are m external connections to each cell and one arterial connection and the cell is capable of producing all $(m+1)$-variable functions on its output. Although the number of functions realizable in such a cascade is increased (obviously all $(m+1)$-variable functions are realizable, for example), it is still inadequate to produce all the functions and becomes increasingly so as the number of variables increases.

The second approach, an increase in the number of interconnections between cells, is shown in Fig, 7 as a generalization of the single-rail cascade discussed in Section II and is called a *two-rail cascade* (Short, 1965). Again, it is assumed that the cell is individually so complex as to produce any 3-variable functions of the cell inputs.

The two-rail cascade is logically complete. The greater flexibility of such a

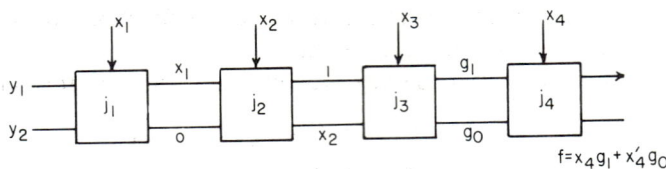

FIG. 7.   A two-rail cascade.

cascade is illustrated by the cascade of Fig. 7 which realizes any 4-variable function. The specialization of the cell is indicated by the notation

$$j : f_1(x, y_1, y_2), f_2(x, y_1, y_2)$$

where f_1 and f_2 are the rail outputs, y_1 and y_2 are the rail inputs, and x is the primary input to the cell. If we set

$$j_1 : x, 0$$
$$j_2 : y_1, x$$
$$j_3 : g_1(x, y_1, y_2), g_0(x, y_1, y_2)$$
$$j_4 : \text{arbitrary}, xy_1 + \bar{x}y_2$$

where g_0 and g_1 are arbitrary 3-variable functions, then the above cascade will realize any 4-variable function expressed as

$$f(x_1, x_2, x_3, x_4) = x_4 g_1 + \bar{x}_4 g_0$$

on the bottom output of the terminal cell. In fact, a two-cell cascade with j_3 and j_4 will realize any function if we allow inputs x_1 and x_2 to be applied as horizontal inputs to cell j_3.

The logical completeness of the cascade is proved as follows: suppose the function is specified as sum-of-products. The two cell outputs are specified as

$$f_1 = (xy_1, \bar{x}y_1, 1)$$
$$f_2 = (xy_1 + y_2, \bar{x}y_1 + y_2, y_2)$$

Then any arbitrary product term can be formed on the f_1 leads, and when completed, can be added to the sum being accumulated on the f_2 lead. At the same point, a new product term can be commenced on the f_1 lead. Of course, the resulting cascade will be redundant. Fig. 8 illustrates one such cascade.

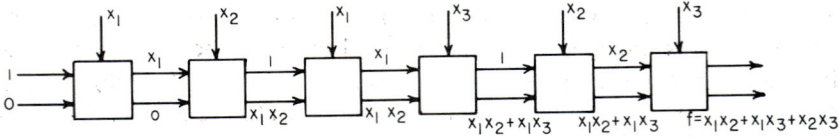

FIG. 8.   A redundant two-rail cascade.

Note the function realized by this cascade is not realizable in a single-rail cascade. It should be obvious that an arbitrary function needs at most $n2^n$ cells. If complementary outputs are also specified for f_2, the number of cells required is at most $n2^{n-1}$.

B. EFFICIENT TWO-RAIL CASCADES

There are three basic problems that can be considered with respect to realization of a function by two-rail cascades. The first one is the derivation of an upper bound on the number of cells in a two-rail universal cascade, that is, a cascade that will realize any arbitrary function $f(x_1, \ldots, x_n)$. The second problem is to develop algorithms for minimal cell cascade synthesis for any given function. The third problem is to find a reduced, if not minimal, set of functions that must be producible by a typical cell in a universal cascade. We will discuss the first and the third problems briefly here. The second problem is still an open one.

Based on the expansion of a function given by

$$f(x_1, \ldots, x_n) = g_1(x_1, \ldots, x_{n-1}) \oplus x_n g_2(x_1, \ldots, x_{n-1})$$

a general form of a two-rail cascade can be conceived as shown in Fig. 9. The

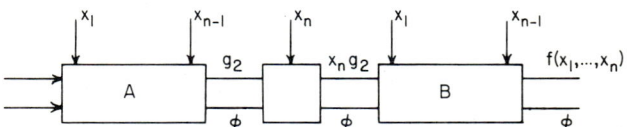

FIG. 9. A general form of a two-rail cascade.

A cascade whose primary inputs are $x_1, \ldots, x_{n-1}$ produces the function g_2. Suppose it needs N_A cells. The middle cell forms $x_n g_2$ on the upper rail (ϕ is arbitrary on the bottom rail). The B cascade adds the function g_1 based on the expansion

$$g(x_1, \ldots, x_{n-1}) = h_0(x_1, x_2) \oplus x_3 h_1(x_1, x_2) \oplus x_4 h_2(x_1, x_2)$$
$$\oplus \cdots \oplus x_3 x_4 \cdots x_{n-1} h_{2^{n-3}-1}(x_1, x_2)$$

Two cells are required for each h-function; one more cell is required per variable occurrence in the expansion. As soon as a term is formed it is added EXCLUSIVE-OR-wise on the top rail and a new term is started there. Thus, in general, the B cascade requires a total of

$$N_B = 2 \cdot 2^{n-3} + \sum_{i=1}^{n-3} \binom{n-3}{i} i = (n+1)2^{n-4}$$

Thus, the number of cells N_n in the complete n-variable cascade is bounded by

$$N_n \leq N_A + N_B + 1 = N_{n-1} + (n+1)2^{n-4} + 1$$

Certain simplifications can be performed to improve the above bound. First, since the expansion of g_1 could be carried with respect to any arbitrary $n-3$ variables, the last variable used as the rightmost variable in the A cascade can be applied to the leftmost input of the B cascade, thus saving one cell. Also, since h_0 is not multiplied by any variable, h_0 can be produced simultaneously with any other h-function from a single cell, thereby saving two more cells. Thus,

$$N_n \leq N_{n-1} + (n+1)2^{n-4} - 2$$

The fact that ϕ outputs are arbitrary could be exploited to obtain further savings. For example, the ϕ output could be specified to be a variable x_i if that variable could be used in the $(i+1)$th cell in the cascade. The key to such savings is to obtain an expansion of g_1 with certain ordering of the variables which occur with the h-functions. For example, if g_1 is a 4-variable function, we can write

$$g_1(x_1, x_2, x_3, x_4)$$
$$= x_1 x_4 h_3(x_2, x_3) \oplus h_0(x_1, x_2) \oplus x_3 h_1(x_1, x_2) \oplus x_4 h_2(x_2, x_3)$$

If the B cascade realizes the terms in the order they occur in the expansion from left to right, then for each term a cell can be saved in the transition from one term to the next in the cascade. Such ordered expansions of g_1 are known to exist for all values of n through $n=8$, but beyond that the problem is an open one (Minnick et al., 1966).

Although two-rail cascades are more flexible compared to single-rail cascades, a typical 3-input, 2-output cell is significantly more complex compared to a 2-input, 1-output cell of single-rail cascades. In the construction of the universal cascade, however, except for the first cell in the cascade, we do not require the complete flexibility in each cell. In fact, we have used only the following restricted cell-types, denoted A, B, C, D and E, in all but the first cell.

$$A : f_1 = x, f_2 = y_2 \oplus xy_1$$
$$B : f_1 = y_1, f_2 = xy_2$$
$$C : f_1 = h(x, y_1), f_2 = y_2$$
$$D : f_1 = xy_1, f_2 = y_2$$
$$E : f_1 = h(x_1, y_1), f_2 = y_2 \oplus h(x_1, y_1)$$

Although the first cell in the cascade needs to be arbitrarily complex, it is convenient to replace the first two cells by a five-cell cascade as shown in Fig. 10, each of whose cells is restricted. This cascade is derived from the expansion

$$g_2(x_1, x_2, x_3, x_4)$$
$$= h_0(x_1, x_3) \oplus x_2 h_1(x_1, x_3) \oplus x_4 h_2(x_1, x_2) \oplus x_3 x_4 h_3(x_1, x_2)$$

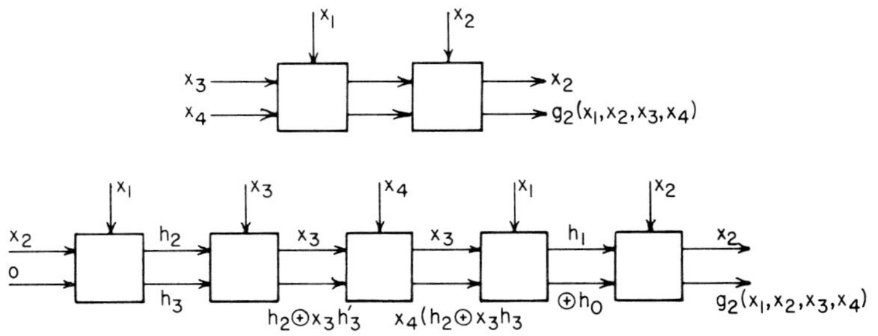

FIG. 10.   A five-cell restricted two-rail cascade for four variables.

Thus an upper bound to the restricted two-rail cascade can be obtained by adding 3 to N_n previously obtained.

It is natural to attempt to extend the study of two-rail cascades to multi-rail cascades. Results of such investigations using group-theoretic approaches are reported in Chapter VIII.

EXERCISES

5. Prove that the growth rate $n2^{n-1}$ can be achieved for two-rail cascades by utilizing Reed–Muller expansion of the function.

6. Show that there exists a 10-cell two-rail cascade which realizes any arbitrary function of four variables.

IV. TWO-DIMENSIONAL ARRAYS

This section describes some of the well-known two-dimensional arrays for realizing arbitrary combinational switching functions. These arrays have been derived by putting some constraints on either the interconnection structure or the cell functions of the array and represent some of the simplest of the arrays which are amenable to mathematical treatment. An interesting class of arrays is obtained by a fixed interconnection structure of the "edge-fed horizontal bus-type" as shown in Fig. 11. Each cell in the main array produces a 2-input, 1-output function. The topmost input to each column is a constant 1 or 0. The bottom row is a collector row which is also a cascade of 2-input, 1-output cells with the leftmost input being a constant 1 or 0. The output of the collector is the desired function $f(x_1, \ldots, x_n)$ of the input variables $x_1, x_2, \ldots, x_n$. In order to exploit some of the canonic expansions of a switching function for the synthesis algorithms, the cells in the collector row are assumed to perform the same function OR or AND, or EXOR, or EQUIV.

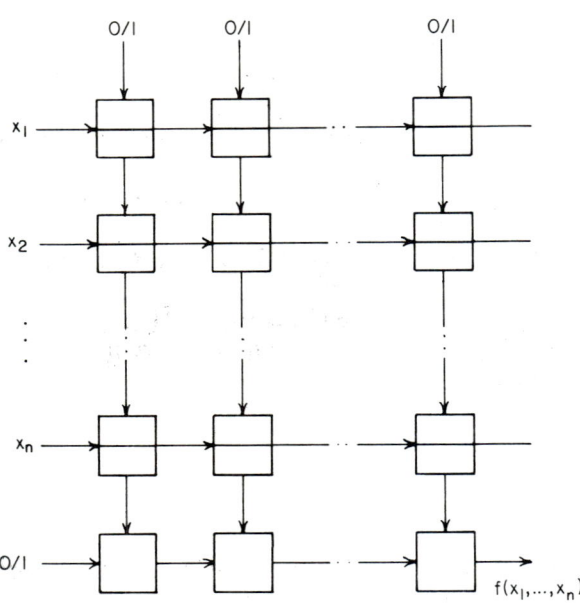

FIG. 11.   Edge-fed horizontal bus type array with 2-input, 1-output cells.

A. CUTPOINT ARRAYS

The cutpoint arrays (Minnick, 1964) are based on Maitra cascades. Each column of Fig. 11 is capable of producing all the functions realizable by a Maitra cascade. The cells in the collector row could have arbitrary functions but usually they are assumed to be OR or AND cells. Any function expressed as a sum-of-products (product-of-sums) can be realized by obtaining the products (sums) by the columns and their sum (product) by the collector row of OR (AND) cells. Obviously, for any arbitrary function of n variables, we need $n + 1$ rows and a maximum of 2^{n-1} columns, and hence a maximum of $(n + 1)2^{n-1}$ 2-input, 1-output cells, which is called the *growth rate* of the array.

Although each column is capable of producing all the Maitra cascade realizable functions, the individual cells need not be as complex as to produce all the 16 functions of two variables. In Section II, we have seen that a canonic form of Maitra cascade need only have cells which can produce AND, OR, and EXOR. If arbitrary complementation of the inputs is not allowed in the different columns of the array, we need to add additional flexibility in the cell. This can be seen by the following theorem.

THEOREM 4.1. If $f(x_1, \ldots, x_n)$ can be realized by a Maitra cascade of n cells in which the input to the ith cell is x_i $(1 \le i \le n)$, then f can also be realized by a cascade with x_i as input to the ith cell which is restricted to produce the following six functions:

1. $f_1 = \bar{x}_i \bar{y}_i$ (NOR)   or $f_2 = \bar{x}_i y_i$ (NIMP)
2. $f_4 = x_i \bar{y}_i$ (NIMP)   or $f_8 = x_i y_i$ (AND)
3. $f_5 = \bar{y}_i$ (NOT)   or $f_{10} = y$
4. $f_6 = x_i \oplus y_i$ (EXOR)   or $f_9 = x_i \oplus \bar{y}_i$ (EQUIV)
5. $f_7 = \bar{x}_i + \bar{y}_i$ (NAND)   or $f_{11} = \bar{x}_i + y_i$ (IMP)
6. $f_{13} = x_i + \bar{y}_i$ (IMP)   or $f_{14} = x_i + y_i$ (OR)

where y_i is the arterial input to the cell and f_j is the ith 2-variable function whose characteristic vector has a decimal equivalent value of j.

Proof: Consider a two-cell cascade as shown in Fig. 12 where θ_1 and θ_2 are the cell functions for the first and the second cells, respectively. If θ_1 is arbitrary, y_2 could be 0, 1, x_1, or $\bar{x}_1$. Allowing θ_1 to be arbitrary, a particular function $f(x_1, x_2)$ can be produced at y_3 for different function values to θ_2. For example, the function $f_6 = x_1 \oplus x_2$ can be produced at y_3, if the function θ_2 is made $x_1 \oplus y_2$ or $x_1 \oplus \bar{y}_2$. By exhaustion it can be easily seen that a

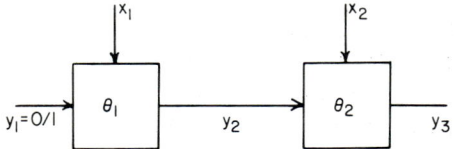

FIG. 12.   A two-cell Maitra cascade.

minimal set of functions that the second cell must generate such that y_3 is an arbitrary function of two variables are the six functions listed in the theorem with $i = 2$. Thus, there are 64 choices for the function θ_2. We can similarly restrict θ_1 because by giving appropriate constant value to y_1 (1 or 0); it is possible to produce $y_2 = 0, 1, x_1$, or $\bar{x}_1$ with the above restricted set of functions. Thus the theorem is true for $n = 2$. We now prove the theorem by induction.

We assert $y_{n+1} = 0, 1, x_n$ or $\bar{x}_n$. This assertion is true for $n = 2$; assume true for $n - 1$, that is, $y_n = 0, 1, x_{n-1}$, or $\bar{x}_{n-1}$. Let $\theta_n = \bar{y}_n$ or y_n. Then $y_{n+1} = 0$ or 1 (setting $y_n = 0$ or 1). Let $\theta_n = \bar{x}_n \bar{y}_n$ or $\bar{x}_n y_n$. If $\theta_n = \bar{x}_n \bar{y}_n$, put $y_n = 0$ so that $y_{n+1} = \bar{x}_n$. If $\theta_n = \bar{x}_n y_n$, put $y_n = 1$, then $y_{n+1} = \bar{x}_n$. Also, $\theta_n = x_n \bar{y}_n$ or $x_n y_n$. If $\theta_n = x_n \bar{y}_n$, put $y_n = 0$, then $y_{n+1} = x_n$; if $\theta_n = x_n y_n$, put $y_n = 1$, then $y_{n+1} = x_n$. Thus, it is proved that y_{n+1} can be 0, 1, x_n or $\bar{x}_n$.

From Lemma 2.1, we observe that if y_n is producible, then so is $\bar{y}_n$. Thus if u is any function produced at the output of a cascade of $n - 1$ cells, allowing θ_n to have the different functional forms, we conclude that y_{n+1} could be any one of the nontrivial 2-variable functions of x_n and u. Thus, y_{n+1} could be any one of the 16 possible functions of x_n and u.

Thus the restricted Maitra cell must be able to produce a set of six functions and therefore needs at least three specification bits or "cutpoints." In the physical realization of the cell, these cutpoints take the form of photocells or switches or just points at which the electrical connection could be mechanically broken at the final of stage specializing the array for a particular function. In the original cutpoint cell a fourth specification bit was included to allow conversion of the cell into a flip-flop with little extra circuitry. A set of basic digital computer elements like decoder add-one circuit, complementor, code translator, shift register, and multiplier using cutpoint cellular arrays have been described by Minnick (1964, 1965b).

Minnick also proposed logical design methods using cutpoint arrays. Except where the logical designer applies his own skill and ingenuity, these methods were found to be inefficient because of large wastage of cells which perform very little or practically no logic. Other reasons for inefficiency are the absence of data paths to cause interaction of information flow from vertical to horizontal directions—a feature which is very desirable in parallel operations, and the excessive use of "jumper" connections from some outputs at the edge to edge-input points of the same array. To remove these difficulties, Minnick proposed a cutpoint-based array with a more complicated interconnection structure, called a *cobweb* array (Minnick, 1965a). A typical cell in a cobweb array has about the same amount of logic as in a cutpoint cell but has additional cutpoints to allow flow of information from right to left and from top to bottom in a diagonal direction. As a result, a cascade function in this array need not be produced by a vertical cascade; it may be formed in a zig-zag path in the array. This gives the logical designer a considerable flexibility in forming his design.

Goldberg (Minnick, *et al.* 1966) proposed two embellishments of cutpoint arrays by complicating the interconnection structure. In one, called a 2-*way array*, the output of a cell can be directed to either or both of two neighboring cells. In another, called a 4-*way array*, the output of the cell can be directed to any one of the four neighboring arrays. The efficiency of these arrays lies intermediate between those of cutpoint and cobweb arrays.

Another approach towards improving the efficiency of cutpoint arrays is to develop optimal synthesis algorithms. This aspect of the problem will be treated in Section V.

B. q-FUNCTION OR ADDER ARRAY

1. The Basic q-Function Array

Another interesting edge-fed horizontal bus-type array proposed by Kautz and Myhill (Minnick, 1966) is based on a cell which can produce either the AND or the OR function of its inputs depending on the value of a single specification bit b. The output z_i of the ith cell with input x_i can be written as

$$z_i = x_i y_i + b_i y_i + b_i x_i$$

where y_i is the arterial input and b_i is 0 or 1. Obviously, when $b_i = 0$ the cell produces an AND function and when $b_i = 1$, the cell produces an OR function. Each cell in the collector cascade performs an EXCLUSIVE OR function. The function z_i is also recognized to be the carry output of a single-bit full adder, where b_i denotes the third input. The function produced by such a cascade is called a q-function and is denoted by q_i if $(b_n b_{n-1} \cdots b_2 b_1)$ is the binary equivalent of the positive integer i $(0 \leq i \leq 2^n - 1)$. The q-functions for $n = 2$ are shown in Table III. Each q-function is a positive unate function

TABLE III
q-Functions For $n = 2$

$b_2 b_1$	Characteristic vector $c_3 c_2 c_1 c_0$	q-Function
00	0 0 0 0	$q_0 = 0$
01	1 0 0 0	$q_1 = x_1 x_2$
10	1 1 0 0	$q_2 = x_2$
11	1 1 1 0	$q_3 = x_1 + x_2$

and has a single run of 1's in the characteristic vector and this statement is true for any arbitrary value of n. Any fundamental product

$$p_i = \dot{x}_n \dot{x}_{n-1} \cdots \dot{x}_2 \dot{x}_1, \qquad i = \sum_{j=1}^{n} f_j 2^{j-1}$$

where

$$f_j = \begin{cases} 1, & \dot{x}_j = x_j \\ 0, & \dot{x}_j = \bar{x}_j \end{cases}$$

can be formed as

$$p_i = q_{2^n - i} \oplus q_{2^n - i - 1}$$

where q_{2^n} is defined as always true. Any arbitrary function f can be specified as $(0 \le i, j, \ldots k \le 2^{n-1})$,

$$
\begin{aligned}
f &= p_i + p_j + \cdots + p_k \\
&= p_i \oplus p_j \oplus \cdots \oplus p_k \\
&= q_{2^n - i} \oplus q_{2^n - i - 1} \oplus q_{2^n - j} \oplus q_{2^n - j - 1} \oplus \cdots \oplus q_{2^n - k} \oplus q_{2^n - k - 1}
\end{aligned}
$$

Thus, any arbitrary function can be synthesized by using a set of cascades with a collector row that forms the EXCLUSIVE OR sum of the cascade outputs. Furthermore,

$$
\begin{aligned}
p_i + p_{i+1} &= q_{2^n - i} \oplus q_{2^n - i - 1} \oplus q_{2^n - i - 1} \oplus q_{2^n - i - 2} \\
&= q_{2^n - i} \oplus q_{2^n - i - 2}
\end{aligned}
$$

and in general

$$p_i + p_{i+1} + \cdots + p_{i+k} = q_{2^n - i} \oplus q_{2^n - (i+k) - 1}$$

Thus, if one counts the *runs* of like bits in the characteristic vector, and if r is the number of these runs, then $r - 1$ columns are needed to realize the function in a q-function array.

To illustrate, let $f = x_3 \bar{x}_2 x_1 + \bar{x}_3 (x_1 + x_2) + x_1 x_2$. The characteristic vector of f is $(c_7 c_6 c_5 c_4 c_3 c_2 c_1 c_0) = (10011110)$. There are 4 runs of like bits. We have $p_1 + p_2 + p_3 + p_4 = q_7 \oplus q_3$ and $p_7 = q_1$. Thus $f = q_7 \oplus q_3 \oplus q_1$. The specification bits for q_7, q_3, and q_1 are $(b_3 = 1, b_2 = 1, b_1 = 1)$, $(b_3 = 0, b_2 = 1, b_1 = 1)$, and $(b_3 = 0, b_2 = 0, b_1 = 1)$, respectively.

The growth rate for the q-function array can be estimated in terms of the bounds on the run measure of switching functions obtained by Elspas and Short (1964). They have proved that the maximum number of runs $r(f)$ for an arbitrary function f of n variables is $[(2^{n+1} + 1)/3]$ when n is even, and $[(2^{n+1} + 2)/3]$ when n is odd. This means that the growth rate for the q-function array is $(n + 1)r(f)$ when arbitrary input variable permutation and complementation is allowed (this is because of the underlying assumptions in the derivation of the run measure).

2. Minimization of q-Function Arrays

The q-function arrays can be simplified by combining together certain q-functions, just as fundamental products or minterms can be combined to produce prime implicants.

First, consider two columns having identical cascades of AND or OR gates, except in row j, where one has an AND gate while the other has an OR gate. This means that the b-numbers of these two columns differ in only a single binary digit in the jth bit position. The situation is shown by the left two columns in Fig. 13. These two columns jointly contribute an output 1 to the collector row if and only if (a) the outputs of the jth row gx_j and $g + x_j$ are themselves different, which means that $g \oplus x_j = 1$; (b) the inputs $x_{j+1}, \ldots, x_n$ are such that any difference detected in row j is propagated all the way down to the collector row. This is true if the inputs, as well as the outputs, from the two columns in the original array are complementary, for all cells $k, j < k \leq n$.

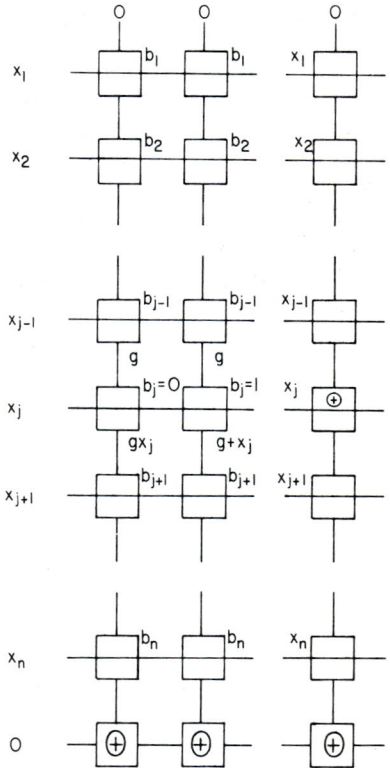

FIG. 13. Simplification of q-function arrays.

This is ensured if and only if $b \neq x$ as shown in Fig. 14. In the single cascade replacement, therefore, we must have $y = 1$ propagated if $b \neq x$, that is, the cell function is $z = y(b \oplus x)$. Thus the pair of columns can be replaced by the single right-most column of Fig. 13, in which the upper cascade is unchanged; the pair of differing gates is replaced by a single EXOR gate, and the AND and OR gates in the lower cascade are transformed into AND and INHIBIT gates, respectively.

The reduction algorithm can be readily extended to pairs of merged columns which originally correspond to four columns whose b-numbers agree in all digits except two. Figure 15 shows the result of a second merging.

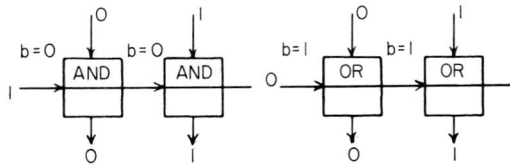

FIG. 14.   Propagation of difference detected in row j.

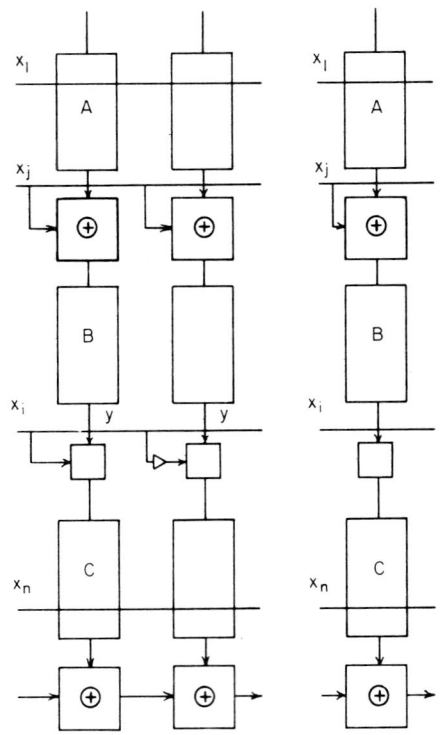

FIG. 15.   Merger of four columns into one.

As before, the least order digit to disagree gives rise to an EXCLUSIVE OR cell, and converts the part of the cascade below it from AND/OR to AND/INHIBIT cells. Referring to the ith row in Fig. 15, the y inputs to the ith cell must be the same since the cascades above the point are identical. However, if the combination of these two cascades is to contribute a 1 to the collector row, their outputs must differ, that is, $x_i y = (\overline{\overline{x}_i y})$, which is true if $y = 1$ independent of x_i. Thus in the single cascade replacement cell on row i, the digit x_i could be skipped, effectively transmitting the signal downward unmodified. Further mergers may proceed in a like fashion, giving rise to large skipped digits and lesser number of columns in the array.

C. 2^n-CELL ARRAYS

These arrays have a fixed number 2^n of cells, but almost all functions require nearly the full-sized array. Thus 2^n-cell arrays should be most useful for those functions which are the more difficult cases by other standards. These arrays have been proposed by Kautz (Minnick, 1966).

1. Minterm Arrays

This array consists of three layers of cells. The first layer produces minterms in variables $(x_1, \ldots, x_{n/2})$ and $(_{n/2+1}, \ldots, x_n)$ in the lower and right hand sides of the array, respectively (assume n is even for the moment), as shown schematically in Fig. 16. The cell logic required for this purpose is either the AND (xy) or the IMP $(\overline{x}y)$ function which can be specified by a single cutpoint. The second layer forms the AND of these outputs by returning them in horizontal and vertical buses. The cell logic is simply an AND gate. The third layer now serves to collect in OR-fashion any desired subsets of these fundamental products or minterms in order to form an arbitrary function. The cell logic is just an OR function and a single specification bit which will select the output of the AND gate of the second layer or apply a 0 to the input. The OR gates are connected in the form of a cascade. Note that when all three layers are superimposed, the specification bits for a cell are different in different regions of the array.

2. Symmetric Function Arrays

A network which realizes the elementary symmetric functions $S_0, S_1, \ldots,$ S_n in the variables $x_1, x_2, \ldots, x_n$ is shown in Fig. 17. This network is derived from the well-known transfer-contact realization of a symmetric function tree (Caldwell, 1958). An arbitrary function f of these variables expressed as

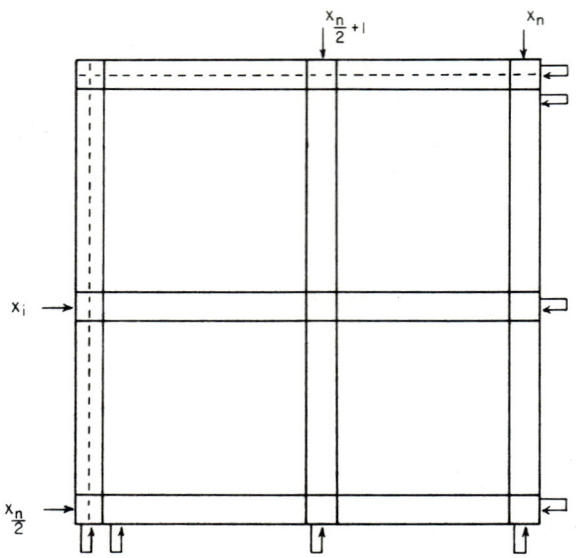

FIG. 16.   Minterm array.

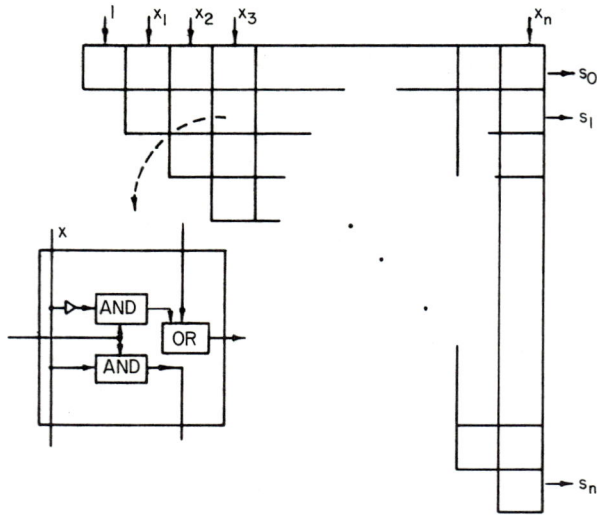

FIG. 17.   Elementary symmetric function array.

a characteristic vector $(c_{2^n-1} c_{2^n-2} \cdots c_j \cdots c_1 c_0)$ may be written as a symmetric function in $2^n - 1$ variables as

$$f = S_{a_1, a_2, \ldots, a_r}(x_1, x_2, x_2, x_3, x_3, x_3, \ldots, x_n)$$

where x_i is repeated 2^{i-1} times and the a-numbers in the subscript to S are the indices j, one for each of the digits f_j in the characteristic vector, which equals 1. It can be easily seen that the elementary symmetric functions in this redundant tree are the 2^n minterms in the variables $(x_1, x_2, \ldots, x_n)$. The resulting triangular array has size $2^n \times 2^n$. An arbitrary function can now be formed by adding a collector column of OR cells at the right, which would form the disjunction of any subset of these minterms as specified by a set of cutpoints, one per collector cell.

The size of the array can be reduced and a square, rather than a triangular, array can be obtained by dividing the set of variables into half, and by superimposing a triangular array for the first half on a corresponding rotated array for the second half. The structure of the array is shown schematically in Fig. 18. The elementary symmetric functions are returned through horizontal

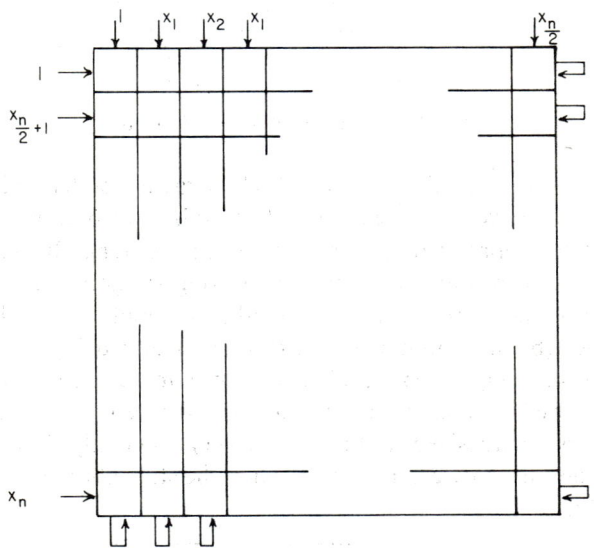

FIG. 18.   Symmetric function array.

and vertical buses; then they are combined by AND gates to form the minterms and a collector cascade of OR gates running through the array is added to form an arbitrary function as in the case of minterm arrays, as selected by cutpoints associated with the OR cells. The array has exactly 2^n cells of only three different types. There are only 2^n cutpoints, one per cell.

D. NOR ARRAYS

The NOR arrays were originated by Brooking in 1961 at the Air Force Cambridge Research Laboratory (Brooking *et al.*, 1965). The base array consists of an 8-*neighbor*, 2-dimensional arrangement of 8-input NOR cells as shown in Fig. 19. Any subset of the eight inputs can be selected by a set of

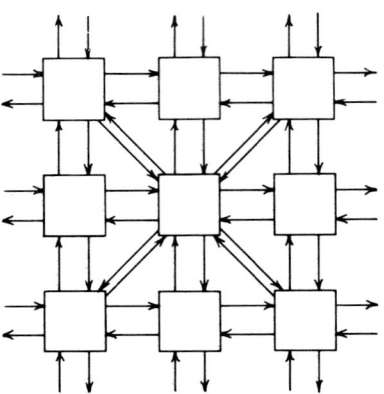

FIG. 19.   The 8-neighbor NOR array.

eight parameters associated with each cell. The array is also termed a "uniform modifiable logic network," because all of the sub- and partial networks can be generated by appropriately selecting the parameters. Brooking and his co-workers studied a number of such interesting subgraphs and found that their behavior could be made to correspond to some of the well-known finite state machines (binary counter, etc.) assuming a unit delay associated with each connection. King (1965) studied the autonomous behavior of such an array and showed that an arbitrary autonomous behavior of a sequential machine can be synthesized, although not very efficiently. Let s_t^i represent the state of the ith NOR-cell at time t. Then the state at time $t + 1$, s_{t+1}^i, can be expressed as

$$s_{t+1}^i = \overline{(a_{i1} s_t^{\,1} + a_{i2} s_t^{\,2} + \cdots + a_{in} s_t^{\,n})}$$

where a_{ij} is 1 if and only if the output of the jth NOR gate is connected to the input of the ith NOR gate. Thus, the state vector S_{t+1}, expressed as a column vector, can be written as

$$S_{t+1} = \overline{(A)S_t}$$

where (A) is the interconnection matrix of the n nodes. The above equation which gives a mapping of S_t to S_{t+1} gives the autonomous state behavior of the array. As can be expected from the general theory of autonomous networks with nonlinear logic (Kautz, 1958; Preparata, 1964), the state diagram consists of a set of cycles or rooted trees or a combination of cycles and rooted trees which unlike those for linear sequential machines (Elspas, 1959) may not be backward deterministic. Note that matrix (A) can represent any arbitrary interconnection and therefore the particular 8-neighbor configuration or any uniform subinterconnection of it can be expressed by suitably choosing the coefficients in (A). Unfortunately, no characterization of state diagrams is known which will correspond to these special cellular interconnections. However, King has proved that allowing arbitrary interconnections and using a redundant encoding of the states, any arbitrary autonomous state behavior can be realized by such arrays.

E. NOR–NAND ARRAYS FOR COMBINATIONAL FUNCTIONS

Spandorfer and Murphy (1963) developed logical design methods for combinational functions on 8-neighbor NAND cell arrays. Minnick (1968) proposed similar design methods using NOR cell arrays. Basic to these arrays is the cascade shown in Fig. 20. If the cell function in the cascade is NOR and if $a_1 = a_2 = \cdots = a_n = 0$, then the output function is $\bar{x}_1\bar{x}_2 \cdots x_n$. If the cell function is NAND and $a_1 = a_2 = \cdots = a_n = 1$, the output of the cascade is $\bar{x}_1 + \bar{x}_2 + \cdots + \bar{x}_n$. Since any arbitrary $f(x_1, \ldots, x_n)$ can be realized in a two-level NOR-to-NOR or NAND-to-NAND network, it is obvious that a cellular array of NOR or NAND cascades with a NOR or NAND collector row can be used to synthesize any arbitrary combinational function. One such array is shown in Fig. 21. The synthesis algorithm consists of expressing the given function in some sum-of-products or product-of-sums form. If it is in sum-of-products form the cell function is NAND and $a = 1$. If it is in product-of-sums form, the cell function is NOR and $a = 0$. Thus, if the array in Fig. 20 is a NAND array, $F = x_1x_2x_3 + \bar{x}_1x_2\bar{x}_3 + \bar{x}_1\bar{x}_2x_3$, but if it is a NOR array, $F = (x_1 + x_2 + x_3)(\bar{x}_1 + x_2 + \bar{x}_3)(\bar{x}_1 + \bar{x}_2 + x_3)$. Note that although the a-line could be bussed, the variables could not be because they depend on F; we may need both complemented or uncomplemented variables to be applied to different columns.

To make both x and $\bar{x}$ available at each input, cascades as shown in Fig. 22 for different variables are interleaved with the main array resulting in a composite array as shown in Fig. 23. The cascades which generate the variable and its complement have been termed horizontal wiggle buses by

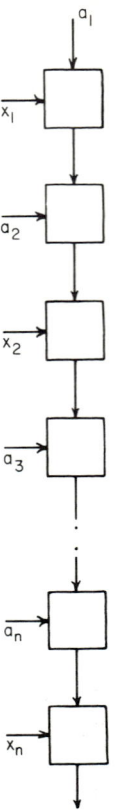

FIG. 20. The basic NOR–NAND cascade.

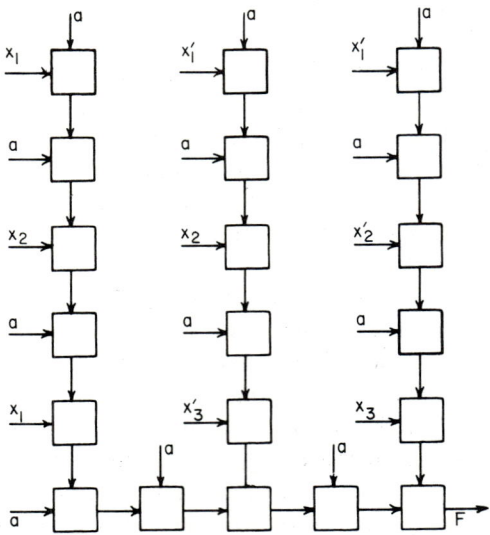

FIG. 21. A 2-dimensional NOR–NAND array.

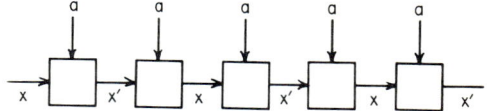

FIG. 22. A "variable" cascade.

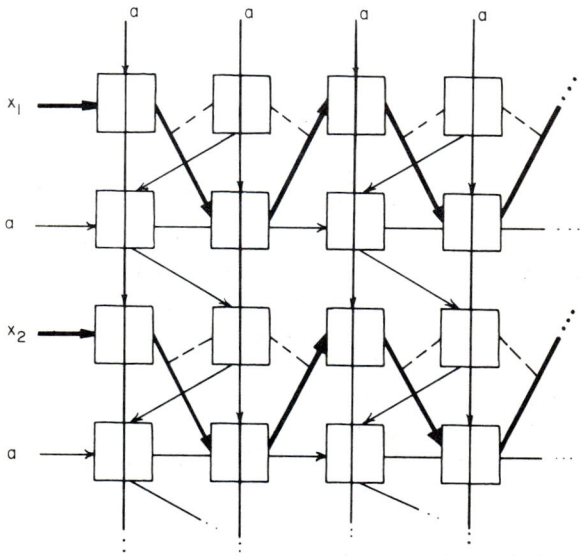

FIG. 23. A 2-dimensional NOR–NAND array interleaved with "variable" cascades.

Spandorfer and Murphy, or simply variable buses by Minnick. The vertical cascades have been termed vertical wiggle buses and they generate the sums or products required in the function. A particular function is synthesized by breaking some of the connections from x_i or $\bar{x}_i$ or both. Actually, these connections may be absent and they could be made by using *synthesizing* arcs. The growth rate of this array is easily calculated to be $2(n + 1)2^{n-1} + n2^{n-1} + n$ which is approximately $n2^{n+1}$ as n increases.

Spandorfer and Murphy also proposed certain variations of the basic NAND array in which they increased the number of inputs to the NAND cells or departed from the base 8-neighbor connections. They have proposed two such specific variations called Diamond I array with growth-rate $n2^n$, and Diamond II array with growth rate $3n2^{n-2}$. Interested readers are referred to original papers or to survey by Minnick (1967).

Certain variations of the basic NOR array have also been proposed by Minnick (1968). The idea in one of them is to obtain a NOR-based cell by treating a 4 × 4 subarray of Fig. 23 as a single cell. The growth rate of this array is reduced at least by a factor of four, although a cell is about four times as complex. Logical simplification on the original NOR array will also be applicable to NOR-based arrays.

F. MAJORITY-GATE ARRAYS

These arrays will be mentioned very briefly here. Each cell function of this array is a majority or minority element, usually of three inputs. These arrays have been studied by Canaday (1969), Short (1965), Amarel *et al.* (1964), and Zschirnt (1966). Canaday's method is based on an algorithm for the synthesis of arbitrary self-dual functions with a regular 2-dimensional special array of majority gates. The synthesis algorithm is then extended to arbitrary functions by showing that any functions can be expressed as a self-dual function in which the number of variables is larger. Short's algorithm is based on a tree network of 3-input majority gates for any arbitrary function developed by Miyata. Short proposes an algorithm to convert this into a rectangular form. Amarel *et al.* considered the synthesis of symmetric functions using majority-gate cellular arrays. Zschirnt proved that the tree-based majority-gate cellular array can be embedded by specializing the " uniformly modifiable logic network" of 8-neighbor NOR array.

V. MINIMIZATION OF CELLULAR ARRAYS

To logic designers, one of the fundamental problems in cellular logic is to be able to design cellular arrays for arbitrary combinational or sequential functions which optimize a set of design parameters or objectives. The parameters involved are numerous and complicated; a partial list of these could be: the size of the array measured in terms of number of cells, the logical complexity of the individual cell as well as the number of specification bits or cutpoints (the cutpoints take a significant fraction of the total cell area, particularly for programmable logic arrays which need active integrated components as well), the interconnection structure, the total number of intracell and intercell connections, speed of signal propagation which depends on maximum and average intercell distance, and numerous other electrical and geometrical parameters which are pertinent in a given technology. It is clear intuitively, and has been demonstrated empirically, that there exist some sort of trade-offs among certain of these parameters. The cobweb arrays illustrate that the total array size for realizing arbitrary function can be reduced by making the interconnection structure more complex at the expense of a number of cutpoints. Research on (n, d, k) problems (Elspas *et al.*, 1964) has yielded some limited analytical results covering the interdependence among some of the purely topological parameters of a cellular array such as total number of connections incident on a cell, maximum and average inter-cell distance, the size and planarity of the arrays. When both logical and topological parameters are considered, the situation becomes extremely diffi-cult to define mathematically. Furthermore, a set of definitions adopted today might become obsolete tomorrow because of the evolving technology. Even if we adopt a set of definitions, it seems to be mathematically a very hard problem to arrive at "minimal" solutions in the more general cases.

As a start on this difficult problem, we consider in this section the import-ant case when the interconnection and logic of the individual cell may be assumed to be fixed as in the case of edge-fed horizontal bus-type arrays. We further assume that the order of the input variables is fixed. The general theoretical problem may be stated as: find a systematic procedure for expres-sing an arbitrary given function as a Maitra function of a minimum number of Maitra functions in the independent variables, taken in a prescribed order. No general solution to this problem is yet known. Taking the primary Maitra function to be a disjunction, a conjunction, or an exclusive-OR or equivalence sum which allows the Maitra functions to be combined in an arbitrary order, some partial results have been obtained. In this section we shall present the minimization algorithm developed by Mukhopadhyay (1969) for unate cellu-lar arrays and then discuss the outline of a more general algorithm proposed by Weiss (1969a,b). A discussion of other minimization problems for cellular arrays is also included at the end.

A. UNATE CELLULAR LOGIC

A 1-dimensional cascade of 2-input, 1-output cells each of which can pro-duce any one of the 14 unate functions of its two inputs (that is, excluding EXOR and EQUIV) is called a *unate cascade*. The following properties of unate cascades can be easily proved: (i) the output of a unate cascade is a unate function; (ii) each unate cascade has a canonic irredundant form in which each cell is either an AND function or an OR function; (iii) a restricted irredundant unate cascade in which each cell function is restricted to produce the five functions (1), (2), (3), (5), and (6) of Theorem 4.1 can realize any arbit-rary unate cascade realizable function; (iv) the sum-of-products form of the cascade output expression of a unate function gives the complete set of prime implicants of the function, all of which are core prime implicants of the function. Property (iv) can be utilized to develop a test and realization procedure of a unate cascade realizable function which needs a maximum of n iterations (Mukhopadhyay, 1969). The test procedure has been further improved by Schmitz (1969).

The minimization problem for cellular array realizing a function f is the following: we have to find a set of unate functions $u_1, u_2, \ldots, u_k$ such that

(a) either $f = \sum_{i=1}^{k} u_i$ or $f = \prod_{i=1}^{k} u_i$;

(b) each of the unate functions u_i is unate cascade realizable with same ordering of the input variables;

(c) k is minimum.

1. Definitions

The terms defined in this section are used in the algorithm given in Section V.A.2.

DEFINITION (D1). If u is a unate function such that $u \subseteq f$ (u is included in f), then u is said to be a *unate component* of f. If $u \neq f$, then u is said to be properly included in f.

DEFINITION (D2). A unate component of f not properly included in any other unate component of f is said to be a *maximal unate component* of f.

DEFINITION (D3). Let $P = \{P_1, P_2, \ldots, P_k\}$ be the set of prime implicants of f. A subset (proper or improper) of P, denoted as $P^j = \{P_{j_1}, P_{j_2}, \ldots, P_{j_s}\}$ is said to be *compatible* if and only if

$$\prod_{i=1}^{s} P_{j_i} \neq 0$$

DEFINITION (D4). A compatible set of prime implicants is said to be *maximal* if it is not a proper subset of any other compatible set of prime implicants.

DEFINITION (D5). Let $\{p_1, p_2, \ldots, p_k\}$ be the set of product terms each of which is an implicant of the function f, but not necessarily a prime implicant of f. Let $P_{u_b} = (p_{i_1}, p_{i_2}, \ldots, p_{i_m})$ be a maximal compatible set such that each $p_{i_j}(1 \leq j \leq m)$ is a prime implicant of the function u_b defined as

$$u_b = \sum_{j=1}^{m} p_{i_j}$$

Then u_b will be called a "basic unate component" of f. Basic unate components include the maximal unate components and, in general, some non-maximal unate components.

DEFINITION (D6). A unate function obtained by dropping only one prime implicant from u_b will be called a "derived" unate component of f with respect to u_b. If there are m prime implicants of u_b, there will be m derived unate components with respect to u_b. The process can be iterated, i.e., there can be derived unate components with respect to a previously derived unate component if necessary.

DEFINITION (D7). Let u_m be a basic or a derived unate component of f realizable by a unate cascade with a specified ordering of the inputs such that there does not exist another basic or derived unate component of f, say u_s, which properly includes u_m, i.e., $u_m \subset u_s$, and is realizable with the same ordering of the inputs. Then u_m is said to be a "prime unate" component of f with respect to the given ordering of the inputs.

   Note that since nonprime cascades can always be replaced by prime cascades without destroying the covering of the function, only prime unate cascades need be considered in the minimization algorithm.

2. Algorithm Steps

   The minimal unate cascade realization of an arbitrary switching function can be found as follows:

   Step 1. *Determine the basic unate components.* Let $\{p_1, p_2, \ldots, p_k\}$ be the set of implicants of the given function f. A compatibility graph G_c is then drawn as follows: G_c has a set of k vertices, one for each implicant term. Two vertices p_i and p_j are connected by an edge if and only if (1) $p_i p_j \neq 0$, i.e., p_i and p_j are compatible; and (2) $p_i \not\subset p_j$, $p_j \not\subset p_i$, i.e., p_i is not included in p_j and vice versa. The cliques[3] or maximal complete subgraphs of G_c correspond to the basic unate components of the given function.
   Step 2. *Set up a partially ordered graph for the basic unate components.* Let $L_1 = \{u_1, u_2, \ldots, u_s\}$ be the basic unate components of f. A partially ordered graph whose vertices correspond to the basic unate components is set up as in Fig. 24. A line is drawn connecting two vertices if and only if one

---

   [3] A clique of a linear nondirected graph is a maximal subgraph such that any pair of vertices within the subgraph is connected by an edge. A maximum clique is a clique having a maximum number of vertices.

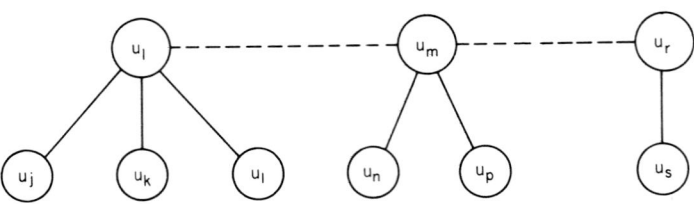

FIG. 24. A partially ordered graph.

of the corresponding basic unate components is included in the other. Components which are not included in any other components correspond to the maximal unate components.

Step 3. *Determine the prime unate components.* For an arbitrarily specified ordering of the variables, the prime unate components of $L_1 = \{u_1, u_2, \ldots, u_s\}$ which appear in the first level of the partially ordered graph of Fig. 24 are tested for unate cascade realizability. Let L_r and L_{nr} denote, respectively, the set of basic unate components on the upper level of the partially ordered graph which are realizable and nonrealizable by unate cascades. Let L_c denote the set of unate components lying on the lower level of the partially ordered graph which are included in at least one element of L_r. Update list L_1 to L_1' as follows:

$$L_1' = L_1 - L_r - L_{nr} - L_c$$

If L_1' is empty, go to Step 4. If not, proceed as follows: Let u_b be an element of L_{nr}. Obtain the set of derived unate components with respect to u_b and consider only those which are not included by any element of L_r. Obtain such derived components for each of the elements of L_{nr}. Let L_d denote all such derived components. Update list L_1' to list L_2 as

$$L_2 = L_1' + L_d$$

A partially ordered graph (as in Fig. 24) is now drawn with respect to L_2 and the above procedure is repeated. At some ith stage, L_1' must be empty since the number of derived components is finite.

Step 4. *Determine a minimal cover.* In Step 3 the list of prime unate components of f was obtained. A cover table for the fundamental products is now set up. A minimal set(s) of prime unate components are picked up which form a cover of the table. Then, for the assumed ordering of the variables, all minimal cascade realizations are obtained.

The principal disadvantage of the algorithm just presented lies in determining the basic unate components. The problem is that the time required in a computer program to find the cliques increases almost exponentially with the number of vertices. For functions with more variables, a greater percentage of the execution time is needed to find the basic unate components. Another disadvantage of the algorithm is that the process of setting up the partially ordered graph of Step 2 may have to be repeated many times, depending on whether the basic unate components or derived components are unate cascade realizable or not.

Schmitz (1969) improved the algorithm by incorporating two important modifications. First, the basic unate components are found by determining the cliques of a number of smaller graphs instead of solving one large graph. Second, the improved algorithm is set up so that only one partially ordered graph need be set up and the prime unate components can be picked from that graph.

B. A MORE GENERAL ALGORITHM

In an independent study by Weiss (1969a), a minimization algorithm applicable to more general type arrays has been proposed. The array structure is very much like the cutpoint array in which the variables are not necessarily bused horizontally. Each column is assumed to be a redundant or irredundant cascade with variable ordering independent of each other. Test and realizability of a function by a cascade is based on a standard algebraic form satisfied by all cascade realizable functions. This standard form is essentially an algebraic equivalent of the canonic form theorem (Theorem 2.1) discussed in Section II. For functions which can be realized by an irredundant cascade, it is possible to derive this standard form working exclusively with the prime implicants of the function. The procedure for doing that is based on the fact that all the prime implicants of irredundant cascade realizable functions are core, which is also true for unate cascades discussed earlier. For a general redundant cascade realizable function, this procedure is much more complicated.

The broad outline of Weiss's algorithm is very similar to the unate array minimization algorithm. One proceeds by first determining all the prime cascade realizable functions which imply the given function. The difference between the prime unate component and the prime cascade realizable function is that the latter is an arbitrary function producible by a cascade without any unateness constraint. In the next step, a minimal or irredundant set of prime cascade realizable functions are picked up which, taken together, cover the given function. The details of the algorithm are significantly different from the unate algorithm and will not be presented here due to space limitation. Weiss further indicates how the algorithm can be modified to find optimal arrays for a set of functions and also for functions having optional states.

C. OTHER MINIMIZATION PROBLEMS

Since cellular arrays can be developed based on expressions involving EXCLUSIVE OR and EQUIVALENCE functions as primary operators, it is natural to pose the following minimization problem: find a systematic procedure to express an arbitrary given function as an EXCLUSIVE OR-sum or EQUIVALENCE-sum of a minimum number of cascade realizable functions. This problem seems to be an inherently difficult one and its solution is not yet known even in the simple case when the cascades produce arbitrary products of variables. Mukhopadhyay and Schmitz (1970) have proposed a minimal solution for the special case when the cascades produce products of variables having a fixed polarity (complementation or no complementation) of its variables. One does not feel very optimistic about the general problem; however, if such algorithms are developed, they will perhaps be applicable to the minimization of "q-function" arrays in which the q-functions are unate, but not necessarily positive unate, or perhaps a general cascade realizable function.

Minimization problems can also be posed with respect to other parameters of the array. Such a problem formulated by Kautz (1965) arose in connection with the minimization of cutpoints in the 2^n-cell array. This problem is relevant because it is sometimes possible to share cutpoint connections between different versions of the same cell. The problem can be formulated as follows: suppose a particular specialization of a cell requires that some electrical connection should be established between subsets of points on the cell. The connectedness between pairs of points for a specialization of that cell defines a binary relation among the points. What is the interconnection structure having the minimum number of physical connections such that each of the binary relations defined by the different specializations of the cell can be derived by eliminating arcs from the interconnection structure? It has been shown that this minimization problem is essentially a minimal cover problem again.

Other minimization problems have been considered on cellular arrays which concern with purely topological parameters of the array. For example, one problem is to find the most uniform interconnection structure for arrays such that signal propagation delay between any pair of cells in the array does not exceed a specified value with minimum fan-in and fan-out of each cell. Extensive research has been done on the graph-theoretic equivalent of this problem, called the (n, d, k) problem, but still many problems remain unsolved and challenging (Elspas et al., 1964).

VI. REVIEW OF OTHER WORKS IN CELLULAR AREA

A. INTRODUCTION

In this section we shall give a brief review of other works done in the cellular area. Except for the work of Hennie, all these works could be classified as macrocellular research because the individual cells are considerably more complex than those usually used in cellular arrays discussed so far, which, by the same token, are called microcellular arrays. The motivations for these works were pattern recognition and detection, parallel computation, and abstract problems of automata theory.

B. WORK OF HENNIE

One of the earliest and systematic studies of cellular networks was conducted by Hennie (1961, 1968). Hennie observed that there is a close relationship between iterative combinational circuits and sequential circuits and that most of the analysis and synthesis techniques for sequential circuits are applicable to iterative combinational circuits having fixed interconnection and cell functions. Hennie first considered the "unilateral one-dimensional" iterative circuits—the multirail cascade equivalent with fixed cell function—and observed that a typical cell performs a fixed transformation between the *intercell signals* entering and leaving that cell, which can be described in the form of a closed "flow-table" in which the states are represented by the combination of values of the intercell signals. If an input sequence of length n is accepted by this "flow table" in the sequential machine sense, an iterative cascade of n cells will also provide an output from the nth cell if this input "sequence" is applied as a temporal sequence of inputs to the n cells with appropriate boundary signals at the edge of the cascade. Thus a system of iterative networks becomes a spatial counterpart of a sequential machine. As can be expected, the classical state minimization algorithm and state assignment procedures applicable to sequential machines, can be used to simplify the logical complexity of a typical cell. The synthesis problem consists of obtaining a typical cell description from the verbal description of the problem (Hennie, 1961).

A major disadvantage of the 1-dimensional cascade considered by Hennie is that being fixed in structure, the number of functions realizable is very small. Flexibility of cell functions was not Hennie's concern, which, as we have seen earlier in this chapter, is a significant factor in cellular logic. Application of Hennie's sequential model to flexible cell cascades will result in a variable "flow-table" model which does not seem to be an easy approach. Furthermore, for bilateral cascades in which signals can be propagated in both directions between adjacent cells, the sequential model breaks down because the outputs have to be considered functions of future inputs. However, Hennie developed analysis procedures using a nondeterministic model and showed that bilateral networks can exhibit both equilibrium or cyclic behavior and that the problem of deciding whether an arbitrary iterative network exhibits a cyclic behavior is difficult in general. Definite iterative cascades which have a unique equilibrium output pattern for every input pattern (each cell of Hennie's cascade has an associated output) has been shown to have canonic form consisting of two unilateral cascades working independently in two directions. Furthermore, a complete characterization of definite bilateral systems (that is, a class of iterative bilateral cascades) has been obtained which says that an input/output transformation can be realized by a definite system if and only if the input and output strings taken together form a regular set. Some of Hennie's ideas have been pursued by Kilmer (1962, 1963). A study on the relationship of iterative systems and abstract computational models has been reported by Cole (1964).

C. WORK OF UNGER

Unger (1958, 1959) described a modular computer oriented toward spatial problems which is specially organized for pattern detection. The computer consists of a master control and a rectangular array of modules. Each module communicates with its four nondiagonal neighbors and receives commands issued by the control which go to all the modules simultaneously. Each module or macrocell has a 1-bit accumulator called the principal register (PR), a small amount of random access memory (six 1-bit memory cells denoted MR1, MR2, ..., MR6), and some associated logic. The PR of each module can be set or reset from the outside world. The master control contains some random access memory for storing instructions, decoding and synchronizing circuitry. The master control reads out instructions from the memory in sequence, decodes them, and sends appropriate command signals to the modules. The instruction set of the machine consists of arithmetic and logical instructions like invert the contents of each PR, add or multiply logically the contents of each PR (MRJ, $1 \leq J \leq 6$) with the contents of

MRJ (PR); register-to-register transfer operations like shift left, right, up or down the content of each PR, write the content of MRJ (PR) into PR (MRJ); control instructions like transfer on zero (indicated by the output of an OR gate whose inputs come from each PR), transfer unconditional or stop, plus two special intructions called LINK and EXPAND which permitted the content of a PR to be directly affected by the contents of arbitrary distant cells. Using flip-flops for PR's and MR's, the logical structure of a module has been given by Unger (1959). Using these instructions several programs for pattern detection and recognition have been written; in particular, Unger describes a detailed program for the recognition of alphanumeric characters. The recognition scheme consists of finding a set of local and global properties which characterize a given letter of the alphabet and the program in the machine will detect such properties in a unique way. In a later paper, Unger (1962) proposed the use of 2-dimensional bilateral iterative networks to extract the feature of the given pattern.

Hawkins and Munsey (1963) considered a 2-dimensional iterative network computer oriented also toward spatial problems. The computer consists of two planes, called the input plane which holds the input pattern, and a resultant plane in which an image of the input plane can be projected via an intervening plane called the mapping mask, using optical techniques. The output plane contains a cellular array of linear threshold elements. The authors have shown that such a system is capable of recognizing certain tiny objects against the background of other large objects. A more flexible and general class of logical systems having a similar structure, called cellular bulk transfer system, has been studied by Roy (1970).

D. WORK OF HOLLAND

Holland (1959) studied a 2-dimensional modular computer capable of simultaneously executing an arbitrary number of subprograms. Each module is a small general-purpose computer containing a storage register, a set of auxiliary registers, and some associated logic. The module can be active or inactive at a given time. When active, the content of its storage register is treated as an instruction and the module executes the instruction. The active status of the module can then be passed to any one of its nondiagonal four neighbors. Thus, an arbitrary number of programs consisting of a sequence of instructions can be stored spatially throughout the array and could be executed simultaneously. The action of an active module during a cycle consists of three phases: In the first phase, called the input phase, module storage and auxiliary registers can be set to desired values by an external source. In the second phase, called a path-building phase, the active modules fetch their operands by setting paths to be gated to them from the locations of operands. In the final phase, called an instruction execution phase, the instructions in the storage register of all the active modules are executed. Although Holland machines would not be practical for reasons of programming and very inefficient use of hardware, it does give a model of parallel computations which apparently motivated several workers to propose modifications of Holland machines. Important among them are the proposals of Comfort (1963), Gonzales (1963), and Slotnick *et al.* (1962). McCormick (1963) proposed a design of a parallel processor for visual information processing. For more information about these machines, the reader is referred to Murtha (1966).

E. WORK OF VON NEUMANN

As early as 1952, von Neumann introduced the notion of a cellular space and discussed universal computing and constructing automata embedded in such a space (von Neumann, 1966). The basic structure of the cellular model is an infinite rectangular grid or tesselation, each cell of which is occupied by the same 29-state automaton. The neighborhood of a cell consists of the cell itself and its four immediate nondiagonal neighbors. The state of a cell at time $t + 1$ is uniquely determined by the state of the neighborhood at time t together with the transition function of the automaton which is associated with each cell. At time $t = 0$, an initial cell assignment, i.e., an assignment of a state to each cell of a finite number of cells in the model is imposed from the " outside world," all cells not in the assignment list being left in a quiescent or unexcitable state. Thereafter, the history of the model for all subsequent time is uniquely determined by the transition rules of the model. By appropriately associating the states of the cells with symbols of an alphabet, configurations of cells with functions over the alphabet, and the history of configurations with computation of a sequence of functions, von Neumann proved that his model is computation universal in the sense of Turing's. He also proved that this model is construction universal in the sense that a constructor can be embedded in the model that can generate any given configuration (with some exceptions) at some place in the space. In particular, a self-reproducing automaton that can duplicate its own configuration can also be embedded in this model. The work of von Neumann inspired investigations in a new field known now as the theory of cellular automata, in which numerous contributions have been made. A discussion of these results is beyond the scope of the present chapter.

ACKNOWLEDGMENTS

The work of the first author (AM) has been supported in part by the National Science Foundation under Grant GJ-723.

REFERENCES

AIKEN, H., *et al.* (1951). Synthesis of electronic computing and control circuits. *Ann. Comput. Lab. Harvard Univ.*, **27**. Harvard Univ. Press, Cambridge, Massachusetts.

AMAREL, S., COOKE, G., and WINDER, R. D. (1964). Majority gate networks. *IEEE Trans. Electron Computers* **13**, pp. 4–13.

ASHENHURST, R. (1959). The decomposition of switching functions. *Ann. Comput. Lab. Harvard Univ.*, **29**, No. 30. Harvard Univ. Press, Cambridge, Massachusetts.

BROOKING, M. E., KING, W. F., III, and GIUSTI, A. (1965). A Survey of In-house Work on Cellular Logic. Unpublished memorandum, AFCRL.

CALDWELL, S. H. (1958). "Switching Circuits and Logical Design." Wiley, New York.

CANADAY, R. H. (1965). Two-dimensional iterative logic. *In* "AFIPS Conference Proceedings" Vol. 27, Part 1, Spartan Books, Washington, D.C.

COLE, S. N. (1964). Real Time Computation by *n*-dimensional Iterative Arrays of Finite Automata. Ph.D. Thesis, Harvard Univ.

COMFORT, W. T. (1963). A modified holland machine. *In* "Proceedings Fall Joint Computer Conference," pp. 481–488.

ELSPAS, B. (1959). The theory of linear sequential networks. *IRE Trans. Circuit Theory* **CT-6**, 45–60.

ELSPAS, B., and SHORT, R. A. (1964). A bound on the run measure of switching function. *IEEE Trans. Electron. Comput.* **EC-13**, 1.

ELSPAS, B., *et al.* (1964). Investigation of Propagation Limited Computer Networks. Stanford Res. Inst. Project 4523 (also republished with same title, 1965).

FANTAUZZI, G. (1968a). A semigroup theory for the Maitra cascade. *In* Proceedings IFIP Congress 1968, Supplement Book I, pp. 54–61.

FANTAUZZI, G. (1968b). Application of Karnaugh maps to Maitra cascades. *In* AFIPS Conf. Proc. **32**, SJCC, Thomson, 291–296.

GONZALES, R. A. (1963). A multilayer iterative circuit computer. *IEEE Trans. Electron. Comput.* **12**, 781–790.

HARVARD Computation Laboratory Staff (1951). Synthesis of Electronic Computing and Control Circuit, *Annals of the Computation Laboratory* **27**, (Harvard Univ. Press, Cambridge, Massachussetts).

HAWKINS, J. K. and Munsey, C. J. (1963). A two-dimensional iterative network computing technique and mechanization. *In* "Proceedings 1962 Workshop on Computer Organization," pp. 93–125.

HENNIE, F. C. (1961). "Iterative Arrays of Logical Circuits." MIT Press, Cambridge, Massachusetts.

HENNIE, F. C. (1968). "Finite-State Models for Logical Machines." Wiley, New York.

HOLLAND, J. H. (1959). A universal computer capable of executing an arbitrary number of sub-programs simultaneously. *In* " Proceedings Eastern Joint Computer Conference," pp. 113–118.

KAUTZ, W. H. (1958). State Logic Relations in Autonomous Sequential Networks. *In* "Proceedings Eastern Joint Computer Conference," pp. 119–127.

KAUTZ, W. H. (1965). Cutpoint Minimization. Unpublished Stanford Res. Inst. Mem. (August 23).

KILMER, W. L. (1962). Iterative switching networks composed of combinational cells. *IRE Trans. Electron. Comput.* **EC-11**, 123–131.

KILMER, W. L. (1963). Topics in the theory of one-dimensional iterative logic networks. *Inform. Contr.* **6**, 180–199.

KING, W. F. III (1965). State-Logic Relations in an Iterative Structure for Autonomous Sequential Machines. AFCRL Rep. 65–439.

LENDARIS, G. G., and STANLEY, G. L. (1963). On the Structure-Dependent Properties of Adaptive Logic Networks. GM Defense Res. Labs., Mathematics and Evaluation Studies Dept., Santa Barbara, California, TR 63-219, July.

LEVY, S. Y., WINDER, R. O., and MOTT, T. H., JR. (1964). A note on tributary switching networks. *IEEE Trans. Electron. Comput.* **EC-13**, 148–151.

McCORMICK, B. H. (1963). The Illinois pattern recognition computer—ILLIAC III. *IEEE Trans. Electron. Comput.* **12**, 791–813.

MAITRA, K. K. (1962). Cascaded switching networks of two-input flexible cells. *IRE Trans. Electron. Comput.* **EC–11**. 136–143.

MINNICK, R. C. (1964). Cutpoint cellular logic. *IEEE Trans. Electron. Comput.* **EC-13**, 685–698.

MINNICK, R. C. (1965a). Application of cellular Logic to the design of monolithic digital systems. *In* " Proceedings Symposium on Microelectronics and Large Systems." Spartan, Washington, D.C.

MINNICK, R. C. (1965b). Cobweb cullular arrays. *In* "Proceedings AFIPS Conference, Fall Joint Computer Conference." Spartan, Washington, D.C.

MINNICK, R. C. (1967). A survey of microcellular research. *J. Ass. Comput. Machinery* **14**, 203–241.

MINNICK, R. C. (1968). "Cellular Networks " (G. Biorci, ed.). Academic Press, New York.

MINNICK, R. C., SHORT, R. A., GOLDBERG, J., STONE, H. S., GREEN, M. W., YOELI, M., and KAUTZ, W. H., *et al.* (1966). Cellular Arrays for Logic and Storage. Stanford Res. Inst., Project 5087, Contract AF 19(628)-4233; AFCRL.

MUKHOPADHYAY, A. (1969). Unate cellular logic. *IEEE Trans. Comput.* **C-18**, 114–121.

MUKHOPADHYAY, A., and SCHMITZ, G. (1970). Minimization of EXCLUSIVE OR and LOGICAL EQUIVALENCE Switching circuits. *IEEE Trans. Comput.*

MURTHA, J. C. (1966). Highly parallel information processing systems. *Advan. Comput.* **7**, 1–116.

PREPARATA, F. P. (1964). State logic relations for autonomous sequential networks. *IEEE Trans. Electronic Comput.* **EC-13**, 542–548.

ROY, K. K. (1970). Cellular Bulk Transfer Systems. Ph.D. Dissertation, Dept. of Electrical Engineering, Montana State Univ., Bozeman, Montana.

SCHMITZ, G. (1969). " Minimization Methods for Cellular Logic Arrays. Ph.D. Dissertation, Montana State Univ., Bozeman, Montana.

SHORT, R. A. (1965). Two-rail cellular arrays. *In* AFIPS Conf. Proc. **27**, pt. 1, Washington, D.C., Spartan, pp. 355–369.

SKLANSKY, J. (1963). General synthesis of tributary switching networks. *IEEE Trans. Electron. Comput.* **EC-12**, 464–469.

SKLANSKY, J., KORENJAK, A., and STONE, H. S. (1964). Canonical tributary networks. *IEEE Trans. Electron. Comput.* **EC-14**, 961–963.

SLOTNIK, D. L., *et al*. (1962). The SOLOMON Computer. *In* Proc. Fall Joint Computer Conf., pp. 97–107.

SPANDORFER, L. M., and MURPHY, J. V. (1963). Synthesis of Logic Functions on an Array of Integrated Circuits. Sci. Rep. No. 1 & 2, UNIVAC Project 4645, AFCRL Contracts -63 -528, Contract AF 19(628)2907, Sperry Rand Corp., Bluebell, Pennsylvania.

STONE, H. S. (1965). On the number of equivalence classes of functions realizable by cellular cascades. *In* Proc. Nat'l Symp. on the Impact of Batch Fabrication on Future Computers, 81–87.

STONE, H. S., and KORENJAK, A. (1965). Canonical form and synthesis of cellular cascades. *IEEE Trans. Electron. Comput.* **EC-14**, 852–862.

UNGER, S. H. (1958). A computer oriented towards spatial problems. *Proc. IRE* **46**, 1744–1750.

UNGER, S. H. (1959). Pattern detection and recognition. *Proc. IRE* **47**, 1737–1752.

UNGER, S. H. (1962). Pattern recognition using two-dimensional bilateral, iterative, combinational switching circuits. *In* Proceedings Symposium on Mathematical Theory of Automata." Polytechnic Institute of Brooklyn, New York.

VON NEUMANN, J. (1966). "Theory of Self-Reproducing Automata" (Edited and completed by A. W. Burks). Univ. of Illinois Press, Urbana, Illinois.

WALTER, C. T., *et al*. (1968). Impact of fourth generation software on hardware design. *Comput. Group News, IEEE*, **2**, 4.

WEISS, C. D. (1969a). Optimal synthesis of arbitrary switching functions with regular arrays of 2-input 1-output switching functions. *IEEE Trans. Comput.* **C-18**, 839–856.

WEISS, C. D. (1969b). "The characterization and properties of Cascade Realizable Switching Functions." *IEEE Trans. Comput.* **C-18**, 624–633.

YOELI, M. (1965). A group theoretical approach to two-rail cascades. *IEEE* TEC, *EC*-14, 6, December, pp. 815–822.

ZSCHIRNT, H. H. (1966). Miscellaneous notes on cellular logic structure and their synthesis in eight-neighbour nets. Unpublished Memorandum.

Chapter VIII

THE THEORY OF MULTIRAIL CASCADES

BERNARD ELSPAS

I.   INTRODUCTION   316
    A. General   316
    B. Historical Preliminaries   317

II. DECOMPOSITION THEORY OF GROUP FUNCTIONS   320
    A. Definitions and Notation   320
    B. Decomposability of Group Functions   323
    C. Characterization of Decomposable Groups   328
    D. Survey of Finite Groups with Respect to Decomposability   331

III. SYNTHESIS OF MULTIRAIL CASCADES   336
    A. Introduction—The Notion of Permutation Cascade   336
    B. Two-Rail Binary Cascades   338
    C. Multirail Synthesis   345
    D. Number of Module Types   350
    E. A Linear-Algebra Approach to Multirail Cascades   354
    F. Permutation Cascades Based on Other Groups   359
    G. Comparisons Among Cascade Varieties   361
    H. Nonstandard Ordering of Input Variables—Equalization
       of Input Variable Loading   365

    REFERENCES   367

ABSTRACT.   This chapter is concerned with systematic techniques for the realization of multioutput combinational logic functions by means of a regular interconnection geometry, that of multirail cellular cascades. As with single-rail (Maitra) cascades, of which multirail cascades are a generalization, interconnections are allowed only between adjacent cells. A mathematical theory, the decomposition theory of group functions, is developed which permits systematic techniques of great generality to be applied to this logic synthesis problem. The synthesis techniques described here permit the exercise of considerable control over the number and internal structure of the individual cells required for arbitrary multioutput function synthesis. Comparisons are made with alternative techniques for the synthesis of multirail cascades.

I. INTRODUCTION

A. GENERAL

The theory presented in this chapter is motivated (as are the theories of
Maitra cascades and cellular arrays) by attempts to find systematic techniques
for the realization of combinational logic functions within regular interconnection geometries. In the situation discussed here, that of multirail cellular
cascades, the interconnection geometry is constrained to be essentially one
dimensional, with only nearest neighbor connections allowed between
"cells," just as with Maitra, or single-rail, cascades. Here, however, we are
concerned with simultaneously realizing several Boolean output functions
(i.e., with multioutput synthesis instead of the synthesis of a single Boolean
function). Moreover, in the multirail case it becomes very important to limit
the number of different basic cell types employed. In the single-rail case, this
limitation hardly arises since there are only 16 different 2-input, 1-output cell
functions possible to begin with (including trivial ones). However, simply by
passing from one to two rails (and hence to 3-input, 2-output cells) the number
of possible cell types rises drastically from 16 to $2^{16} = 65,536$. Clearly, serious
inventory problems ensue if any significant fraction of this number of functions is needed for use as "standard" cell types. This problem becomes
increasingly significant and complex as the number of rails is further increased.

The basic problem to which multirail cascade synthesis, as described here,
addresses itself is to find systematic techniques for the efficient realization of
multioutput (say, r-output) switching function networks with the following
properties:

(1) the network is to be a cascade of $(r + 1)$-input, r-output cells, each
cell being connected only to its immediate predecessor in the cascade
via the r rails,

(2) the rail outputs of the last cell in the cascade realize r arbitrarily specified Boolean functions

$$y_i = F_i(x_1, \ldots, x_n); \qquad i = 1, \ldots, r$$

of the n Boolean input variables, only one variable x_j being supplied
to each cell (though each variable may be applied to a number of
different cells, if necessary).

(3) the cells of the cascade are to be chosen from a fixed roster (or "inventory") of standard cell types, where the inventory is constrained
to be of "reasonable" size.

(4) the cells of the inventory should preferably not be unduly complex in their internal logical structure,

(5) the length (number of cells) of the cascade should be "reasonable" in magnitude (minimal, if possible).

Some comments and discussion are now in order. Requirements (1) and (2) are straightforward enough, though one might argue that they are, perhaps, too narrow. For example, one might conceivably allow more than one external Boolean input x_j to be applied to each cell. This would lead to a multirail generalization of the Lendaris-Stanley (1963) cascades. Also, the whole idea of using only a 1-dimensional cascade interconnection geometry is a considerable restriction. One might allow some reasonable amount of "treeing" of cells in order to enhance their logical power in the network without thereby greatly complicating the interconnection geometry. Generalized approaches such as this are quite feasible, and have been pursued by a number of researchers, though it should be clearly pointed out that the ensuing mathematical complications are considerable. The simpler assumptions that we have chosen to follow here turn out to be self-justifying to a remarkable extent in that they appear to provide just enough mathematical regularity to permit the application of powerful mathematical tools—the theory of finite groups and linear algebra—to the synthesis problem. These assumptions thus permit some rather elegant and satisfying results to be obtained within this formulation. The resulting treatment also provides an excellent basis for further extensions of these mathematical techniques to other switching problems.

The reader will note that requirements (3), (4), and (5) are hedged with such words as "reasonable" and "unduly." What is "unreasonable" with one fabrication technology today may rapidly become state-of-the-art tomorrow. We shall not, therefore, attempt to be any more specific about what might constitute "reasonable" complexity or size for cells or networks of cells. The requirements as stated above are still useful as qualitative guidelines and as criteria for comparing different synthesis techniques. Even more important, trade-offs will inevitably exist between and among network parameters such as the cell complexity, the inventory size, and the length of cascade required. The elucidation of such trade-offs, is in one sense, the real goal of studies such as those reported here.

B. HISTORICAL PRELIMINARIES

The earliest work on multirail cascades was exclusively concerned with *binary two-rail* cascades, as reported by Short (1965) and Yoeli (1965), and was itself motivated by the desire to generalize the still earlier work on Maitra (or single-rail) cascades done by Stone and Korenjak (1965) and others. (See Chapter VII.) The salient fact about single-rail cascades is their logical incompleteness: the fractional number of n-variable Boolean functions realizable at the end of a (single-rail) cascade of two-input gates goes rapidly toward zero as n grows, even when unlimited repetitions of the input variables are allowed.

In contrast, Short (1965) showed that the addition of a second rail yields a logically complete structure for the realization of an arbitrary Boolean function on *one* of the two rails. The essence of Short's argument is that one rail may be used to "collect" minterms (or implicants), while the second rail is employed successively to build up the products forming these minterms. As each product is completed, it is added (by Boolean addition) to the sum already accumulated on the first rail. Thus, an arbitrary function can be realized on the first rail; but little if any control can be exercised in determining which function is produced on the second rail in this process. It is clear that Short's procedure leads to canonical cascades (*canonical* in that the input variables may be applied in a fixed order (with repetitions allowed) to the cells of the cascade, independent of the particular output function being realized). The logical function realized by each cell is, of course, dependent on the desired overall output function. Short's work displays considerable ingenuity in obtaining canonical orderings of the input variables for various values of n, which appear to minimize the number of cells required. The Reed expansion (ring-sum or EXCLUSIVE OR expansion) and variants thereof play a large role in obtaining these efficient canonical orderings. In Short's synthesis procedure, the realization of an arbitrary n-variable Boolean function requires N_n 3-input, 2-output logical cells, where $N_n \cong (1/8)n2^n$. Short also treated the question of cell complexity in his two-rail cascades, showing that 3-variable general-function cells could be dispensed with by merely adding three additional cells to the cascade. Three of the five remaining generic cell types are fairly simple, but the other two still involve general functions in two variables. The salient features are summarized in Sec. III,G of this chapter. For further details of Short's procedure, the reader should see the already referenced work of Short (1965) and Minnick *et al.* (1966).

In contrast to Short's procedures for the cascade synthesis of switching functions, which while they are highly ingenious are not dependent on sophisticated mathematical techniques, Yoeli (1965) has described a set of techniques which use the theory of groups and semigroups for the synthesis of what he originally called "generalized Maitra cascades." (See also Minnick *et al.*, 1966.) Yoeli's treatment made use of abstract group theory and graph theory, particularly results depending on the theory of homomorphisms of directed graphs. Certain questions were left unresolved in this treatment, particularly since the best decomposition results contained therein did not apply to arbitrary two-rail Boolean functions, but only those for which certain restrictive conditions could be satisfied. Also, specific results were obtained for only 4-variable and 5-variable functions.

Attempts to remove these restrictions led Yoeli and Turner (1966) to a considerably sharpened theory for two-rail cascades, which also contained the seeds of a number of powerful extensions (described in Section III). The Yoeli–Turner theory not only achieves all the results of the earlier work by Yoeli, it does so in a simpler manner. In addition, the Yoeli–Turner theory shows that two-rail cascades are *functionally complete* for realizing an arbitrary *pair* of Boolean output functions in an arbitrary number of variables. It also establishes an upper bound YT(n) for the number of cells required to realize an arbitrary pair of n-variable functions in a two-rail cascade,

$$\text{YT}(n) = 3 \cdot 2^{n-1} - n - 8$$

And, of course, with one more cell, one can get an arbitrary single function of $n + 1$ variables, permitting a comparison[1] to be made with Short's earlier results. This will be discussed in more detail in Section III,G, but it should be mentioned here, that for $n = 7$, YT(n) = 177, while Short's cascades require S(n) = 190 cells. For $n > 7$, the Yoeli–Turner cascades continue to improve slightly, relative to Short's.

Now, the Yoeli–Turner bound, YT(n), given above depended on a decomposition technique which uses occasional cells to accomplish an Ashenhurst (1959) decomposition of the two-rail function to be realized. These "Ashenhurst cells" completely extract the functional dependence on a given input variable, x_i ; but as a consequence they must be *general function cells*. That is, they must be capable of realizing an arbitrary pair of 3-variable functions. For some purposes, this amount of intracell complexity may be deemed undesirable. The other cells, which are employed in the Yoeli–Turner decomposition to make possible an Ashenhurst extraction, may be called *permutation cells*.

In one sense, cascades contructed solely from such permutation cells will not be functionally complete. This is the case when the input-rail signals to the first cell of the cascade are considered to be among the input variables of the functions to be realized (a basic assumption used, e.g., in Yoeli (1965) and in Yoeli and Turner (1966)). However, if one drops this assumption and regards these ultimate rail inputs as fixed parameters of the cascade, not to be used as free variables, then functional completeness is regained with permutation cell cascades. This striking observation (Stone) has permitted us to exploit the Yoeli–Turner decomposition theory of permutation functions for general

---

[1] Of course, this manner of comparing is somewhat unfair to the Yoeli–Turner technique, but even so it eventually comes out ahead of the Short cascades.

cascade synthesis. These results are discussed in Section III,C. Somewhat surprisingly, even when one gives up the Ashenhurst (general function) cells in favor of the permutation approach using only relatively simple cells, the cascade length $P(n) = 3 \cdot 2^{m-1} - 2$, is increased by only $n + 6$ over the value, YT(n). It still crosses over the Short cascade length, S(n), at the same point, $n = 7$.

The work of Yoeli and Turner (1966) also contains several other results in the decomposition of group functions, which while they are of considerable mathematical interest, are possibly of less potential value to applications in the synthesis of logical cascades. One such result gives sufficient conditions for a six-fold decomposition analogous to the four-fold decomposition we have chosen to concentrate on. Another set of results relates to the decomposability of what might be called *factor group* functions. These latter results will probably be of value in further studies of group-theoretic methods of cascade synthesis, although thus far we have not had to make direct use of these results. For further details on these questions, the reader should see the cited work of Yoeli and Turner (1966).

EXERCISES

1. Show that the number of possible $(r + 1)$-input, r-output cell types is given by 2^M, where $M = r2^{r+1}$.

II. DECOMPOSITION THEORY OF GROUP FUNCTIONS

A. DEFINITIONS AND NOTATION

In this section we attempt to summarize and interpret the most pertinent results obtained by Yoeli and Turner (1966) regarding the decomposition of functions from the Boolean m-cube into a group. Following their approach, we will apply these results to the synthesis of cascade networks, but we defer this application to Section III. Moreover, there will be a significant shift in emphasis from that of Yoeli and Turner, in that we use only *group* functions (eschewing semigroup functions) and thereby arrive at syntheses for the class of *permutation cascades*, which form a logically complete class of cascade nets to be defined below. Comparisons of these networks, those of Yoeli–Turner, and those of Short are given in Section III,G.

Our notation will follow that of Yoeli and Turner (1966) for the most part. First, some definitions and conventions. We let $X = \{0, 1\}$ = the Boolean 1-cube, and X^m = the cartesian product of m copies of X = the Boolean m-cube. Occasionally we also write $X^m = \{0, 1, \ldots, 2^m - 1\}$ in accordance with the usual numbering for vertices of the Boolean m-cube, whereby vertex $(x_m \cdots x_2 x_1)$ is denoted by the integer $i = \sum_{j=1}^{m} x_j 2^{j-1}$, whose binary code representation is the m-tuple $x = (x_m, \ldots, x_1)$.

Let G be an arbitrary group. Much of our discussion will concern functions from X^m into G, i.e., functions with domain X^m and range G, which we shall call *group functions* (into G). Thus:

DEFINITION.   *A group function f is a function which associates with each point x of the Boolean m-cube an element $f(x)$ in a group G.*

$$f: X^m \to G$$

In engineering terms we may visualize a group function f as a "box" (i.e., a logical network) with two kinds of inputs—x-inputs and group-element inputs—and group-element outputs. The signals on the group-element inputs and outputs will also be called the *rail signals*, which represent (according to some binary coding scheme) elements of the group G. (Several group elements may, in general, be represented by the same signal.) Thus, if g is the group element corresponding to the input rail signals, and the x-inputs (also called "vertical inputs") are supplied with the logical signals $x_1, \ldots, x_m$, our convention will be that the rail outputs (for the box performing group function f) represent the group elements $g \cdot f(x)$. In particular, if the rail inputs represent the group identity I, the output will be (the encoded form of) group element $f(x)$. Figure 1 shows these conventions.

We may define the *composition* of group functions in such a way as to correspond to the cascade interconnection of their associated networks. Thus, if $f_1: X^m \to G$ and $f_2: X^n \to G$ are two group functions into the same (range)

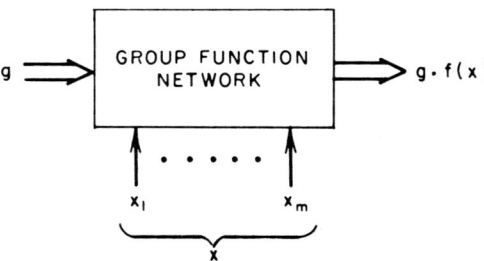

FIG. 1.   Conventions for group function networks.

group G, we define their composition $f = f_1 \circ f_2$ as the group function from $X^m \times X^n$ into G, given by:

$$f(x, y) = f_1(x) \cdot f_2(y)$$

for all x in X^m and y in X^n, where the dot represents group multiplication. This relation is illustrated in Fig. 2.

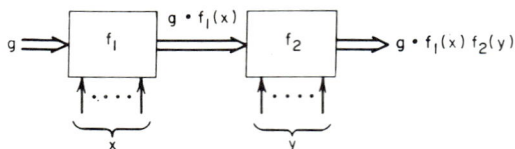

FIG. 2.   Cascade connection of group function networks.

A *cell* will be defined as a network realizing a group function $f: X \rightarrow G$, i.e., as a network with only one binary x-input[2] (though the number of (binary) rail inputs is arbitrary).

A *permutation cascade* (or group cascade) is a cascade connection of cells; i.e., it realizes a composition of group functions $f_i : X \rightarrow G$. The reason for using the term "permutation" will become clear later on when we discuss various ways of coding the rail states into group elements.

It is important to realize that the Boolean cubes X^m and X^n, appearing in the definition of composition of group functions, need not be disjoint. That is, some of the x-inputs to the f_1-box may be repeated as components of the vector y (the "x-input" to the f_2-box). In particular, a permutation cascade may have repeated occurrences of the variables x_i forming vertical inputs to its cells, so that the number of cells may exceed the number of distinct variables x_i. As with Maitra cascades, we shall call such cascades *redundant cascades*. A convenient notational formalization for this situation is as follows:

Consider a cascade consisting of n cells, realizing the group functions $f_i: X \rightarrow G$, where $i = 1, \ldots, n$ and the vertical input variable to the ith cell is $x_{\lambda(i)}$, where $\lambda(i)$ is some integer from 1 to m for each value of $i = 1, \ldots, n$, i.e., $\lambda: I_n \rightarrow I_m$ where $I_n = \{1, 2, \ldots, n\}$ = the first n positive integers. The overall network formed by this cascade connection of cells is shown in Fig. 3, and the group function $f: X^m \rightarrow G$ it realizes is given by:

$$f(x) = f(x_1, \ldots, x_m) = \prod_{i=1}^{n} f_i[x_{\lambda(i)}] \tag{1}$$

---

[2] Thus this treatment excludes, e.g., the Lendaris–Stanley cascades, where each "cell" may have several x-inputs. A generalization to cells with $m > 1$ is certainly feasible, and probably interesting, but we have not pursued this particular line of thought. (See also Yoeli (1967).)

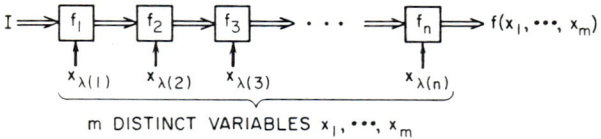

FIG. 3. Permutation cascade notation.

where $x_i \in X$ and the product sign denotes group multiplication in G. Thus, f is the group function formed by composition of the n group functions f_i, with specialization of their vertical inputs in accordance with the rule $i \to \lambda(i)$. A general notation occasionally useful for this kind of combination is

$$f = (f_1 \circ f_2 \circ \cdots \circ f_n)^\lambda \qquad (2)$$

The theory of cascade decomposition of group functions is concerned with the question: Under what circumstances can one decompose a given group function $f: X^m \to H$ in the form of Eq. (2) for suitable component functions $f_i: X \to G$ and a suitable input variable assignment function $\lambda: I_n \to I_m$? When such a decomposition exists, we say that f is *decomposable over G* (where G may be a larger group than H, and usually H is a proper subgroup of G). When interest centers on finding a decomposition with minimum value of n (for given m), we say that f is *n-decomposable*.

B. DECOMPOSABILITY OF GROUP FUNCTIONS

We have already said that the principal concern of the decomposition theory for group functions is with the problem of determining under what circumstances a given group function f can be decomposed in the form of Eq. (1) or Eq. (2). This subsumes several subquestions: (1) What group pairs (G, H) are amenable to such decomposition? (2) How can one determine the minimum number of cells (elementary functions f_i) required? (3) How can one determine the minimum number of *distinct types* of cells required? (4) Under what conditions are there canonical decompositions which are valid for *all* functions mapping X^m into H, for all m? Most of these questions have now been answered in considerable detail. The answers will be presented in the remainder of this section and in Section III.

To fix our ideas we first examine a fairly special case, which is nevertheless instructive and contains the seeds of greater generality. Suppose we wish to decompose a group function of two variables, x_1, x_2; i.e., we have $f: X^2 \to H$. How many cells do we need? For instance, let f be defined by the table:

x_1	x_2	$f(x_1, x_2)$
0	0	h_0
0	1	h_1
1	0	h_2
1	1	h_3

where h_i is a specified element of some group H for $i = 0, \ldots, 3$. It is clear that two cells are not sufficient, in general, since the relation $f(x_1, x_2) = f_1(x_1)f_2(x_2)$ implies that whenever $h_0 = h_1$, we also have $h_2 = h_3$, but the h_i are independent quantities. Likewise, it can easily be seen that three cells do not suffice, in general. First of all, any repetition of a variable x_1 or x_2 in adjacent cells reduces the cascade essentially to one with fewer cells. Hence, apart from the interchange of the roles x_1 and x_2, the only 3-cell cascade will have the form:

$$f(x_1, x_2) = f_1(x_1)f_2(x_2)f_3(x_1)$$

From which we find

$$h_0 = f_1(0)f_2(0)f_3(0)$$
$$h_1 = f_1(0)f_2(1)f_3(0)$$
$$h_2 = f_1(1)f_2(0)f_3(1)$$
$$h_3 = f_1(1)f_2(1)f_3(1)$$

But these three equations imply that whenever we happen to have $h_0 = h_1$, we also have $f_2(0) = f_2(1)$, which implies that $h_2 = h_3$ contradicting the specified independence of the h_i. This contradiction shows that three cells are not, in general, adequate for the cascade decomposition of 2-variable group functions. On the other hand, we will shortly show that for certain pairs of groups (G, H) (where $G > H$), *all* functions $f \colon X^2 \to H$ are decomposable into a 4-cell cascade. This 4-cell canonical decomposition[3] becomes the basis of an iterative synthesis technique for the realization of a very wide variety of group functions of any number of variables. Yoeli and Turner have discussed the two-rail case in detail; their methods are refined and generalized here to any number of rails.

THEOREM 2.1 (Generalization of Theorem 3. Yoeli and Turner (1966)). Let G be a finite group and H a subgroup of G, with the Property P that

---

[3] By Yoeli and Turner (1966). In their Theorem 3 this decomposition is stated and demonstrated for the case where $H =$ Klein 4-group and $G =$ alternating group on four objects, a case applicable directly to two-rail cascades.

there exists an element τ in G such that the mapping $U_\tau : h \to \tau h \tau^{-1}$ is an automorphism[4] of H leaving no point of $H - \{I\}$ fixed.

Then every group function $f : X^2 \to H$ is 4-decomposable over G in the form:[5]

$$f(x_1, x_2) = f_1(x_1)\tau^{x_2}f_2(x_1)\rho^{x_2}$$

where $\tau, \rho \in G$. Conversely, if every 2-variable group function $f : X^2 \to H$ is 4-decomposable over $G > H$, then an element τ exists in G with the above Property P.

Finally, the function f_2 is *normalized* (i.e., $f(0) = I$); f_1 is likewise normalized whenever $f(0, 0) = I$ holds, and τ^{x_2} and ρ^{x_2} are normalized by their form, since $\tau^0 = \rho^0 = I$. The functions f_1 and f_2 have range H.

Theorem 2.1 is a powerful generalization of Theorem 3 of Yoeli and Turner (1966), where essentially this result is given for the case $H = K$ (Klein group) and $G = A_4$ = alternating group of degree four. Our proof, which follows that of Yoeli and Turner very closely, is deferred until the following preliminary lemmas have been discussed.

LEMMA 2.1. Let G be a finite group with subgroup H. The following assertions are then equivalent:

(1) G contains an element τ such that the mapping $U_\tau : h \to \tau h \tau^{-1}$ is an automorphism[4] of H leaving no point of $H - \{I\}$ fixed.

(2) G contains an element τ such that the mapping $T_\tau : h \to h^{-1}\tau h \tau^{-1}$ maps H onto itself in one-to-one fashion.

(3) G possesses an inner automorphism[4] under which H is invariant as a set, but which leaves no element of H other than the identity fixed.

Proof of Lemma 2.1. If Assertion (1) holds, then so does Assertion (2), because if U_τ has the desired property, then the mapping $T_\tau : h \to h^{-1}U(h)$ certainly takes H into itself. Moreover, if $h_1^{-1}\tau h_1 \tau^{-1} = h_2^{-1}\tau h_2 \tau^{-1}$, then $\tau(h_1 h_2^{-1})\tau^{-1} = h_1 h_2^{-1}$, so that the element $h_1 h_2^{-1}$ in H is left fixed by U_τ. By Assertion (1), therefore, $h_1 h_2^{-1} = I$, or $h_1 = h_2$, so that the mapping T is one-one (*onto* H, since H is finite).

---

[4] An *automorphism* of a group H is a one-to-one mapping U of H onto H such that $U(xy) = U(x)U(y)$ and $U(x^{-1}) = U(x)^{-1}$ for all x, y in H. An *inner* automorphism of H is an automorphism of the form $U(h) = aha^{-1}$ for some a in H. All other automorphisms are said to be *outer* automorphisms. An inner automorphism of a group G may, when restricted to a subgroup H, be an outer automorphism of H, as in Lemma 2.1(3).

[5] The notation τ^x appearing here is defined (for $x = 0, 1$) by the obvious conventions: $\tau^0 = I, \tau^1 = \tau$.

Conversely, if Assertion (2) holds, then so does Assertion (1), because if T_τ maps H onto itself one-to-one, then the mapping $U_\tau : h \to \tau h \tau^{-1} = hT_\tau(h)$, which is an inner automorphism of G by its form, also maps H into H. Therefore, U_τ is also an automorphism of H. If U_τ maps some element h of H onto itself, then T_τ maps that element onto the identity I. But T_τ is one-one by assumption, and T_τ maps I onto I; therefore, $h = I$.

Finally, Assertion (3) is equivalent to Assertion (1), because U_τ is precisely the inner automorphism of G required by Assertion (3).

Proof of Theorem 2.1:   We are now ready to give a proof for Theorem 2.1. First, the sufficiency portion of the theorem: We have a group G, a subgroup H of G, and an element τ in G such that $\tau H \tau^{-1} = H$, and such that $\tau h \tau^{-1} = h$ implies $h = I$. For an arbitrary group function $f: X^2 \to H$, we wish to determine four group functions $f_i : X \to G$ such that f is the composition of these functions. We assert that this is possible in the form:

$$f(x_1, x_2) = f_1(x_1)\tau^{x_2} f_2(x_1)\rho^{x_2} \tag{3}$$

where in fact, f_1 and f_2 are group functions into the *subgroup* H, $f_2(0) = I$, and the other two factors are normalized group functions into G:

$$\tau^{x_2} = \begin{cases} I & \text{when} \quad x_2 = 0 \\ \tau & \text{when} \quad x_2 = 1 \end{cases}$$

and similarly for ρ.

First of all, we may assume without loss of generality that $f(0, 0) = I$ (and hence also, $f_1(0) = I$). If f is not a normalized function, we may replace it by $f^* = [f(0, 0)]^{-1}f$, realize f^* as shown below, and finally incorporate the unnormalizing factor $f(0, 0)$ into the first component function, $f_1(x_1)$.

Thus, we seek solutions for τ, ρ, f_1 and f_2 in the equations:

$$\begin{aligned} f(0, 0) &= f_1(0)f_2(0) &= I \\ f(0, 1) &= f_1(0)\tau f_2(0)\rho = \tau\rho \\ f(1, 0) &= f_1(1)f_2(1) \\ f(1, 1) &= f_1(1)\tau f_2(1)\rho \end{aligned} \tag{4}$$

Now let,

$$w = [f(1, 0)]^{-1}f(1, 1)[f(0, 1)]^{-1} \tag{5}$$

This clearly represents an element of H. Moreover, in terms of the sought-for representation, we have

$$w = [f_2(1)]^{-1}\tau f_2(1)\tau^{-1} \tag{6}$$

At this point the hypotheses regarding the properties of the automorphism U_τ come into play; because, by hypothesis and Lemma 2.1, Eq. (6) must have a unique solution $f_2(1)$ in H, for each possible value of w in H. Moreover, we can also find an element $f_1(1)$ in H, such that $f(1, 0) = f_1(1)f_2(1)$, namely, the element $f_1(1) = f(1, 0)f_2(1)^{-1}$. The remaining unknown quantity, ρ, in Eq. (4) is found to be

$$\rho = \tau^{-1}f(0, 1) \qquad (7)$$

and is consequently in G along with τ. Note that it is *essential* that the element τ be in G and not in H because if τ were in H, it would necessarily commute with some elements of H (e.g., τ and its powers) so that the automorphism U_τ would leave these (nonidentity) elements of H fixed, contrary to hypothesis.

We have demonstrated the sufficiency of the hypothesis to guarantee the existence of the four-cell decomposition (Eq. (3)) for $f(x_1, x_2)$. Consider now the converse assertion: that $f(x_1, x_2)$ will have such a decomposition only if some element τ in G makes U_τ an automorphism of H leaving no point in $H - \{I\}$ fixed (Property P). This is clearly not true if only a single, particular group function f is to be decomposed. For example, the constant function $f(x_1, x_2) = I$ has the trivial decomposition with $\tau = \rho = I$ and $f_1(x_1) = I$. However, the converse to be proved asserts that if *every* function $f: X^2 \to H$ has a decomposition (Eq. (3)), then τ exists in G with the desired Property P. We have actually demonstrated this already in the course of solving the Eqs. (4) which result from Eq. (3). For, if Eqs. (4) are to have solutions for arbitrary f, then Eq. (6) must have a solution for all possible[6] values of w in H. But this means that there must exist an element τ in G such that the mapping $h \to h^{-1}\tau h \tau^{-1}$, $h \in H$, is one-to-one. By Lemma 2.1, this is equivalent to Property P for τ.

We have just shown that four cells are necessary and sufficient, in a certain sense, for cascade decomposition of 2-variable group functions over a certain class of groups. Later, we study the question of characterizing those groups for which this holds. At the moment, we wish to consider group functions $f: X^m \to H$ of an arbitrary number m of Boolean variables. Theorem 2.2 shows that the four-cell decomposition iterates nicely to yield a canonical decomposition for such f's.

THEOREM 2.2.   Let $f: X^m \to H$ be an arbitrary group function into a subgroup H of a finite group G. If G contains an element τ with the Property P

---

[6] Observe that the definition for w, Eq. (5), is sufficiently general that each value for w in H can actually be realized with some group function f.

that $U_\tau : h \to \tau h \tau^{-1}$ is an automorphism of H leaving no point of $H - \{I\}$ fixed, then f is k_m-decomposable[7] over G, where $k_m = 2^m + 2^{m-1} - 2$.

Proof: Write $x = (x_1, \ldots, x_{m-1}, x_m) = (\hat{x}, x_m)$, where $\hat{x} = (x_1, \ldots, x_{m-1}) \in X^{m-1}$. The proof is by induction on m. Assume the theorem is true for $m - 1$; then the decomposition process used in proving Theorem 2.1 shows that we may write

$$f(x) = F_1(\hat{x})\tau^{x_m}F_2(\hat{x})\rho^{x_m} \tag{8}$$

where $F_2(\hat{x})$ is found, independently for each value of $\hat{x}$ in X^{m-1}, from the equation

$$F_2(\hat{x})^{-1}\tau F_2(\hat{x})\tau^{-1} = w(\hat{x})$$

and where $w(\hat{x})$ is defined as:

$$w(\hat{x}) = [f(\hat{x}, 0)]^{-1}f(\hat{x}, 1)[f(0, 1)]^{-1} \tag{9}$$

in H^*, where $\mathbf{0}$ denotes the vector $\hat{x}$ for which all components are zero.
   The other quantities are also determined, as in Theorem 2.1, to be

$$F_1(\hat{x}) = f(\hat{x}, 0)[F_2(\hat{x})]^{-1}$$
$$\rho = \tau^{-1}f(\mathbf{0}, 1)$$

   By the inductive hypothesis, we may decompose each of the group functions F_1 and F_2 over G into a product (composition) of $k_{m-1} = 2^{m-1} + 2^{m-1} - 2$ elementary functions. Thus, we have a total of $2k_{m-1} + 2$ elementary factors in the decomposition of the original m-variable function, f. Furthermore, $2k_{m-1} + 2 = k_m$ as required.
   The theorem is true for $m = 2$ by Theorem 2.1 since $k_2 = 4$. Since we have shown it to be true for m if it holds for $m - 1$, this completes the induction.

C. CHARACTERIZATION OF DECOMPOSABLE GROUPS

   Next, we consider the question of characterizing those finite groups which lend themselves to the decomposition of group functions as in Theorems 2.1 and 2.2. As Lemma 2.1 and the two theorems clearly show, a group H will be decomposable over *some* group G (of which H is a subgroup) only if H possesses an outer automorphism U which leaves no element of H fixed except the identity element. It follows that whenever H, for example, has no outer automorphisms, then *some* functions $f: X^m \to H$ will be indecomposable over

---

[7] We shall refer to this kind of decomposition hereafter as a Yoeli–Turner decomposition.

every group G. Some groups H of this sort are readily identified. They are, of course, poor candidates for cascade synthesis of Boolean functions, but it is useful to know which groups to avoid. We formalize the above conclusions as follows:

LEMMA 2.2. Every group function from X^m into a finite group H is decomposable (as in Theorems 2.1 and 2.2) only if H possesses an outer automorphism U which leaves no element of H fixed except the identity.

DEFINITION. A group H is said to be *complete* if and only if (1) All automorphisms of H are *inner* automorphisms, i.e., are of the form $h \rightarrow aha^{-1}$ for some a in H; and (2) The only element k in H which commutes with *every* element h in H, $hk = kh$, is the element $k = I = $ identity.

The concept of a complete group is a standard one (see, e.g., Hall (1959)). The property of interest to us is part (1) of the definition, which excludes the presence of outer automorphisms. Lemma 2.3 follows directly from what we have already said:

LEMMA 2.3. If H is a complete group, then regardless of what group $G(G > H)$ we may consider, some functions $f: X^m \rightarrow H$ will fail to possess a Yoeli–Turner decomposition over G.

Thus, complete groups are indecomposable in a particularly strong sense. We shall refer to them as being *completely indecomposable*. In particular, the symmetric groups, S_n, are known to be completely indecomposable for $n \neq 6$. Other examples of such groups are discussed in Sec. II,D.

In general (i.e., except for complete groups) the question of decomposability is a function not only of the group H, but also of the group G over which the decomposition is to be carried out. That is, decomposability is a function of the group pair (G, H), with $G > H$. The following discussion is based largely on work of Stone and succeeds in characterizing those group pairs (G, H) for which Yoeli–Turner decompositions exist for all functions $X^m \rightarrow H$.

THEOREM 2.3. Let G and H be finite groups $(G > H)$, such that all group functions $f: X^m \rightarrow H$ have Yoeli–Turner decompositions over G. Then:
(1) There is a subgroup G_1 of G, such that H is a *normal* subgroup of G_1.
(2) Every group function $f: X^m \rightarrow H$ is Yoeli–Turner decomposable over G_1 as well as over G.
(3) The factor group G_1/H is cyclic.

(4) G_1 is a union of cosets of the form $\tau^i H$, each distinct power of τ in a different coset, where τ is an element of G such that $\tau H = H\tau$, and for all $\sigma \in H$, $\tau\sigma\tau^{-1} = \sigma$ if and only if $\sigma = I$.

Proof: If every group function $f: X^m \to H$ is decomposable over G, then Theorem 2.1 guarantees the existence of a $\tau \in G$ that serves as a candidate to satisfy part (4) of Theorem 2.3. We now show that every $\tau \in G$ that satisfies Property P (of Theorem 2.1) will satisfy part (4) of Theorem 2.3.

Select any τ that satisfies Property P. Element τ and all of its powers τ^i form a cyclic group of order r, where r is the smallest integer such that $\tau^r = I$. Every power of τ has the property that $\tau^i H = H\tau^i$.

Let G_1 be the group generated by τ and H. Clearly, $H < G_1 < G$. Consider the left cosets of G_1 relative to H. Two such cosets are $\tau^i H$ and $\tau^j H$. It is well known that any two left cosets relative to a subgroup are either disjoint or identical. We now show that $\tau^i H$ and $\tau^j H$ are identical if and only if $\tau^i = \tau^j$. It follows immediately that if $\tau^i = \tau^j$, then $\tau^i H = \tau^j H$. In the other direction, if $\tau^i H = \tau^j H$, then there must be an element h of H such that $\tau^i = \tau^j h$. Hence, $h = \tau^{i-j}$. But then $\tau h \tau^{-1} = h$, which (from Property P) implies that $h = I$. But then $\tau^i = \tau^j$.

So far we have proved that the distinct powers of τ generate distinct left cosets and that $G_1 \supseteq \bigcup_i \tau^i H$. To complete the proof of part (4) we have to show that the cosets generated by powers of τ exhaust G_1; i.e., $G_1 \subseteq \bigcup_i \tau^i H$. Since τ and H generate G_1, every element g of G_1 can be written as a product of the form $\tau^j h_a \tau^k h_b \tau^l h_c \cdots$. Since $\tau^i H = H\tau^i$ for all i, it follows that for every g there is an $h \in H$ such that g can be rewritten in the form $\tau^{j+k+1+\cdots} h$. Then, every g in G_1 is in some coset of the form $\tau^i H$, so that $G_1 \subseteq \bigcup_i \tau^i H$. This proves part (4) of Theorem 2.3.

Part (1) of Theorem 2.3 follows because the left cosets of G_1 relative to H all have the form $\tau^i H$; but $\tau^i H = H\tau^i$ for all i, so that the left cosets equal the right cosets. Hence H is a normal subgroup of G_1.

Part (2) follows because the Yoeli–Turner decomposition requires only those group functions derived from compositions of τ and H; but the domain of such functions lies in the group G_1, because G_1 is the group generated by τ and H.

Finally, Part (3) is easily derived from the fact that the factor group G_1/H (the group of cosets of G_1 relative to H under coset multiplication), is isomorphic to the cyclic group of order r, where r is the order of τ in G. Distinct cosets of G_1 relative to H are associated with distinct powers of τ; furthermore, under coset multiplication, $(\tau^i H) \cdot (\tau^j H) = \tau^{i+j} H$, so that the mapping that carries $\tau^i H$ into τ^i establishes the isomorphism between G_1/H and C_r, the cyclic group of order r.

This concludes the proof of Theorem 2.3.

D. SURVEY OF FINITE GROUPS WITH RESPECT TO DECOMPOSABILITY

1.   The Symmetric Groups, S_n

The symmetric groups S_n, $n \geq 3$ are (with the exception of S_6) known to be *complete groups* (see Burnside (1955)). Hence, they have only inner auto-morphisms, and are therefore strictly indecomposable. The group S_2 (of order 2) is isomorphic to the cyclic group C_2, which is shown to be strictly indecomposable below.

The case of S_6 is still somewhat clouded. It is known (Burnside (1955)) that S_6 has 6! outer automorphisms (in addition to its 6! inner automorphisms), and could, therefore, conceivably be decomposable. However, it is strongly suspected that every such outer automorphism leaves some element of $S_6 - \{I\}$ fixed, in which case, S_6 would also be shown to be indecomposable.

2.   The Alternating Groups, A_n

The group A_n is the group of all $\frac{1}{2}n!$ even permutations of n objects. It is a normal subgroup of S_n. A_2 is the trivial group of order one, and is (trivially) decomposable. A_3 is isomorphic to the cyclic group C_3, shown to be decom-posable below.

For $n \geq 4$, and $n \neq 6$, the following argument shows that A_n is strictly indecomposable.

For $n \neq 6$ it is known that the automorphism group $\mathscr{A}(A_n)$ is isomorphic to S_n, with each distinct element τ in S_n inducing a distinct automorphism $U_\tau : a \rightarrow \tau a \tau^{-1}$ of A_n. Hence, these $n!$ automorphisms account for all auto-morphisms of A_n. Thus, if $A_n(n \neq 6)$ is decomposable at all, it must be decom-posable over S_n. However, for $n \geq 4$, it can be shown (by a rather lengthy argument due to Elspas *et al.* (1967)) that each automorphism of A_n leaves some element of $A_n - \{I\}$ fixed. Thus, for $n \geq 4$ and $n \neq 6$, A_n is strictly indecomposable, and A_6 is (at least) indecomposable over S_6 (possibly strictly indecomposable).

3.   The Cyclic Groups, C_n

LEMMA 2.4.   Let $C_n = \{I, a, a^2, \ldots, a^{n-1}\}$ be the cyclic group of order n. Then C_n is decomposable over the dihedral group D_n of order $2n$, provided n is odd. When n is even, C_n is strictly indecomposable.

Proof: Since $C_n = \{I, a, a^2, \ldots, a^{n-1}\}$ is abelian, all its automorphisms are outer ones. In fact, every mapping, $a \rightarrow a^t$ with t relatively prime to n, defines an (outer) automorphism of C_n, and all automorphisms of C_n are found in this way.

If n is even, say, $n = 2q$, then t must be odd, $t = 2s + 1$. Then the element, a^q is mapped into $(a^q)^t = a^{q(2s+1)} = (a^{2q})^s a^q = a^q$, i.e., into itself. Thus, all automorphisms of C_{2q} leave $a^q \neq I$ fixed, showing that C_{2q} is strictly indecomposable.

Next, suppose that n is odd. The mapping $x \rightarrow x^{-1}$ is easily seen to be an automorphism of C_n. Moreover, if $(a^i)^{-1} = a^i$, then $a^{2i} = I$, i.e., $2i \equiv 0$ (mod n). Since n is odd, this can happen only for $i \equiv 0$. Hence, only the identity is left fixed by this automorphism.

We have shown that a suitable (outer) automorphism exists for n odd. The next step is to find a group G in which C_n is a subgroup, and to find an element τ in G such that the mapping $x \rightarrow \tau x \tau^{-1}$ is the desired automorphism $x \rightarrow x^{-1}$ of C_n. To do this, consider the Cayley representation of C_n, whereby the generator a of C_n takes the form of the permutation $(0, 1, 2 \cdots n - 1)$. Let τ be the permutation (a product of transpositions), $\tau = (0)(1, n - 1)(2, n - 2) \cdots ((n - 1)/2, (n + 1)/2)$. A simple calculation shows that $\tau a \tau^{-1} = a^{-1}$, hence also $\tau x \tau^{-1} = x^{-1}$ for all $x = a^i$. The group G generated by C_n and τ is a subgroup of S_n, and it is of order $2n$, since $\tau^2 = I$. This group has the two generators, a and τ, satisfying the relations $a^n = \tau^2 = I$ and $\tau a = a^{-1}\tau$. It is known as the *dihedral group* D_n of order $2n$ (the group of rotations and reflections of a regular n-sided polygon). This completes the proof of the lemma.

Observe that since $A_3 \cong C_3$ and $S_3 \cong D_3$, we have shown that A_3 is decomposable over S_3. Likewise, since $S_2 \cong C_2$, this group is strictly indecomposable.

4.   The Dihedral Groups, D_n

The dihedral groups, D_n, introduced above, may themselves be considered for decomposability. Although the details are omitted here (see Elspas *et al.*, 1967) it is not difficult to show that D_n is strictly indecomposable for all $n \neq 2$. The group D_2 is isomorphic to $C_2 \times C_2$ (also known as the Klein group). The decomposability of the Klein group (over A_4) was first demonstrated by Yoeli and Turner (1966), and it is dealt with in detail in Section III,B of this chapter.

5.   Other Indecomposable Groups

Other well-known families of finite groups which can be shown to be strictly indecomposable are: the *dicyclic groups*, of order $4m$, defined by the relations

$$A^{2m} = I, \qquad A^m = (AB)^2 = B^2$$

(the case $m = 2$ gives the quaternion group of order 8); the *metacyclic* groups of order $p(p - 1)$, where p is an odd prime, generated by S and T, where

$$S^p = T^{p-1} = I, \qquad ST = TS^k$$

and k is a primitive root modulo p.

6. Survey of all Groups Through Order Twelve

Table I below summarizes the decomposability properties of all groups of order twelve or less (comprising 23 groups, not counting the trivial 1-element group, which is, of course, decomposable over itself). This table embodies the

TABLE I
Decomposability of Groups Through Order Twelve

ORDER	GROUP H	TYPE	$\mathscr{A}(H)$	$\mathscr{I}(H)$	DECOMPOSES OVER G	$|G|$
2	$C_2 \cong S_2$	Abelian, cyclic	I	I	Indecomposable	—
3	$C_3 \cong A_3$	Abelian, cyclic	C_2	I	$D_3 = S_3$	6
4	C_4	Abelian, cyclic	C_2	I	Indecomposable	—
	$K \cong C_2{}^2 \cong D_2$	Abelian	S_3	I	A_4	12
5	C_5	Abelian, cyclic	C_4	I	D_5	10
6	$C_6 \cong C_2 \times C_3$	Abelian, cyclic	C_2	I	Indecomposable	—
	$S_3 \cong D_3$	Nonabelian	S_3	S_3	Completely indecomposable	—
7	C_7	Abelian, cyclic	C_6	I	D_7	14
8	C_8	Abelian, cyclic	$C_2{}^2$	I	Indecomposable	—
	$C_4 \times C_2$	Abelian	D_4	I	Indecomposable	—
	D_4	Nonabelian (dihedral)	D_4	$C_2{}^2$	Indecomposable	—
	Q_2	Nonabelian (quaternion)	$S_3 \times C_2{}^2$	$C_2{}^2$	Indecomposable	—
	$C_2{}^3$	Abelian	$L[V_3(2)]$	I	$<$ Affine group on V_3	56
9	C_9	Abelian, cyclic	C_6	I	D_9	18
	$C_3{}^2$	Abelian	$L[V_2(3)]$	I	$D_3{}^2$	36
10	C_{10}	Abelian, cyclic	C_4	I	Indecomposable	—
	D_5	Nonabelian (dihedral)	$\mathscr{A}(D_5)^a$	D_5	Indecomposable	—
11	C_{11}	Abelian, cyclic	C_{10}	I	D_{11}	22
12	C_{12}	Abelian, cyclic	$C_2{}^2$	I	Indecomposable	—
	$C_2{}^2 \times C_3 \cong C_2 \times C_6$	Abelian	$C_2 \times S_3$	I	$A_4 \times D_3$	72
	D_6	Nonabelian (dihedral)	D_6	S_3	Indecomposable	—
	Q_3	Nonabelian	?	?	Indecomposable	—
	A_4	Nonabelian (alternating)	S_4	A_4	Indecomposable	—

[a] $\mathscr{A}(D_5)$ is the metacyclic group of order 20.

decomposability results on cyclic groups and direct products given elsewhere in this section.

Table I also includes some pertinent information on the automorphism group $\mathscr{A}(H)$ and the inner automorphism group $\mathscr{I}(H)$, for most of the groups listed. Notation follows the scheme: S_n = symmetric group of order $n!$; A_n = alternating group of order $n!/2$; C_n = cyclic group of order n; D_n = dihedral group of order $2n$; Q_n = dicyclic (generalized quaternion) group of order $4n$; K = Klein four-group = $C_2 \times C_2$; $L[V_r(q)]$ = linear group on the vector space of dimension r over $GF(q)$; $\mathscr{A}(H)$ = the group of all automorphisms of H; $\mathscr{I}(H)$ = the group of all inner automorphisms of H; $|G|$ = the order of G.

7.   The Groups $C_2{}^r$

As we shall see in Section III,C, the groups $C_2{}^r$ (i.e., the direct products of r copies of C_2) play a central role in the synthesis of r-rail binary cascades. We shall defer detailed analysis of these groups until that point, so that we may tie the mathematics in directly with the switching application. However, for the sake of completeness, we state the pertinent result here.

The group $C_2{}^r$ is decomposable for all values of $r > 1$. We have already seen that C_2 is indecomposable, above.

8.   Theorems on Decomposability of Direct Products

LEMMA 2.5.   If group H_1 is decomposable over G_1 and group H_2 is decomposable over G_2, then the direct product $H_1 \times H_2$ is decomposable over $G_1 \times G_2$.

Proof: Let $U_i : h_i \to \tau_i h_i \tau_i^{-1}$; $i = 1, 2$ be the respective automorphisms of H_1 and H_2, whose existence is guaranteed by Lemma 2.2. We have τ_i in G_i for $i = 1, 2$; moreover, $U_1(h_1) = h_1$ only if $h_1 = I$ in H_1, and likewise for U_2, H_2, and h_2. We define $U_1 \times U_2 : H_1 \times H_2 \to H_1 \times H_2$ as the mapping that takes (h_1, h_2) in $H_1 \times H_2$ into the element $[U_1(h_1), U_2(h_2)]$ in $H_1 \times H_2$. It is easily seen that $U_1 \times U_2$ is an automorphism of $H_1 \times H_2$ that is induced by the element (τ_1, τ_2) in $G_1 \times G_2$. Moreover, $[U_1(h_1), U_2(h_2)] = (h_1, h_2)$ only if $h_1 = h_2 = I$, so that $U_1 \times U_2$ has no fixed points in $H_1 \times H_2$ except the identity element. Hence, $H_1 \times H_2$ is decomposable over $G_1 \times G_2$ by Theorem 2.2.

The reader is cautioned not to assume the converse of this theorem to be true. The group $C_2 \times C_2$, though decomposable, is the product of two indecomposable groups. Similarly, $C_2 \times (C_2{}^2 \times C_3)$ is the product of an indecomposable group (C_2) and a decomposable one. But, it can also be written as $C_2{}^3 \times C_3$, which is decomposable by the above theorem.

Nevertheless, Lemma 2.5 taken together with our previously stated results on the decomposability of $C_2{}^r$ for $r \geq 2$, and of C_n for n odd, shows that the following result holds:

THEOREM 2.4.   The group $C_2{}^r \times H_0$ is decomposable for $r \geq 2$, $H_0 =$ product of odd cyclic groups. It is indecomposable for $r = 1$.

Proof: As already stated, the positive assertion follows from Lemmas 2.4 and 2.5. The direct product of C_2 with any product H_0 of odd cyclic groups is indecomposable because H_0 (being of odd order) has no element of order two. Since all elements are of the form h, or bh (with h in H_0 and $b \neq I$ in C_2), the only element in $C_2 \times H_0$ of order two is b itself. But the order of an element is preserved under automorphism. Thus, all automorphisms of $C_2 \times H_0$ map b into itself. By Lemma 2.2, then, $C_2 \times H_0$ is indecomposable.[8] Incidentally, this establishes the following corollary:

COROLLARY 2.1.   For every integer $n \not\equiv 2 \bmod 4$, there exists at least one decomposable abelian group of order n.

Proof: For $n \equiv 1$ or $3 \bmod 4$, i.e., for odd n, C_n is decomposable by Lemma 2.4. For $n \equiv 0 \bmod 4$, let $n = 2^r q$, where q is odd, and $r \geq 2$. Then $C_2{}^r \times C_q$ is an abelian group that is decomposable by the above theorem.

We thus have available to us decomposable groups of all odd orders and all orders divisible by 4. We have not been able to find even one example of a decomposable group of singly-even order. Nor have we been able to prove that none exist, although not much attention was given to this question.

Also surprising is the following observation: All of the decomposable groups we have been able to find are abelian. There seems, however, to be no fundamental reason why this property should be necessary for decomposability.

---

[8] Essentially the same argument suffices to show the stronger result that $C_2 \times H_1$ is indecomposable for any group H_1 of odd order (whether abelian or not).

EXERCISES

2. Investigate the group S_6. Try to show that each of the 6! outer automorphisms of S_6 leaves some element of $S_6 - I$ (or even of $A_6 - I$) fixed.

3. Prove that D_n is strictly indecomposable when $n \neq 2$. (Hint: Consider separately the cases n odd and n even. The even case is easy.)

4. Show that $C_2 \times C_4$ is (strictly) indecomposable as asserted in Table I. Since this fact does not follow from any of the general theorems proved above, you will have to go through a detailed analysis of the automorphisms of this group. (Hint: represent the group by means of the generators, A and B, where $A^4 = B^2 = I$ and $AB = BA$; find all eight automorphisms in terms of their action on A and B; then find an element of $C_2 \times C_4$ left fixed under all automorphisms. Use the fact that the order of an element is fixed under any automorphism.)

5. Show that the automorphism group found in the preceding exercise is isomorphic to D_4.

6. Prove the assertion left undemonstrated in Section II,D,2 about A_n, viz that for each automorphism of A_n, $n \geq 4$, there exists an element of A_n (other than the identity permutation) which is fixed under that automorphism.

III. SYNTHESIS OF MULTIRAIL CASCADES

A. INTRODUCTION—THE NOTION OF PERMUTATION CASCADE

In Section III we shall apply the theory of decomposition of group functions developed in Section II to the synthesis of multirail, multifunction cascade networks. An indication of one way to make such application was given in Section I,B, where the approach of Yoeli and Turner (1966) to this problem was summarized. As noted there, the role played by group functions (i.e., by their hardware counterpart, which we call *permutation cascades*) is simply to provide a recoding of rail states so that a subsequent Ashenhurst decomposition can extract a single Boolean variable. The cells involved in realizing the Ashenhurst decomposition are necessarily general-function cells (e.g., for the case of two rails they must have general function capability for any 3-input, 2-output function). This seems a high price to pay for a small (asymptotic) reduction in the cascade length relative to that achievable with the permutation cascades to be described here.

At first glance it may seem surprising—even paradoxical—that synthesis of arbitrary m-variable Boolean functions might be achievable with cascades of that special type we have called *permutation cascades*. We repeat and amplify the definition from Section II,A.

DEFINITION.   A *permutation cascade* network is a cascade of logical cells, each cell's action being under the control of a single binary variable x_i, where this action is the transformation of an input state y (on r binary rails) to an output state y' (on r rails). The mapping $y \rightarrow y'$ is required (for each fixed x_i) to be a *permutation* of the set of 2^r rail states.

Thus, permutation cascades are special, in that each cell performs (for fixed vertical input x_i) a group function (permutation) of the possible input states. So, for example, a cell with constant output (independent of its input rail signals) is ruled out, since this is not a one-to-one function (permutation). A cascade connection of such cells is likewise restricted to one-to-one mappings (permutations) of the input rail states into the output rail states. How, then, can we claim logical completeness (generality) for such networks? The answer to the apparent contradiction lies in the fact that we are not interested in whether or not the overall function is a permutation between rail inputs and rail outputs for fixed vertical inputs. We *are* interested in the dependence of the rail outputs on the vertical inputs for fixed rail inputs. As we shall see shortly, we can easily guarantee that this latter functional dependence is arbitrary, leading to synthesis of arbitrary sets of Boolean functions of the vertical inputs. Thus, the crux of the matter is that we be willing to give up the use of the ultimate rail inputs as "live" inputs on a par with the vertical inputs x_i. Instead, we tie the left-hand rail inputs to a convenient source of logical constants, thus providing a boundary condition on the rail signals. This matter will be further elucidated below when we carry through application of the Yoeli–Turner decomposition to a number of different groups under different rail-state coding assumptions.

In the remainder of this section we shall apply the Yoeli–Turner decomposition first to the two-rail case and then to its natural generalization for an arbitrary number of rails, using the affine group on the rail states. Finally, we consider the use of other groups, whose order is not equal to a power of 2, together with alternative encodings of the rail states into group elements. The realization of incompletely specified Boolean functions (the "don't cares" case) is treated in this context.

B. TWO-RAIL BINARY CASCADES

It is desirable, for both historical and pedagogical reasons, to discuss two-rail cascades separately before advancing to the general, r-rail, case. The first multioutput synthesis technique of any generality for the binary two-rail cascades was developed by Yoeli and Turner (1966). The earlier work of Short (1965) was concerned only with single-output function synthesis as discussed in Section I,B. It is essentially the Yoeli–Turner technique and its underlying mathematical theory (based on permutation groups) that we cover in this section. A few simplifications have been introduced to make the material more easily assimilable by the student, and also to enhance the relationship with the general theory for multirail cascades to be given in Section III,C.

It is desired to realize an arbitrary pair of Boolean functions $F_1(x_1, \ldots, x_m)$, $F_2(x_1, \ldots, x_m)$ of m Boolean variables $(x_1, \ldots, x_m) = x \in X^m$. That is, we wish to realize the two mappings $F_i : X^m \to X$, $i = 1, 2 \cdots$ or, what amounts to the same thing, a single mapping $F = (F_1, F_2) : X^m \to X^2$. In order to achieve this realization, we identify a certain group of order four and place its four elements in one-to-one correspondence with the four rail-states $0 = (0, 0)$, $1 = (0, 1)$, $2 = (1, 0)$, and $3 = (1, 1)$, i.e., the vertices of the Boolean 4-cube, X^2. A natural group of order four for this purpose turns out to be the Klein 4-group, $K = \{\sigma_0, \sigma_1, \sigma_2, \sigma_3\}$, defined by the relations $\sigma_i \sigma_j = \sigma_j \sigma_i$, $\sigma_0 = I = \sigma_i^2$, $\sigma_1 \sigma_2 = \sigma_3$, $\sigma_2 \sigma_3 = \sigma_1$, and $\sigma_3 \sigma_1 = \sigma_2$. Thus, K is isomorphic to the additive group (modulo two) on the vectors (y_1, y_2); $y_i = 0, 1$, and $K \cong C_2 \times C_2$ (where C_2 is the cyclic group of order two).

We associate with each of the four rail-states $j = 0, 1, 2, 3$, a corresponding element σ_j in K. That is, if we regard K as a permutation group acting on the four rail-states, we have the (Cayley) representation:

$$\sigma_0 = (0)(1)(2)(3)$$
$$\sigma_1 = (01)(23)$$
$$\sigma_2 = (02)(13)$$
$$\sigma_3 = (03)(12)$$

Thus, rail-state j is associated with that permutation σ_j which permutes[9] rail-state $0 = (0, 0)$ into rail-state j. In terms of the truth tables for the output

---

[9] It is important to note that there must be at least one permutation σ in the group K that permutes rail-state 0 into each arbitrary state j. Thus, the group must be transitive on the rail-states. The group chosen, K, is in fact *simply transitive* (there is *exactly one* element that permutes any state i into state j).

functions F_1 and F_2, this means that we may replace each tabular entry $[F_1(x), F_2(x)]$ with the corresponding element of K. For example, the truth table

x_3	x_2	x_1	F_1	F_2	$\sigma = f(x)$
0	0	0	0	0	σ_0
0	0	1	0	1	σ_1
0	1	0	0	1	σ_1
0	1	1	1	0	σ_2
1	0	0	1	1	σ_3
1	0	1	1	0	σ_2
1	1	0	0	0	σ_0
1	1	1	0	1	σ_1

shows how a particular pair of three-variable Boolean output functions is made to correspond to a group function f. In general, if for minterm $x = (x_1, \ldots, x_m)$ we have output function values $F_1(x), F_2(x)$ forming the output rail state

$$j(x) = [F_1(x), F_2(x)] \in \{0, 1, 2, 3\}$$

then

$$f(x) = \sigma_{j(x)} \in K$$

Thus, in order to realize the functions F_1, F_2 in cascade form, we need merely decompose the group function f in a permutation cascade, apply input rail-state $(0, 0)$ at the left, and the output rail-state will be (F_1, F_2) as desired.

By Theorem 2.2 we know that an outer automorphism of K must be found that leaves no element of $K - \{I\}$ fixed. The mapping defined by $\sigma_0 \to \sigma_0, \sigma_1 \to \sigma_3, \sigma_2 \to \sigma_1, \sigma_3 \to \sigma_2$ is such an automorphism (it is outer because K, being abelian, has no nontrivial inner automorphisms). There is one other such automorphism of K, viz., the inverse of the above one, permuting the nonidentity elements in the cycle (123) instead of (132). The other automorphisms of K comprise the identity and the three that leave one element besides I fixed, i.e., (23), (31), and (12), in the above notation. The automorphism group of K is thus a group of order six; it is easily seen to be isomorphic to S_3, the symmetric group on three objects.

Fixing on the automorphism which permutes (132) in that order as the one suitable for our purposes, we next identify a group G of which K is a subgroup, and an element τ in G such that the inner automorphism (of G) corresponding to τ is the outer automorphism U_τ of H to be used in the Theorem 2.2 decomposition. The group G may be chosen to be S_4 (although this is not the smallest group suitable for this purpose, as we shall see). The element τ may then be selected[10] to be $\tau = (0)(123)$, since then $U_\tau : \sigma \to \tau\sigma\tau^{-1}$ is the desired automorphism. Now, since τ is an element of order three in S_4, the group G_1 generated by K and τ together is only of order $12 = 3 \times 4$. In fact, G_1 must be the alternating group A_4, since both τ and K consist of even permutations. It follows that all group functions from X^m into K are decomposable over A_4 (a result first obtained by Yoeli and Turner (1966)).

The detailed general design equations given in Theorem 2.2 take a considerably simpler form here, because (1) the group $H = K$ is abelian, and (2) all elements of K are self-inverse, so that exponents "-1" disappear on K operations. Thus, $F_2(\hat{x})$ is obtained by solving the equation

$$F_2(\hat{x}) \cdot (123) \cdot F_2(\hat{x}) \cdot (132) = w(\hat{x})$$

where

$$w(\hat{x}) = f(\hat{x}, 0)f(\hat{x}, 1)f(0, 1)$$

The other quantities, at each stage of the reduction, are found from

$$F_1(\hat{x}) = f(\hat{x}, 0)F_2(\hat{x})$$

$$\rho = (132) \cdot f(0, 1)$$

The solution of the equation for $F_2(\hat{x})$, which might appear to be the most difficult part of the whole algorithm, is actually quite simple once $w(\hat{x})$ has been evaluated. A simple calculation shows that the mapping $\sigma \to \sigma \cdot (123) \cdot \sigma \cdot (132)$ takes $\sigma_0 \to \sigma_0$, $\sigma_1 \to \sigma_2$, $\sigma_2 \to \sigma_3$, and $\sigma_3 \to \sigma_1$, and solving for $\sigma = F_2(\hat{x})$ thus involves a table look-up in a 4-entry table.

We next carry through the details of the synthesis procedure for realizing the particular 3-variable, 2-rail function specified earlier in this section. In doing this we make use of an abbreviated notation where members of the group $K = \{\sigma_0, \sigma_1, \sigma_2, \sigma_3\}$ are denoted simply by their corresponding integer subscripts, 0, 1, 2, or 3. All of the calculations are in K except for determining

---

[10] This choice is, of course, not unique. The reader may wish to examine for himself in detail what the other possibilities are.

the element ρ at each stage of the reduction, therefore this abbreviated notation causes no ambiguity. Multiplication of elements in K then follows the rules:

$$1 \cdot 2 = 3, \quad 2 \cdot 3 = 1, \quad 3 \cdot 1 = 2, \quad j \cdot j = 0, \text{ etc.}$$

$(x_3 x_2 x_1)$	$f(\hat{x})$	$f(\hat{x}, 0)f(\hat{x}, 1)$	$w(\hat{x})$	$F_2(\hat{x})$	$F_1(\hat{x})$
0	0	3	0	0	0
1	1	3	0	0	1
2	1	1	2	1	0
3	2	3	0	0	2
4	3	$f(0, 1) = \sigma_3 = (03)(12)$			
5	2				
6	0	$\rho = (132)(03)(12) = (103)(12)$			
7	1				

$x_2 x_1$	F_2	$F_2(x_1, 0)F_2(x_1, 1)$	$w(x_1)$	$F_{22}(x_1)$	$F_{21}(x_1)$
0	0	1	0	0	0
1	0	0	1	3	3
2	1	$\rho_2 = (132)(01)(23) = (120)(3)$			
3	0				

$x_2 x_1$	F_1	$F_1(x_1, 0)F_1(x_1, 1)$	$w(x_1)$	$F_{12}(x_1)$	$F_{11}(x_1)$
0	0	0	0	0	0
1	1	3	3	2	3
2	0	$\rho_1 = (132)I = (132)(0)$			
3	2				

The cascade network resulting from this synthesis procedure is shown in symbolic form in Fig. 4. Here a symbolic notation for cascade cells has been introduced, whereby a diagonal slash across a cell separates the group function the cell performs when the vertical input is zero (shown above the slash) from the group function it performs when the vertical input is one (shown below the slash). Two kinds of cells are indicated. Square boxes designate cells performing permutations in K; round cells designate operations in the larger

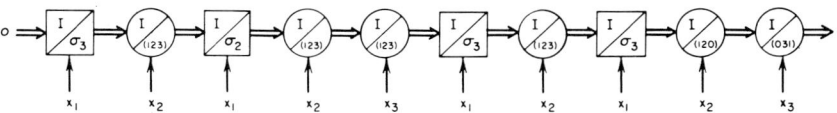

FIG. 4.   Cascade realization of a particular two-rail function.

group A_4. The A_4 operations are written out as permutations on the rail-states 0, 1, 2, 3, while the K operations are shown simply in the form, σ_0, σ_1, σ_2, or σ_3. In this particular network, all the cells happen to be normalized cells (i.e., the element I lies above the slash in each case). The desired output is realized with left-hand rail input (0, 0). The only cell which might happen to come out nonnormalized is that performing F_{11}. In that event, F_{11} could be normalized by altering the rail-inputs supplied to it accordingly.

Having achieved the synthesis of this particular two-rail cascade by way of illustrating the general procedure, it is pertinent to ask some questions: (1) How many different types of cells are required to realize any two-rail function in cascade form? (2) How complex are these cells?

Clearly, there are two *general* types of cells involved, as we have already mentioned. First, there are the K-type cells, of which there are just four (one of them is the identity cell, which cannot really be counted as a "cell" but which we include just for the sake of uniformity). Second, there are the cells (in $A_4 - K$) which perform the functions $\tau = (123)$ and $\tau^{-1}f(0, 1) = (132) \cdot \sigma_j$, where $j = 0, 1, 2, 3$. Thus, there are five cells of this second type, making a total of nine cell types for synthesis of arbitrary two-rail functions in cascade form. Later we shall show that it is possible to eliminate one of the A_4-type cells to bring the total down to eight (still including the trivial (identity) cell). This saving becomes proportionately larger for the case of r-rail synthesis where $r > 2$, and it seems appropriate to defer the discussion until r-rail synthesis has been covered.

What do these two-rail cells "look like" in terms of logical hardware? The group-theoretic description is still too abstract to permit one to evaluate this by inspection. So let us have a closer look, using the specific encoding of rail-states introduced above. First, the K cells. Consider the (normalized) cell realizing σ_3. When its vertical input is active this cell performs the following transformation on the rail states:

Input	Output
(0, 0)	(1, 1)
(0, 1)	(1, 0)
(1, 0)	(0, 1)
(1, 1)	(0, 0)

since $\sigma_3 = (03)(12)$. Hence, in the active mode this cell simply complements both rail signals. In the inactive mode, it passes the rail signals unaltered. Hence, this cell consists of two EXCLUSIVE OR gates, as shown in Fig. 5c. Likewise, the realizations for the (normalized) cells σ_1 and σ_2 simply complement one or the other of the two rail signals, as shown in Fig. 5a–b.

Consider now the cells in $A_4 - K$; e.g., the normalized cell for the permutation $\tau = (123)$ which has the state-permutation table (for $x = 1$):

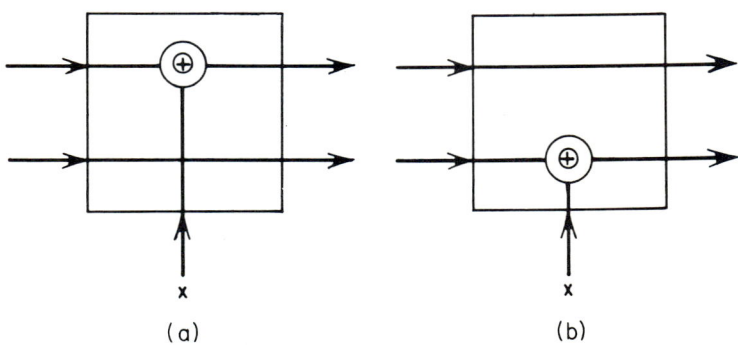

(a)   (b)

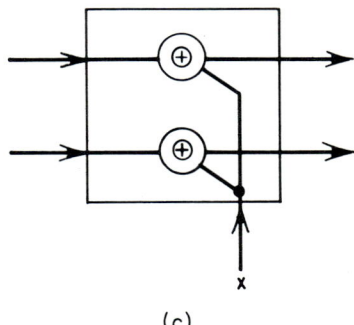

(c)

FIG. 5.   K-type affine two-rail cells. (a) Cell σ_1. (b) Cell σ_2. (c) Cell σ_3.

Input		Output	
y_2	y_1	y_2'	y_1'
0	0	0	0
0	1	1	0
1	0	1	1
1	1	0	1

thus yielding the functions:

$$y_1' = \bar{x}y_1 + xy_2$$
$$y_2' = \bar{x}y_2 + x(y_1 \oplus y_2) = y_2 \oplus xy_1$$

The other four required permutations in $A_4 - K$ are similarly found to have the input/output functional expressions:

$$\tau^{-1} = (132): \begin{cases} y_1' = y_1 \oplus xy_2 \\ y_2' = \bar{x}y_2 + xy_1 \end{cases}$$

$$\tau^{-1}\sigma_1 = (012): \begin{cases} y_1' = y_1 \oplus x\bar{y}_2 \\ y_2' = \bar{x}y_2 + xy_1 \end{cases}$$

$$\tau^{-1}\sigma_2 = (023): \begin{cases} y_1' = y_1 \oplus xy_2 \\ y_2' = \bar{x}y_2 + x\bar{y}_1 \end{cases}$$

$$\tau^{-1}\sigma_3 = (031): \begin{cases} y_1' = y_1 \oplus x\bar{y}_2 \\ y_2' = \bar{x}y_2 + x\bar{y}_1 \end{cases}$$

One sees that all five of these cells are very similar. They use only two symmetry types of 3-variable Boolean functions, in fact, they differ only in complementation of one or the other of the input variables, and interchange of the output variables. The whole cell for realizing τ, for example, requires only five logical gates: three ANDs, one OR, and one EXCLUSIVE OR, and

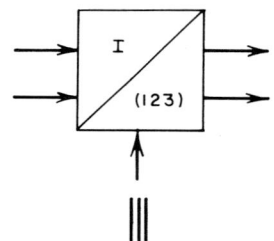

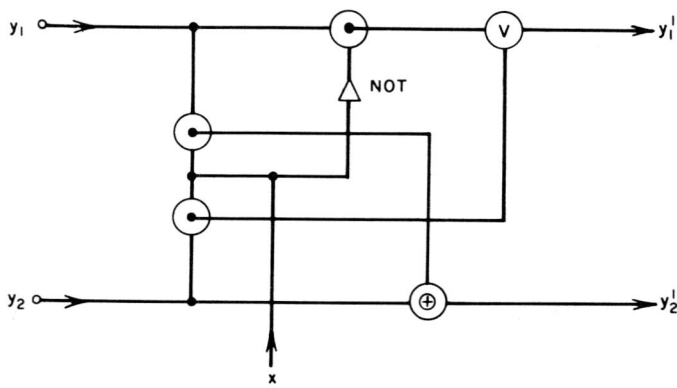

FIG. 6.   Typical A-type two-rail cells.

the other cells are the same except for possible inverters (depending on the technology used). See Fig. 6.

Later, in Section III,G, we will make some comparisons between the cell types (and cascade lengths) required above and those required for Short's synthesis of two-rail cascades.

C. MULTIRAIL SYNTHESIS

We next treat in detail the synthesis of multirail permutation cascades, using the Yoeli–Turner decomposition technique applied to group functions into the group $C_2{}^r$. This group turns out to be the natural generalization of the group K used for the two-rail case. The two-rail theory will then be seen to take its place as a special case of r-rails for $r = 2$.

The difficult part of the generalization (although it appears simple once the trick is revealed) is to guess at an appropriate group G, for which $C_2{}^r$ is a normal subgroup, and such that G contains an element τ inducing an automorphism $U_\tau : \sigma \to \tau\sigma\tau^{-1}$ which leaves no element of $C_2{}^r$ fixed, except for the identity in $C_2{}^r$. First, though, let us justify the selection of $C_2{}^r$. Following the basic viewpoint developed in the preceding section, we must choose for H, the range of our group functions, a group which is transitive on the set of rail states X^r. Hence, H must contain at least 2^r elements (at least one to correspond to each output rail state of the cascade). We may visualize X^r as the linear vector space of dimension r over GF(2), instead of simply as the Boolean r-cube. Thus, the rail state space is $V_r(2)$, and it consists of binary r-tuples which can be added (componentwise) by modulo two addition. The set K_r of all translations $y \to y + b$ of this vector space is an additive group isomorphic to $C_2{}^r$, and isomorphic to the additive group of $V_r(2)$ itself. This group is simply transitive on the 2^r rail states, since there is precisely one translation which takes a given r-tuple into another given r-tuple. In binary terms, this translation group, K_r, consists just of the selective complementations of the binary rail signals. So far, this is a clear generalization of the two-rail case.

Now, how does one choose a larger group G in which to embed K_r? The symmetric group on 2^r objects would certainly suffice but it is much too large and unwieldy for one to do anything with it. A clue is given by the automorphisms of K_r. Since $K_r \cong C_2{}^r$ is an abelian group, it has no inner automorphisms (except identity). Its (outer) automorphisms are the set L_r of all nonsingular linear transformations of the vector space $V_r(2)$, for any such nonsingular linear transformation can be regarded as a change of coordinates in the vector space, leaving all its abstract properties unchanged. However, the translations K_r do not form a subgroup of the linear group L_r;

we need to extend L_r to include translations by going over to the affine group, which allows as operations: $y \rightarrow Ay + b$, where A is a linear transformation on V_r and b is a (constant) vector in V_r; that is, the affine group, $\mathscr{A}_r$ contains linear transformations followed by translations. Clearly, $\mathscr{A}_r$ contains both L_r and K_r as subgroups. It is also well known that the group of translations is a normal subgroup of the affine group (over any vector space).

The preceding argument suggests (but does not prove) that it may be possible to find an appropriate element τ in $\mathscr{A}_r$ such that the mapping $U_\tau : \sigma \rightarrow \tau\sigma\tau^{-1}$ is an automorphism of K_r that leaves only the element I fixed in K_r.

LEMMA 3.1. K_r is a normal subgroup of the affine group $\mathscr{A}_r$, i.e., $\alpha^{-1}\sigma\alpha$ is in K_r for any α in $\mathscr{A}_r$ and any σ in K_r (see also Birkhoff and MacLane (1953)).

Let σ be an arbitrary element of K_r, and α be an arbitrary element of $\mathscr{A}_r$. We must show that $\alpha^{-1}\sigma\alpha$ is in K_r; that is, $\alpha^{-1}K_r\alpha = K_r$. If $y \in V_r$ and $\sigma(y) = y + c$, and $\alpha(y) = Ay + b$, then

$$\alpha^{-1}\sigma\alpha(y) = \alpha^{-1}\sigma(Ay + b) = \alpha^{-1}(c + Ay + b) = A^{-1}(c + Ay) = y + A^{-1}c$$

Thus, $\alpha^{-1}\sigma\alpha$ takes y into $y + c'$, where $c' =$ the fixed vector $A^{-1}c$, and hence $\alpha^{-1}\sigma\alpha$ is in K_r. (Note that $\alpha^{-1}\sigma\alpha$ is equal to σ only if $Ac = c$.)

The next lemma shows that a suitable automorphism-inducing element τ can be found, not only in $\mathscr{A}_r$, but even in L_r, the group of (homogeneous) linear transformations on V_r. Curiously enough, this proof has points of contact with the theory of binary shift-register sequence generators. See Elspas (1959).

LEMMA 3.2. There exists in L_r $(r \geq 2)$ an element τ having the following properties: (1) The mapping $U_\tau : \sigma \rightarrow \tau\sigma\tau^{-1}$ is an automorphism of K_r. (2) U_τ leaves no point of K_r fixed, except the identity.

Proof: U_τ is an automorphism of $\mathscr{A}_r$ for each choice of τ in $\mathscr{A}_r$. Also, each such mapping U_τ leaves K_r invariant by Lemma 3.1; hence, each such U_τ is an automorphism of the subgroup K_r of $\mathscr{A}_r$. This guarantees Property (1). In order to satisfy Property (2), a little care must be exercised in the choice of τ. Let T be any nonsingular matrix of order r over GF(2). Moreover, specify that the characteristic polynomial of T, i.e., the polynomial

$$\varphi(\lambda) = \det(\mathsf{T} - \lambda\mathsf{I}) = \lambda^r + \cdots + 1$$

does not have $\lambda = 1$ as a root. Thus, $\varphi(1) \neq 0$. Since our arithmetic is over the field GF(2), this means merely that $\varphi(\lambda)$ must have an odd number of (non-zero) terms, and thus $\varphi(1) = 1$. We can insure this, for example, by taking $\varphi(\lambda) = \lambda^r + \lambda + 1$ and choosing T to be the companion matrix of this polynomial. Then it follows that $y\mathsf{T} = y$ implies $y = 0$ (i.e., T does not have eigenvalue 1).

Now define τ to be that element of L_r which maps $y \in V_r$ into $y\mathsf{T}$; i.e., τ is the linear transformation on V_r represented by the matrix T. Suppose that for some element σ in K_r we had the relation $\mathsf{T}\sigma\mathsf{T}^{-1} = \sigma$. This would mean that for every y in V_r we would have

$$(y)\tau\sigma = (y)\sigma\tau$$

If σ is the element of K_r which maps $y \in V_r$ into $y + c$ (i.e., $(y)\sigma = y + c$), then $(y)\tau\sigma = y\mathsf{T} + c$, while $(y)\sigma\tau = (y + c)\mathsf{T} = y\mathsf{T} + c\mathsf{T}$. Hence, if $\tau\sigma = \sigma\tau$ we would have $y\mathsf{T} + c = y\mathsf{T} + c\mathsf{T}$, or $c\mathsf{T} = c$. But by the way T was chosen, this can happen only if $c = 0$, i.e., if $\sigma = \mathsf{I} =$ identity mapping of V. Hence, $\tau\sigma\tau^{-1} = \sigma$ with σ in K_r only if $\sigma = \mathsf{I}$. This concludes the proof.

Note that there was considerable freedom of choice regarding T in the above proof. We might equally well have taken $\varphi(\lambda)$ to be a *primitive* polynomial of degree r over GF(2). Then T corresponds to a maximal-period feedback shift register of r binary stages, counting to $2^r - 1$. This would in turn make τ an element of order $2^r - 1$ in L_r. That is, the cycle structure of τ considered as a permutation on 2^r objects would then be $(1, 2^r - 1)$, an even permutation. However, this choice has the drawback of making the group G_1 generated by K_r and τ (see Theorem 2.3) rather large. In fact, then order $(G_1) = (2^r - 1)2^r$, and this is also the number of different types of cells required for synthesis with this choice of τ.

The best choices for T from this point of view are those which make $\mathsf{T}^p = \mathsf{I}$ for some small value of $p > 1$. For $r = 1$, this is impossible, since only $\mathsf{T} = \mathsf{I}$ is available among the nonsingular "matrices" (another way of seeing the incompleteness of single-rail (Maitra) cascades). For $r = 2$, the best (and only) choice is $p = 3$, obtained with the matrix T satisfying $\mathsf{T}^2 + \mathsf{T} + \mathsf{I} = 0$ (corresponding to a maximal-period feedback register), because $p = 2$ is excluded by the eigenvalue condition. For larger values of r, various possibilities occur. In general, there are 2^{r-2} ways of choosing the characteristic polynomial $\varphi(\lambda)$ so that it has an odd number of nonzero terms. Thus, for $r = 4$, we have the choices:

$$\lambda^4 + \lambda + 1$$
$$\lambda^4 + \lambda^3 + 1$$
$$\lambda^4 + \lambda^2 + 1 = (\lambda^2 + \lambda + 1)^2$$
$$\lambda^4 + \lambda^3 + \lambda^2 + \lambda + 1$$

THEOREM 3.1. Let $F(x_1, \ldots, x_m)$ be an arbitrary r-rail function of m-Boolean variables $F: X^m \rightarrow X^r, r \geq 2$. Then there exists an r-rail canonical cascade decomposition which realizes F at the output rails. The cascade contains k_m cells, where $k_m = 2^m + 2^{m-1} - 2$.

Proof: Identify elements of $C_2{}^r = K_r$ with the rail-states X^r in the natural way. This identification transforms F into a group function $f: X^m \rightarrow K_r$. Let K_r and $\mathscr{A}_r$ be the groups H and G, respectively, in Theorem 2.2. Then Lemmas 3.1 and 3.2 above show that the hypothesis of this theorem is satisfied. Thus, f has a Yoeli–Turner-type decomposition over $\mathscr{A}_r$, with k_m cells. This decomposition of f, when reinterpreted in terms of the r-rail function F, leads to a k_m-cell r-rail cascade realizing the function F at its output rails, when suitable input rail signals are applied.

Several matters relating to this decomposition now merit more detailed discussion. First, note that the length of cascade k_m required for this canonical realization of r Boolean functions is independent of the number of rails. Second, it should be observed that the two-rail case, $r = 2$, is covered by Theorem 3.1 as a special case, where K_2 is the Klein four-group $C_2 \times C_2$, and τ is represented by the matrix

$$\mathsf{T} = \begin{pmatrix} 1 & 1 \\ 1 & 0 \end{pmatrix}$$

with characteristic polynomial $\lambda^2 + \lambda + 1$. We have $\mathsf{T}^3 = \mathsf{I}$, hence also $\lambda^3 = 1$ and the group generated by K_2 and τ has order $4 \times 3 = 12$. In fact, the permutation induced on K_2 by T is $(0)(123)$, corresponding exactly to the description given in Section III,B. Hence, the group G_1 generated by τ and K_2 is just A_4, the alternating group on four letters.

Another matter of interest is that of the freedom of choice in selecting τ (or what amounts to the same thing, selecting the matrix T). We have already pointed out that there are 2^{r-2} different ways of picking a characteristic polynomial $\varphi(\lambda)$ for T, and we have explicitly written out these four choices for the case $r = 4$. In three of these four choices, φ is an irreducible polynomial, and hence T is uniquely[11] determined by φ as the corresponding companion matrix. Thus, we have the (nonderogatory) matrices:

$$\mathsf{T}_1 = \begin{bmatrix} 1 & 1 & 0 & 0 \\ 1 & 0 & 1 & 0 \\ 1 & 0 & 0 & 1 \\ 1 & 0 & 0 & 0 \end{bmatrix} \qquad \mathsf{T}_2 = \begin{bmatrix} 1 & 1 & 0 & 0 \\ 0 & 0 & 1 & 0 \\ 0 & 0 & 0 & 1 \\ 1 & 0 & 0 & 0 \end{bmatrix} \qquad \mathsf{T}_3 = \begin{bmatrix} 0 & 1 & 0 & 0 \\ 0 & 0 & 1 & 0 \\ 1 & 0 & 0 & 1 \\ 1 & 0 & 0 & 0 \end{bmatrix}$$

$$\lambda^4 + \lambda^3 + \lambda^2 + \lambda + 1 \qquad\qquad \lambda^4 + \lambda^3 + 1 \qquad\qquad\quad \lambda^4 + \lambda + 1$$

---

[11] Uniquely determined up to equivalence.

belonging to the polynomials indicated. In the case of the fourth possible choice for $\varphi(\lambda)$, namely the polynomial $\lambda^4 + \lambda^2 + 1$, T can be either the (nonderogatory) companion matrix

$$T_4 = \begin{bmatrix} 0 & 1 & 0 & 0 \\ 1 & 0 & 1 & 0 \\ 0 & 0 & 0 & 1 \\ 1 & 0 & 0 & 0 \end{bmatrix}$$

or the reducible matrix

$$T_5 = \begin{bmatrix} 1 & 1 & 0 & 0 \\ 1 & 0 & 0 & 0 \\ 0 & 0 & 1 & 1 \\ 0 & 0 & 1 & 0 \end{bmatrix}$$

The difference arises because this last polynomial is factorable, and hence the minimal polynomial[12] of T need not be its characteristic polynomial. We have, in fact, $m(T_4) = \varphi(\lambda) = \lambda^4 + \lambda^2 + 1$, while $m(T_5) = \lambda^2 + \lambda + 1$. The case of T_5 is interesting because it partitions into two 2×2 matrices, each identical to the choice of T for the two-rail case.[13] Thus, this four-rail decomposition corresponds to simply putting two two-rail cascades in parallel next to each other, with no "mixing" of the rail signals. This leads to somewhat more complex cells (considering a pair of two-rail cells as a four-rail cell) than does the choice of T_2, T_3, or T_4, because each four-rail cell corresponding to T_5 must contain two EXCLUSIVE OR's, while the other four-rail cells each need only one EXCLUSIVE OR. On the other hand, T_1 involves three EXCLUSIVE OR gates and is thus the worst choice from this standpoint.

The four-rail cells implied by the choice of $\lambda^4 + \lambda + 1$ as characteristic polynomial for τ in the case $r = 4$ are of the type shown in Fig. 7 and they obey logical equations of the form:

$$y_1' = \bar{x}y_1 + x(y_1 \oplus y_4) = y_1 \oplus xy_4$$
$$y_2' = \bar{x}y_2 + xy_1$$
$$y_3' = \bar{x}y_3 + xy_2$$
$$y_4' = \bar{x}y_4 + xy_3$$

The four-rail cells corresponding to group elements in K_4 are, of course, independent of the choice of τ. They are the simpler sort, involving selective

[12] The minimal polynomial of T, $m(T)$ is the polynomial of lowest degree such that $m(T) = 0$.

[13] In more complicated cases, and in general, matrices with the same characteristic polynomial may even have the same minimal polynomial but be inequivalent because their elementary divisors are different. See Birkhoff and MacLane (1953).

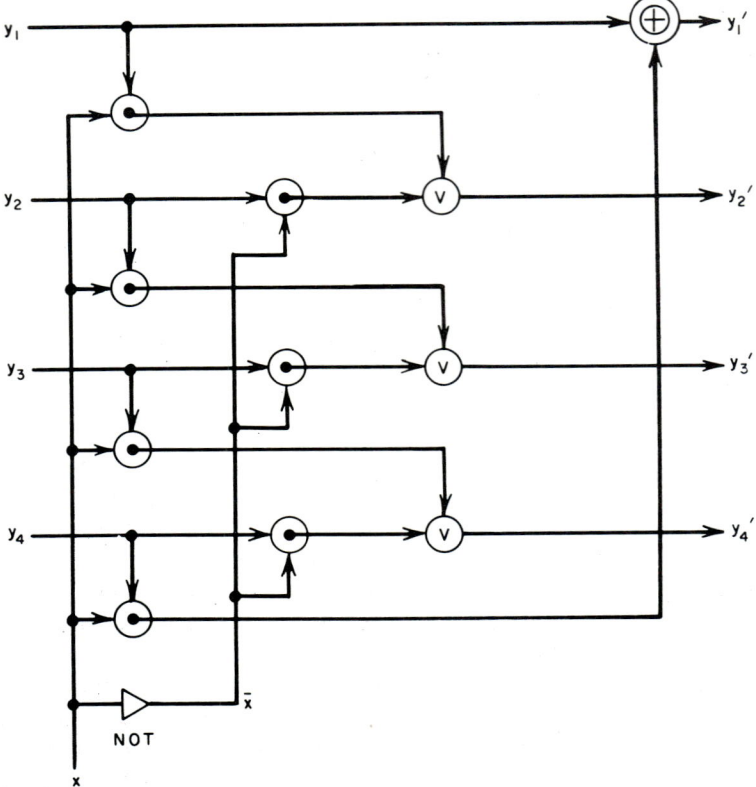

FIG. 7. Typical A-type four-rail cell.

complementation of one or more of the rail signals by the vertical input variable x, just as in the two-rail case. In the limit, as the number m of vertical input variables increases, one-third of the cells in a cascade are of the simple K_4 type, while two-thirds are in the affine group.

D. NUMBER OF MODULE TYPES

We return now to the question of the number of module types required for cascade synthesis, a question which has been given some consideration already in the two-rail case (see Section III,B). We now show that some reductions are possible in the r-rail case, although these save only one cell type for $r = 2$; the savings become proportionately greater as r increases, becoming asymptotic to a 25% saving in cell types as r becomes large.

Recall that we need 2^r K-type cells plus (apparently) $2^r + 1$ cells of the affine type (because the decomposition uses a cell realizing τ and 2^r cells of the type $\tau^{-1}\sigma$, where $\sigma \in K_r$). It turns out that fixing τ to be the same at each stage of the iterative decomposition is wasteful, because it forces the second $\mathscr{A}$-cell to run through a possible total of 2^r distinct types, depending on the value of $f(0, 1)$ in group K_r (see the proof of Theorem 2.2). A more efficient procedure (by Stone) consists in using both τ and σ flexibly at each stage of the decomposition. This poses the following problem: Find a minimum set of elements $g_1, g_2, \ldots, g_t$ in the affine group $\mathscr{A}_r$, such that the set of pair-wise products $\{g_i g_j : i, j = 1, \ldots, t\}$ includes all 2^r elements of K_r, and such that each g_i has the property that $g_i \, \sigma g_i^1 = \sigma$ only if $\sigma = I$. Let $C_a(r)$ be the size of the minimal set of g_i's.

A solution (perhaps not the optimum one) is given by the following inductive algorithm, which we first define for $r = 2$ and 3 before indicating (informally) how the induction goes.

Case $r = 2$: We wish to express all elements of $K = C_2{}^2 = \{\sigma_0, \sigma_1, \sigma_2, \sigma_3\}$ as pairwise products of elements of $\mathscr{A}$. Here four elements suffice, for let a be any element of $\mathscr{A}$ such that $T_a : \sigma \to a\sigma a^{-1}$ leaves only $\sigma_0 = I$ fixed in K. Let $b = \sigma_1 a^{-1}$, and $c = a\sigma_2$. Thus,

$$a \cdot a^{-1} = \sigma_0$$
$$b \cdot a = \sigma_1$$
$$a^{-1} \cdot c = \sigma_2$$
$$b \cdot c = \sigma_3 = \sigma_1 \sigma_2$$

Case $r = 3$: The minimum number of elements $C_a(3) \leq 7$, since we may append to the above solution for $r = 2$ the additional elements c^{-1}, d, and e, defined by $d = c\sigma_4$ and $e = \sigma_5 d^{-1}$. Thus

$$c^{-1} \cdot d = \sigma_4$$
$$e \cdot d = \sigma_5$$
$$a^{-1} \cdot d = \sigma_6$$
$$b \cdot d = \sigma_7$$

Hence, $a, a^{-1}, b, c, c^{-1}, d$ and e form a set of seven elements of $\mathscr{A}$ whose pairwise products cover $K = C_2{}^3$.

Case $r = 4$: Each element of $C_2{}^4$ not in $C_2{}^3$ can be written as $\sigma_8 \sigma_i$ for $i = 0, \ldots, 7$, and we have already factored the elements σ_i of $C_2{}^3$ into products of two elements of $\mathscr{A}$. By introducing an element $f = d\sigma_8$ into our list, we automatically get the desired factorizations for the elements $\sigma_{12}, \sigma_{13}, \sigma_{14}$, and σ_{15}, since the expressions for $\sigma_4, \sigma_5, \sigma_6$, and σ_7 all end in a factor d. We need merely introduce (besides f) the elements $d^{-1}, g = \sigma_9 f^{-1}, h = \sigma_{10} f^{-1}$, and $j = \sigma_{11} f^{-1}$ in order to express the remaining elements of C_2^4. Also, we are ready for the next stage of this recursive process, since eight of the factorizations just obtained (those for $\sigma_8, \ldots, \sigma_{15}$) end in the same factor f. Thus, $C_a(4) \leq 7 + 5 = 12$.

General Case r: Half of the elements in C_2^{r-1} having been expressed as two-factor products with the same second factor, say z, we need only append $2^{r-2} + 1$ new elements to the list $g_1, g_2, \ldots$ in order to express all elements of C_2^r in the desired form. (The extra "1" is required because we must introduce two new elements, z^{-1} and one other element, to express σ_{2r-1} as a pair product.) Hence, if $C_a(r) = $ minimum length of the list $g_1, g_2, \ldots$, we have

$$C_a(r) \leq C_a(r-1) + 2^{r-2} + 1$$

$$C_a(2) \leq 4$$

Combining these, we get $C_a(r) \leq 2^{r-1} + r$, for $r \geq 2$. In order to complete the argument, we need to show that each new element g_i introduced in the algorithm lies in $\mathscr{A}_r$ and has the property that for all σ in K_r, we have $g_i \sigma g_i^{-1}$ only if $\sigma = I$. The variables introduced are in one of the three forms: p^{-1}, $\sigma_0 p$, or $p\sigma_0$, where p is already known to be an element of $\mathscr{A}_r$ with the desired property, and σ_0 is in K_r. Clearly, all three forms lie in $\mathscr{A}_r$, since K_r is a subgroup of $\mathscr{A}_r$. Next, $p^{-1}\sigma p = \sigma$ implies $p\sigma p^{-1} = \sigma$, so that p^{-1} is admissible if p is. Finally, from either of the relations $p\sigma_0 \sigma(p\sigma_0)^{-1} = \sigma$ or $(\sigma_0 p)\sigma(\sigma_0 p)_i^{-1} = \sigma$, we conclude that $p\sigma p^{-1} = \sigma$, by making use of the fact that K_r is an abelian group. Hence, $p\sigma_0$ and $\sigma_0 p$ are admissible if p is. This proves that all the elements g_i required in the algorithm are admissible.

Let $C_k(r)$ be the number of K-type modules required in r-rail cascade synthesis $= 2^r$; $C_a(r)$ be the number of $\mathscr{A}$-type modules required $\leq 2^{r-1} + r$, as shown above; then the total number of module types $C(r) = C_k(r) + C_a(r) \leq 2^r + 2^{r-1} + r$. These values are shown for $r \leq 5$ in Table II.

TABLE II
Numbers of Module Types Required
Versus Number of Rails

Number of rails, r	$C_k(r)$	$C_a(r)$	$C(r)$
2	4	4	8
3	8	7	15
4	16	12	28
5	32	21	53

What do these modules "look like" in the general case? We have already examined some cases for $r = 4$. Observe now that once the initial element, a, of the construction algorithm has been selected, all the subsequently selected $\mathscr{A}$-cells (or their inverses) have the form $a\sigma_i$ for some $i = 0, \ldots, 2^{r-1}$. That is, they may be represented by the specific a-cell followed by (gated) complementation of some output rail signals. We have noted that a good choice for a is the linear transformation with characteristic polynomial, $\lambda^r + \lambda + 1$.

Using this choice for a, we readily find the canonical equations for the a-cell to be:

$$a: \begin{cases} y_1' = \bar{x}y_1 + x(y_1 \oplus y_r) = y_1 \oplus xy_r \\ y_2' = \bar{x}y_2 + xy_1 \\ \vdots \\ y_i' = \bar{x}y_i + xy_{i-1} \quad \text{for} \quad i = 2, 3, \ldots, r \\ \vdots \\ y_r' = \bar{x}y_r + xy_{r-1} \end{cases}$$

The inverse equations (forward equations for the cell a^{-1}) are seen to be

$$a^{-1}: \begin{cases} y_1' = \bar{x}y_1 + xy_2 \\ \vdots \\ y_{r-1}' = \bar{x}y_{r-1} + xy_r \\ y_r' = \bar{x}y_r + x(y_1 \oplus y_2) \end{cases}$$

Now let σ_c correspond to the rail-complementation vector, $[c_1 c_2 \cdots c_r]$ where $c_i = 0$ or 1. Then the normalized cell performing the operation, $a\sigma_c$ has the canonical equations obtained by adding $(\oplus)c_i x$ to the ith equation:

$$a\sigma_c: \begin{cases} y_i' = \bar{x}y_i \oplus xy_{i-1} \oplus xc_i \quad \text{for} \quad i = 2, \ldots, r \\ y_1' = y_1 \oplus xy_r \oplus c_1 x \end{cases}$$

or generically,

$$a\sigma_c: \begin{cases} y_i' = \bar{x}y_i \oplus x\dot{y}_{i-1}, \quad \text{for} \quad i = 2, \ldots, r \\ y_1' = y_1 \oplus x\dot{y}_r \text{ (where } \dot{y} = \text{either } y \text{ or } \bar{y}). \end{cases}$$

Similarly, the inverse equations (forward equations for a^{-1}) may be augmented to yield the equations for the normalized cell $\sigma_c a^{-1}$ by selectively complementing the *input* variables. Thus,

$$\sigma_c a^{-1}: \begin{cases} y_i' = \bar{x}\dot{y}_i + x\dot{y}_{i+1} \quad \text{for} \quad i = 1, 2, \ldots, r-1 \\ y_r' = \bar{x}\dot{y}_r + x(\dot{y}_1 \oplus \dot{y}_2) \end{cases}$$

Another possibility suggested by the foregoing discussion is that of simply using a (redundant) cascade of two cells, σ_c and a^{-1} or a and σ_c, to realize each element $\mathscr{A}$ needed. This cuts down the total inventory of cell types to $2^r + 2$, but it also increases the lengths of the cascades required by as much as two-thirds in unfavorable cases.

E. A LINEAR-ALGEBRA APPROACH TO MULTIRAIL CASCADES

In the preceding portions of this section, we have given a general treatment of multirail cascade synthesis based on group-theoretic notions and results. A rather general treatment was followed in various places so that the results might be applicable under a variety of circumstances, e.g., to nonbinary as well as to binary cascades, or to multioutput function synthesis with *don't care* conditions. However, in the simplest situation—that of completely specified binary rail outputs—it was observed that an appropriate group H for encoding group functions from the Boolean m-cube $\{0, 1\}^m$ into H is the group C_2^r. It was also found that the affine group of dimension r over GF(2) constitutes a suitable (larger) group $G > H$ for the representation of the cell permutation functions.

Thus, the operations required for realizing r-rail cascades under the above assumptions are entirely within the realm of linear algebra (i.e., operations with vectors and matrices). However, we have not made full use of this fact in the development of the results. An alternative treatment of r-rail cascades can be based entirely on linear-algebraic concepts and notation, without making use of the more general (and more powerful) group-theoretic tools developed in Section II.

Below we outline the essential features of such an alternative treatment. Although no new cascades are thereby made available, this alternative approach has the distinct advantages of being easier to understand and apply. It is doubtful, however, if we could have arrived at this treatment independently, i.e., apart from the specialization from the general group-theoretic treatment. This fact suggests also that the linear-algebraic view may have intuitive value in suggesting other applications for group-theoretic decompositions and possibly for improving our understanding of networks, such as the Short (1965) cascades that were not developed by group-theoretic methods.

Let $F_i(x_1, \ldots, x_m)$; $i = 1, \ldots, r$ be r independently specified Boolean functions (each depending on m variables), which are to be realized on the output rails of a cascade network. That is, the output signals from the final cell of the cascade are specified to be

$$y_i' = F_i(x_1, \ldots, x_m), \qquad i = 1, \ldots, r \tag{10}$$

For convenience we lump the Boolean variables x_j together to form a "vector" $x = (x_1, \ldots, x_m)$ in the Boolean m-cube $\{0, 1\}^m$. Similarly, the signals y_i' are treated jointly as a vector $\mathbf{y}' = (y_1', \ldots, y_r')$ in the r-dimensional vector space $V_r(K)$, where $K = $ GF(2). Thus,

$$\mathbf{y} = \mathbf{F}(x) \in V_r(K) \tag{11}$$

We note that **F** denotes a function from the Boolean m-cube $\{0, 1\}^m$ into $V_r(K)$, i.e.,

$$\mathbf{F}: \{0, 1\}^m \to V_r(K) \tag{12}$$

We now propose to decompose **F** into a cascade of functions represented in Fig. 8. In vector equation form, we have the desired decomposition:

$$\{[\mathbf{y}_0 \oplus \mathbf{H}_1(\hat{x})]\mathsf{T}^{-x_m} \oplus \mathbf{H}_2(\hat{x})\}\mathsf{T}^{x_m} = \mathbf{y}' = \mathbf{F}(x) \tag{13}$$

where $\oplus$ deonotes addition in the vector space $V_r(K)$, and $\hat{x} = (x_1, \ldots, x_{m-1})$. The quantity T^{x_m} may require some explanation; T is a nonsingular square

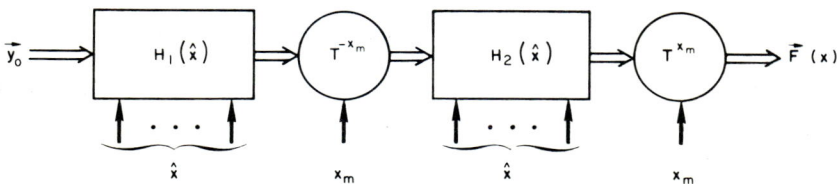

FIG. 8.   Basis for cascade decomposition.

matrix of order r over GF(2), i.e., a matrix containing 0's and 1's only. The quantity T^{x_m} represents $\mathsf{T}^1 = \mathsf{T}$ when $x_m = 1$, and $\mathsf{T}^0 = \mathsf{I} =$ the identity matrix when $x_m = 0$. Of course, $\mathsf{T}^{-x_m} = (\mathsf{T}^{x_m})^{-1} = (\mathsf{T}^{-1})^{x_m}$. Thus, $\mathsf{T}^{x_m}\mathsf{T}^{-x_m} = \mathsf{I}$ for all x_m. It is important to note also that T^{x_m} may be written as

$$\mathsf{T}^{x_m} = \mathsf{I} \oplus (\mathsf{I} \oplus \mathsf{T})x_m \qquad \text{for} \quad x_m = 0 \text{ or } 1$$

We inquire as to when the decomposition, Eq. (13), will be possible, i.e., when there will exist $(m-1)$-variable functions $\mathbf{H}_1(\hat{x})$, $\mathbf{H}_2(\hat{x})$ and a matrix T for which Eq. (13) holds. Expanding Eq. (13) and using the relations given below, we easily find that

$$\mathbf{y}_0 \oplus \mathbf{H}_1(\hat{x}) \oplus \mathbf{H}_2(\hat{x}) \oplus \mathbf{H}_2(\hat{x}) \cdot (\mathsf{I} \oplus \mathsf{T})x_m = \mathbf{F}(x) \tag{14}$$

Now an arbitrary (vector) Boolean function $\mathbf{F}(x)$ can always be written as a Reed decomposition (ring-sum decomposition):

$$\mathbf{F}(x) = \mathbf{F}_A(\hat{x}) \oplus \mathbf{F}_B(\hat{x})x_m \tag{15}$$

with respect to the variable x_m, where both $\mathbf{F}_A$ and $\mathbf{F}_B$ are independent, arbitrary $(m-1)$-variable-vector Boolean functions.

Hence, comparing Eq. (14) with Eq. (15), we see that if Eq. (14) is to be a valid decomposition for **F**, then we must have

$$\begin{aligned}
\mathbf{F}_A(\hat{x}) &= \mathbf{y}_0 \oplus \mathbf{H}_1(\hat{x}) \oplus \mathbf{H}_2(\hat{x}) \\
\mathbf{F}_B(\hat{x}) &= \mathbf{H}_2(\hat{x}) \cdot (\mathsf{I} \oplus \mathsf{T})
\end{aligned} \tag{16}$$

The second of these equations must be solvable for the unknown function $H_2(\hat{x})$. In order to guarantee the possibility of such solution, we impose the following condition: (*Condition C*) The matrix $I \oplus T$ is nonsingular, i.e., $(I \oplus T)^{-1}$ exists.

It is obvious, of course, that a necessary and sufficient condition for Condition C to hold is that $\det(I \oplus T) \neq 0$, or equivalently, that the matrix T does not possess any unit eigenvalues.

Under this assumption we have the immediate solution:

$$\begin{aligned} \mathbf{H}_1(\hat{x}) &= \mathbf{y}_0 \oplus \mathbf{F}_A(\hat{x}) \oplus \mathbf{F}_B(\hat{x}) \cdot (I \oplus T)^{-1} \\ \mathbf{H}_2(\hat{x}) &= \mathbf{F}_B(\hat{x}) \cdot (I + T)^{-1} \end{aligned} \tag{17}$$

Observe that according to the above equations the input vector $\mathbf{y}_0$ to the cascade of Fig. 8 is specifiable independently of the function $\mathbf{F}(x)$ to be realized. We might, for example, simply set $\mathbf{y}_0 = 0$ (all rail inputs equal to zero). Alternatively, we might set $\mathbf{y}_0 = \mathbf{F}(0)$ in order to allow the use of normalized cell functions throughout the cascade, as discussed below.

The above decomposition realizes an arbitrary r-rail function $\mathbf{F}(x)$ of m Boolean variables as a cascade involving two elementary functions and two functions of $m - 1$ variables. The latter two functions may be similarly decomposed into functions of $m - 2$ variables (plus two elementary functions), and so forth. Thus, with $\mathbf{y}_0 = 0$, for convenience, we have

$$\begin{aligned} \mathbf{H}_1(\hat{x}) &= [\mathbf{H}_{11}(\hat{x}) \cdot T^{-x_{m-1}} \oplus \mathbf{H}_{12}(\hat{x})] \cdot T^{x_{m-1}} \\ \mathbf{H}_2(\hat{x}) &= [\mathbf{H}_{21}(\hat{x}) \cdot T^{-x_{m-1}} \oplus \mathbf{H}_{22}(\hat{x})] \cdot T^{x_{m-1}} \end{aligned}$$

where $\hat{\hat{x}} = (x_1, \ldots, x_{m-2})$, $\hat{x} = (\hat{\hat{x}}, x_{m-1})$, and $\mathbf{H}_{ij}$; $i, j = 1, 2$ are (vector) functions of $m - 2$ Boolean variables. This iterative decomposition terminates when the resultant H-functions have been reduced to elementary (one-variable) functions $\{0, 1\} \to V_r(K)$.

As described above, the final form of the decomposition contains 2^{m-1} cells represented by H-functions, while the remaining $2^m - 2$ cells realize the linear transformations $y \to yT$ and $y \to yT^{-1}$. The H-function cells simply perform transformations of the type $y \to y \oplus H(x_j)$, i.e., selective complementation of the rail signals under control of a Boolean input variable x_j. In other words, the H-cells perform translations of the input vector y.

The T-cells, which realize the transformations $y \to yT^{x_j}$ or $y \to yT^{-x_j}$, have the property that when $x_j = 0$ the transformations reduce to the identity $T^0 = I$, a property we have referred to as normalization. Note that the H-cells do not necessarily possess this property. That is, we do not necessarily have $H(0) = 0$. The burden of achieving a realization of $F(x)$ in a cascade consisting entirely of normalized cells can be placed entirely on suitable selection of the input rail signal vector $\mathbf{y}_0$ to be applied at the extreme input end of the cascade (as was shown in Section III,C). However, when this is done, we must also allow the use of T-cells other than the (homogeneous) linear trans-formations $y \to yT^{\pm x}$ used above.

In order to see this, consider the case of a cascaded combination of a T-cell (driven by x_1) and a (nonnormalized) H-cell (driven by x_2). The relation be-tween input vector $\mathbf{y}$ and output vector $\mathbf{y}'$ for this two-cell cascade is simply

$$\mathbf{y}' = \mathbf{y}T^{x_1} \oplus H(x_2) = \mathbf{y}T^{x_1} \oplus H(0) \oplus [H(1) \oplus H(0)]x_2$$

where $H(0) \neq 0$, since the H-cell is not normalized, and where $H(0) \neq H(1)$, unless the H-cell is vacuously dependent on its variable x_2. We propose to replace the H-cell by the normalized H-cell with function $H^*(x_2) = x_2 H_s$, where H_s is the constant (0 or 1):

$$H_s = H(0) \oplus H(1)$$

Consider the cascade shown in Fig. 9. We claim that this cascade realizes the

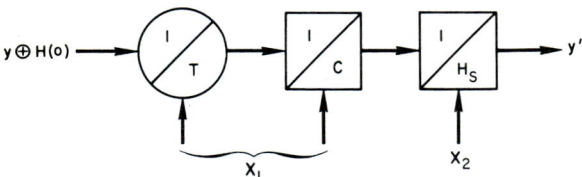

FIG. 9.   Normalization of a cascade.

same output function y' as the original cascade provided that the (vector) constant C is properly chosen. For, indeed,

$$[y \oplus H(0)]T^{x_1} \oplus x_1 C \oplus x_2 H_s$$

is its output function, and this reduces to

$$yT^{x_1} \oplus H(0)[1 \oplus x_1(1 \oplus T)] \oplus x_1 C \oplus x_2 H_s$$
$$= yT^{x_1} \oplus H(0) \oplus x_2 H_s \oplus x_1[H(0)(1 \oplus T) \oplus C]$$
$$= yT^{x_1} \oplus H(0) \oplus x_2 H_s = y'$$

provided that the constant C is selected to be

$$C = H(0)(1 \oplus T)$$

We have, in effect, "moved" a constant cell function $H(0)$ through the T-cell in order to normalize the final H-cell and place the normalizing term at the input end. In doing this, however, the T-cell is changed from a homogeneous linear transformation into an affine transformation (homogeneous linear transformation followed by a translation, both controlled by the same Boolean variable, x_1). The two cells making up this affine transformation, since they are both controlled by x_1, may be combined into a single affine cell realizing the transformation,

$$y_{out} = y_{in} \, T^{x_1} \oplus x_1 C$$

which is itself normalized. Typical cells of this sort are shown in Figs. 6 and 7.

It follows that, in using this normalization procedure, one apparently must provide for an inventory of 2^{r+1} distinct affine cells, i.e., T, T^{-1}, and the cells derived from them by cascading a constant vector function C, which has $2^r - 1$ distinct nonzero values. By using a slight modification of this procedure, one can see that only 2^r affine cells are actually needed, i.e., cells T and (T^{-1}, C), for 2^r possible values of C. However, there is a tremendous advantage in using normalization, because the stockpile of H-cells required is reduced from 2^{2r} cells, realizing any pair $[H(0), H(1)]$, to the 2^r cells required to realize any normalized function $[0, H(1)]$. Thus, a saving of $2^r(2^r - 1)$ H-cells is paid for by the smaller price of stockpiling extra $2^r - 1$ affine cells.[14] The net saving in inventory size obviously grows rapidly with the number of rails r.

The reader will have observed that the only conditions imposed on the matrix T in Eq. (13) and on the subsequent decomposition arguments are that T be nonsingular, which is required by the presence of T^{-xm} in Eq. (13), and that the matrix $I \oplus T$ also be nonsingular (Condition C). A few words are in order here to relate these conditions to the more general assumptions made in the group-theoretic decompositions used in Section II.

We first recall the statement of Property P used in Section II. *Property P*: There exists in group G an element τ such that the mapping $U_\tau : h \rightarrow \tau h \tau^{-1}$ is an automorphism of H leaving no element of $H - \{I\}$ fixed.

We now show that this condition is equivalent (in the case of the group $H = C_2^r$) to the existence of a nonsingular matrix T with $T \oplus I$ nonsingular. For the group G can be taken to be the affine group over $V_r(K)$. If the element h of H [considered as a permutation of $V_r(K)$] takes vector y of $V_r(K)$ into $y \oplus c$, i.e.,

$$h: y \rightarrow y + c$$

and τ corresponds to the matrix T over GF(2) = K, then τ (considered as a permutation of $V_r(K)$) takes y into yT. Hence, the mapping U_τ takes y into $(y)\tau h \tau^{-1}$, i.e.,

$$U_\tau : y \rightarrow (yT \oplus c)T^{-1} = y \oplus cT^{-1}$$

---

[14] The actual saving is even better than the above argument indicates, because a more sophisticated procedure for assigning the affine cells in a cascade decomposition (described in Section III,D) shows that only $2^{r-1} + r$ affine cells are needed.

The result of this mapping can coincide with the result of applying h, i.e., a translation by amount c if and only if we have

$$c = c\mathsf{T}^{-1}$$

Equivalently,

$$c(\mathsf{T} \oplus \mathsf{I}) = 0$$

This equation always holds, of course, if $c = 0$. But that solution corresponds to the case $h = I$ (identity transformation). If a solution $c \neq 0$ is to exist, then it is necessary and sufficient (by elementary matrix theory) that the matrix $\mathsf{T} \oplus \mathsf{I}$ be singular, i.e., that $\det(\mathsf{T} \oplus \mathsf{I}) = 0$.

F. PERMUTATION CASCADES BASED ON OTHER GROUPS

The cascade synthesis procedure for r-rail cascades which was described in the preceding section used the very natural choice of C_2^r as the subgroup H for decomposition. This is by no means the only possible choice, however. In principle, we could use for H any decomposable finite group with at least 2^r elements, such that with some fixed input rail-state y_0, at least one of the group elements maps y_0 into an arbitrary rail-state at the output. The most difficult part of this requirement to satisfy is the decomposability. Lemma 2.2 and Theorem 2.3 give a characterization of decomposable groups. However, the mathematical literature does not supply us with a great deal of ready information as to which groups possess the right kind of outer automorphisms. On the negative side, certain groups are known to be complete, and hence these are indecomposable (over any G). By detailed investigation we determined in Section II,D that a number of other classes of finite groups are indecomposable and hence are unsuitable for cascade decompositions.

One may legitimately ask, "What is the point in seeking other decomposable groups H for cascade synthesis once one has found an eminently suitable group of just the right order, 2^r, to encode signals on r binary rails?" There are several good reasons for doing this, as we shall now explain. First, it is always a good idea to have several alternative design techniques at hand for a problem, whenever possible. One or another of the alternatives may prove considerably more efficient (e.g., in numbers of cells required or in the simplicity of the cells) than the other, for the *particular* problem at hand. Second, for the synthesis of functions involving "don't cares," or when certain rail states can be disallowed for other reasons, it would appear that advantages might accrue from exploiting these restrictions by using a smaller decomposition group than the full 2^r-element group, C_2^r. A third potential reason for exploring other decomposition groups is simply that combinations of several cascades based on different groups might lead to alternative canonical cascades or arrays better than the straight affine-group-based cascades.

In order to illustrate some of the above ideas, we shall consider a cascade synthesis technique based on the use of the cyclic group $C_3 = \{I, a, a^2\}$. This might find application, for example, in two-rail synthesis where one of the four possible rail-states, say the state $(1, 1)$, is excluded by the use of "don't cares," or by other constraints inherent in the application.[15] We shall identify group element I with rail state $(0, 0)$, a with $(1, 0)$, and a^2 with $(0, 1)$, somewhat arbitrarily. For simplicity, we can refer to these states as 0, 1, and 2, respectively. Considered as permutations, we have $a = (012)$, $a^2 = (021)$. C_3 has been shown to be decomposable over D_3 (by Lemma 2.4), with the element $\tau = (0)(12)$ generating the required outer automorphism: $I \rightarrow I$, $a \rightarrow a^2$, $a^2 \rightarrow a$. Also, C_3 is transitive on the set of admissible rail states.

Now let us see what kinds of cells are required for this restricted two-rail synthesis. First, the (normalized) cell which performs the rail-state permutation (012) when its x-input is on: We have, with $x = 1$, $y_1' = \bar{y}_1 \bar{y}_2$ and $y_2' = y_1$. Hence, for arbitrary x,

$$a: \begin{cases} y_1' = \bar{x}y_1 + x\bar{y}_1\bar{y}_2 \\ y_2' = \bar{x}y_2 + xy_1 \end{cases}$$

constitute the logical equations for cell a. Similarly, the a^2 cell is given by the equations:

$$a^2: \begin{cases} y_1' = \bar{x}y_1 + xy_2 \\ y_2' = \bar{x}y_2 + x\bar{y}_1\bar{y}_2 \end{cases}$$

which are seen to differ from the equations for the a cell in that the rails are transposed.

Next, let's examine the τ-cell and the cells derived from it, which perform the (normalized) functions τa and τa^2 (since $\tau^{-1} = \tau$). As permutations, these group elements are: $\tau = (0)(12)$, $\tau a = (01)(2)$, and $\tau a^2 = (02)(1)$. The corresponding logical equations are readily found to be:

$$\tau: \begin{cases} y_1' = \bar{x}y_1 + xy_2 \\ y_2' = \bar{x}y_2 + xy_1 \end{cases}$$

$$\tau a: \begin{cases} y_1' = \bar{x}y_1 + x\bar{y}_1\bar{y}_2 \\ y_2' = y_2 \end{cases}$$

$$\tau a^2: \begin{cases} y_1' = y_1 \\ y_2' = \bar{x}y_2 + x\bar{y}_1\bar{y}_2 \end{cases}$$

Note that the τ-cell is a simple 2-input permutation cell which either transposes its rail inputs or passes them through unchanged, depending on the

---

[15] For example, the autosynchronous logic of Sims and Gray (1959) propagates signals 0, 1 and $\emptyset$ (null signal) on two binary rails, with $(1, 1)$ excluded as an invalid signal.

value of its vertical (x) input. The other two cells are like the cell realizing a, except that in each case one of the rail signals is unaltered.

The following comments are pertinent to this synthesis. First, there are only five basic cell types required (six, if we include the identity cell), as opposed to the eight types needed for canonical two-rail synthesis using the group C_2^2. Second, these cells are somewhat simpler than the affine-based cell shown in Fig. 6 because they lack EXCLUSIVE OR gates.

The above example illustrates the potential utility of cascade decompositions employing groups other than C_2^r. It also points up another somewhat arbitrary assumption—the way in which group elements are identified with (or coded into) rail states. Here we arbitrarily set $I \leftrightarrow (0, 0)$, $a \leftrightarrow (1, 0)$, and $a^2 \leftrightarrow (0, 1)$. There is no reason why some other encoding might not have been used; in fact, even the unused state $(1, 1)$ above, might have been different.

More generally, it is suggested that one need not even use a one-to-one identification between rail states and group elements. There might be more than one group element assigned to each rail state, in which case we might have a permutation group that is multiply transitive on the rail states, instead of simply transitive as all our examples have been. This redundancy in group encoding could conceivably be exploited in a manner similar to that of "don't-care" synthesis, since the transition from a truth table to a group function table would leave some flexibility of choice. At this point we do not know whether this flexibility really represents a useful gain. Further study may clarify this and other related questions.

G. COMPARISONS AMONG CASCADE VARIETIES

Thus far in this chapter we have presented a generalization of cascade decomposition theory that leads to the synthesis of multirail functions in structures we call "permutation" cascades. Generally, all types of multirail cascades have a length which depends only on the number of input variables and not on the number of outputs, and their length grows exponentially with the number of input variables. In this section we compare the permutation cascades to cascades of Short (1965) and Yoeli–Turner (1966), with regard to their lengths, cell complexity, and number of cell types required.

Short exhibited two basic synthesis procedures. The first and more efficient of the two algorithms was rigorously demonstrated to be valid only for $n \leq 8$ input variables. This algorithm leads to two-rail cascades of length $S_1(n)$ cells, where $S_1(n)$ satisfies the recursion:

$$S_1(n) = S_1(n-1) + (n-1)2^{n-4} \quad \text{for} \quad n > 4$$
$$S_1(3) = 1, \quad S_1(4) = 2$$

It is not difficult to show that an exact closed-form expression for $S_1(n)$ is given by:

$$S_1(n) = n2^{n-3} - 2^{n-2} - 2 \qquad \text{for} \quad n \geq 4$$

Short conjectured that this bound, $S_1(n)$, was valid also for $n > 8$, although he did not succeed in demonstrating this conjecture.

Short's second algorithm, which was shown to be valid for *all* values of n, leads to two-rail cascades which are somewhat longer for $n \geq 9$ variables. Letting $S_2(n)$ be the length of this somewhat less efficient cascade, we have

$$S_2(n) = S_2(n-1) + (n+1)2^{n-4} - 2 \qquad \text{for} \quad n > 8$$
$$S_2(8) = 190 = S_1(8)$$

A corresponding closed-form expression for $S_2(n)$ is readily shown to be:

$$S_2(n) = n2^{n-3} - 2n - 50, \quad n \geq 8$$

It is convenient also to assume that $S_2(n) = S_1(n)$ for $n < 8$.

The Yoeli–Turner synthesis procedure, on the other hand, requires $YT(n) = 2^n + 2^{n-1} - n - 8$ cells in a canonical two-rail cascade to realize an arbitrary *pair* of n-variable functions. It is difficult to make a completely fair comparison between a cascade which realizes a pair of (independent) functions and a cascade which realizes only a single function. However, we may add one cell to the Yoeli–Turner (Y–T) cascade to combine the two independent functions with a new $(n+1)$th variable, in order to realize (albeit somewhat inefficiently) a single function of $n+1$ variables. On this basis we are led to compare the cascade lengths, $S_2(n)$ and $YT(n-1) + 1$ for the realization of a single function of n variables:

$n =$	3	4	5	6	7	8	9	10
$S_2(n) =$	1	2	10	30	78	190	508^{16}	1210^{16}
$YT(n-1) + 1 =$	1	2	13	36	83	178	369	752

We note that the Y–T cascades become shorter than (both kinds of) the Short cascades, for $n \geq 8$. That such a crossover occurs is not surprising, since both forms of $S(n)$ grow as $n2^{n-3}$, while $YT(n)$ grows as $3 \cdot 2^{n-1}$. It should be remembered also that the Y–T cascades have a flexibility that is not fully exploited in the above comparison, since they are capable of realizing *two* arbitrary functions of n variables with $YT(n)$ cells.[17] This is a capability which is lacking in Short's synthesis procedure.

---

[16] Even if we replace these (firm) values by Short's conjectured values of $S_1(9) = 446$ and $S_1(10) = 1022$, the Yoeli–Turner cascade is still superior.

[17] For $n \geq 13$, the Y–T cascade uses fewer cells to realize *two* functions than the Short needs for a *single* function.

Next, consider the *permutation cascades,* discussed in Section III,C. As shown there, permutation cascades capable of realizing r arbitrary functions of n variables on r binary rails can be constructed using at most $P(n) = (2^n + 2^{n-1} - 2)$ cells. This number is independent of r, for $r \geq 2$, and in particular it is valid for $r = 2$. $P(n)$ exceeds the Y–T cascade length by $(n + 6)$ cells (again independent of r). The cost of $n + 6$ additional cells is the price paid for dispensing with the general-function cells used in the Yoeli–Turner synthesis. Thus we have:

$$n = \quad 3 \quad 4 \quad 5 \quad 6 \quad 7 \quad 8 \quad 9 \quad 10$$
$$P(n) = 10 \quad 22 \quad 46 \quad 94 \quad 190 \quad 382 \quad 766 \quad 1534$$

Again, we have a crossover in the neighborhood of $n = 7$, as compared to the Short cascades. However, to make proper comparison between these two types of cascades, we should not penalize the permutation cascade for doing without general-function cells. One way to compensate for this factor is to eliminate the initial general-function cell in the Short cascade—which can be done, as Short showed, by simply adding three more cells. Let $S_3(n) = S_2(n) + 3$ be the length of this *reduced cell* Short cascade. Then, comparing $S_3(n)$ and $P(n - 1) + 1$, we find:

$$n = 3 \quad 4 \quad 5 \quad 6 \quad 7 \quad 8 \quad 9 \quad 10$$
$$S_3(n) = 4 \quad 5 \quad 13 \quad 33 \quad 81 \quad 193 \quad 511 \quad 1213$$
$$P(n - 1) + 1 = 5 \quad 11 \quad 23 \quad 47 \quad 95 \quad 191 \quad 383 \quad 767$$

Note that, again, the permutation cascade becomes relatively shorter than the corresponding Short cascade for $n = 8$ and beyond.

An alternative way to make such a comparison is by introducing a *truncated permutation cascade,* where we replace the first ten cells of the permutation cascade by a single general-function cell. After all, the first ten cells in a permutation cascade accomplish nothing more than realization of two arbitrary functions of three input variables, since $P(3) = 10$. If we are willing to permit the use of a single general-function cell in a permutation cascade, then we can save nine cells at the beginning of the cascade. Thus, the length $P_1(n)$ of a truncated permutation cascade is given by $P_1(n) = P(n) - 9 = 2^n + 2^{n-1} - 11$. We may compare this length directly with that of a Short cascade, $S_2(n)$, in the same fashion already employed, *viz.,* by adding a cell and another input variable to $P_1(n - 1)$ so as to get a single function of n variables. Thus,

$$n = 3 \quad 4 \quad 5 \quad 6 \quad 7 \quad 8 \quad 9 \quad 10$$
$$S_2(n) = 1 \quad 2 \quad 10 \quad 30 \quad 78 \quad 190 \quad 508 \quad 1210$$
$$1 + P_1(n - 1) = 1 \quad 2 \quad 14 \quad 38 \quad 86 \quad 182 \quad 374 \quad 758$$

As in previous comparisons, we see that a crossover occurs between $n = 7$ and $n = 8$.

We close our comparisons on the basis of cascade lengths with the observations that, in a certain sense, Short's conjecture that his bound $S_1(n)$ is valid for $n > 8$ has been validated by the Yoeli–Turner results, since they lead to shorter cascades than $S_1(n)$ for $n > 8$. However, the structure of the Yoeli–Turner cascades, and also of the related permutation cascades, is quite different from those of Short. Thus, we are still not sure that Short's construction for $S_1(n)$ cascades is valid when n exceeds 8. However, this question is now academic, since the Y–T cascades, and even the permutation cascades, are shorter than the $S_1(n)$ cascades in this range. It would be very interesting to see whether any of the new techniques (i.e., group decomposition) could be applied to the Short cascades, or vice versa, whether portions of Short's cascades can be used to shorten, say, a permutation cascade. Preliminary efforts in this direction have not been very promising. Apparently the two kinds of cascades are so different from each other that hybrids are unlikely to succeed. For example, we have not gained any new insight into the choice of cells for Short-like cascades from experience with permutation cascade cells.

We examine next the question of the number and complexity of cell types required for the various kinds of two-rail cascades which we have been discussing. The Short cascades are of two varieties: (1) the cascades using (3-variable) general-function cells, (2) the reduced-cell cascades, which eschew these general-function cells but are three cells longer. The reduced-cell cascades employ just five basic types. If the rail input variables are denoted as y_1, y_2, with x representing the vertical input variable, then the function pairs realized in the Short reduced cells are:

(a) x, $y_2 \oplus xy_1$

(b) y_1, xy_2

(c) xy_1, y_2

(d) $h(x_1, y_1)$, y_2

(e) $h_1(x, y_1)$, $y_2 \oplus h_2(x, y_1)$

This sixth cell type is the 3-variable, 2-general-function cell.

(f) $g_1(x, y_1, y_2)$, $g_2(x, y_1, y_2)$,

which is not used on Short's reduced cell cascades.

Note that cell types (a), (b), and (c) are extremely simple. Certainly (b) and (c) are simpler than any of the cell types used in our permutation cascades (see Figs. 5, 6). Cell type (a) is, perhaps, comparable in complexity to the permutation cells. Cell types (d) and (e), however, involve arbitrary 2-input functions, so that these two types really comprise a large number of special cells. They are not complex, however; any particular 2-variable functions which one might insert for h in (d), or for h_1 and h_2 in (e), lead to cells of the same complexity as the permutation cells of Section III,B. Also, types (a), (c), and (d) are really just special cases of (e), so that if we count individual cell types, there are only type (b) and the pertinent cases of type (e). It is not clear exactly how one should count the varieties of individual cell types needed under basic type (e). Potentially, there are $(16)^2 = 256$ types subsumed under this category. Certainly this is an upper bound, but quite probably not all 256 types would be needed. Detailed investigation on this point has not been carried out. Using this (perhaps pessimistic) upper bound, one finds that at most, 257 individual cell types are needed in Short's reduced-cell cascades. One compares this with a total of eight individual cell types used in permutation cascades. The relative complexities of these cells are roughly comparable, with the Short cells being somewhat simpler, but not markedly so. Of course, the Short cascades are shorter than the permutation cascades, at least for fewer than eight independent variables.

The original Y–T cascades, employing Ashenhurst cells (see Section I,B) have cells of greater complexity than either the Short reduced-cell cascades or the permutation cascades derived from the Y–T decomposition. The reason for this is that the Ashenhurst step in the Y–T decomposition introduces a general-function cell of the same type as Short's type (f). Since these cells occur anywhere embedded in the middle of the Y–T cascades, it is not clear that they can be replaced by short strings of less complex cells, as Short does in his reduced-cell cascades where they occur only at the very beginning of the cascade. In any case, it seems clear that the permutation cascades, although, they are slightly longer than the Y–T cascades, have cells of less complexity and variety than the Y–T cascades. At the present state of knowledge, the permutation cascades also have cells of slightly less complexity than the cells of Short's cascades. However, future work may even the score here somewhat.

H. NONSTANDARD ORDERING OF INPUT VARIABLES—EQUALIZATION OF INPUT VARIABLE LOADING

The cascades derived above are *canonical* cascades in that, regardless of what output functions are being realized, one may use a fixed, standard ordering of the input variables. Moreover, the cascades are *redundant* in that every input variable is applied to more than one cell. However, as we shall see, the number of different cells driven by a given input variable may vary greatly from one variable to another. Since such a wide variation in loading may cause difficulties in some circuit technologies, it is appropriate to investigate whether the loading may be equalized.

If one follows the decomposition procedure described in Section II with decomposition centering on the variables $x_n, x_{n-1}, \ldots, x_2, x_1$ in that order, for the realization of a multirail Boolean function $F(x_1, \ldots, x_n)$, then one arrives at the *standard canonical ordering* for cell inputs. For the case $n = 4$, this is (in terms of subscripts):

$$1212\,3 \quad 1212\,3\,4 \quad 1212\,3 \quad 1212\,3\,4$$

This ordering requires that in each of the subfunctions obtained during the decomposition process, one follows the same variable ordering. The number of cells driven by variable x_1 is then exactly 2^{n-1} (these are the "H-cells" of the cascade). The other $n-1$ variables drive only affine cells. Thus, x_2 drives 2^{n-1} affine cells, x_3 drives 2^{n-2} affine cells, $\ldots$, x_j drives 2^{n-j+1} affine cells for $j = 2, \ldots, n$. (In the case $n = 4$ given above, the loading varies from two cells to eight cells.) This constitutes a large variation in loading on the input Boolean variables that should be avoided for obvious circuit reasons. Fortunately, a remedy is found easily.

To see this, let a function $F(x_1, \ldots, x_n)$ be realized as a decomposition in terms of the $n-1$ variable subfunctions $H_1(x)$ and $H_2(x)$. Let H_1 be realized as a standard canonical cascade, i.e., with the ordering described above in its variables, $x_1, \ldots, x_{n-1}$. Further, let H_2 be realized with the *reverse* variable ordering, i.e., in decomposing H_2, extract x_1 first, then x_2, etc. Of course, variable x_n will drive only two cells (and it is clear that we cannot prevent that from happening for the first variable to be extracted), but all the other variables will then be subjected to cell loadings differing at most by two. For example, with $n = 5$ the standard canonical ordering yields a load distribution of

$$16 + 16 + 8 + 4 + 2$$

over the 46 cells in the cascade. The new load distribution is obtained by superposing $8 + 8 + 4 + 2$ and its reverse, $2 + 4 + 8 + 8$, and then appending a term "2" for the fifth variable. Thus,

$$
\begin{array}{r}
8 + 8 + 4 + 2 \\
2 + 4 + 8 + 8 \\
\hline
10 + 12 + 12 + 10 + 2
\end{array}
$$

The situation is somewhat similar to the rearrangement of relay contacts in a complete decoding tree in order to obtain optimally uniform loading on the relay armatures. There, however, the optimum distribution has all loads within one unit of each other (except for one of the variables, which has unit loading).

ACKNOWLEDGMENTS

The author wishes to express his debt to the following people who have contributed greatly to the development of the subjects reported in this chapter: Harold S. Stone, James B. Turner, and especially Michael Yoeli, whose ideas led directly to the whole development of permutation cascades. He is also grateful to William H. Kautz, Karl N. Levitt, and Robert A. Short for their participation in many useful discussions. Most of the research described herein was supported under contract with the Air Force Cambridge Research Laboratories, Office of Aerospace Research, and this support is gratefully acknowledged.

REFERENCES

ASHENHURST, R. (1959). The decomposition of switching functions. *Ann. Comput. Lab. Harvard Univ.*, **29**, No. 30. Harvard Univ. Press, Cambridge, Massachusetts.

BIRKHOFF, G and MACLANE, S. (1953). "A Survey of Modern Algebra," p. 254. Macmillan, New York.

BURNSIDE, W. (1955). "Theory of Groups of Finite Order," Dover, New York.

ELSPAS, B. (1959). The theory of autonomous linear sequential networks. *IRE Trans. Circuit Theory* **CT-6**, 45–60.

ELSPAS, B., GOLDBERG, J., JACKSON, C. L., KAUTZ, W. H., and STONE, H. S. (1967). Properties of Cellular Arrays for Logic and Storage, Section II–Theory and applications of cascade decompositions. Sci. Rep. 3, SRI Project 5876, AFCRL-67-0463, Stanford Res. Inst., Menlo Park, California.

HALL, M., JR. (1959). "The Theory of Groups." Macmillan, New York.

LENDARIS, G. G., and STANLEY, G. L. (1963). On the Structure-Dependent Properties of Adaptive Logic Networks. TR 63-219; GM Defense Research Labs., Mathematics and Evaluation Studies Dept., Santa Barbara, California.

MINNICK, R. C., GOLDBERG, J., GREEN, M. W., KAUTZ, W. H., SHORT, R. A., STONE, H. S., and YOELI, M. (1966). Cellular Arrays for Logic and Storage. Final Rep., Contract AF 19(628)-4233, SRI Project 5087, Stanford Research Institute, Menlo Park, California (April 1966) AFCRL-66-613; pp. 45–61 and 140–155.

SHORT, R. A. (1965). Two-rail cellular cascades. *AFIPS Conf. Proc.*, Washington, D.C., **27**, Part I, 355–369.

SIMS, J. C., JR., and GRAY, H. J. (1958). Design criteria for autosynchronous circuits. *Proc. Eastern Joint Comput.* **T-114**, 94–99. Philadelphia, Pa., December 3–5, 1958.

STONE, H. S., and KORENJAK, A. J. (1965). Canonical form and synthesis of cellular cascades. *IEEE Trans. Electron. Comput.* **EC-14**, 852–862.

YOELI, M. (1965). A group-theoretic approach to two-rail cascades. *IEEE Trans. Electron. Comput.* **EC-14**, 815–822.

YOELI, M. (1967). Ternary Cellular Cascades. EE Publication 64, Technion, Israel Institute of Technology.

YOELI, M., and TURNER, J. B. (1966). Decompositions of Group Functions with Applications to Two-Rail Cascades. Sci. Rep 2, Contract AF 19(628)-5828, SRI Project 5876, AFCRL 66-472. Stanford Res. Inst., Menlo Park, California.

Chapter IX

PROGRAMMABLE CELLULAR LOGIC

WILLIAM H. KAUTZ

I. INTRODUCTION 369

II. PROGRAMMABLE CELLULAR ARRAYS 373

III. ARRAYS FOR ARBITRARY LOGIC 374
    A. Minterm and Prime-Implicant Arrays 375
    B. Reed and Modified Reed Arrays 380
    C. NOR Gate Arrays 387
    D. Sequential Arrays 393

IV. SPECIAL-PURPOSE ARRAYS 394
    A. Threshold Array 394
    B. Sorting Array 399
    C. Coding Array 406
    D. Interconnection Arrays 409

V. CONCLUSION 420

REFERENCES 421

ABSTRACT.   This chapter is concerned with the analysis and design of efficient micro-cellular logic arrays whose behavior is controlled by electronic programming (one or two flip-flops) in each cell. In addition to the usual advantages of cellularity, these arrays are characterized by their versatility—they can exhibit a very wide range of behavior from a single type of array. Both combinational and sequential arrays are discussed. Each is capable of completely arbitrary behavior but can also realize with a higher gate utilization efficiency many commonly encountered functions such as arithmetic units, counters, encoders, decoders, comparitors, checking logic, etc. Various theoretical techniques are employed in the derivation of these arrays and in the logical design problems associated with their application. From a practical point of view the arrays presented are highly compatible with large-scale-integrated (LSI) semiconductor technology.

I. INTRODUCTION

The synthesis problem of classical switching theory begins with the specification of a desired switching function, or set of functions, and one or more types of logical elements or " building blocks " out of which the network is to

369

be synthesized. The synthesis procedure or algorithm itself generates from this information all of the necessary interconnections between and among a set of copies of these elements and the inputs and outputs of the network. Note here that the synthesis itself takes place mainly in the topological domain since it is the interelement connections that vary from one problem to another, while the logical elements themselves are assumed to be fixed.

There is another equally valid approach that is gaining increased importance in present-day digital systems, which tend to be highly interconnection-limited. In this approach, it is the interconnections that are assumed to be fixed, and the logic of the individual elements is allowed to vary from one problem to the next. For example, the assumed interconnection pattern might take the form of a simple tree, as shown in Fig. 1. In this case the synthesis

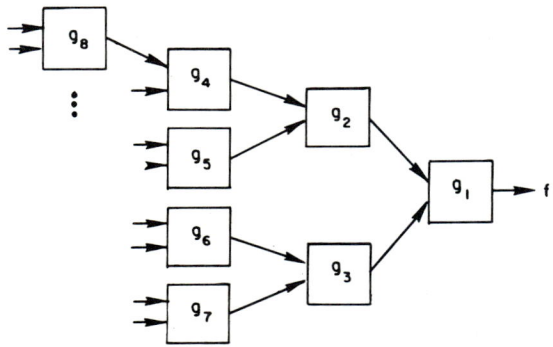

FIG. 1.   A fixed-interconnection network in the form of a tree.

problem consists of determining, for a specified output function f, the entire set of 2-variable element functions $g_1, g_2, \ldots$, corresponding to the nodes in the tree. While there is certainly some topological content in this problem, the synthesis itself really takes place entirely within the switching-functional domain. In contrast to the classical approach, this fixed-interconnection approach to synthesis can be expected to be relatively free of those difficulties that arise from the awkwardness of representing complex network topologies in algebraic form.

One of the simplest interconnection patterns for a fixed-interconnection switching network is provided by a cellular array that has connections between neighboring cells only. The type of array that is the most common and that turns out to be potentially the most useful is the 2-dimensional rectangular array with adjacent-cell interconnections in just two directions (that is, no diagonal connections), as shown in Fig. 2. The cell connections are usually,

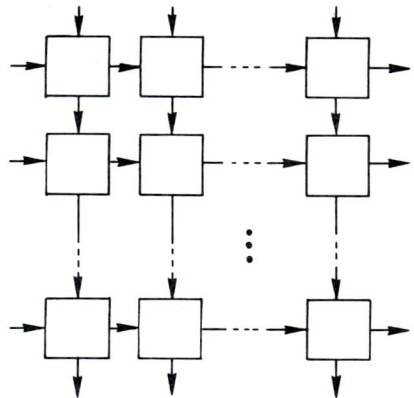

FIG. 2.   A cellular fixed-interconnection network.

but not always, assumed to be unilateral, and the input variables are usually, but not always, applied along just one of the edges—for example, the left-hand edge. The cells are endowed with sufficient logical variability so that a cellular array of suitable dimensions can realize a wide range of functional behavior.

The problem studied first is that of obtaining completely arbitrary combinational behavior for a given number n of input variables. By analogy with conventional switching structures, the most crucial questions underlying the synthesis of such combinational cellular logic arrays are:

1. What types of cell logic are logically most efficient (in some sense) for the realization of arbitrary functional behavior in cellular arrays?
2. How does the required size of the array depend upon the number of input variables, the number of terminals per cell, the function being realized, and the cell complexity? Also, what are the trade-offs among these parameters?
3. Can usable and practical synthesis algorithms be found to generate logically efficient cellular arrays, and what are they?

The corresponding questions can be asked for sequential networks, but it will be well to achieve some answers in the combinational case before becoming involved in the many intricacies of sequential phenomena.

In addition to the purely theoretical interest, there is actually considerable practical motivation for the study of cellular arrays. Modern batch fabrication techniques for digital circuitry produce circuit modules or "chips" that are characterized by very low cost per logic gate, very high reliability, and very large size (100's to 1,000's of gates per chip). However, they are inherently unrepairable and have a relatively small number of terminals in comparison with conventional digital modules and subsystems of comparable logical complexity. As a result, for unstructured logical circuitry there exist serious problems of fault testing and accommodation, circuit standardization, and logical design for which no good solutions are in sight at the present time.

Many of these problems can be solved or circumvented by arranging the chip circuitry in the form of a cellular array (Kautz, 1969; Minnick, 1967; Wahlstrom, 1967). These arrays, especially those in which each individual cell function is electronically programmable (as described in Section II) offer the following advantages:

1. Flexibility in function; that is, the number of different array types needed in a system is much reduced.
2. Testability—the ability to test a chip for faults, working entirely from the set of terminals on its edges (Kautz, 1967).
3. Fault accommodation—the capability for making use of arrays known to contain a small number of faults.
4. Subarray interconnectability—the capability for connecting many small arrays together to form large arrays.
5. High logical performance per chip (good gate utilization).
6. Ease of logical design.
7. Low power levels and high speed.
8. Ease of functionally decomposing a digital system into chip-size modules.

As a result of these engineering advantages of cellularity, it is important to determine what are the inherent capabilities of cellular arrays and how to synthesize them. From the digital systems point of view, the task of the logical designer can then shift from conventional gate-manipulation procedures, toward the selection of compatible and useful large-circuit functions and the derivation of techniques for realizing these functions in array form.

Following this motivational background, the next section introduces the notion of programmed cellularity and describes its implementation and its limitations. Section III contains a detailed description of the design and operation of several types of programmable cellular arrays for the realization of arbitrary logical behavior. In Section IV, four different types of special-purpose cellular arrays are offered as examples of how certain restricted types of functional behavior can be achieved naturally and efficiently through cellularity.

II. PROGRAMMABLE CELLULAR ARRAYS

In fixed-interconnection networks, such as those illustrated in Figs. 1 and 2, there are fundamentally two different ways in which one may actually achieve in practice the desired degree of node or cell variability. On one hand, one may maintain an inventory of all different cell types that might be required, and during fabrication of a particular array insert in each cell location the appropriate one of these types as specified by the synthesis algorithm or the logic designer. Clearly, this approach requires a large inventory of cell types or masks to achieve a significant variability. Alternatively, one might employ a single " universal " cell that contains enough gatery to realize all of the cell functions that might be needed. The choice of cell function or " mode " of cell behavior at a particular array position for a particular design must then be externally imposed. This approach uses a completely uniform array, but one in which the cells are more complex.

The first cellular arrays that were proposed assumed the use of " cutpoints " —small photocells or switches, or merely points at which selected conductors could be mechanically broken—in order to specialize each cell to its selected function (Minnick, 1964). Later proposals assumed that each cell contained a small number of flip-flops, possibly in shift-register form, that are externally loaded prior to use of the array, in order to "program" each cell to its proper mode and the array as a whole to a desired functional behavior (Minnick, 1967; Wahlstrom, 1967). This latter arrangement has several advantages. Not only does such a programmable array have a very wide range of possible behavior, but if the cells are selected to have a sufficiently rich set of modes, then the array as a whole can realize arbitrary behavior and is truly universal, subject only to limitations on its overall size. Thus, one could even build an entire digital processor out of this single type of universal array. (However, it would probably not be economical to do so.) Another advantage arises from the fact that a programmable logic array can also be regarded as a memory array, each of whose storage-digit locations (cell programming registers) is augmented with a small amount of logic to permit some limited processing to take place on the digits or words stored in the memory. As a result, some programmable logic arrays can also be employed as content-addressed memories or memories that are specialized to carry out a special function, such as sorting, counting, rounding, etc. As an example, a particularly efficient array for sorting is described in Section IV,B. Finally, programmability of a cellular array renders it much more susceptible to techniques of fault accommodation, inasmuch as the same physical array can normally be programmed in more than one way to realize a given function. In some cases, a defective cell can be bypassed completely by appropriate programming. In others, an entire row or column containing the faulty cell must be taken from use (as in a conventional core memory). In still other cases, appropriate reprogramming can utilize the cell in one mode even though it has failed in another.

In the next two sections we offer several examples of cellular arrays that have been found to be particularly efficient for the realization of either arbitrary combinational behavior (Section III) or well-defined and useful particular functions (Section IV). Greatest emphasis has been placed on the use of very simple cells. In each case, we describe the cell logic, the cell complexity, the size of the array required, and (where known) an effective synthesis procedure.

Some of the arrays intended for the realization of arbitrary combinational logic can also be used for arbitrary sequential logic. These possibilities are indicated, but no really good solutions to the general sequential problem are currently known.

III. ARRAYS FOR ARBITRARY LOGIC

Arrays will be treated first in which one of the cell inputs (the left-hand input or x-input) is connected directly through the cell to the opposite side (the right-hand output or $\hat{x}$-output). Therefore, the left-edge input variables to an array of such cells are bussed horizontally through the array to the right-hand

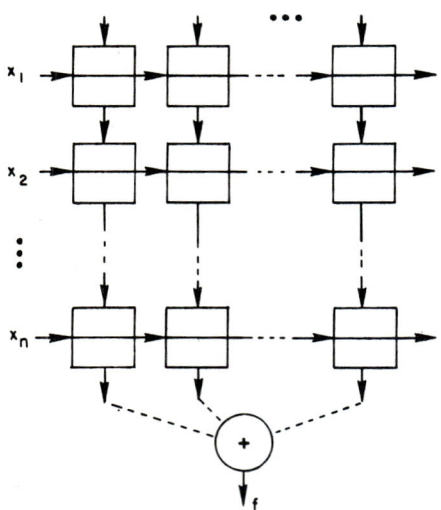

FIG. 3.   A row-bussed cellular array.

edge, as indicated in Fig. 3, and all logical subfunctions are formed in a down-ward direction in the columns of the array. Next treated are cellular arrays without such row bussing of the input lines, so that functions and subfunctions may be formed within the array in both horizontal and vertical directions.

A. MINTERM AND PRIME-IMPLICANT ARRAYS

The simplest and most naïve way to make use of an array of the form shown in Fig. 3 is to employ the canonical expansion of an arbitrary switching function $f(x_1, x_2, \ldots, x_n)$ of n input variables in terms of the fundamental products, or *minterms* in these variables; namely:

$$f = a_0 \bar{x}_n \cdots \bar{x}_2 \bar{x}_1 + a_1 \bar{x}_n \cdots \bar{x}_2 x_1 + a_2 \bar{x}_n \cdots x_2 \bar{x}_1 + \cdots + a_{N-1} x_n \cdots x_2 x_1$$

$$(1)$$

The set of binary coefficients $a_0, a_1, a_2, \ldots, a_{N-1}$, where $N = 2^n$, completely and uniquely specify the function to be realized. For the cellular realization, each minterm needed (that is, each minterm for which $a_i = 1$) for a given function will be realized by a separate column of cells in which the cell in the jth row performs the AND operation in the downward direction between either x_j or $\bar{x}_j$ and the partial-product signal received from the cell above. Designating the vertical input and output of a typical cell by z and $\hat{z}$, respectively, and the generic horizontal input and output of the cell by x and $\hat{x}$, respectively, the equations for the two cell modes may be written:

$$\hat{z} = z\bar{x} \qquad \hat{z} = zx$$
$$\hat{x} = x \qquad \hat{x} = x \qquad (2)$$

The corresponding cell logic circuits for these two modes are illustrated in Fig. 4a. Indicating the single programming digit, which is held in a flip-flop, by y, the entire cell is shown in Figure 4b (exclusive of the circuitry for initially clearing and for setting up the flip-flop). As a result, if the input variables $x_1, x_2, \ldots, x_n$ are applied in any fixed order to the left edge of an array (as in Fig. 3) of such cells then by suitable programming each column of the array generates an arbitrary minterm in these variables.

Probably the simplest way to sum a set of these minterms, in order to form the function f, is to combine all of the lower-edge column outputs in a large OR gate that is exterior to the cellular array proper, as shown in Fig. 3. In many device technologies this OR action may be achieved simply by wiring directly together all of the column output lines. An alternative arrangement, one that is especially appropriate when several functions are being formed

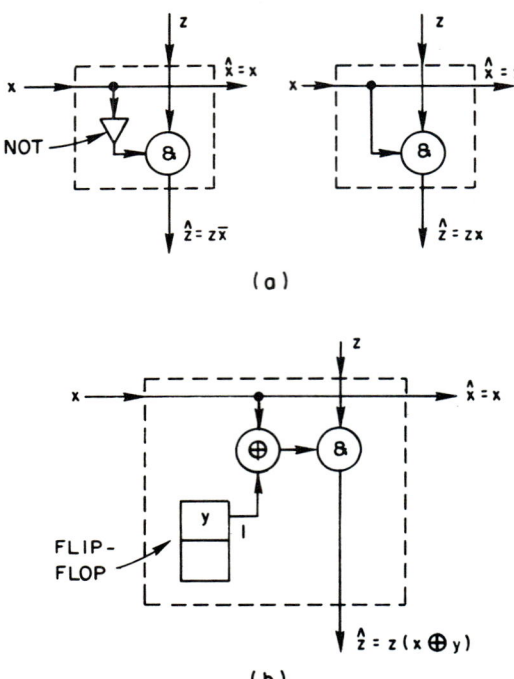

FIG. 4. Cell modes and logic for the minterm cell, corresponding to Eq. (2). (a) The two modes. (b) The combined cell.

simultaneously from the same set of input variables, is to employ in addition to the original "minterm" array a second "summation" array. This second array has *column*-bussed cells, and its upper-edge inputs are connected directly to the corresponding lower-edge outputs from the minterm array. This combined configuration, shown in Fig. 5, employs in the summation array a type of two-mode cell that allows an arbitrary subset of column inputs (minterms) to be summed in each row. Thus, each cell of the summation array may be described by the pair of mode equations

$$\hat{z}^* = z^* \qquad \hat{z}^* = z^*$$
$$\hat{x}^* = x^* \qquad \hat{x}^* = x^* + z^* \tag{3}$$

where the asterisks distinguish the summation array. These two modes are illustrated in Fig. 6a, and the combined cell employing the programming flip-flop y^* in Fig. 6b. Consequently, any m functions of n variables can be realized with one array pair, composed of simple cells having together $n + m$

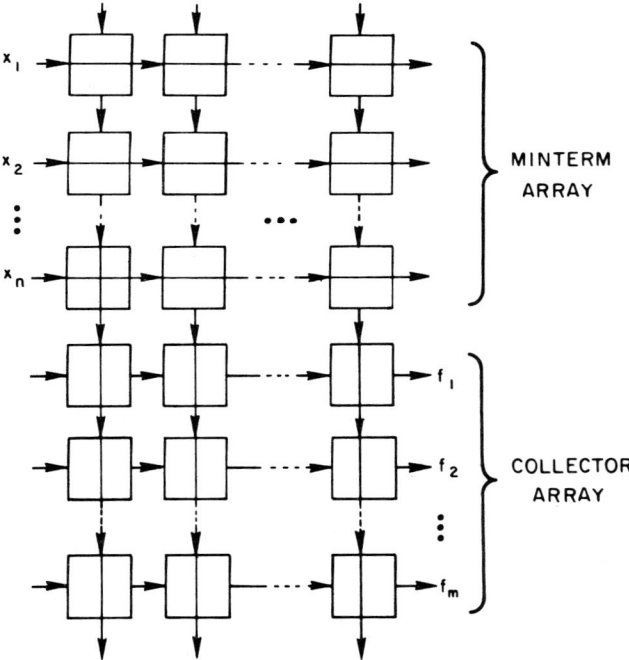

FIG. 5.   Augmentation of the row-bussed array for the simultaneous realization of many functions.

rows and a number of columns equal to the largest number of minterms required for the set of all m functions (N at most).

If only a single type of array rather than an array pair is desired, then the minterm and summation arrays may be merged. This may be done in either of two ways, as indicated in Figs. 7 and 8. In Fig. 7, the summation array has been folded over and superimposed on the minterm array so that each resulting cell contains the circuitry of both constituent arrays. Each cell then contains two programming flip-flops y and y^*, which generate four modes [Eqs. (2) and (3)], and has two intercell lines in each direction. The cell complexity has now doubled, and the height of the array is MAX(m, n). The arrangement of Fig. 8 also employs a cell with four modes, but no row merging is assumed. Thus, all cells in the same row participate *either* in minterm formation or in summing, but not both. The cell equations are again (2) and (3) together, but the asterisks are now removed from Eq. (3).

$$
\begin{array}{cccc}
\hat{z} = z\bar{x} & \hat{z} = zx & \hat{z} = z & \hat{z} = z \\
\hat{x} = x & \hat{x} = x & \hat{x} = x + z & \hat{x} = x
\end{array}
\qquad (4)
$$

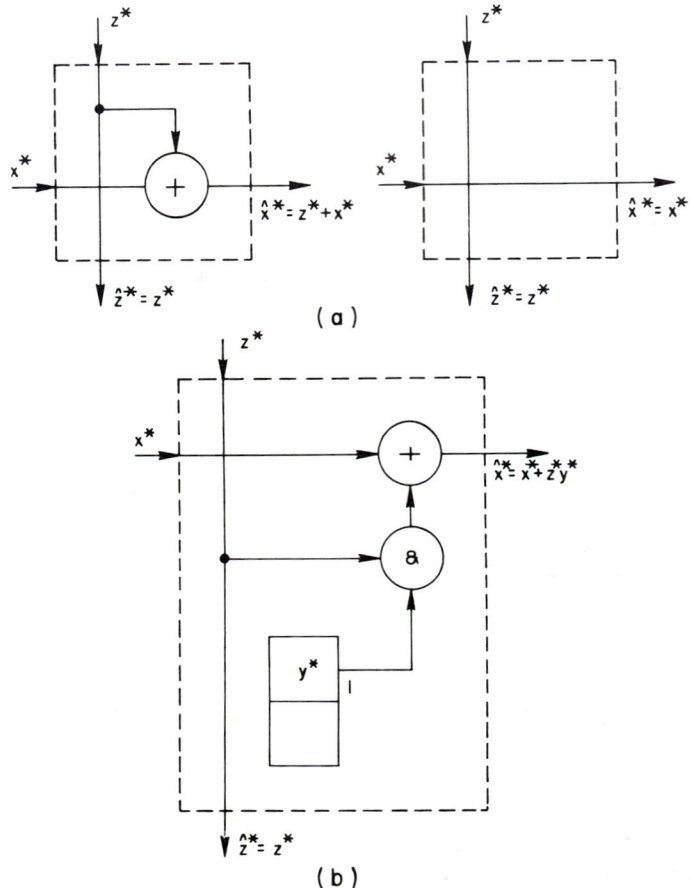

FIG. 6.   Cell modes and logic for the summation cell, corresponding to Eq. (3). (a) The two modes. (b) The combined cell.

The cell complexity is about the same as in Fig. 7, but the required number of rows is now $n + m$. Since the second programming digit y^* is the same for all cells in each row, it is most conveniently introduced along a bus supplied from the left-hand edge of the array. In both Fig. 7 and Fig. 8 the number of columns has remained the same as before.

If, instead of bussing one of the programming digits along rows, all four cell modes that arise in the merging are retained for each individual cell, then an array results that is narrower than before, since a shorter functional expansion can now be used. The four modes are still given by Eq. (4), and the corresponding cell is shown realized in NOR gate form in Fig. 9. It is apparent from Eq. (4) that the first three modes may be used together to form products

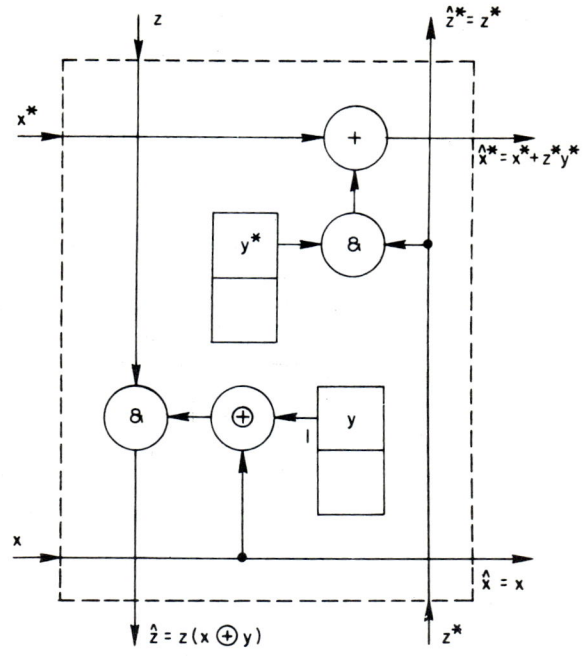

FIG. 7. First version of merged minterm and summation cells.

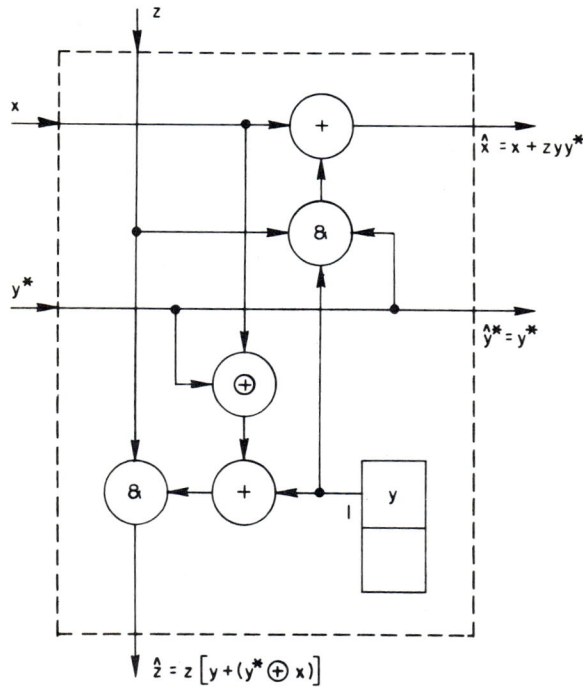

FIG. 8. Second version of merged minterm and summation cells.

379

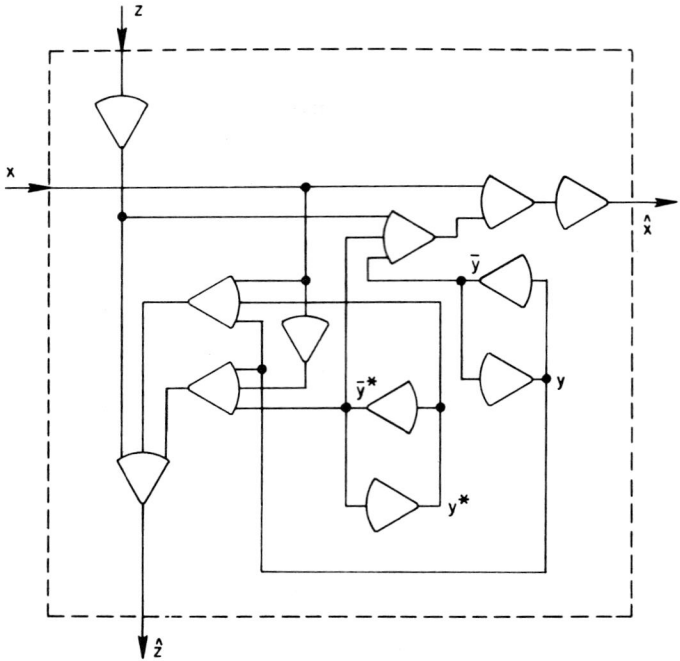

FIG. 9.   NOR gate realization of the second version of merged minterm and summation cells.

of natural and complemented variables in which, unlike minterms, some of the input variables are missing, i.e., products that have the form of *prime implicants*. Techniques for deriving minimal-length expansions of a switching function or set of functions in terms of prime implicants are very well-known. These expansions almost always have many fewer terms than do minterm expansions of the same functions. Consequently, the width of an array composed of cells as shown in Fig. 9 will normally be much less than an array of cells as shown in Fig. 7 and 8.

B. REED AND MODIFIED REED ARRAYS

An alternative but analogous family of arrays may be based on the well-known *Reed* expansion of an arbitrary switching function, namely:

$$f = b_0 \oplus b_1 x_1 \oplus b_2 x_2 \oplus b_3 x_1 x_2 \oplus b_4 x_3 \oplus \cdots \oplus b_{N-1} x_1 x_2 \cdots x_n \quad (5)$$

In comparison with the minterm expansion, Eq. (1), the Reed expansion employs EXCLUSIVE OR ($\oplus$) rather than INCLUSIVE OR ($+$) summation, and each complemented input variable $\bar{x}_j$ is replaced by a binary 1. Thus, in the Reed array the cell mode equations (2) for the minterm array are replaced by the set

$$\hat{z} = z \qquad \hat{z} = zx$$
$$\hat{x} = x \qquad \hat{x} = x \qquad (6)$$

and the summation array equations (3) now become

$$\hat{z}^* = z^* \qquad \hat{z}^* = z^*$$
$$\hat{x}^* = x^* \qquad \hat{x}^* = x^* \oplus z^* \qquad (7)$$

Combination of Eqs. (6) and (7) to yield a single set of equations for the 4-mode array now yields two identical modes—the first modes of (6) and (7). To avoid this duplication, one might just as well employ for the fourth mode a different pair of logical functions $\hat{z}$ and $\hat{x}$, chosen to allow the width of the array to be reduced as much as possible through the use of a shorter functional expansion. There appear to be three interesting possibilities.

First of all, one might employ for the fourth mode the first mode of Eq. (4), yielding the set:

$$\hat{z} = z \qquad \hat{z} = zx \qquad \hat{z} = z \qquad \hat{z} = z\bar{x}$$
$$\hat{x} = x \qquad \hat{x} = x \qquad \hat{x} = x \oplus z \qquad \hat{x} = x \qquad (8)$$

In this case, the appropriate functional expansion is an EXCLUSIVE OR sum of prime-implicant-like terms, i.e., each term is a single product of selected input variables, natural and complemented. Simple functions can be readily manipulated algebraically into short expansions of this type. Although some bounds are available on the length of such " modified Reed " expansions of arbitrary functions (Cohn, 1962; Even *et al.*, 1967; Mukhopadhyay and Schmitz, 1970), good synthesis algorithms are not currently available. The cell logic of the four modes and the combined cell are shown in Fig. 10a and 10b, respectively.

For the second possibility, let the fourth mode be employed for vertical INCLUSIVE OR summation:

$$\hat{z} = z \qquad \hat{z} = zx \qquad \hat{z} = z \qquad \hat{z} = z + x$$
$$\hat{x} = x \qquad \hat{x} = x \qquad \hat{x} = x \oplus z \qquad \hat{x} = x \qquad (9)$$

Each column function generated in this array is a mixed AND and INCLUSIVE OR concatenation of an ordered subset of the input variables. The terms in the functional expansion are therefore all positive unate functions of a

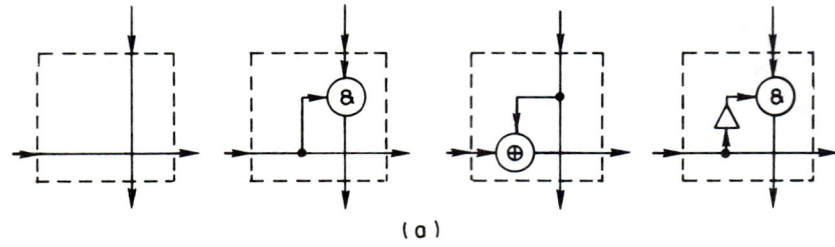

(a)

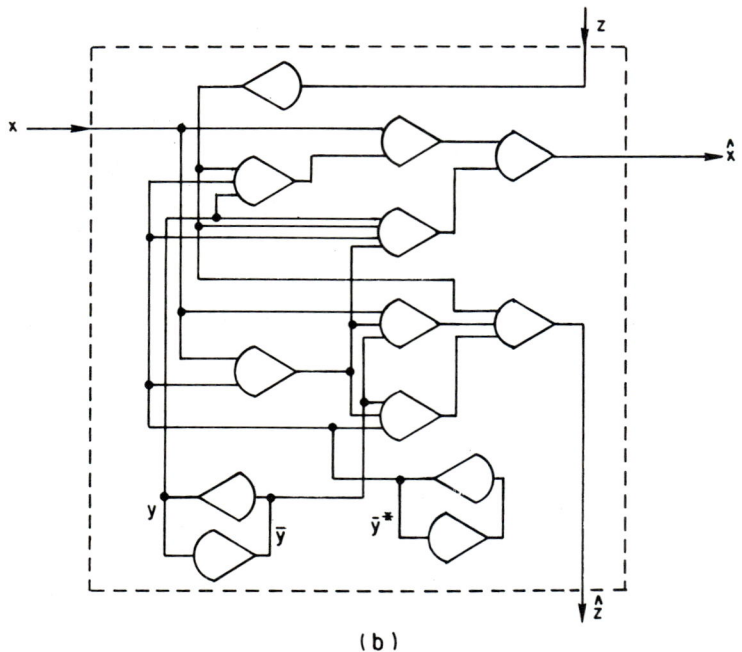

(b)

FIG. 10.   Cell modes and logic for the first Reed cell, corresponding to Eq. (8). (a) The four modes. (b) the combined cell.

certain type. To obtain a functional interpretation of this expansion, suppose first that only the second and fourth modes in Eq. (9) are used to form column functions (expansion terms). In this case, each column of the array behaves as a concatenation of majority gates, the free input of each of which is the programming digit y, as indicated in Fig. 11a. The majority gate behaves as an AND gate when $y = 0$ and as an OR gate when $y = 1$. Such a cascade may be regarded as the carry chain for a parallel binary adder of the two n-digit

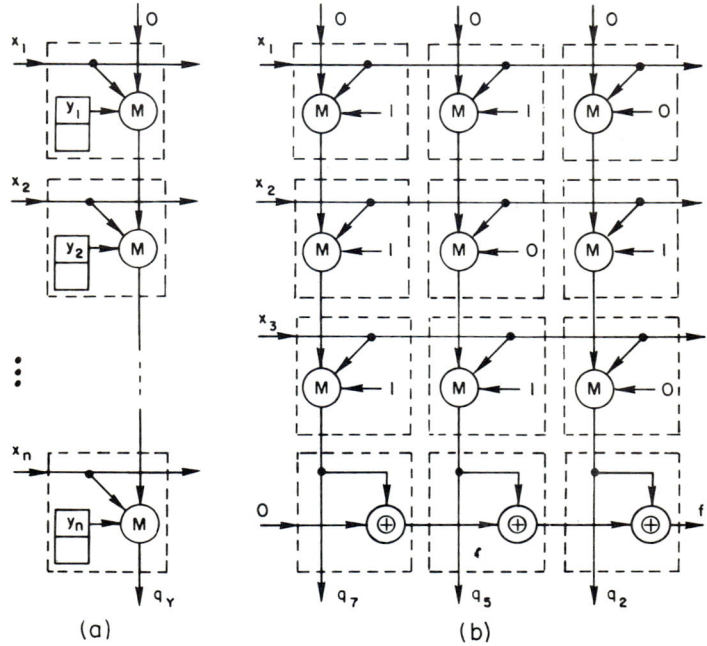

FIG. 11.   Illustration of column-function formation with the second Reed cell (without first mode). (a) One column of the array. (b) The entire array.

binary numbers $X = (x_n \cdots x_2 \, x_1)$ and $Y = (y_n \cdots y_2 \, y_1)$. An overflow will be created during addition when and only when

$$X + Y \geq N$$

In other words, the output column of the table of combinations for the column function q_Y, say, consists of a string of $N - Y$ 0's in rows $0, 1, 2, \ldots, N - Y - 1$, followed by a string of Y 1's in rows $N - Y, N - Y + 1, \ldots, N - 1$. More compactly, the function q_Y can be expressed in the form of its characteristic vector, which is the column of the table of combinations turned on its side; namely,

$$q_Y = [\underbrace{000 \cdots 0011}_{\substack{N - Y \\ 0\text{'s}}} \underbrace{\cdots 111}_{\substack{N \\ 1\text{'s}}}] \tag{10}$$

Thus, by injecting the appropriate n-digit binary number Y into the programming flip-flops in a column, one may realize as the output from that column any function whose characteristic number has the form (10).

It is now easy to see that an arbitrary function can be expressed as an EXCLUSIVE OR sum of several of these functions of the form (10), each of which has a Y value corresponding to the start of a 0-string or 1-string in the characteristic vector of f. For example, the function $f = [0110\ 0011]$ may be expressed as $f = [0111\ 1111] \oplus [0001\ 1111] \oplus [00000011] = q_7 \oplus q_5 \oplus q_2$, leading to the majority-gate array shown in Fig. 12. The left-end boundary digit in the summation row [whose cells are in the third mode in Eq. (9)] is simply the value of f when $X = 0$, i.e., the first digit in the characteristic vector for f.

Consequently, an array, each of whose cells can realize either the second, third, or fourth mode of Eq. (9), can be used to realize an arbitrary function f, using a number of columns that is one less than the total number of 0-strings and 1-strings in the characteristic vector for f. The number of strings of a given function certainly depends upon the order in which the input variables are applied to the rows of the array. One should, therefore, select an input-variable permutation that minimizes the total number of strings. For functions chosen at random from the set of all possible functions of n variables, Elspas and Short (1964) have shown that the expected number of strings is about $N/3$, and the maximum number is $2N/3$. For most common functions, however, this number is certainly much smaller.

If the first mode in Eq. (9) is now included, considerable additional flexibility is offered in the formation of the column functions. Again, algebraic manipulation of simple functions presents no real problem, but no systematic approach is presently known for minimizing the width of the array in the general case. The four cell modes and the complete cell logic corresponding to Eq. (9) are shown in Fig. 12.

The third alternative for the fourth mode corresponds to a simple interchange of leads (Elspas *et al.*, 1967):

$$\begin{array}{llll} \hat{z} = z & \hat{z} = zx & \hat{z} = z & \hat{z} = x \\ \hat{x} = x & \hat{x} = x & \hat{x} = x \oplus z & \hat{x} = z \end{array} \qquad (11)$$

These four modes are shown in Fig. 13a, and the corresponding combined cell in Fig. 13b. This new mode may be used in an array based on the Reed expansion in order to effect an exchange of some of the row and column variables, thereby allowing subfunctions to be formed that can then play the role of additional inputs. However, the capabilities of this type of array are actually much greater, in terms of the range of logical behavior possible from a limited number of cells. As evidenced by many examples of simple functions that have been worked out, a large variety of functional decompositions can be directly accommodated. Subarrays for the realization of a few commonly occurring functions and subfunctions are shown in Fig. 14. For the synthesis

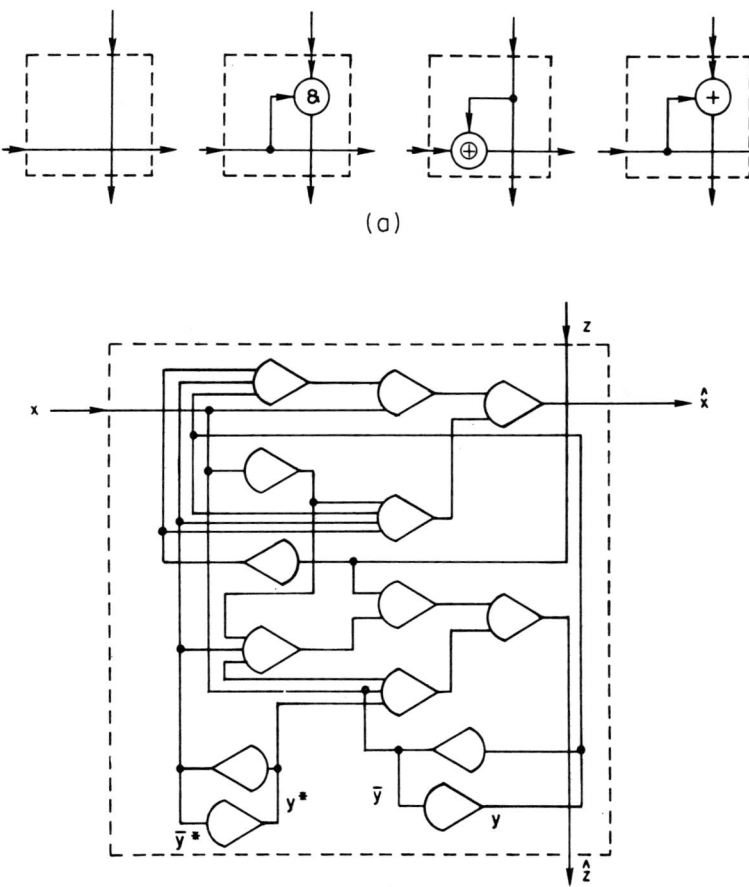

FIG. 12. Cell modes and logic for the second Reed cell, corresponding to Eq. (9). (a) The four modes. (b) The combined cell.

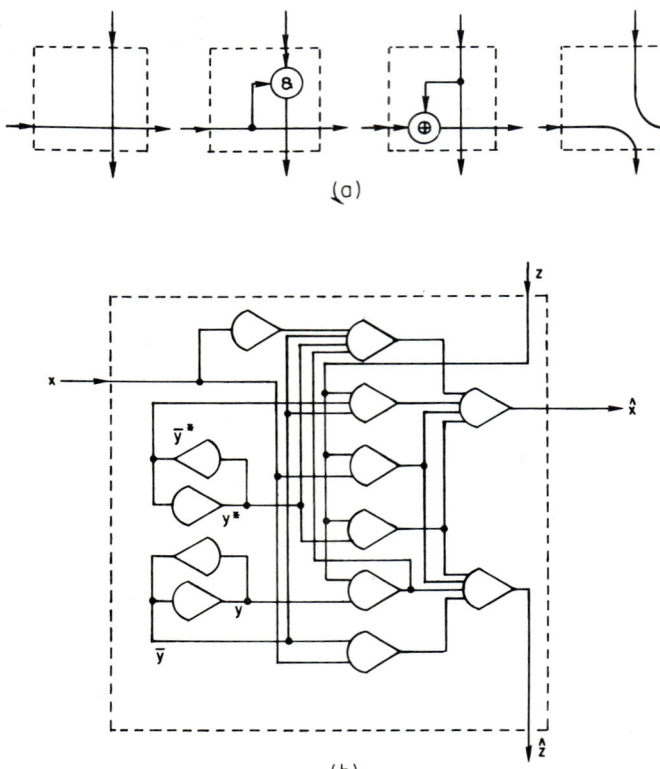

FIG. 13.   Cell modes and logic for the third Reed cell, corresponding to Eq. (11). (a) The four modes. (b) The combined cell.

of arbitrary functional behavior, it is perhaps best to start with an awareness of how blocks of variables may be combined to form product or sum terms, and to then make use of familiar functional expansions, or of algebraic expressions that are prescribed or easily derived for a given function. Several such subarrays are displayed in Fig. 14d–g. Subarrays are shown for forming simple variable products and EXCLUSIVE OR sums, for masking and complementation, and for the permutation of leads. The subarray shown in Fig. 14h suggests by example how a set of leads may be permuted within an array. Arrays for arbitrary permutation are discussed in Section IV,D.

C. NOR GATE ARRAYS

It is sometimes more convenient to form the cellular realization of a given function directly from an already available noncellular network composed of conventional gates, instead of basing the array upon some functional expansion. This approach often allows the designer to make greater use of his prior experience, much of which may be intuitive, in the synthesis of switching networks. The 4-mode cell depicted in Fig. 15 is proposed for this situation, it being assumed that the prior realization employs simple NOR gates as the primitive logical elements.

As may be seen from Fig. 15a, the cell equations are now

$$\hat{z} = \bar{z}\bar{x} \qquad \hat{z} = x \qquad \hat{z} = \bar{z} \qquad \hat{z} = z$$
$$\hat{x} = \bar{z}\bar{x} \qquad \hat{x} = z \qquad \hat{x} = x\bar{z} \qquad \hat{x} = x \tag{12}$$

The second and fourth modes of this "NOR network" cell provide simple line permutation and no logic. It is shown in Section IV,D that an arbitrary permutation of leads can be achieved with a sufficient number of cells in just these two modes. The first mode in Fig. 15a represents the fundamental NOR gate itself in the network that is being distorted into a cellular form. Therefore, these three modes are adequate for the representation in cellular form of any loop-free network of 2-input NOR gates.

Actually, however, the inclusion of the third mode in Fig. 15a allows a substantial reduction to be made in the total number of cells required for a realization, and also simplifies considerably the process of distorting the given network into a cellular array, each cell of which is in one of these four modes.

The subarrays shown in Fig. 15c–f illustrate how some important subfunctions can be formed locally within a larger array. Note in particular the use of the third mode for forming variable products along rows, as in

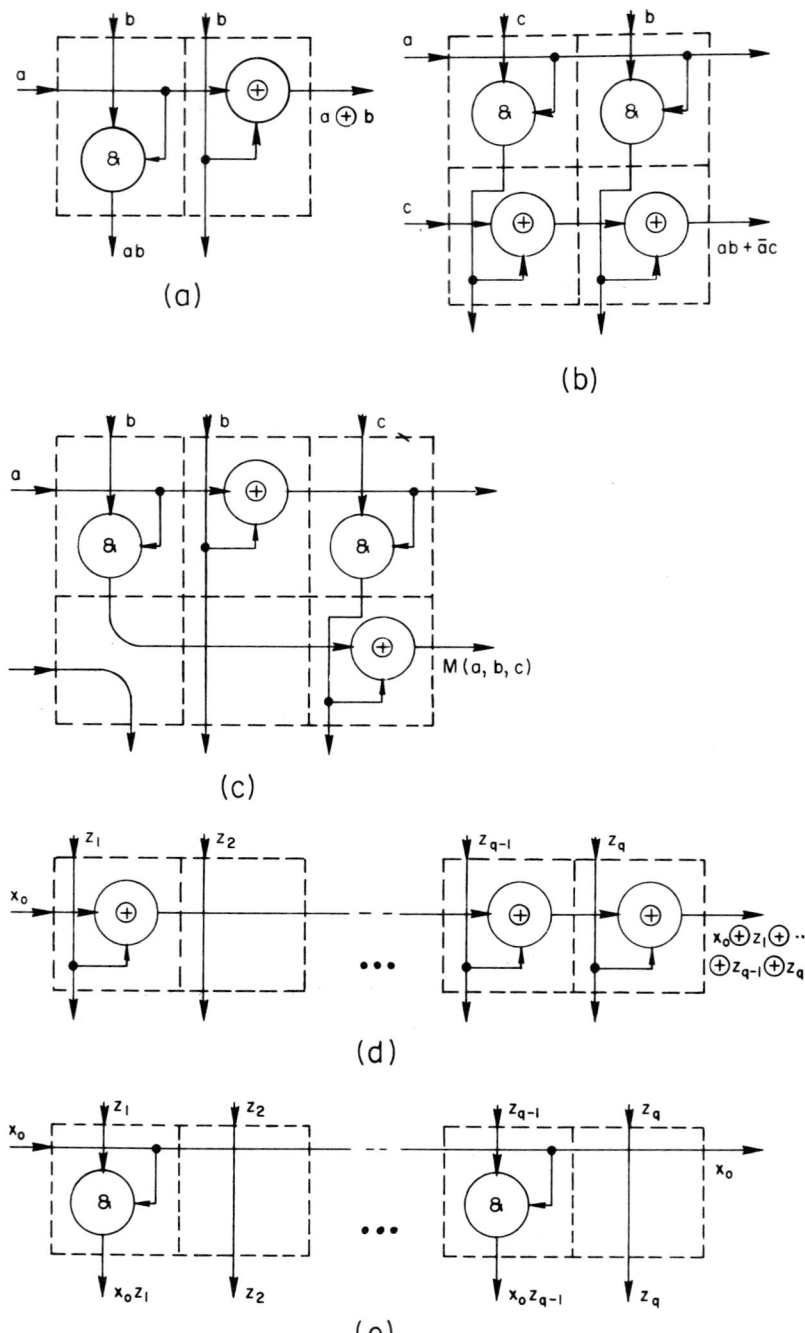

FIG. 14. Subarrays for the realization of some common functions, using the cell of Fig. 13. (a) Half adder. (b) Switch. (c) Majority gate. (d) Parity sum. (e) Masking.

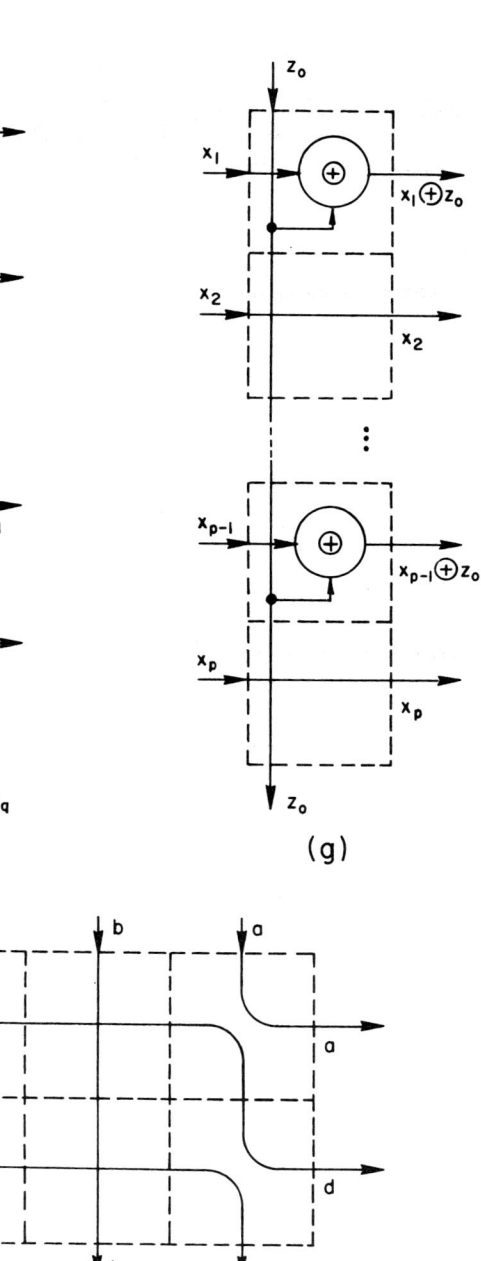

(f)

(g)

(h)

FIG. 14.   (Continued). (f) Product. (g) Complementation. (h) Permutation.

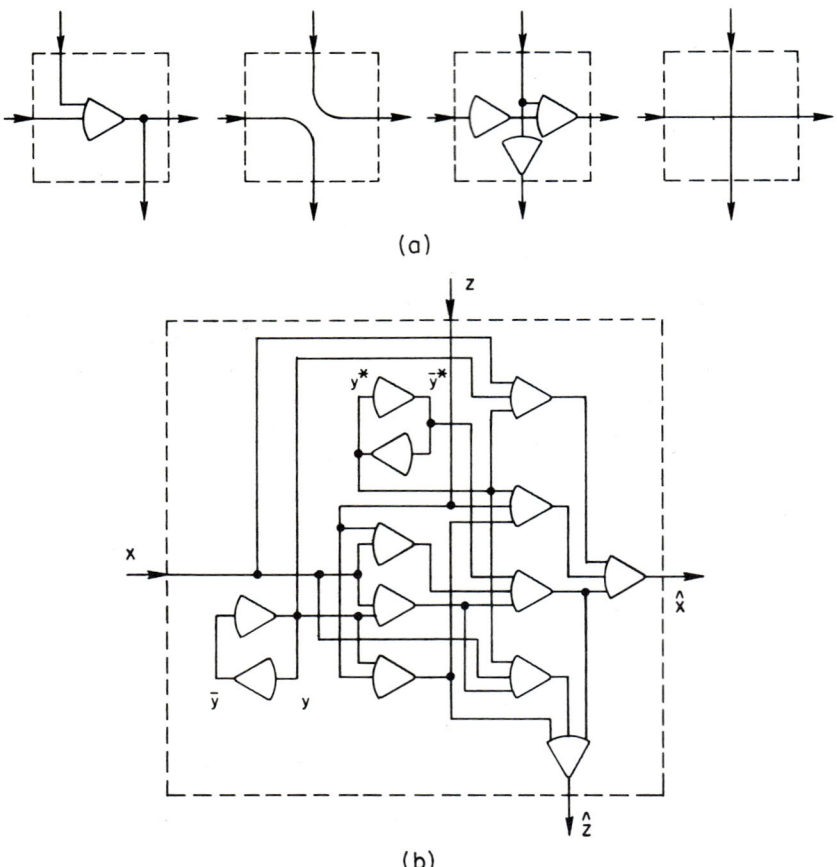

FIG. 15.   The NOR-gate array. (a) Cell modes. (b) Cell logic.

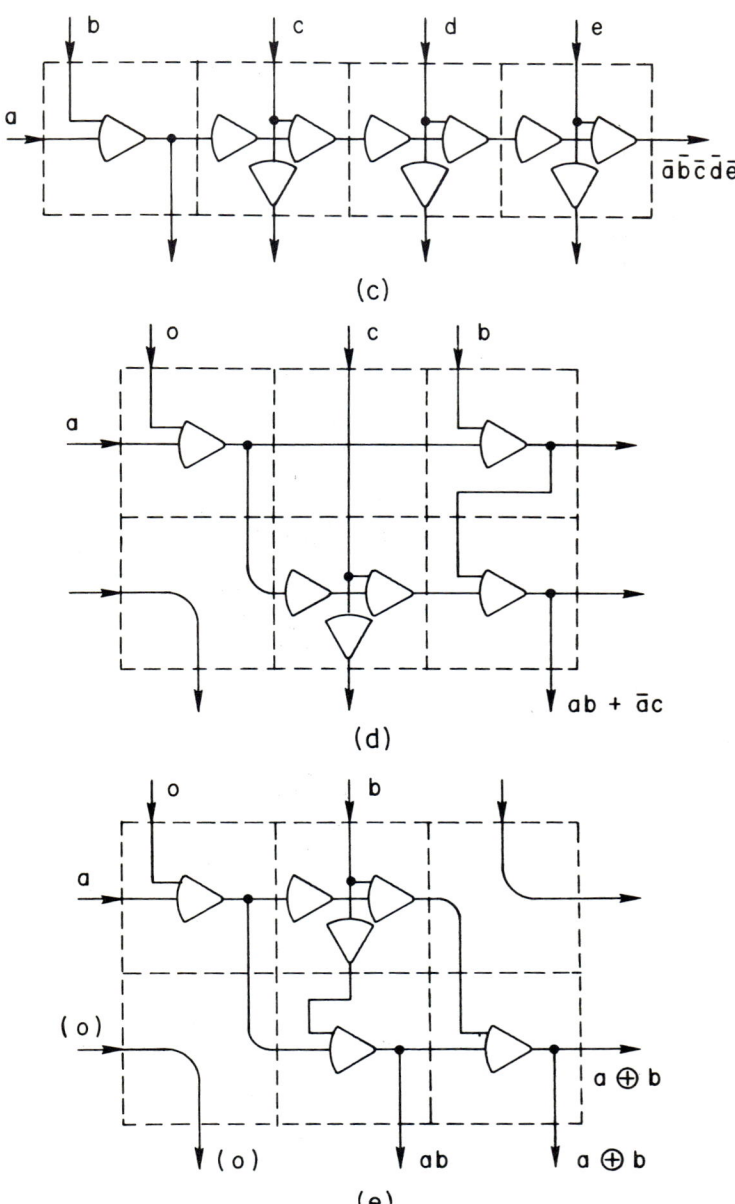

FIG. 15 (Continued). (c) NOR product. (d) Switch. (e) Half adder.

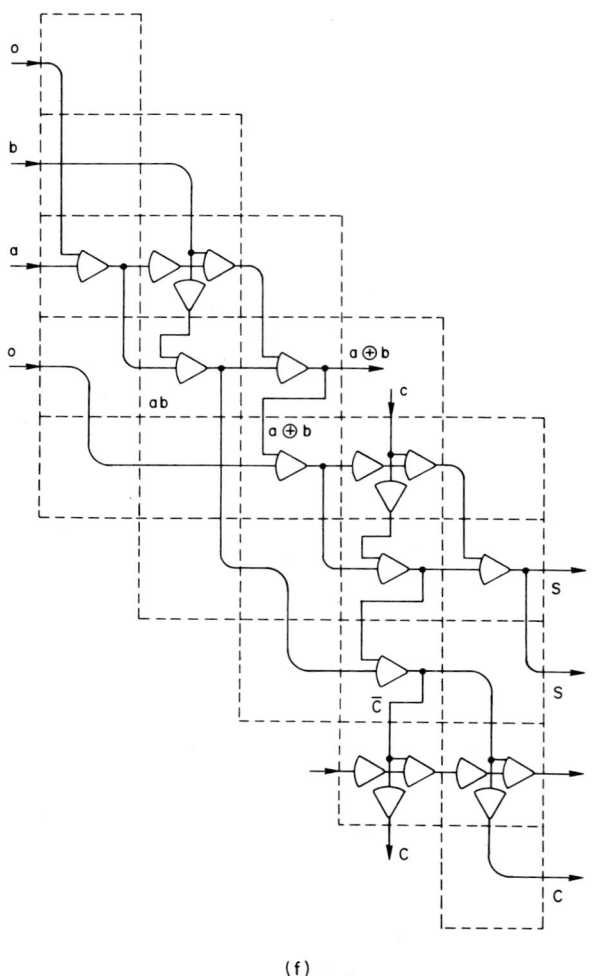

(f)

FIG. 15 (Continued). (f) Full-adder stage.

Fig. 15c, so that a row of cells can behave as a single NOR gate with fan-in greater than 2. Figures 15e and 15f show subarrays that can be iterated vertically to form an add-one network (binary counter) and a parallel binary adder, respectively.

D. SEQUENTIAL ARRAYS

The most direct realization of arbitrary sequential behavior employs a multioutput combinational array such as that shown in Fig. 5, one of its folded versions employing the cell of Fig. 7 or Fig. 8, or any of the other arrays described in the previous two subsections that are based on cells described by Eqs. (8), (9), (11), or (12). The unit delays (single register stages) or flip-flops that provide the storage essential to sequential behavior are exterior to the array proper and are connected in feedback fashion between some of the array outputs and some of the array inputs. The balance of the array outputs and inputs serve as external outputs and inputs, respectively, of the sequential network as a whole.

A typical configuration is shown in Fig. 16, although in general the array outputs and inputs need not be confined to the right and left sides of the

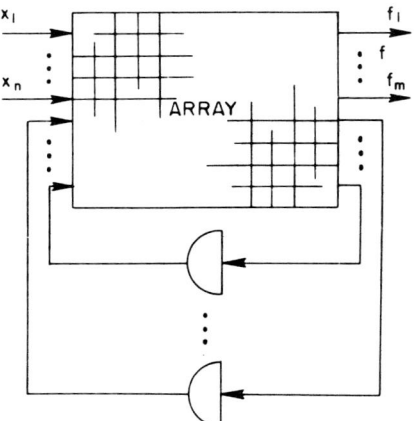

FIG. 16.   Standard network form for the realization of sequential behavior.

array and the order of these signal lines is not normally important. It is well-known that completely arbitrary sequential behavior can be achieved with this form of realization, even without special restrictions on the particular state assignment used. In most cases, however, a much smaller array results if one chooses the state assignment advantageously. To a first approximation the criteria normally employed in solving the state assignment problem lead to functional expansions for the sequential (next-state) functions that have a reduced number of terms and these lead in turn in the general direction of a minimal-width array. However, no systematic method is available for picking a judicious state assignment to minimize directly the total or the maximum number of terms.

Other cellular realizations of sequential machines have been proposed by Low and Maley (1961), Ferrari and Grasselli (1969), and Arnold *et al.* (1970).

IV. SPECIAL-PURPOSE ARRAYS[1]

The subsections that follow describe four arrays that are capable of realizing restricted classes of functional behavior instead of completely arbitrary behavior. They all tend to operate with a high logical efficiency in that simple cells and small arrays are required in comparison with competitive alternatives.

A. THRESHOLD ARRAY[2]

A threshold switching function $f(x_1, x_2, \ldots, x_n)$ of the n input variables $x_1, x_2, \ldots, x_n$ is a switching function which can be represented in the form

$$f = 1 \quad \text{iff} \quad \sum_{i=1}^{n} w_i x_i - w_0 \geq 0$$

in which the summation is arithmetic rather than Boolean, the w_i are *weights* (which may be assumed to be positive or negative integers, without loss of generality), and the integer w_0 is called the *threshold*. While not all switching functions are threshold functions, any switching function can be composed of threshold functions, this composition corresponding to a network interconnection of two or more functional elements, each of which realizes a particular threshold function.

Figure 17 illustrates (a) the main array, (b) a typical cell, and the cell logic equations of a cellular array capable of realizing any threshold function. The binary input variables $x_1, x_2, \ldots, x_n$ are distributed on buses along the rows of the array, one input to a row, from drivers external to the array proper.

---

[1] The results reported in this section are extracted from four technical papers (Kautz, 1969; Kautz, 1967b; Levitt and Kautz, 1969; Kautz *et al.*, 1968) reproduced by permission of the Institute of Electrical and Electronics Engineers.
[2] From Kautz, 1967b.

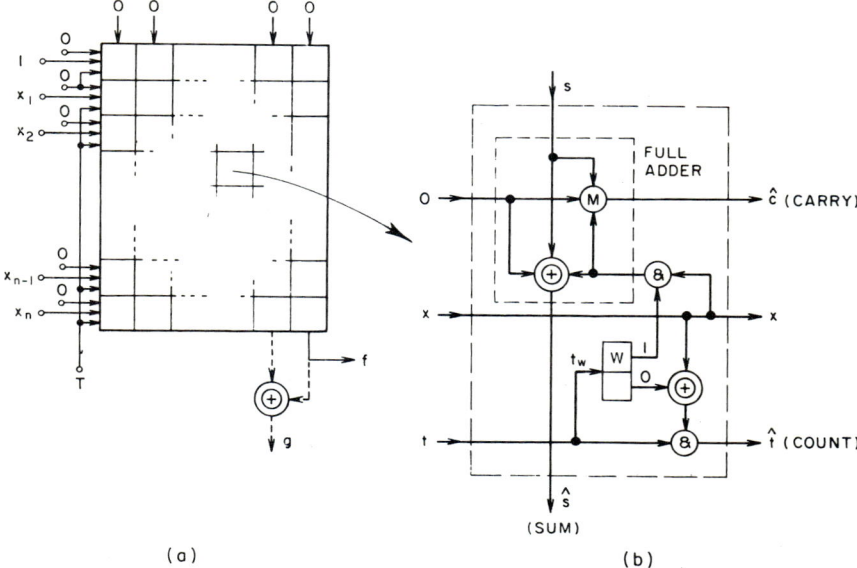

(a)   (b)

FIG. 17. Cellular threshold array. (a) The main array. (b) A typical cell. The cell
logic equations are

$$\hat{s} = s \oplus c \oplus wx \qquad \hat{t} = t(\overline{w} \oplus x)$$
$$\hat{c} = M(s, c, wx) \qquad w' = w \oplus t$$

The single output f emanates from the lower right-hand corner of the array.
Observe also that each main cell consists of a flip-flop with contents w; a full
binary adder, with outputs $\hat{s}$ (the sum) and $\hat{c}$ (the carry), and inputs s (sum
digit from the cell above), c (carry digit from cell to the left), and wx (the flip-
flop contents, gated by the input variable x); and counter circuitry, which
connects the set of w flip-flops in each row as a binary up/down counter, to be
described below. (Subscripts will be appended to these cell signal variables
when it becomes necessary to distinguish the position of the cell in the array.)
The array has $n + 1$ rows and m columns.

The m flip-flops in the ith row of this array perform the function of a
register that stores the ith weight w_i, which is an integer and is encoded as an
m-digit binary number $(w_{im} \cdots w_{i2} w_{i1})$, with the least significant digit w_{i1} at
the left side of the array. Each negative weight $w_i < 0$ is represented in two's-
complement form—that is, as a binary number with value $2^m - |w_i|$. The top
row stores the threshold in the form $(2^{m-1} - w_0)$.[3] The set of m full adders in
the ith row of the array forms a full parallel binary adder that adds the stored
weight w_i, gated by the value of x_i, to the number received in parallel from

[3] To obtain $\bar{f}$ instead of f from array, the threshold should be stored as $2^m - w_0$ instead.

the row above, to form a sum that is passed down to the row below. As a result, the number appearing at the bottom of the array is

$$S \equiv \sum_{i=1}^{n} w_i x_i - w_0 + 2^{m-1} (\text{mod } 2^m)$$

Provided only that m is large enough to accommodate the extreme excursions of value of S, as the binary inputs x_i are varied, this sum S will be greater or less than 2^{m-1}, depending only on whether that portion of the sum involving weights is greater or less than zero. That is, S will fall in the range $2^{m-1} \leq S \leq 2^m - 1$ if and only if

$$\sum_{i=1}^{n} w_i x_i - w_0 \geq 0 \qquad (13)$$

which is the defining inequality for the output condition $f = 1$ for a threshold element. This condition on the range of S is contained in the value of its most significant digit, which equals 1 when and only when S is at least as large as 2^{m-1}. Therefore,

$$f = \hat{s}_{nm}$$

Thus, an array storing the weights $w_1, w_2, \ldots, w_n$ (and the threshold value $2^{m-1} - w_0$ in the top, or zeroth row), and having the input and output connections shown in Fig. 17, realizes the threshold function given by the defining inequality (13) above. Note that the gate circuitry for forming $f(x_1, x_2, \ldots, x_n)$ is completely combinational. The longest propagation path through the array includes $m + n$ gates. The threshold array has a total of $n + 1$ rows, one for each of the input variables, plus one more for the threshold. (Alternatively, the fixed threshold may simply be injected as the top-edge boundary condition of an n-rowed array.) The number m of columns of the array must be large enough to accommodate both the positive and negative extremes of cumulative weight sums which might arise in any row, as the input variables $x_1, x_2, \ldots, x_n$ are allowed to vary through all possible values. These extremes will be reached at the bottom of the array, so m must satisfy the two inequalities

$$2^{m-1} > \sum_{+} w_i - w_0, \qquad 2^{m-1} \geq \sum_{-} |w_i| + w_0 \qquad (14)$$

where the first and second sums are taken over positive and negative weights, respectively.

As an example, consider the realization of the switching function $f_1(x_1, x_2, x_3)$ defined by the linear inequality

$$f_1 = 1 \qquad \text{iff} \quad 3x_1 - 2x_2 + x_3 \geq 2$$

The limits (14) give $2^{m-1} > 3 + 1 - 2 = 2$, and $2^{m-1} \geq 2 + 2 = 4$, thus requiring that $m = 3$. Therefore, the row counters in rows 0, 1, 2, and 3 should store the binary numbers $2^2 - w_0 = 2 = (010)$, $w_1 = 3 = (011)$, $w_2 = -2 = (110)$, and $w_3 = 1 = (001)$, respectively. This pattern of counter contents is shown in Fig. 18; the array realizes the function f_1. (Note that the least significant digits of the weights are stored in the left-most columns of the array.)

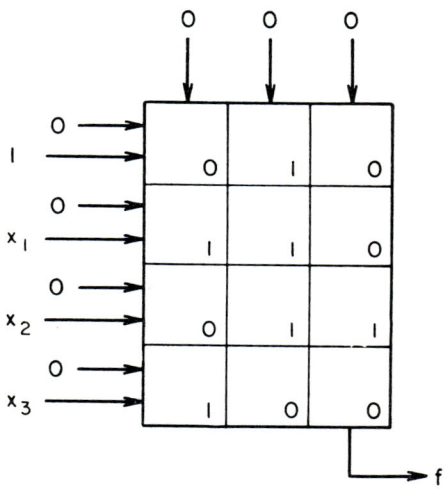

FIG. 18.   Contents of the cell flip-flops for the example worked out in the text.

As may be seen from Fig. 17, the m flip-flops in the ith row of the threshold array are also wired as an up/down counter, with common pulse input T. When T is applied, the counters in all rows of the array are changed at the same time; the count increases by *one* in every row in which $x_i = 1$, and decreases by *one* in every row in which $x_i = 0$. Note that transitions between positive and negative weight values occur smoothly, without requiring any special control, by virtue of the use of the two's-complement number system. In this manner the threshold array may be used adaptively, by allowing each weight to be repeatedly modified ($T = 1$), in a positive ($x_i = 1$) or negative ($x_i = 0$) direction, in order to reinforce favorable behavior during a "teaching" phase of the operation. Teaching algorithms other than the one assumed here can also be realized by changing the cell logic associated with the counter advance. Descriptions and analyses of various algorithms are available in the literature (Nilsson, 1965).

Figure 19 shows how a set of several threshold arrays may be stacked and interconnected to form a two-layer pattern-classification network of "Perceptron," "Minos," or "Madaline" type. The possible extension to more than two layers is obvious.

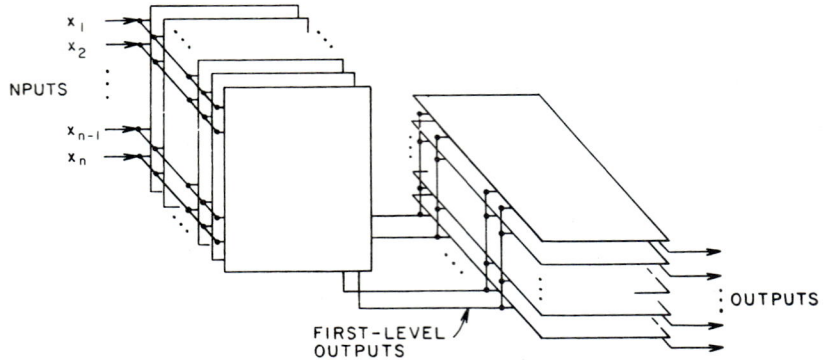

FIG. 19. Manner of stacking threshold arrays to form a pattern-recognition machine.

Also, note that by applying the proper set of control signals to the x-inputs, prior to normal use of the array, the weights may be programmed to a set of desired initial values (at least, any set of all even or all odd values). A set up schedule of applied signals may be derived without difficulty.

For adaptive use the threshold w_0 would normally be set equal to zero. The number of rows in the array may then be reduced to n, and the number m of columns may be selected to be the smallest integer that satisfies the inequality

$$m > \log_2(n|w_{max}|) + 1$$

form (14), above. It may also be desirable for adaptive use to provide additional circuitry to prevent (directly or indirectly) the magnitudes of the individual weights from becoming so large during adaptation that the capacity of the array is exceeded.

The simplest way to limit weight excursion is to generate an output g that becomes 0 as soon as the weight sum S exceeds prescribed limits, in order to signal the termination of the adaption period. A convenient range is one-half the maximum weight excursion—namely, $g = 1$ iff $2^{m-2} \leq S < 3 \cdot 2^{m-2}$. The limits of this range may be detected with a simple logical circuit applied to the two most significant sum digits at the bottom of the array, as shown dotted in Fig. 17.

$$g = \hat{s}_{n,m} \oplus \hat{s}_{n,m-1}$$

Note that this approach does not limit the individual weights directly, but only their sum. It will be effective only if the mode of teaching used allows one to assume a sufficient degree of statistical uniformity among the weights, so that the probability is negligible that one weight may become impermissibly large while the sum remains less than 2^{m-2} in magnitude.

If a cellular threshold array is to be used without the need for adaptation, then the bidirectional counter may be replaced by a low-speed shift register or a unidirectional counter. One of these alternatives would be preferred if it leads to a simpler circuit realization, or if it allows the weights to be set initially at completely arbitrary values.

B. SORTING ARRAY[4]

1. Basic Operation

We now describe an array that behaves as a single-address, multiple-word memory that keeps in sorted order all data words that are fed into it.[5] Words that are read out are obtained in order of size, with the largest word first (or alternatively, with the smallest first, as desired). The words are assumed to be of maximum length n. This memory could find use as a functional unit that can be attached to the central processor of a general-purpose computer, or to a special-purpose computing system for which sorting capability is needed. As explained below, the array may also be used for various other purposes having nothing to do with sorting.

Figure 20 displays the elementary sorting array, the logic circuitry and equations of a typical cell, and the registers that would normally be used (an

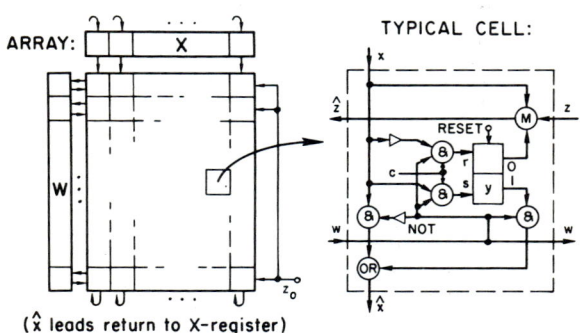

FIG. 20. Cellular sorting array. The cell logic equations are

$$\hat{x} = \bar{w}x + wy$$
$$s_y = wcx, \qquad r_y = wc\bar{x}$$
$$\hat{z} = M(x, \bar{y}, z) = x\bar{y} + z(x + \bar{y})$$

[4] From Kautz, 1969.
[5] A similar array based on cryogenic technology was described by Seeber (1960), but his elementary cell design is needlessly complex when converted to gate-type logic.

input/output register X, and a word select register W). All cell input terminals on the right-hand side of the array are connected to a logical signal (z_0, normally fixed at 1), and the outputs labeled $\hat{z}$ from the left edge of the array serve as inputs to the stages of the W-register.

It may be noted from the figure that each cell of the array contains one flip-flop, whose contents is designated y, so that the set of n flip-flops in any one of the m rows of the array may be employed to store one n-digit word. This set of words is assumed to be encoded in a uniform digit-weighted code, such as the conventional binary or binary-coded decimal number system, with the most significant digits at the left ends of the words. A one's or two's complement representation is assumed for negative numbers, with any minus sign encoded as a 0 at the left end of the word. (The case in which sorting is performed with other number representations or over only a portion of the digits in the word is treated later.) An entire word is handled as a unit during the input, output, and sorting operations.

One cycle of operation of the array consists of two steps.

(1) A *comparison* step, in which the word X in the X-register is simultaneously compared with all m words stored in rows of the array; a 1 is injected into the W-register in those rows whose words (including blank words) are smaller than or equal to the word X, and a 0 is injected into the W-register in those rows whose words are larger than the word X.

(2) An *execution* step, in which: (a) the set of all words that are stored in the subset of rows having a 1 in the W-register are collectively moved downward one row within the set, while the word in the X-register is copied into the uppermost such row; and (b) the lowermost such word is copied into the X-register. Words in rows having a 0 in the W-register are not moved. Only one clock is needed, but if desired, the two substeps (a) and (b) can be executed simultaneously with two separate clocks, so that readout may be carried out without concurrent write-in.

Sorting with this array is accomplished by maintaining a sorted file of previously entered words, with the largest at the top of the array, and the smallest and any blank rows (rows containing all 0's) at the bottom. Each new word that is to be sorted is inserted into this file in a single operational filing cycle. In step (1) every word smaller than or equal to the new word is marked with a 1 in the corresponding position of the W-register, and in step (2) all marked words are shifted down one word position, with the new word being inserted in the uppermost marked row. Unless the array is already full prior to the filing cycle, the X-register will contain all 0's at the end of the cycle. Otherwise, it will contain the smallest word in the array.

To read out in order the words in the file, largest to smallest, place a single 1 in the uppermost stage of the W-register. Now carry out step (2) of the cycle repeatedly, shifting the 1 downward in the W-register, one row with each step. If the words are desired in the opposite order, the single 1 may be started at the bottom of the W-register and shifted upward, although this procedure will produce an initial string of all-0 words if the array is not full. (See the next subsection for a way to avoid this.) This sequence of operations gradually empties the file; if it is desired to merely copy the file into the output channel, without clearing it, then only step (2b) should be used.

An alternative method of readout is to enter repeatedly the number $(1\ 1\ 1\ \cdots\ 1)$ from the X-register. This forces the contents of the array out of the bottom, one word at a time. This method avoids the use of the W-register as a shift register, but leaves the array full of 1's, which must then be cleared by some other means.

The detailed operation of the array during the comparison step (1) proceeds as follows. With reference to the cell circuitry and equations shown in Fig. 20, note that the majority gate at the top of the cell forms part of a chain of n such gates along each row of the array. The inputs x and $\bar{y}$ of this gate allow it to play the role of a size comparator, so that the leftmost $\hat{z}$-output in each row takes on the value 1 when and only when the number represented on the set of x-lines entering this row is greater than or equal to the number represented in the cascade of y-flip-flops in the same row.

Normally, step (1) is carried out with the W-register initially empty, so that the w busses in all rows of the array carry the value 0. In this case the $\hat{x}$-output in each cell carries the same value as the x-input; $\hat{x} = x$. That is, the contents of the X-register is passed downward to all rows of the array. As a result, the comparisons in step (1) are made between the word X in the X-register and every word Y stored in the array.

The detailed operation of the array during the execution step (2) proceeds as follows. Within each row for which the W-register contains a 0, we have $w = 0$, so that each cell in this row behaves according to

$$\hat{x} = x, \qquad y' = y$$

That is, the row is static, and behaves as if it were not even present. Within each row for which the W-register contains a 1, we have $w = 1$, so each cell behaves according to

$$\hat{x} = y, \qquad y' = cx$$

where c is the clock. Thus, the word stored in the flip-flops in this row is transferred onto the set of $\hat{x}$-lines that pass downward from this row. Also, with the application of the clock, the word received on the x-lines from the row above is transferred into the flip-flops in this row. For the array as a whole, therefore, all words in the subset of rows consisting of the X-register and those rows marked with a 1 in the W-register shift down cyclically one row position within this subset. The contents of the X-register fills the top position, and the contents of the lowest marked row passes back into the X-register.

For some special uses of the array, it may be desired to use the W-register during step (1). Suppose that some one row containing word Y_i has been marked with a 1 in the W-register. The comparison process will then be modified so that all words Y in rows below the marked row will now be compared with the word Y_i in the marked row, instead of the word X. If several rows are so marked, then each word in the array will be compared against the first marked word above it (or against the word X, if there is no marked word above it).

2.   Simplifications and Embellishments

In most computation and data-processing problems, words are not sorted on the basis of just their relative magnitudes, but of their magnitudes taken over only a subset of the digits, usually called the *key*, of the words. That is, certain nonkey digit positions within the words carry auxiliary data (or pointers to the location of auxiliary data) that should not enter into the comparison process, but that must be retained for other computation purposes not related to the sorting. To inhibit the comparison on these digit positions, the array shown in Fig. 20 may be augmented, as depicted in Fig. 21, to include a *mask register M*, the stage outputs of which are bussed vertically along columns to all rows of the array. Each cell of the new array contains the additional gatery to inhibit the comparison operation in a column

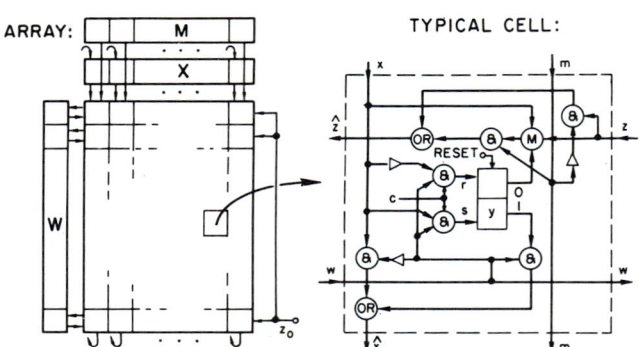

FIG. 21.   Cellular sorting array with masking. The cell logic equations are

$$\hat{x} = \overline{w}x + wy$$
$$s_y = wcx, \qquad r_y = wc\overline{x}$$
$$\hat{z} = \overline{m}z + mM(x, \overline{y}, z)$$

whenever $m = 0$ for that column. When $m = 1$, the cell behaves normally. With this arrangement, the position of the key within the data words may be selected externally by injecting a string of 1's into the proper positions in the M-register. In fact, multiple keys may be employed, and their positions need not be contiguous, but it is assumed as before that the significance of the digits in the keys increases from right to left.

A less flexible but simpler arrangement is shown in Fig. 22. Here the key and nonkey portions of the words are handled in two separate sorting arrays, the latter of which is simplified over the former by having its comparison circuitry (the majority gate and associated wiring) removed. Corresponding w-busses of the two arrays are directly connected, however, so that the two portions of each word undergo the same transfers.

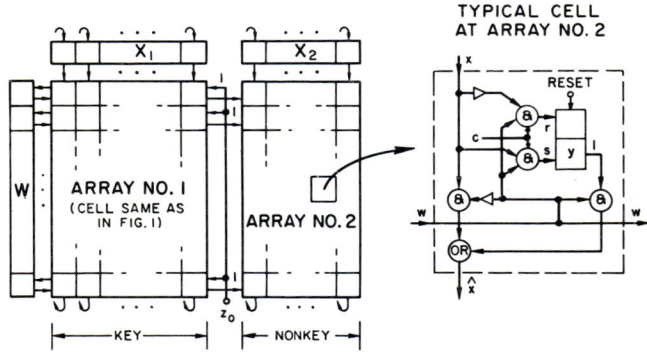

FIG. 22.   Cellular sorting array with separate nonkey array.

When a mask register is used, it may be desirable to change to a new key for the entire set of words in an already ordered file, in order to re-sort the file on the basis of the new key. Probably the simplest way to achieve this re-sorting capability is to reserve the leftmost digit position within each word as a *tag* digit to indicate "re-sort" status. This digit will normally have the value 0. When the key is changed, the array is cycled $(m + 1)$ times, starting with the W-register full of 0's, but clearing it as usual at the end of each cycle. A 1 is held in the leftmost digit position in the X-register through-out the re-sorting. In this way, the file is gradually pushed out of the bottom of the array in successive cycles, and is reinserted into the top for re-sorting, one word at a time. The extra 1 causes all words in the new file to be treated as if they were larger than all words in the original file, thereby keeping the new file on top of the old one. The operation may be stopped as soon as the first word having an extra 1 in its most significant digit position appears at the output of the array.

If this process were going to be repeated with still another key, one could either: (a) first circulate the file again for $m + 1$ cycles, holding all 1's in the W-register but forcing the leftmost x-digit to be 0 before repeating the above process (in order to clear the leftmost column); or (b) augment the array with a reset line attached to all flip-flops in the leftmost column, to clear this column before repeating the re-sort operation; or (c) reserve a second digit position to indicate the next "re-sort" status.

If two or more words equal in magnitude are filed in the array described in the previous section, they will occupy adjacent rows in the array, with each entry located *above* all earlier equal-sized entries. If the opposite ordering of equal-sized words is desired for readout, this may be achieved by changing the signal value on the boundary input z_0 on the right-hand edge of the array of Fig. 20 from 1 to 0. This modification causes the comparisons to be executed according to a strict inequality $(X > Y)$ instead of a simple inequality $(X \geq Y)$, so that a word equal in size to a previous word is treated as if it were larger rather than smaller than the previous word.

Floating-point numbers are handled with no special provisions required, provided only that the exponent is placed to the left of the mantissa, and the representation is normalized before being injected into the array proper. Negative numbers represented in a "magnitude-plus-sign" form must be complemented (in the X-register, for example) before comparison, and the sign must be moved to the left end of the word and complemented (to be 0 for a minus and 1 for a plus), if necessary.

Actually, it is sometimes possible to dispense with one or both of the X- and W-registers, depending upon the computing environment in which the sorting array is used. The X-register is used only as a buffer. If the signals and timing on the x- and $\hat{x}$-lines are compatible with those of the input-output channel connecting the sorting array to the rest of the digital system, then this register can be eliminated. Even during resorting, the $\hat{x}$-lines can be fed back directly to the x-lines. The W-register may also be eliminated by tying each $\hat{z}$-output directly to the x-bus in the same row. To see that this simplification is valid, recall that the effect of step (1), which is purely combinational and is unclocked, is to force w to take on the value 1 in a lower group of rows, each of whose words Y is less than or equal to the word X supplied to the top of the array. This change in w now changes the comparison in all of these rows (except the uppermost of them), so that each word Y is compared with the word immediately above it instead of with the word X. In an ordered file, however, the magnitude of the words decreases downward, so the value of $\hat{z}$, hence w, will never change to 1 and then back to 0 again, but in fact, will be held latched at the value 1, if it changes to 1 at all. Step (2) proceeds normally. When the input word X changes, the boundary between the 0 and 1 strings of *w*-values will ripple upward or downward until the insertion point for the new *x*-word is located. Consequently, this registerless array may operate somewhat more slowly, but still carries out the filing operation properly.

From a computing system point of view, this array is probably best treated as a single-address multiword memory having a storage capacity of *m* words of *n*-digits each, and having the property that a readout command will always retrieve the largest word in the memory. If a mask register is used, it should certainly be addressable as well. It might also be desirable to make the *W*-register addressable, so that prescribed blocks of words or individual words can be selected or inhibited during the sorting and readout operations. Even a small degree of external control of the *W*-register offers the possibility of employing some rather sophisticated selection criteria, such as: selection of all words whose magnitudes fall between given limits; selection of the *k*th largest word; selection on the basis of multiple keys, possibly improperly ordered within the words; and selection using combined inequality and equality testing.

If it is desired to use a sorting memory in which the roles of "smallest" and "largest" are interchanged, it is only necessary to modify the cell so that the equation for $\hat{z}$ has the form:

$$\hat{z} = M(\bar{x}, y, z) = \bar{x}y + z(\bar{x} + y)$$

The cell required has substantially the same complexity as before.

If ordering based upon binary inclusion ($X \subseteq Y$) rather than size comparison is desired, only the second term in the above expression should be used—a rather special case since binary inclusion normally generates only a partial ordering.

3.   Content-Addressed Memory

The sorting array may also be used as a content-addressed (associative) memory. For inequality searching ($X \geq Y$ or $X > Y$, the selection between these choices depending upon the value of z_0), step (1) is carried out as first described. This leaves 1's in the *W*-register in just those rows containing words *Y* that satisfy the inequality. This entire subset of words may now be shifted out of the array by executing step (2) repeatedly without resetting the *W*-register but with the input register empty, until an all-0's word is encountered on the output lines from the array.[6]

---

[6] For nondestructive readout, start with the *X*-register full of 0's, but cycle back to the *X*-register the words read out, for reentry into the array. Stop when the all-0's word reappears in the *X*-register.

If the circuitry shown in Fig. 20 is modified so that the $\hat{z}$-line inputs to the stages of the W-register are arranged to enter the *trigger* inputs rather than the *set* inputs of these flip-flops, then the associative search may be carried out on the basis of certain additional searching conditions without giving up any of the capabilities discussed so far. For example, assuming that the words in the array are in order, a search for all numbers Y in the range $X_1 \leq Y \leq X_2$ can be conducted as follows.

(1) Lead X_1 into the X register.
(2) Compare $X_1 > Y$ [i.e., execute step (1) with $z_0 = 0$], apply clock to W-register.
(3) Load X_2 into the X register.
(4) Compare $X_2 \geq Y$ [execute step (1) with $z_0 = 1$], and apply clock to W-register.

As a result, the W-register will contain a 1 in every row whose word Y satisfies *one* of the two tests, but not both (since two $\hat{z}$-signals will return the W flip-flop to its initial state). This can happen only when $X_1 \leq Y \leq X_2$. When $X_1 = X_2$, this constitutes an equality test, and step 3) is unnecessary. Readout of the selected words is conducted as before.

This supplementary use of the sorting array is offered merely as an example. Many other interesting uses can be devised by the ingenious designer. In particular, the array may also be utilized as a push-down memory (stack) or as a queue memory (buffer), and the multiple-comparison feature just described could lead to procedures for handling complex testing conditions involving several inequalities and equalities.

C. CODING ARRAY[7]

Another example is provided by a special-purpose array for encoding and decoding any single-error-correcting plus multiple-error-detecting linear binary code.

This array is shown in Fig. 23, along with the logical circuitry of a typical cell of the array. The cells (stages) of the input register X and the output register Z are also depicted in Fig. 23. Each array cell is seen to consist of a single flip-flop with binary contents q, plus a small amount of cascade logic (a total of about 13 elementary NOR gates, as illustrated in Fig. 24).

For such a code, the encoding and decoding processes may be thought of as being applied to an n-digit codeword (X, Z) that consists of a k-digit

[7] From Levitt and Kautz, 1969.

$H = [I \vdots Q]$

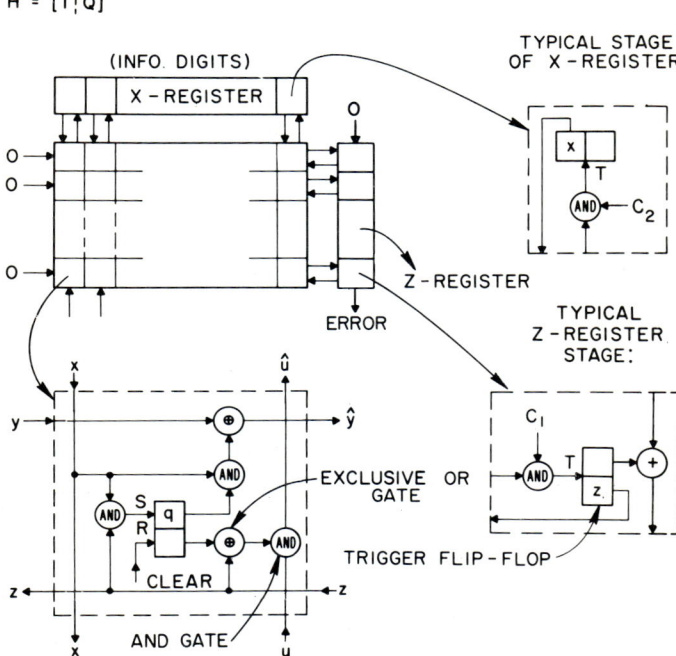

FIG. 23. Array for encoding and decoding single-error-correction codes. The cell equations are

$$\hat{u} = u(z \oplus \bar{q}) \qquad s_q = xzc_3$$
$$\hat{y} = y \oplus qx \qquad r_q = \text{CLEAR}$$

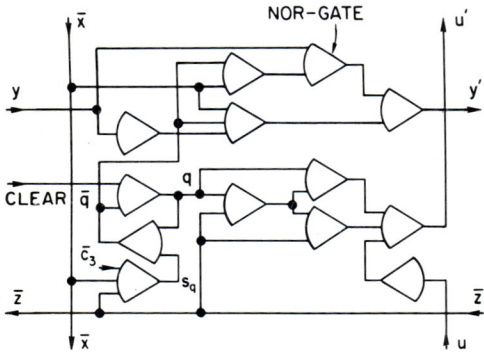

FIG. 24. NOR gate realization of main-array coding cell.

information portion X and an $r(=n-k)$-digit check portion Z. The task of the encoder is the calculation of the Z vector from the X vector, assumed given. For a parity check matrix H in echelon canonical form, namely

$$H = [Q_{r \times k} : I_{r \times r}]$$

the computation of Z may be expressed as

$$Z = QX,$$

where all vectors are treated as column vectors (Peterson, 1961).

For encoding, then, the array and its two associated registers operate as follows. For a given code, each digit q_{ij} of the Q portion of the matrix H is placed in the cell flip-flop in row i, column j of the cellular array (by a setup process to be described subsequently). The block of k information digits of a particular codeword is placed in the k-digit register X, and the r check digits are computed combinationally by the array and inserted in the r-digit check register Z in a single clock time. This computation proceeds by means of the chain of EXCLUSIVE OR gates along each row, independently of the other rows, on the basis of the x-digits that are bussed vertically down each column. Each x-digit x_i contributes to the sum in a particular row i if and only if the corresponding digit q_{ij} of the Q matrix equals 1. Thus, after the clock has been applied to the Z register, this register contains the block of check digits that are to be associated with the given block of information digits.

In general, decoding of a received, possibly erroneous codeword (X^*, Z^*) may be carried out by first recomputing the check digits QX^* from X^* and adding them to the received check digits Z^*, to obtain the error *syndrome*

$$S = Z^* \oplus QX^*$$

An error has occurred if and only if S is nonzero. For single-error correction, the digit position in error is identified by the corresponding column in H that is identical to S; that is, by the corresponding column of Q (stage of the X register) if the error is in the *information* portion of the codeword, and the corresponding column of I (stage of the Z register) if the error is in the *check* portion. As soon as the digit in error is identified, it may be corrected by complementing it.

For decoding, then, let the data portion X^* and the check portion Z^* of the received codeword be entered into the X and Z registers, respectively. When clock c_1 is applied, exactly the same operation is carried out as in encoding, except that the calculated check digits QX^* are added (digitwise) to the received check digits Z^*, thereby leaving the syndrome S in the Z register. The digits of this syndrome are now passed back along the rows of the array on the z busses, for digitwise comparison with the flip-flop contents (columns of the Q matrix) in each column of the array. When clock c_2 is applied, any column in which exact coincidence is found causes a 1-signal to be applied to the corresponding column of the X register, thereby complementing it. If no such column generates such a coincidence signal, then either there was no error, or the error occurred in the check portion Z^* of the received codeword and not in the data portion X^*.

Note that error detection may be provided by attaching a simple OR gate having r inputs onto the Z register, as indicated in Fig. 23. A 1-output from this gate following the application of clock c_1 during decoding then indicates the presence of one or more errors.

For loading the flip-flops of the main array, assume first that the flip-flops are all reset initially. To load a 1 into the flip-flop in row i and column j, inject 1's into registers Z and X in this row and column, and apply clock $c_3 (i = 1, 2, \ldots, n - k, j = 1, 2, \ldots, k)$. The contents of these flip-flops are not changed during the encoding and decoding operations.

Various extensions can be made to this basic array to permit the correction of burst errors, erasure errors, multiple independent errors, and others (Levitt and Kautz, 1969).

D. INTERCONNECTION ARRAYS[8]

One of the most fundamental functions in data processors is data switching (or line-switching), a function which is achieving increasing importance in modern digital systems. This section describes a class of such switching arrays organized as cellular logic arrays. Each cell in these interconnection arrays is very simple—it behaves as a reversing switch, under the control of a single memory flip-flop. Various ways are shown below of programming an array of these simple cells to perform an arbitrary permutation of a set of input lines onto a set of output lines. The solutions offered differ in the number of cells needed, the shape of the array, and the number and type of external connections required around the periphery of the array proper. Also discussed are ways of setting up the memory flip-flops in an array to achieve a desired permutation.

A simple cell capable of performing an elemental line-switching operation is shown in Fig. 25. In essence, it is a double-pole double-throw reversing switch controlled by a storage element (e.g., a flip-flop), with some means provided for setting the storage element to the desired state. The two modes consist of a "crossing" [Fig. 25a] and a "bending" [Fig. 25b] of the pair of input leads to the pair of output leads. Figure 25c shows a NOR gate realization of the cell in question. This cell by itself therefore provides for a

---

[8] From Kautz *et al.*, 1968.

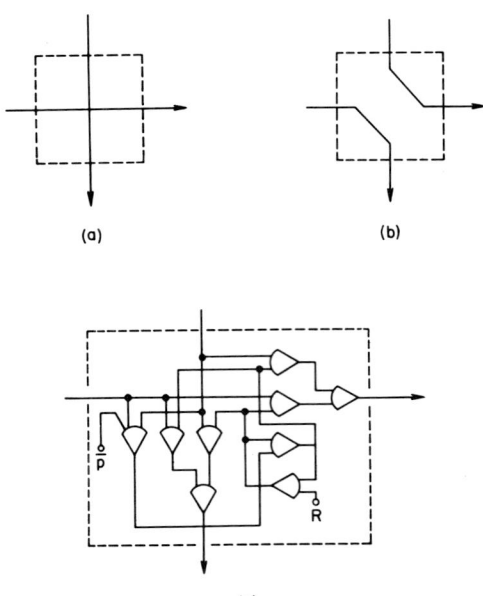

FIG. 25.   The permutation cell. (a) The crossing mode. (b) The bending mode. (c) A NOR gate realization.

simple interchange on two lines, and constitutes a solution for the case when $n = 2$ to the general problem of synthesizing an n-input n-output interconnection network.

Even before any constraints on the regularity of intercell connections are imposed, a lower bound may be obtained on the number $M(n)$ of elemental cells required for a complete n-input n-output interconnection network. The network must contain enough two-state storage elements (cells) to specify all possible permutations; thus

$$M(n) \geq \log_2(n!)$$

or, asymptotically (from Stirling's formula),

$$M(n) \geq n \log_2(n) - 1.443n + 0.5 \log_2(n)$$

It is noted that the interconnection network problem discussed in this paper is a generalized form of the problem of interconnecting telephone lines in a telephone central office (Beneš, 1965). In the latter case it is necessary to design a so-called "rearrangeable" contact network which permits arbitrary connections from p subscribers to t trunks. Clearly, this connection capability is achieved by a $p \times t$ crossbar switch, requiring pt crosspoint contacts. It has been shown by Beneš (1965), however, that for the case $p = t = n$ (when n is a power of 2 and $n \geq 8$) it is possible to design a rearrangeable network requiring only $4n \log_2(n) - 8n$ crosspoints. Waksman (1968) and independently Goldstein and Leibholz (1967) have shown that the Beneš network can be visualized as one form of an interconnection network composed of $n \log_2(n) - n + 1$ cells of the type described above. Various additional solutions have been derived which require somewhere between about $n^2/2$ and $n \log_2(n) - n + 1$ cells.

Figure 26a displays for $n = 8$ the form of an interconnection array that is probably the simplest of the arrays to be described, both in form and operation. Its capability for performing an arbitrary permutation of n inputs may easily be proved inductively as follows. (Proof for this and succeeding arrays will be illustrated in the figures for the case $n = 8$.) The leftmost cell by itself obviously performs the permutation for $n = 2$. Suppose that the leftmost $n - 2$ columns (all but the last column in Fig. 26a) are capable of permuting the set of input lines $X_1, X_2, \ldots, X_{n-1}$ into any desired order at the right side of the $(n - 2)$nd column. The nth input line X_n entering at the top of the last column may then be switched into this sequence of lines at any point, by using just this column, as depicted in Fig. 26b. This is done by setting the last-column cells above this point to the "crossing" mode, and the remaining last-column cells to the "bending" mode. By induction, then, a total of $n - 1$ columns is adequate to achieve an arbitrary permutation of n input lines.

The total number of cells in the triangular array is

$$M(n) = 1 + 2 + \cdots + (n - 1) = \tfrac{1}{2}(n^2 - n)$$

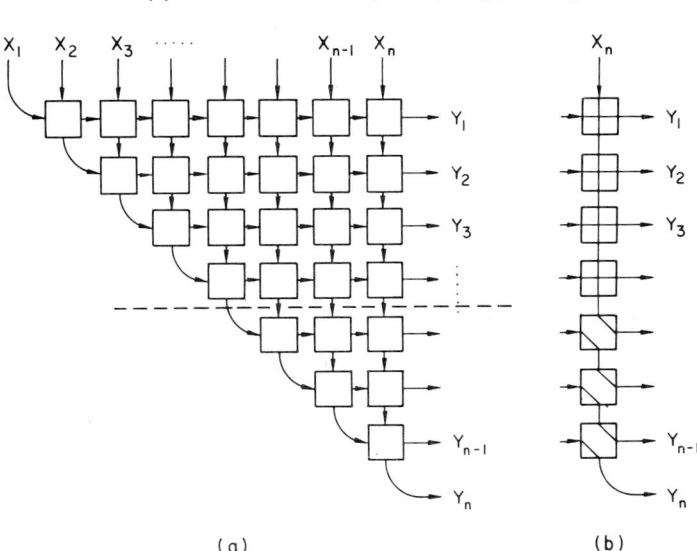

(a)   (b)

FIG. 26.   The triangular array, $n = 8$. (a) The array. (b) The last column.

The diamond-shaped array of Fig. 27a, shown for $n = 8$, can also perform an arbitrary permutation, and also has $M(n) = \frac{1}{2}(n^2 - n)$ cells. Its permutation capabilities are readily established as follows. Let an arbitrary input X_i be routed to output Y_1 in an n-input array by the type of path shown in Fig. 27b: first horizontally, then following the upper right-hand edge. In addition, the cells on the upper left-hand edge below the point of entry should all be set to the "bending" mode. It is now easily seen that the set of unspecified cells and

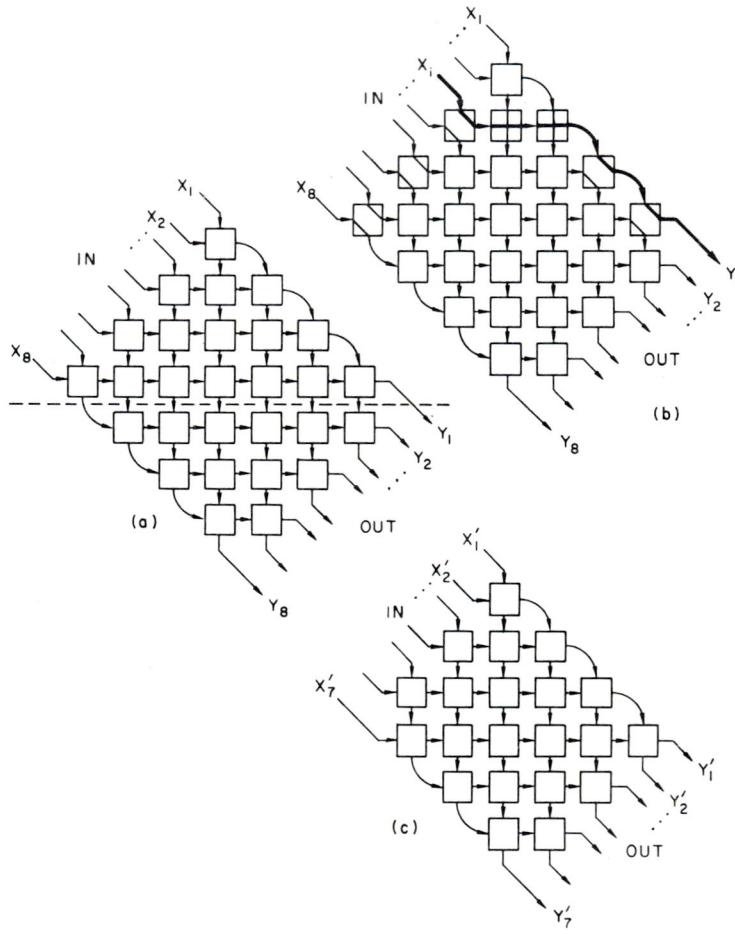

FIG. 27.   The diamond array, $n = 8$.

[9] For n-odd (as displayed in Fig. 27c for $n = 7$) the left and bottom corners of the diamond are "clipped."

connections which remain constitute an $(n - 1)$-input array of exactly the same form, as shown in Fig. 27c. This same process may be repeated with this resultant array, and so on, until only two inputs remain to be permuted by a single cell.

If a rectangular array is desired, to simplify fabrication or for some other reason, several alternatives are feasible. One may be obtained from the triangular array of Fig. 26, by reflecting about its hypotenuse the smaller triangular array below the dotted line, and placing this subarray to the left of the remainder, allowing a diagonal file of isolating cells set to the "bending" mode. The resulting array is shown in Fig. 28 for n even. It has a few external

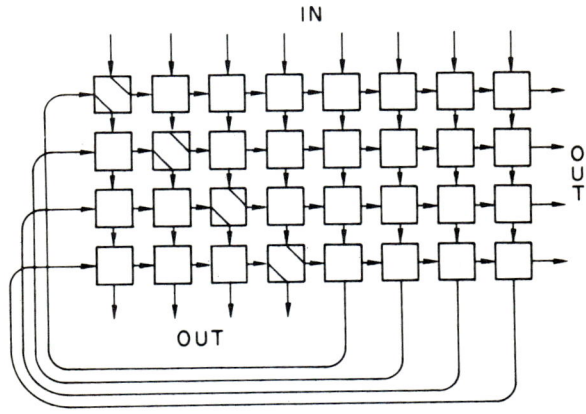

FIG. 28.   The rectangular array with external connections, $n = 8$.

boundary connections, and a total of $n^2/2$ cells. [When n is odd, $(n^2 - 1)/2$ cells are required.]

Another type of rectangular array is shown in Fig. 29a. The proof that this array, which has the same number of cells as that of Fig. 28, can actually effect all permutations is again inductive and, for n even, proceeds as follows. First, reflect the right-hand half about its $-45°$ diagonal, and redraw the array as two symmetrically interconnected square arrays, as shown in Fig. 29b. We will show that any two inputs, say, X_i and X_j, of this array can be connected to outputs Y_1 and $Y_n(= Y_8)$, respectively, in such a manner as to leave the remaining cells in the form of an $(n - 2)$-input array of the same shape. Assume for the first case that X_i enters along the upper edge and X_j along the left edge. The pattern of cell settings shown in Fig. 29c achieves the desired connection: X_i to output 1, X_j to output n. Only the uppermost row and leftmost column are used in each square, and the rest of the array is in the form of a $(n - 2)$-input network. Clearly, this same result would have been

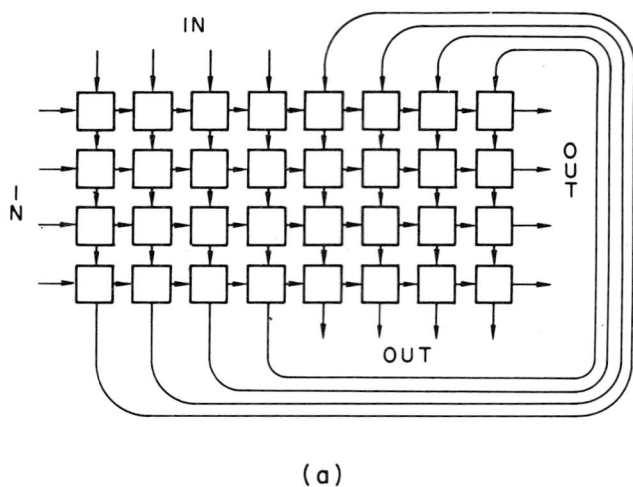

(a)

FIG. 29.   Another rectangular array with external connections, $n = 8$. (a) Original form.

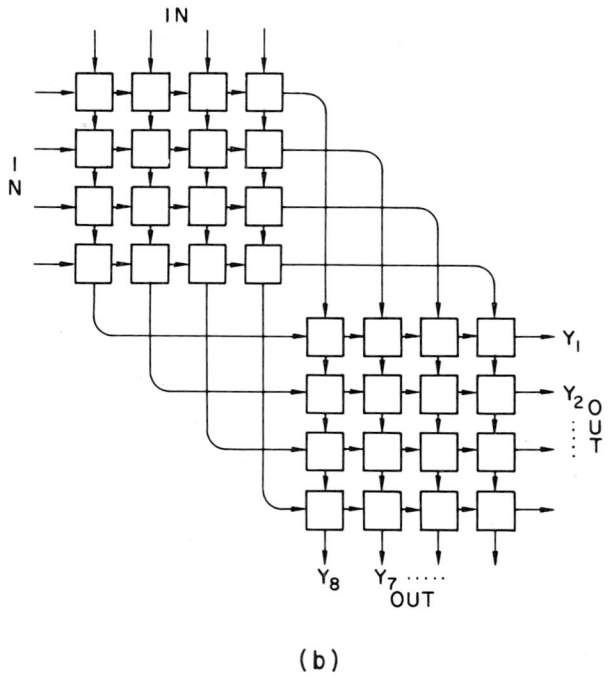

(b)

FIG. 29 (Continued). (b) Redrawn to be symmetrical.

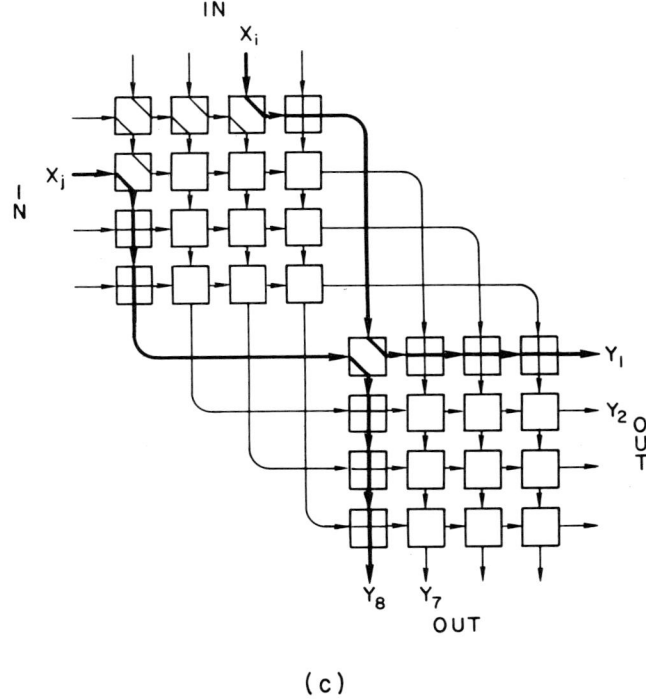

(c)

FIG. 29 (Continued). (c) A pair of typical paths.

obtained if X_i were introduced anywhere along the upper edge, and X_j were introduced anywhere along the left edge.

If the positions of X_i and X_j are interchanged, the desired output connection can still be obtained by changing the upper-leftmost cell in the lower square in Fig. 29c from the "bending" mode to the "crossing" mode. An almost identical argument pertains when both X_i and X_j are applied along the same edge of the upper square, and for the corresponding cases when n is odd. Consequently, any array of the form of Fig. 29a has complete permutation capability.

Actually, the upper-left triangle of cells may be pruned from the rectangular array of Fig. 29a, and one full column may be deleted without reducing its permutation capability. The input connections follow the pattern shown in Fig. 30. The number of cells in this "pruned" rectangular array is found to be

$$M(n) = 3(n^2 - 1)/8 \quad \text{for} \quad n \text{ odd}$$
$$M(n) = n(3n - 2)/8 \quad \text{for} \quad n \text{ even}$$

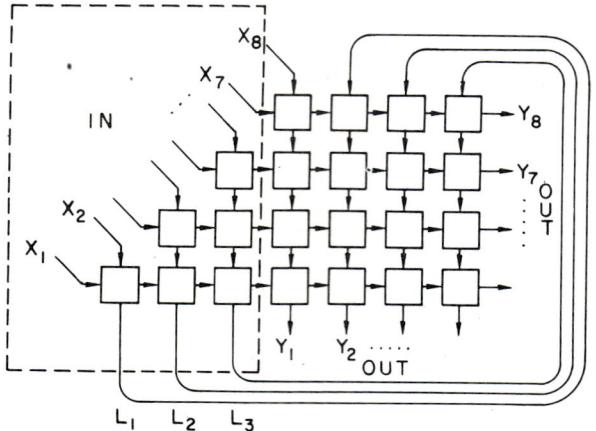

FIG. 30.   The pruned rectangular array, $n = 8$.

Alternatively, the lower-right triangle of cells could also be removed, with the same results.

A simple $p \times p$ square array composed of p^2 cells can be used for inter-connections by applying inputs to the upper and left sides, and by taking outputs from the right and lower sides. However, not all such side terminals can be used for inputs and outputs if the array is to realize all possible per-mutations of the inputs into the outputs. By feeding back enough of the left-over bottom and right-side terminals to the top and left sides, respectively, the array may be made to have complete permutation capability. An example of this form of array is shown in Fig. 31 for $n = 8$. In general, the number of

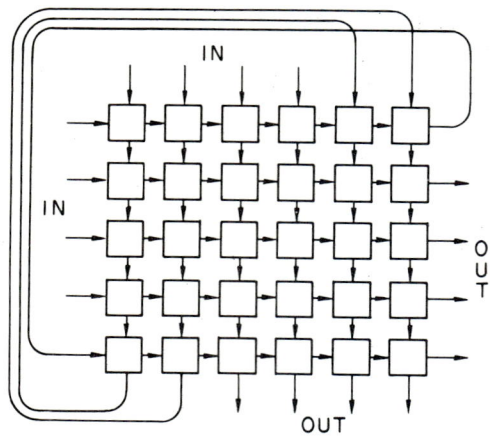

FIG. 31.   The almost-square array, $n = 8$.

cells required for a square or almost square array of this type is given by:

$M(n) = (3n - 2)^2/16$ when n is even but $n/2$ is odd;

$M(n) = 3n(3n - 4)/16$ when n is even and $n/2$ is even;

$M(n) = (3n + 1)(3n - 3)/16$ when n is odd and $(n - 1)/2$ is even;

$M(n) = (3n + 3)(3n - 5)/16$ when n is odd and $(n - 1)/2$ is odd.

Other arrays can be found that achieve complete permutability with a smaller number of cells, provided the condition of adjacent-cell connections is relaxed to allow longer intercell lines. Such arrays fall outside of the scope of this discussion, but one example of such an array is illustrated in Fig. 32 for $n = 8$. In general, this family of arrays requires $n \log_2 (n) - n + 1$ cells (when n is a power of two) (Kautz *et al.*, 1968).

In order to set up or " program " an interconnection array to realize a desired permutation, it is desired to apply a sequence of binary input signals to the edges of the array so as to insert the desired pattern of binary digits into all of the cell flip-flops. Special set-up buses can be used for programming the array in a fairly direct fashion. For the type of cell assumed here and shown in Fig. 25a, however, each cell flip-flop is *set* from a coincidence of 0-signals on its horizontal and vertical inputs, simultaneous with a globally

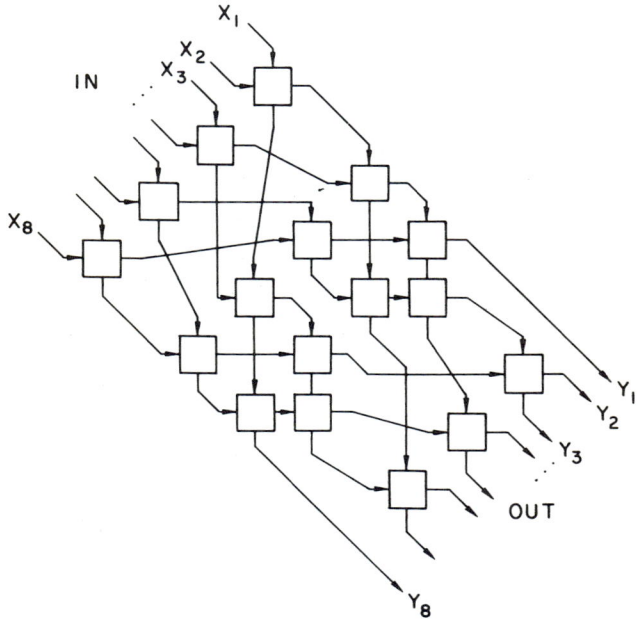

FIG. 32.   The rearrangeable array, $n = 8$.

applied "clock" or "set-up" pulse. We assume, by means not specifically shown in the figure, that all cell flip-flops may be initially cleared to the 0 state, so that each cell is in the "crossing" mode at the start of the set-up process. This might be done through control of the dc power applied to the array, or with a special reset line.

The array will be programmed one cell at a time by applying to the array boundary at each step just two 0's, one of which is to find its way to the horizontal input and the other to the vertical input of the cell in question. The other boundary inputs are held at 1. Such paths are guaranteed to exist, because of the fact that each cell merely permutes its input leads, so that no paths are ever terminated or combined. However, the cells must be set up in such an order that the 0-signal paths through an array intersect only in the cell whose flip-flop is being set. If this were not the case, then an attempt to set one cell flip-flop could actually cause two or more to be set, resulting in a loss of independent control.

Consider first the rectangular array of Fig. 28 (without the external feed-back connections). The entire array may be set up one row at a time, top to bottom, and proceeding from right to left in each row. A typical condition during set-up is depicted in Fig. 33. Assume that the cells above the dotted line

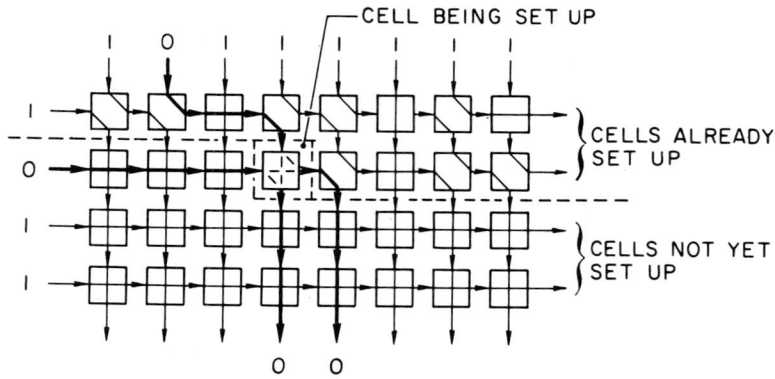

FIG. 33. Typical signal paths used on a rectangular array during programming.

have already been set up, and those below the dotted line have not. Note that one of the applied 0-signals starts at the left end of the row being set up, while the other starts at a boundary terminal higher up in the array. Thus, the two 0-signal paths (shown in heavy line) cannot intersect before they converge at the cell in question. On leaving this cell, one 0-signal path goes directly downward in the same column, while the other is passed onto a column farther to the right. Thus, the two 0-signal paths cannot intersect after they leave the cell in question. Therefore, by following the sequence stated above, all cells may be set up in succession and independent of one another.

This same argument also applies unmodified to all of the other arrays shown in prior figures, provided that the jumper and feedback connections around the edges of the array are removed, since all of these arrays can be imagined to be imbedded in a sufficiently large rectangular array.

Even with the short corner-jumping connections present, arbitrary programming is usually possible. For the triangular array of Fig. 26, for example, the diagonals may be set up in succession, longest to shortest. The cells within each diagonal may be set up in any order. It may be seen from the typical condition depicted in Fig. 34 that, for each cell on the diagonal being set up

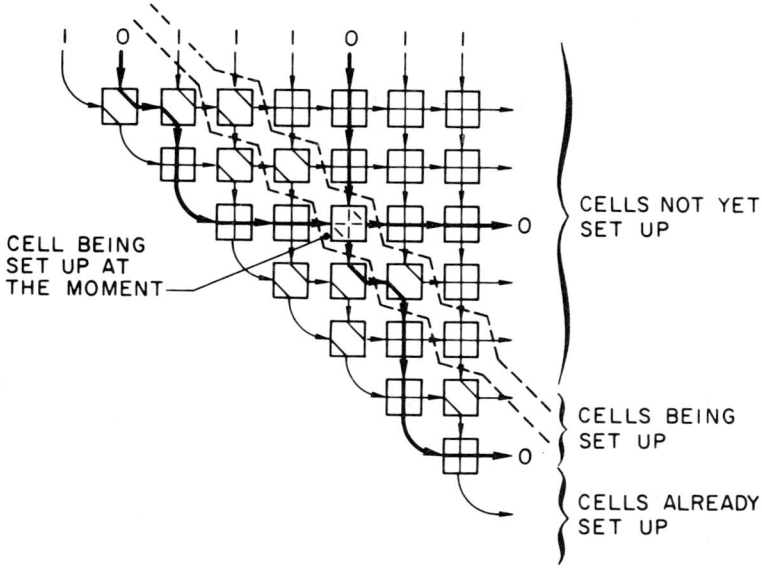

FIG. 34. Typical signal paths used on a triangular array during programming.

(the third diagonal, between the dotted lines), one of the incoming 0-signals comes directly from above, while the other is further to the left. Similarly, one of the outgoing 0-signals goes directly to the right, while the other follows a lower path to the right-hand edge. Therefore, intersection of the two 0-signals cannot occur except at the cell in question. This same argument applies to all cells of an arbitrary diagonal, so the entire triangular array may be set up in any desired pattern.

The array of Fig. 28 which was derived from the triangular array may also be set up along diagonals. In this case the long external jumpers need not be broken. The single diagonal of isolation cells to be preset to the "bending" mode may be set up first, before the procedure described for the triangular array is started.

In summary, various cellular arrays have been described that have the capability of permuting a set of n digital input lines onto a set of n digital output lines in an arbitrary manner. All of these arrays employ the same basic cell, which is essentially a flip-flop controlled reversing switch, requiring about ten elementary gates for its realization. Table I summarizes the results derived on the number $M(n)$ of cells required for the various types of interconnection arrays. The various arrays differ from one another in their shape, the total number of cells required, and the number and length of peripheral connections. It has also been shown how to set up cell flip-flops in these arrays in a systematic fashion, so as to achieve a desired permutation.

TABLE I
Number of Cells in Interconnection Arrays

Array form	Fig.	Number M of cells required													
		$n=2$	3	4	5	6	7	8	9	10	11	12	16	32	Asymptotic
Triangular, diamond	26, 27	1	3	6	10	15	21	28	36	45	55	66	120	496	$n^2/2$
Rectangular	28, 29	2	4	8	12	18	24	32	40	50	60	72	128	512	$n^2/2$
Pruned rectangular	30	1	3	5	9	12	18	22	30	35	45	51	92	376	$3n^2/8$
Approximately square	31	1	3	6	12	16	24	30	42	49	63	72	132	552	$9n^2/16$
Rearrangeable	32	1	3	5	8	11	14	17	21	25	29	33	49	129	$n \log_2 n$
$\langle \log_2 (n!) \rangle$	—	1	3	5	7	10	13	16	19	22	26	29	45	119	$n \log_2 n$

V. CONCLUSION

In this chapter the principal engineering design features of programmable cellular logic arrrays have been presented and discussed by means of numerous examples. Most of these examples are very efficient from a synthesis point of view, i.e., they require for a prescribed, typical functional behavior a number of cells that appears to be close to the minimum number that would be required using any cell of comparable logical complexity.

From a practical point of view, programmable cellular arrays offer an attractive alternative to customized arrays in many applications of large-scale-integration technology to the design of digital computers and digital information-handling systems. As the per-gate costs of integrated circuitry drop and the demand for more special-purpose hardware increases, programmed cellular arrays should become more and more attractive. From a theoretical point of view, the design of these arrays raises many minimization problems arising from unconventional functional expansions—those based on Eqs. (8), (9), and (11), for example. In addition, design with the NOR gate array raises some interesting problems in the geometrical reconfiguration of gate networks into a cellular form.

ACKNOWLEDGMENTS

The author would like to express his gratitude to K. N. Levitt and A. Waksman for their direct collaboration on some of the work reported in Section V, to B. Elspas and H. S. Stone for their long-standing participation and assistance in many aspects of cellular research, and to S. Wahlstrom for his basic contributions in designing good implementations of programmable cellular arrays.

The research that led to the results reported in this chapter was conducted at Stanford Research Institute and supported partly by the Information Systems Branch, Office of Naval Research, Contract Number Nonr-4833(00), and partly by the Air Force Cambridge Research Laboratories, Office of Aerospace Research, under Contract Numbers AF-19(628)-418 and AF-19(628)-5828.

REFERENCES

ARNOLD, T. F., TAN, C. J., and NEWBORN, M. M. (1970). Iteratively realized sequential circuits. *IEEE Trans. Comput.* **C-19**, 54–66.

BENEŠ, E. V. (1965). "Mathematical Theory of Connecting Networks and Telephone Traffic." Academic Press, New York.

COHN, M. (1962). Inconsistent canonical forms of switching functions. *IRE Trans. Electron. Comput.* **EC-11**, 284–285.

ELSPAS, B., GOLDBERG, J., JACKSON, C. L., KAUTZ, W. H., and STONE, H. S. (1967). Properties of Cellular Arrays for Logic and Storage. Sci. Rep. 3, Project 5876, AFCRL Report 67-0463, pp. 103–107. Stanford Res. Inst., Menlo Park, California.

ELSPAS, B., and SHORT, R. A. (1964). A bound on the run measure of switching functions. *IEEE Trans. Electron. Comput.* **EC-13**, 1–4.

EVEN, S., KOHAVI, I., and PAZ, A. (1967). On minimal modulo-2 sums of products for switching functions. *IEEE Trans. Electron. Comput.* **EC-16**, 671–674.

FERRARI, D., and GRASSELLI, A. (1969). A cellular structure for sequential networks. *IEEE Trans. Comput.* **C-18**, 947–956.

GOLDSTEIN, L. J., and LEIBHOLZ, S. W. (1967). On the synthesis of signal switching networks with transient blocking. *IEEE Trans. Electron. Comput.* **EC-16**, 637–641.

KAUTZ, W. H. (1967a). Testing for faults in combinational cellular logic arrays. *In* Proc. 8th Ann. Symp. on Switching and Automata Theory, pp. 161–174, October 1967.

KAUTZ, W. H. (1967b). A cellular threshold array. *IEEE Trans. Electron. Comput.* **EC-16**, 680–682.

KAUTZ, W. H. (1969). Cellular logic in-memory arrays. *IEEE Trans. Comput.* **C-18**, 717–727.

KAUTZ, W. H., LEVITT, K. N., and WAKSMAN, A. (1968). Cellular interconnection arrays. *IEEE Trans. Comput.* **C-17**, 433–445.

LEVITT, K. N., and KAUTZ, W. H. (1969). Cellular arrays for the parallel implementation of binary error-correcting codes. *IEEE Trans. Inform. Theory* **IT-1**, 597–607.

LOW, P. R., and MALEY, G. A. (1961). Flow-table logic. *Proc. IRE* **49**, 221–228.

MINNICK, R. C. (1964). Cutpoint cellular logic. *IEEE Trans. Electron. Comput.* **EC-13**, 685–698.

MINNICK, R. C. (1967). Survey of microcellular research. *J. Ass. Comput. Machinery* **14**, 203.

MUKHOPADHYAY, A., and SCHMITZ, G. (1970). Minimization of EXCLUSIVE OR and LOGICAL EQUIVALENCE switching circuits. *IEEE Trans. Comput.* **C-19**, 132–140.

NILSSON N. (1965). "Learning Machines." McGraw-Hill, New York.

PETERSON, W. W. (1961). "Error-Correcting Codes." Wiley, New York.

SEEBER, R. R. (1960). Associative self-sorting memory. *In* Proc. Eastern Joint Comput. Conf., Vol. 18, pp. 179–188.

WAHLSTROM, S. E. (1967). Programmable arrays and networks. *Electron.* **40**, 91–95.

WAKSMAN, A. (1968). A permutation network. *J. Ass. Comput. Machinery* **15**, 159–163.

AUTHOR INDEX

Numbers in italics refer to the pages on which the complete references are listed.

A

Abramowitz, M., 233, 234, *253*
Aiken, H., *225*, 243, *253*
Amarel, S., 300, *311*
Ankerlin, R. A., 244, *253*
Arnold, T. F., 394, *421*
Ashenhurst, R., 171, 179, *225*, 260, 261, 280, *311*, 319, 367

B

Bachman, G., *225*
Batree, T. C., *225*
Beckenback, E. F., *120*
Bellman, R., 133, *225*
Beneš, E. V., 410, *421*
Berge, C., 72, *83*
Bergland, G. D., 132, 133, *225*
Birkhoff, H., 174, *225*, 346, 349, *367*
Brooking, M. E., 296, *311*
Burks, A. W., 1, *26*, 58, *83*
Burnside, W., 331, *367*

C

Caldwell, S. H., 293, *311*
Calingairt, P., 135, 136, *225*
Canaday, R. H., 300, *311*
Cohn, M., *225*, 381, *421*
Cole, S. N., 308, *311*
Comfort, W. T., 310, *311*
Cooke, G., 300, *311*
Curtis, H., *225*

D

Daykin, D. E., 179, *225*
DeBruijn, N. G., 113, *120*, 180, *225*
Dietmeyer, D. L., *225*
Duley, J. R., *225*
Dunham, B., 252, *253*

E

Elpas, B., 124, 151, 200, 210, *225*, 244, 245, 252, *254*, 290, 297, 301, 306, *311*, 331, 332, 346, *367*, 384, *421*
Even, S., 381, *421*

F

Fantauzzi, G., 280, *311*
Ferrari, D., 394, *421*
Forslund, D. C., 244, 252, 253, *254*
Friedman, A. D., *225*

G

Gilbert, E. N., 30, *55*
Gill, A., 179, *225*
Guisti, A., 296, *311*
Glushkov, V. M., 5, *26*
Goldberg, J., 124, 200, 210, *225*, 244, 245, 252, *254*, 318, 331, 332, *367*, 384, *421*
Goldberg, R. R., *225*
Goldstein, L. J., 411, *421*
Golomb, S. W., *120*, 132, 133, 149, *225*
Gonzales, R. A., 310, *311*
Good, I. J., 132, 133, *226*

Gould, R., 18, *26*
Grasselli, A., 394, *421*
Gray, H. J., 360, *367*
Green, M. W., 318, *367*
Guinn, D. F., 139, *227*

H

Hale, H. W., 132, 133, *225*
Hall, M., *120*
Hall, M. Jr., 329, *367*
Harrison, 243, 246, 251, *254*
Harrison, M. A., 58, *83*, 87, 105, 109, 110, 112, 113, *120*, 124, 178, *225*
Hartmanis, J., 221, 222 *226*
Hawkins, J. K., 309, *311*
Hellerman, L., 178, 222, *226*
Hennie, F. C., 306, *311*
Hewitt, E., 140, *226*
High, R. G., 105, *120*
Holland, J. H., 309, *312*
Huffman, D. A., 30, *55*
Hwa, H. R., 219, *226*

I

Ibuki, K., 3, *26*

J

Jackson, C., 175, 177, 180, *227*
Jackson, C. L., 124, 200, 210, *225*, 244, 245, 252, *253*, *254*, 331, 332, *367*, 384, 421
Jacobson, N., *226*

K

Karp, R. M., *227*
Kautz, W. H., 124, 200, 210, *225*, 244, 245, 252, *254*, 297, 306, *312*, 318, 331, 332, *367*, 372, 384, 394, 399, 406, 409, 417, *421*, *422*
Kautz, W. H., 15, *26*, 28, *55*, 63, *83*
Kellerman, E., *226*
Kilmer, W. L., 308, *312*
King, W. F., III, 253, *254*, 296, *311*, *312*
Kobayashi, K., 3, 16, 17, *26*

Kuhavi, S., 381, *421*
Korenjak, A., 258, 280, *313*, 318, *367*

L

Lawler, E. L,, *226*
Lechner, R. J., 87, 100, 112, *120*, 132, 137, 148, 177, 178, 184, 189, 191, 223, *226*, 251, 252, *254*
Leibholz, S. W., 411, *421*
Lendaris, G. G., 281, *312*, 317, *367*
Levitt, K. N., 394, 406, 409, 417, *422*
Levy, S. Y., 264, 280, *312*
Littlewood, D., 127, 142, *226*
Loomis, H. H., Jr., 16, *26*, 140, 142, 146, *226*
Low, P. R., 394, *422*
Luccio, F., *226*
Lui, C. L., *120*
Lupanov, O. B., 58, *83*

M

Maclane, S., 174, *225*, 346, 349, *367*
McCluskey, E. J., 168, 170, *226*
McCormick, B. H., 310, *312*
McKeller, A. C., 172, *221*
McNaughton, R., 58, *83*, *226*
Maitra, K. K., 259, 264, 280, *312*
Maley, G. A., 394, *422*
Markov, A. A., 17, 20, *26*, 30, 33, *55*
Menger, K. L., Jr., 139, *226*
Mev, A. R., 170, *226*, 253 *254*
Mileto, F., *226*
Miller, R. E., 157, *226*
Minnick, R. C., *227*, 256, 280, 286, 288, 289, 293, 297, 299, *312*, 318, *367*, 372, 373, *422*
Moore, E. F., 58, *83*, 230, *254*
Morreale, E., 165, 168, *227*
Mott, T. H., J., 264, 280, *312*
Mukhopadhyay, A., 22, *26*, 83, *83*, 301, 302, 306, *422*
Muller, D. E., 3, 20, *26*, 30, *55*, 138, *227*, 235, 253, *254*
Munsey, C. J., 309, *311*
Murphy, J. V., 297, *313*
Murtha, J. C., 310, *312*

N

Naemura, K., 3, *26*
Nechiporuk, E., 179, 189, 190, *227*
Necula, N. N., 165, 168, *227*
Newborn, M. M., 394, *421*
Nilsson, N., 397, *422*
Ninomiya, I., 124, 132, 134, 139, 148, 174, 175, 177, 180, 183 187, 192, 203, 204, *227*, 251, 252, *254*
Nozaki, A., 3, *26*

P

Patt, Y. N., 253, *254*
Patterson, G. W., *226*
Paz, A., 381, *421*
Peterson, W. W., 62, *83*, 159, *227*, 408, *422*
Pollmer, C. H., 58, *83*
Post, E. L., 2, 3, *26*
Pólya, G., 87, 93, 97, 105, *120*, 179, *227*
Prather, R. C., *227*
Preparata, F. P., 230, *254*, 297, *312*
Putzolo, G., *226*

Q

Quine, W. V., 18, *26*, 168, 170, *227*

R

Reddy, S. M., 22, *26*
Reed, J. S., 3, *26*, 139, *227*
Ross, K. A., 140, *226*
Roth, J. P., *227*
Roy, K. K., 309, *312*
Rudin, W., 140, *227*

S

Schmitz, G., 302, 305, *312*, 381, *422*
Seeber, R. R., 399, *422*
Semon, W. L., 18, 26
Shannon, C. E., 2, *26*, 58, 60, *83*
Sheng, C. L., 219, *226*
Shen, V. Y., 172, *227*
Shestopal, H. A., 3, 7, 15, *26*

Short

Short, R. A., 58, 63, 72, *83*, 281, 300, *312*, 317, 318, 338, 354, 361, *367*, 384, *421*
Sims, J. C., Jr., 360, *367*
Singer, T., 179, *227*
Sklansky, J., 260, 264, 280, *312*
Slepian, D., 105, *120*, 178, 179, 189, *227*
Slotnik, D. L., 310, *313*
Smith, D. R., *227*
Spandorfer, L. M., 297, *313*
Stanley, G. L., 281, *312*, 317, *367*
Stearns, R. E., 221, 222, *226*
Stegun, I. A., 233, 234, *253*
Stone, H. S., 175, 177, 180, *227*, 244, 245, 252, *254*, 258, 280, *313*, 318, 331, 332, *367*, 384, *421*
Stone, W. S., 124, 200, 210, *225*

T

Tan, C. J., 394, *421*
Tang, C. K., 249, 253, *254*
Thrall, R., *228*

U

Unger, S. H., 308, 309, *313*

V

Vasilev, Yu. L., 83, *83*
Von Neumann, J., 230, *254*, 310, *313*

W

Wahlstrom, S. E., 372, 373, *422*
Waksman, A., 394, 409, 411, 417, *422*
Walter, C. T., 257, *313*
Warren, D. W., 58, *83*
Warman, R., 244, 252, 253, *254*
Weiner, P., 172, *227*
Weiss, C. D., *228*, 280, 301, 305, *313*
Wells, W. W., 139, 223, *227*
Whelchel, J. E., Jr., 139, *228*
Winder, R. D., 264, 280, 300, *311*, *312*
Wright, J. B., 1, *26*, 58, *83*
Wyman, R. H., Jr., 16, *26*

Y

Yablonskii, S. V., 3, 12, *26*
Yau, S. S., 249, 253, *254*
Yoeli, M., 317, 318, 319, 320, 321, 322,
   324, 325, 332, 336, 338, 340, 361, *367*

Z

Zhegalkin, I. I., 3, *26*
Zierler, N., 139, *227*
Zschirnt, H. H., 300, *313*

SUBJECT INDEX

A

Abelian group, 93, 333, 335, 345
Abstract Fourier transform, 130
  pair, 131
Add-one network, 323
Adder array, 289
Affine cell, 358
Affine group, 101, 124, 333, 346
Affine group based cascade, 360
Affine operators, 123
Affine transformation, 251, 358
AG, 189, 222
Algebraic expression of residual functions,
  76
Algorithm
  for cascade realizability, 263, 270, 271
  for unate arrays, 303
Almost complete sets, 16
  characterization of, 17
  nontrivially, 19
Almost square array for interconnection,
  416
"Almost" universal function, 243, 244
Alternating Group A_n, 331, 333, 334, 348
AND-OR-type function, 18, 20
AND-type function, 18, 20
Application of Polya's theorem to switching
  functions, 28
Approximate number of classes under G_n,
  115
Arbitrary logic by arrays, 376
Ashenhurst cell, 319, 320, 365
Associativity, 87
Asymptotic behavior
  of $C(m, n)$, 233
  for $M(n)$, 231
Asymptotic formula, 234
Asynchronous sequential logic circuitry, 174
Automata theory, 307
Automorphism, 185, 325, 326, 328, 329, 339,
  345, 346

group, 334
  inducing element, 346
Autosynchronous logic, 360
Auxiliary array, 164

B

Bases, 15
  simple, 15
  strong, 15
  weak, 15
Basic two-rail cascade, 281
Basic unate component, 303
Basis functions, 129
  of Fourier transform, 127
Batch fabrication technique, 256, 371
Bending mode, 409–411, 415
Bessel inequality, 148
Best upper and lower bounds for $M(n)$, 248
Bidirectional counter, 399
Bilateral iterative circuits, 308
Bipartite graph, 72
Binary counter, 393
Binary encoding, 237
Binary operation, 87
Binary two-rail cascade, 317
Binary valued signal, 51
Binomial distribution, 193
Block, 235, 237, 238
  of minterms, 236
Boolean function, 86, 316, 354
Boolean n-cube, partition of, 235
Bound
  for complete tree
    Lupanov's, 63
    Moore's, 61
    Short's, 64
  for $M(n)$, 231
  on minimum number of terminals, 230
Building blocks, 219, 369
Bulk transfer system, 309

Burst errors, 409
Bypassing defective cell, 374

C

$C(m, n)$, 234
$C(n)$, maximum length bound on redundant
    cascade, 278
C_2^n, 86
C_2^n genera, 113
Canonical
  cascades, 365
  decomposition, 348
  two-rail, 318
Canonical decoding tree, 172
Canonical decomposition, 324, 327
Canonical form, 220
  of redundant cascade, 259, 260
Canonic form theorem for cascade, 262
Canonic expansion of switching function, 2
Cardinality, 88, 205
Cartesian product, 123, 173
Cascade, 257
Cascade
  connection of group function network,
    322
  decomposition, 268
    basis for, 355
  length, 320
    by Short's method, 362
    Yoeli–Turner method, 362
  realizability algorithm, 263
  realizable, 269
  realizable symmetry type $W(n)$, 278
Cayley representation, 338
Cell, 316, 322
  complexity, 371
    in two-rail cascade, 318
  logic, 371
  types, 257, 323
Cellular array, 256, 370, 372
Cellular cascades, 316
Cellular logic, 230, 256
Cellular sorting array, 299, 403
Cellular space, 310
Cellular threshold array, 395
Character group, 147
Characteristic function, 145, 160, 166
Characteristic number, 383
Characteristic polynomial, 346–349

Characterization
  of almost complete sets, 17
  of decomposable groups, 328
  of group pair for decomposition, 329
Check digits, 408
Chip, 256, 371
Circuit equilibrium, 38
Classification
  of Boolean functions, 86, 100
  of switching functions, 86, 98
  of 1- and 2-variable, 6
  of 3-variable, 8, 9
Clique, 303
Closed system of function, 3, 5
Closely packed codes, 82
Cobweb array, 288
Code, 61, 82
  conversion, 57
Coding
  array, 406
  theory, 100
Column bussed cells, 376
Combinational, 156
  circuits, 1
  recursive definition of, 1
  logic function, 316, 371
  logic net, 230
Commutative diagram, 102
  group, 91
Comparison among cascade varieties, 361,
    362
Comparison matrix, 347, 349
Complimentary functions, 273, 283
Complimentary group C_2^n, 104
Complimentary outputs, 230
Complimentation, 33, 231, 233, 389
Complete decoding network, 58, 69
Complete group, 329, 331
Complete tree, 63, 64, 366
Completely indecomposable, 329
Complex modules, 229, 230
Complexity
  of cell types, 364
  of inversion, 54
  of switching circuits, 57
Composition, 3
  of group functions, 321, 326
Computational bounds, 168
Conjecture on true identity projection, 80
Connection types $C(m, n)$, 232

Content-addressed memory, 373, 405
Convolution sum, 140, 161
  theorem, 122, 140
Core
  memory, 170
  prime implicant, 166
Coset, 89, 147, 158, 222, 237, 240, 330
  leader, 159
  multiplication, 330
Costs of decoding networks, 64
Counting, 373
  problem under $C_2{}^n$, 104
  theorems, 84, 96
Covering of shell center, 62
Cross-bar switch, 410
Cross-operation, 103
Crossing mode, 409–411, 415
Cubical complex, 204
Custom-designed array, 257
Cutpoint, 286, 373
  array, 286
Cycle, 92, 96
  of permutation, 95
  in iterative circuits, 308
Cycle index, 92, 98, 101
  of $A_n(\mathbb{Z}_2)$, 109, 111
  of $C_2{}^n$, 104
  of G_n, 105, 108
  of $GL_n(\mathbb{Z}_2)$, 109, 110
  of linear group on $\{0, 1\}^n$, 110
  of S_n, 105
  of $S_n \otimes B$, 107
Cycle index polynomial, 92, 96
  for $A_n(\mathbb{Z}_n)$, 119
  for G_n, 105, 117
  for $GL_n(W_2)$, 119
  for S_n, 116
Cycle length, 93
Cycle notation, 99
Cycle structure
  of group, 91, 92
  of product of permutation, 102
Cyclic cover, 158
Cyclic group, 92, 329–331, 333, 334, 360

D

De Brujn's theorem, 184
Decoding, 57

  by cellular array, 407
  complete, 58
  disjunctive, 58
  nondisjunctive, 58
  tree, 58
Decomposable finite group, 359, 360, 333
Decomposable function, 329
Decomposability
  of direct products, 334
  of group (table), 333
Decompositions, 235, 319
  disjunctive, 122, 123, 156, 171
  of function, 266, 268
    over group, 323
  of group functions, 323, 331
    theory of, 320
Degenerate function, 142, 196
Degree of polynomials, 109
Delay element, 39
Δ-forms, 135
Δ-sum canonical form, 138
Diamond array for interconnection, 412
Dicyclic group, 332
Dihedral group D_n, 332, 333
Dimension, 157
Direct product, 174
  of permutation groups, 102
Direct product groups, 124
Direct sum, 123, 161, 173
Discrete mathematics, 100
Disjoint block, 240
Disjunctive canonical form, 126
Disjunctive decomposition, 122, 123, 156, 171
DNF, 158
Domain of boolean function, 98, 102
Domain encoding, 149
"Don't care" states, 337, 340, 359
Double pole double-throw reversing switch, 409
Dummy terminal, 233, 344

E

Edge-fed array, 286
Efficient two-rail cascade, 283
Encoded input-logic, 122, 124, 125, 177, 196, 204, 210
Encoded input threshold logic, 219

Encoding by cellular array, 407
Encoding transformation, 123, 124, 223
Erasure errors, 409
Essential prime implicant, 166, 170
Equalization of input-variable loading, 365
Equivalent, 89, 91, 93, 113
Equivalent form, 201
Equivalence
  classes, 86, 90, 98, 184, 175
  of inversion and inhibition, 33
  relation, 86, 89
  singleton, 99
Exact number of classes under G_n, 115
Excitation, 31, 43
Exclusive or canonical form, 2, 135

F

Factor group, 329, 330
Factor group function, 320
Factoring variable, 172
Fast-Fourier transform, 132, 134, 163, 164, 186, 193, 206, 210, 219
Fault accommodation, 372
Fault testing, 372
Feed forward, 176
Feed forward network, 28
Feedback, 28, 393
  loops, 2, 38
  role in circuit minimization, 31
  shift register, 347
Finite group theory, 317, 324
Fixed interconnection approach, 370
Flow-table, 308
4-decomposable functions, 325
4-way array, 289
Fourier basis vectors, 139
Fourier coefficients, 204
Fourier spectrum, 186, 221
Fourier transform kernel, 127, 129, 138
Froebenius theorem, 90
  generalization of, 94
Full adder, 392
Function counting series, 96
Function matrices, 256
Function selection, 230
Functional equivalence, 175
Functionally complete two-rail cascade, 319
Fundamental invariance theorem, 148, 154, 155

G

g-function, 33, 36
G-genus, 113
G_n, 86, 99
$GF(2)$, 250, 346, 354
Galois field, 123
Gedanken experiments on multiinversion circuits, 50, 51
Genera, 113, 242, 186
General cell types, 342
General form of two-rail cascade, 283
General function cell, 319
General linear group, 100
Generalization
  of group K, 345
  of shell function, 65
Generalized matrix cascade, 318
Generalized switching function, 102
Generating function, 91, 93, 98
Generic class, 187, 174, 189
Genus, 231, 233, 187
Genus equivalence, 231
Graph, 184
Graph theory, 318
Group, 89, 320
  $C_2^r \otimes H_0$, 335
  C_2^r, 334
  characters, 127
  function, 321, 325, 327
  G, 87
  G_n, 231
  $S_n \otimes B$, 105
  homomorphism, 127
  theory, 318
  structure of G_n, 105
Greatest common divisors, 102
Growth rate of array, 287

H

H-cells, 357, 358, 366
h-function, 36, 33
Hadamard transform matrix, 133
Hamming code, 61
Harmonic analysis, 121–124
Harvard class, 273
  H_{87}, H_{76}, and H_{77}, 243
Harvard function, 273

Heuristics, 173
Hill-climbing technique to find ULM, 244, 245
Homogeneous array, 199, 200
Homogeneous linear transformation, 346, 357
Homomorphic image, 91, 94
Homomorphism, 91, 94
  of directed graph, 318
Horizontal bus type array, 286

I

IC circuit package, 257
Identification of inputs, 8, 10–12
Identity, 87
  coset, 159
  group, 92, 98, 102, 321
  projection of switching function, 77
Implicant extraction, 161
Implicants, 147, 156, 157
Incompletely specified Boolean function, 337, 340, 359
Incompleteness of single-rail cascade, 347
Indecomposable groups, 333
Index of subgroup, 88, 89
Indistinguishable subsets, 233
Induced permutation group, 94
Inductive algorithm for affine group, 351
Information digits, 408
Inhibition, 33
Inner automorphism, 181, 329, 331
Inner transform, 131
Input/output terminal of ULM, 230
Integer valued function, 131
Integrated circuit technology, 229
Intercell signals, 307
Interconnection
  approach, to ULM, 249
  array, 409
  limited network, 370
Inverse, 87
Inverse transform, 131
Inversion
  complexity, 54
  of three variables, 33
  of 2^{k-1} variables, 35
  of two variable by one NOT element, 39
Invertible linear transformation, 100

Irreducible polynomial over $\mathbb{Z}_2$, 109
Irredundant cascade, 259
Isomorphic, 91, 102, 182
Isomorphic group, 330
Isomorphic machines, 221
Iteratively closed system of functions, 3, 5
Iterative circuits, 307, 308
Iterative construction of ULM, 240
Iterative counting, 203
Iterative decomposition technique, 324, 356
Iterative network, 230
Iterative circuit computer, 309
  input phase of, 309
  path building phase of, 310

K

K_m-decomposable function, 328
k-cell, 157
K-type cell, 351
K-type affine two-rail cells, 343
Karnaugh map, 59
Kautz's network with feedback, 29
Kernel functions, 129
  of Fourier transform pairs, 127
Klein four-group $C_2 \times C_2$, 348
Kronecker product, 163, 165
  of matrices, 133

L

LG, 222
LSI, 197, 217, 256
Large scale integration, 256
Largest factor, 106
Lattice, 178
Least common multiple, 102
Left coset, 88
Length of redundant cascade in canonical form, 273
Level set, 157, 222
Lexicographic, 193, 203
Linear algebra, 317
  approach to multirail cascade, 354
Linear domain transformation, 207
Linear function, 4, 9, 91
Linear group, 124, 345
  on n-dimensional vector space, 100

Linear mappings, 100
Linear operation, 100, 123
Linear transformation, 346, 347
Logic-in-memory, 257
Logical completeness of two-rail cascades, 281, 282, 318
Logical incompleteness of single-rail cascade, 318
Lower bound
on cells in interconnection network, 410
on number of classes, 114
$M(n)$, 231, 232, 234

M

m-cube, 321
$M(n)$, 231, 410
Macrocellular, 121, 124, 199–203
Macrocellular logic, 307
Madaline, 397
Maitra cascades, 259, 287, 316
Maitra cell, 259
Majority element, 39, 388
Majority function, 33, 46, 242
Majority gate array, 300, 384, 385
Many valued functions, 125
Many valued logic function, 105
Mappings, 98, 325
Markov circuits, 37, 38
Markov's work on minimum NOT-problem, 30
Mask register, 402
Masking, 388
Matrix A, 236
Matrix H_s, 237
Matrix formulation of transform pair, 131
Maximal-period feed back shift register, 347
Maximal unate component, 302
Mean square error minimization, 148
Memory, 28
Merger of column in q-function array, 291
Metacyclic group, 333
Microcellular, 121, 124, 199–201, 203
Microprogrammed read only store, 257
Minimal cover, 304, 158
Minimal nontrivially almost complete, 19
Minimal polynomial, 349
Minimization
of Adder array, 291

of cellular array, 300, 305
of q-function array, 291
of minterm and summation cell, 379
Minimization problem in cellular array, 306
Minimum NOT-problem, 30
Minority function, 33
Minos, 397
Minterm, 164
array, 293, 375, 376
Minterm cell
minimization, 379
NOR-gate realization of, 380
Mixing of rail signals, 349
Mobius function, 106
Mobius inversion formula, 109
Mode of cell, 373
Modified Reed array, 380, 381
Modular computer, 308, 309
Monte-Carlo analysis, 251
Monotonicity, 4
Moore bound for complete tree, 61
Muller's work on minimum-NOT problem, 30
Multiinput-output functions, 125
Multiinversion circuits, 43, 50, 51
Multioutput switching functions, 316
Multioutput synthesis technique, 338
Multiple independent errors, 409
Multirail Boolean function, 366
Multirail cascades, 316, 281
Multirail synthesis, 345, 328
Mutually independent encoding, 175

N

n-decomposable, 323
n-variable Boolean function, 231
(n, d, k) problem, 306
N_n, upper bound on cells in two-rail cascade, 284
$N(m, n)$, number of connections to terminals 233
Natural equivalence relation, 86, 98
Necessity of unstable equilibrium, 48
Negation, 231
synthesis, 6
Negative weights, 33
Nesting of Markov circuits, 37, 38

Network topology, 310
Neutral functions, 116, 264
Next-state function, 394
Ninomiya class N_{92}, N_{90}, and N_{91}, 243
Ninomiya transform, 194
Nonabelian group, 333
Noncellular array, 257
Nondegenerate Boolean functions, 234
Nondegenerate genera, 232
Nondisjunctive decoding network, 58
Nonexistence of (4, 3) ULM, 242
Nonisomorphic groups with same cycle
    index, 93
Nonlinear closely-packed codes, 82
Nonlinear function, 6, 10
Nonmonotonic function, 6, 7
Non-one-preserving functions, 12
Non-self-dual function, 8, 11, 12
Nonsingular matrix, 100, 346, 347, 358
Nonzero-preserving function, 12
NOR
    array, 296, 390, 391, 387
    based array, 300
    function, 229
    gate realization of coding cell, 407
    NAND array, 297, 298, 299
Normal form, 135, 206, 208
Normal subgroups, 329, 346
Normalization of cascade, 357
Normalized cell, 342, 343, 356
Null space, 158, 160, 239
Number genera, 114
Number
    of cascade realizable functions, 278, 279
    of cell types, 364
    of cells in almost square array, 417
        in interconnection array, 420
    of classes
        under A_n ($\mathbb{Z}_2$), 112
        under GL_n ($\mathbb{Z}_n$), 112
        under G_n, 109
        induced by RAG_n, 251
        under S_n, 108
    equivalance classes under $C_2{}^n$, 104
    genera, 114
    module types, 350, 352
    self-complementing classes, 114
    shell center patterns, 73
    shell functions, 73, 82
    terminals in You–Tang ULM, 249

O

Odd cyclic groups, 335
One-NOT realization, 31
One preservation, 4
Optional terms in simplification of shell
    function, 67, 68
OR-type function, 18, 20
Orbits, 222
Order of group, 88, 89
Ordered pairs, 175
Ordering of input variable, 365
Oscillation, 40, 41, 43
One-to-one homomorphism, 91
Outer automorphism, 329, 331, 340, 360
Output-incrementing circuit, 44, 45
Overlapped shell functions, 64

P

$P(x_1, X_2, \ldots, x_6)$, 96
Parallel computation, 307, 310
Parallel logic, 258
Parallel processor, 310
Parity
    check matrix, 408
    function, 265, 270
    sum, 388
Parsival identity, 148
Partial decoder, 58, 60
    6-variable, 68
    7-variable, 79
    8-variable, 71
Partial decomposition, 268
Partial symmetry of function, 151
Partially ordered, 202
Partially ordered graph, 303, 304
Partition, 89, 92, 235, 338
    of Boolean function, 86
Partitioned matrix, 177
Pattern
    classification network, 397
    detection, 308
    recognition, 307, 398
Perceptron, 397
Permutation, 231, 233, 389
    cascade, 320, 322, 336, 337, 359, 363
    cell, 319, 410
    group, 90, 92, 94, 101, 102, 338

Permutationally equivalent group, 102
Poisson summation theorem, 146, 148
Polya's couting theorem, 180, 184, 185
Polya's theorem, 96
  application to switching function, 98
Positive T-elements, 33
Positive unate, 49
Post's theorem, 13
Prime implicant, 122, 123, 157, 163, 165, 380
Prime implicant array, 375
Prime unate component, 303, 304
Primitive, 1
  polynomials, 347
Probability, 219
Programmable cellular logic array, 257, 373, 375
Programming
  flip-flop, 377, 383
  interconnection array, 417
Projection, 160, 242, 244
  of switching function, 76, 77
Proof of almost completeness theorem, 21
Prototype, 251
  classes, 124, 174, 183, 187
  equivalence, 177
  equivalence class, 183
  equivalence relation, 173, 174
  functions, 189
  of Ninomiya, 250
  transformation, 122, 123, 220
Pruned rectangular array, 415

Q

q-function array, 289
Quaternion group, 333
Quasitriangular matrix, 181
Quotient group, 142, 333
Quotient group character theorem, 142

R

RAG_n, 250
RAG (restricted affine group), 173–175, 180, 182–186, 189, 193, 219–222, 250
Rademacher–Walsh transform, 139
Radix-two expansion, 165
Rail-complementation vector, 353

Rail-signal, 321
Range, 113
  encoding, 152
  enumerating function, 93
  enumerating series, 96, 98
  transformation, 176
  translation, 134
Rank, 193, 208
Rational canonical forms, 123
Rearrangable contact network, 410
Rectangular array for interconnection, 413, 414
Recursive definition of cascade, 261
Reduced cell short cascade, 363
Reduction theorems, 7
Redundancies in cycle structures, 92
Redundant cascade, 259, 322, 365
Redundant decomposition, 270
  of 4-variable function, 274
Reed arrays, 380, 381
Reed cell, 382
Reed decomposition, 355
Regular interconnection geometries, 316
Regular set, 308
Residual function, 75, 269
  algebraic expression of, 76
Restricted affine group, 122, 123, 177, 181, 250
Restricted cell types in two rail cascade, 284
Restricted linear group, 181
Restricted Maitra cascade, 287
Restricted Maitra cell, 287, 288
Reverse variable ordering, 366
Reversing switch, 405
Right coset, 88
Ring-sum decomposition, 355
Ring-sum expansion of switching function, 2
RLG, 189, 193
Rounding, 373
Row-bussed cellular array, 286, 374

S

S_n, 86
  genera, 113
Search technique to find (7, 4) ULM, 244
Selection, 57
Selective complementation of rail signals, 356

Self-complementing equivalence class, 114, 116
Self-duality, 4
Self-inverse, 138
Self-reproducing automata, 310
Self-reproduction, 230
Semiconductor technology, 256
Sequence-generator, 193
Sequential array, 393
Sequential cells, 230
Sequential circuits, 28, 31, 38
Sequential functions, 394
Sequential model for iterative circuits, 308
Sequential network, 371
Shell center, 59, 62
Shell function, 60
  general theory of, 72
  generalization of, 65
Shift-register, 373
  sequence generator, 346
Signal paths during programming
  in rectangular array, 418
  in triangular array, 419
Signal propagation delay around feedback loop, 47
Signal replication trick, 41, 43
Simple bases, 15, 16
Simple function, 15
Simple transitive group, 338, 345
Simple tree, 370
Single rail cascade, 258, 316
Singleton equivalence class, 99
Sorting, 373
  array, 399, 400, 402
Source
  pair, 233
  wires, 230, 232
Special purpose array, 394
Spectral coefficients, 208, 219
Spectrum, 192, 194, 196, 210, 211
  invariance, 122
Split class, 192, 193, 194,
Stable equilibrium, 47
Stable multiinversion circuits, 51
Stability groups, 222
Stacked threshold array, 396
Standard canonical ordering, 366
Standard cell types, 316
Standardization in LSI, 372
  of cell, 257

State assignment for sequential machines, 125, 221, 294
State-permutation table, 343
Stirling number of second kind, 233
Strong bases, 15
Strong completeness, 2
Strongly enumeratively equivalent, 97
Storage bound, 170
Structure theorems for permutation groups, 101
Subarray interconnectability, 372
Subcubes, 157, 410
Subgroup, 88, 89, 326, 345
Submatrix, 239
Subspace, 158, 160, 165, 237
Sum, mod 2, 99
Summation cell, 378
  minimization of, 379
  NOR gate realization of, 380
Switching function, 86, 98
Switching theory, 100, 369
Symmetric group, 87, 173, 331, 334, 345
Symmetric function, 60
Symmetric function array, 293–295
Symmetric transform matrix, 131
Symmetry group of n-cube, 100
Symmetry operation on connection of ULM, 250
Symmetry type, 231
Symmetry type approach to ULM, 249
Syndrome, 408
Synthesis
  of AND, 6, 7
  of negation, 6
  of OR, 6, 7
  of multirail cascade, 336

T

t-tuple of natural numbers, 93
$T(n)$, 249
Ternary valued signal, 51
Terminal count $M(n)$, 231
Tessellation, 310
Testing cascade realizability, 263, 270
Theory
  of finite groups, 317
  of shell functions, 72
Threshold, 31, 394

Threshold array, 394
Threshold element, 31
Threshold logic, 124
Transform coefficients, 187, 218
Transform pair, 131
Transformation, 98
Transformation groups, 123, 175
Translation, 159
Translation groups, 345
Translation vector, 175
Truth table, 339
Triangular array for interconnection, 411
True projection of switching function, 77
Truncated permutation cascade, 363
Trunks, 410
Truth table, 339
Turing machine, 230
Two-dimensional array, 285
Two-rail binary cascade, 338, 281
Two-rail universal cascade, 283
2-way array, 289
2^n-cell array, 293

U

ULM, 230, 257
  for $n = 3$, 241
  for n, 241
  of Yau–Tang type, 250
(8, 4) ULM, 244
(5, 3) ULM, 243
(m, n) ULM constructive design, 235
(7, 4) ULM, 244, 245, 246
(7, 4) ULM, connection, 246, 247, 248
Unate cellular logic, 301
Unate component, 302
Unate functions, 32, 49, 54
Uniform digit weighted code, 400
Uniform loading tree, 366
Unilateral iterative circuit, 307
Unit delays, 393
Universal cell, 373
Universal computing automata, 310
Universal constructing automata, 310

Universal function, 235
Universal logic module, 230, 257, 241
Universal module, 231
Unrepairable network, 372
Unstable equilibrium, 30, 48
Upper bound on cells in-two-rail cascade, 284
Upper bound on $M(n)$, 231, 241, 246

V

Vn, 237
Vector space of n-tuples, 237, 345
Vector sum, 237
Visual information, 310

W

Weak completeness, 2, 5, 215
Weak completeness theorem, 7
Well-formed combinational circuits, 1
Weight, 394
  of function, 95, 98, 264
  of vector, 159
Weight function, 95
Weight "preserving" equivalence, 101
Weighted sum, 31
Weighted transform, 174

Y

Y-function, 45, 52
$YT(n)$, 319
YT-decomposition, 365
Yoeli–Turner bound on cells in two-rail cascade, 319
Yoeli–Turner decomposition, 324, 345, 348

Z

Zero preservation, 4

ELECTRICAL SCIENCE

A Series of Monographs and Texts

Editors

Henry G. Booker
UNIVERSITY OF CALIFORNIA AT SAN DIEGO
LA JOLLA, CALIFORNIA

Nicholas DeClaris
UNIVERSITY OF MARYLAND
COLLEGE PARK, MARYLAND

Joseph E. Rowe. Nonlinear Electron-Wave Interaction Phenomena. 1965

Max J. O. Strutt. Semiconductor Devices: Volume I.
  Semiconductors and Semiconductor Diodes. 1966

Austin Blaquiere. Nonlinear System Analysis. 1966

Victor Rumsey. Frequency Independent Antennas. 1966

Charles K. Birdsall and William B. Bridges. Electron Dynamics of Diode Regions. 1966

A. D. Kuz'min and A. E. Salomonovich. Radioastronomical Methods of Antenna
  Measurements. 1966

Charles Cook and Marvin Bernfeld. Radar Signals: An Introduction to Theory and Application.
  1967

J. W. Crispin, Jr., and K. M. Siegel (eds.). Methods of Radar Cross Section Analysis. 1968

Giuseppe Biorci (ed.). Network and Switching Theory. 1968

Ernest C. Okress (ed.). Microwave Power Engineering:
  Volume 1. Generation, Transmission, Rectification. 1968
  Volume 2. Applications. 1968

T. R. Bashkow (ed.). Engineering Applications of Digital Computers. 1968

Julius T. Tou (ed.). Applied Automata Theory. 1968

Robert Lyon-Caen. Diodes, Transistors, and Integrated Circuits for Switching Systems. 1969

M. Ronald Wohlers. Lumped and Distributed Passive Networks. 1969

Michel Cuenod and Allen E. Durling. A Discrete-Time Approach for System Analysis. 1969

K. Kurokawa. An Introduction to the Theory of Microwave Circuits. 1969

H. K. Messerle. Energy Conversion Statics. 1969

George Tyras. Radiation and Propagation of Electromagnetic Waves. 1969

Georges Metzger and Jean-Paul Vabre. Transmission Lines with Pulse Excitation. 1969

C. L. Sheng. Threshold Logic. 1969

Dale M. Grimes. Electromagnetism and Quantum Theory. 1969

Robert O. Harger. Synthetic Aperture Radar Systems: Theory and Design. 1970

M. A. Lampert and P. Mark. Current Injection in Solids. 1970

W. V. T. Rusch and P. D. Potter. Analysis of Reflector Antennas. 1970

Amar Mukhopadhyay. Recent Developments in Switching Theory. 1971

A. D. Whalen. Detection of Signals in Noise. 1971

In Preparation

J. E. Rubio. The Theory of Linear Systems